PARASITIC NEMATODES
Molecular Biology, Biochemistry and Immunology

Parasitic Nematodes
Molecular Biology, Biochemistry and Immunology

Edited by

M.W. Kennedy
Division of Infection and Immunity
Institute of Biomedical and Life Sciences
University of Glasgow
Glasgow
UK

and

W. Harnett
Department of Immunology
University of Strathclyde
Glasgow
UK

CABI *Publishing*

CABI *Publishing* is a division of CAB *International*

CABI Publishing	CABI Publishing
CAB International	10 E 40th Street
Wallingford	Suite 3203
Oxon OX10 8DE	New York, NY 10016
UK	USA
Tel: +44 (0)1491 832111	Tel: +1 212 481 7018
Fax: +44 (0)1491 833508	Fax: +1 212 686 7993
Email: cabi@cabi.org	Email: cabi-nao@cabi.org
Web site: http://www.cabi.org	

A catalogue record for this book is available from the British Library, London, UK.

A catalogue record for this book is available from the Library of Congress, Washington DC, UK.

ISBN 0 85199 423 7

Typeset by AMA DataSet Ltd, UK.
Printed and bound in the UK by Biddles Ltd, Guildford and King's Lynn.

Contents

Part III   Specialist Products and Activities

Contributors

Ralf Adam, Department of Molecular Parasitology, Institute of Biology, Humboldt University Berlin, Invalidenstrasse 43, 10115 Berlin, Germany. Present address: Proctor & Gamble European Services Gmbh, Industriestr. 30–34 G. T. 3, 65733 Eschborn, Taunus, Germany. adam.r@pg.com

Timothy J.C. Anderson, Instituto di Patologia Generale Veterinaria, Università di Milano, Via Celoria 10, 20133 Milan, Italy. Present address: Department of Genetics, Southwest Foundation for Biomedical Research, PO Box 760549, San Antonio, TX 78245-0549, USA. tanderso@darwin.sfbr.org

Judith A. Appleton, James A. Baker Institute for Animal Health, College of Veterinary Medicine, Cornell University, Ithaca, NY 14853, USA. jaa2@cornell.edu

David Artis, Immunology Group, School of Biological Sciences, Stopford Building, Oxford Road, University of Manchester, Manchester M13 9PT, UK. Present address: School of Veterinary Medicine, Department of Pathobiology (Rosenthal 205), University of Pennsylvania, 3800 Spruce Street, Philadelphia, PA 19104-6008, USA. dartis@vet.upenn.edu

Claudio Bandi, Istituto di Patologia Generale Veterinaria, Università di Milano, Via Celoria 10, 20133 Milano, Italy. cbandi@imiucca.csi.unimi.it

Alan F. Bird, CSIRO Land and Water, Private Bag 2, Glen Osmond, SA 5062, Australia.

David McK. Bird, Plant Nematode Genetics Group, Department of Plant Pathology, Box 7616, North Carolina State University, Raleigh, NC 27695-7616, USA. david_bird@ncsu.edu

Mark L. Blaxter, Institute of Cell, Animal and Population Biology, University of Edinburgh, Ashworth Laboratories, King's Buildings, West Mains Road, Edinburgh EH9 3JT, UK. Mark.Blaxter@ed.ac.uk

Anne Dell, Department of Biochemistry, Imperial College of Science, Technology and Medicine, London SW7 2AY, UK. a.dell@ic.ac.uk

Birgit Drabner, Department of Molecular Parasitology, Institute of Biology, Humboldt University Berlin, Invalidenstrasse 43, 10115 Berlin, Germany.

Paul Garside, Department of Immunology, University of Glasgow, Western Infirmary, Glasgow G11 6NT, UK. paul.garside@clinmed.gla.ac.uk

Robin B. Gasser, Department of Veterinary Science, The University of Melbourne, 250 Princes Highway, Werribbee, Victoria 3030, Australia. r.gasser@vet.unimelb.edu.au

Claudio Genchi, Istituto di Patologia Generale Veterinaria, Università di Milano, Via Celoria 10, 20133 Milano, Italy.

Richard K. Grencis, Immunology Group, School of Biological Sciences, Stopford Building, Oxford Road, University of Manchester, Manchester M13 9PT, UK. rgrencis@fsl.scg.man.ac.uk

David W. Halton, School of Biology and Biochemistry, Medical Biology Centre, The Queen's University, Belfast BT7 1NN, UK. D.Halton@queens-belfast.ac.uk

Margaret M. Harnett, Department of Immunology, University of Glasgow, Western Infirmary, Glasgow G11 6NY, UK. m.harnett@bio.gla.ac.uk

William Harnett, Department of Immunology, University of Strathclyde, Glasgow G4 0NR, UK. w.harnett@strath.ac.uk

Stuart M. Haslam, Department of Biochemistry, Imperial College of Science, Technology and Medicine, London SW7 2AY, UK. s.haslam@ic.ac.uk

Siân M. Henson, Department of Biochemistry, Imperial College of Science, Technology and Medicine, London SW7 2AY, UK.

Ayman S. Hussein, Department of Biochemistry, Imperial College of Science, Technology and Medicine, London SW7 2AY, UK.

Douglas P. Jasmer, Department of Veterinary Microbiology and Pathology, Washington State University, Pullman, WA 99164-7040, USA. djasmer@vetmed.wsu.edu

Malcolm W. Kennedy, Division of Infection and Immunity, Institute of Biomedical and Life Sciences, Joseph Black Building, University of Glasgow, Glasgow G12 8QQ, UK. malcolm.kennedy@bio.gla.ac.uk

David P. Knox, Moredun Research Institute, Pentlands Science Park, Bush Loan, Penicuik, Midlothian EH26 0PZ, UK. knoxd@mri.sari.ac.uk

Patricia R. Komuniecki, Department of Biology, University of Toledo, Arts & Sciences, 2801 E. Bancroft St, Toledo, OH 43606-3390, USA. pkomuni@uoft02.utoledo.edu

Richard Komuniecki, Department of Biology, University of Toledo, Arts & Sciences, 2801 E. Bancroft St, Toledo, OH 43606-3390, USA. rkomuni@uoft02.utoledo.edu

Catherine E. Lawrence, Division of Infection and Immunity, Joseph Black Building, University of Glasgow, Glasgow G12 8QQ, UK. c.lawrence@strath.ac.uk

Alex Loukas, Institute of Cell, Animal and Population Biology, University of Edinburgh, Ashworth Laboratories, King's Buildings, West Mains Road, Edinburgh EH9 3JT, UK. alex@panther.qimr.edu.au

C. Stewart Lowden, Department of Preclinical Veterinary Sciences, Royal (Dick) School of Veterinary Science, Summerhall, The University of Edinburgh, Edinburgh EH9 1QH, UK.

Richard Lucius, Department of Molecular Parasitology, Institute of Biology, Humboldt University Berlin, Invalidenstrasse 43, 10115 Berlin, Germany. richard.lucius@rz.hu-berlin.de

Rick M. Maizels, Institute of Cell, Animal and Population Biology, University of Edinburgh, Ashworth Laboratories, King's Buildings, West Mains Road, Edinburgh EH9 3JT, UK. r.maizels@ed.ac.uk

Nikki J. Marks, School of Biology and Biochemistry, Medical Biology Centre, The Queen's University, Belfast BT7 1NN, UK.

Richard J. Martin, Department of Veterinary Medicine, College of Veterinary Medicine, Iowa State University, Ames, IA 50011-1250, USA. rjmartin@iastate.edu

Aaron G. Maule, School of Biology and Biochemistry, Medical Biology Centre, The Queen's University, Belfast BT7 1NN, UK. a.maule@qub.ac.uk

Howard R. Morris, Department of Biochemistry, Imperial College of Science, Technology and Medicine, London SW7 2AY, UK.

George F. Newlands, Moredun Research Institute, Pentlands Science Park, Bush Loan, Penicuik, Midlothian EH26 0PZ, UK.

Charles H. Opperman, Plant Nematode Genetics Group, Box 7616, Departments of Plant Pathology and Genetics, North Carolina State University, Raleigh, NC 27695-7616, USA. charlie_opperman@ncsu.edu

Antony P. Page, Wellcome Centre of Molecular Parasitology, The Anderson College, The University of Glasgow, 56 Dumbarton Road, Glasgow G11 6NU, UK. A.Page@udcf.gla.ac.uk

Jenny Purcell, Department of Preclinical Veterinary Sciences, Royal (Dick) School of Veterinary Science, Summerhall, The University of Edinburgh, Edinburgh EH9 1QH, UK.

Diane L. Redmond, Moredun Research Institute, Pentlands Science Park, Bush Loan, Penicuik, Midlothian EH26 0PZ, UK.

Alan P. Robertson, Department of Biomedical Sciences, College of Veterinary Medicine, Ames, IA 50011-1250, USA.

Wayne S. Russell, Department of Biochemistry, Imperial College of Science, Technology and Medicine, London SW7 2AY, UK.

Murray E. Selkirk, Department of Biochemistry, Imperial College of Science, Technology and Medicine, London SW7 2AY, UK. m.selkirk@ic.ac.uk

Philip J. Skuce, Moredun Research Institute, Pentlands Science Park, Bush Loan, Penicuik, Midlothian EH26 0PZ, UK.

Momtchil A. Valkanov, Department of Preclinical Veterinary Sciences, Royal (Dick) School of Veterinary Science, Summerhall, The University of Edinburgh, Edinburgh EH9 1QH, UK.

Mark E. Viney, School of Biological Sciences, University of Bristol, Woodland Road, Bristol BS8 1UG, UK. Mark.Viney@bristol.ac.uk

Preface

Dramatic advances are being made in the study of parasitic nematodes, progress made possible by the wealth of new molecular biological, biochemical and immunological tools which have recently become available. This exciting situation has been accelerated by the genetic and biological information gleaned from work on the free-living nematode, *Caenorhabditis elegans*, and there is increasing evidence of scientists working on this worm and parasitic nematodes in parallel. It is also becoming clear that animal and plant parasitologists may have much to learn from one other, particularly now that the mechanisms whereby both classes of parasite control their tissue environments and the defence reactions of their hosts are receiving particular attention. It remains to be seen whether animal- and plant-parasitic nematodes use similar mechanisms, but there is clearly a possibility that they do, and new information on, for example, lipid signalling in plants and animals may be pertinent. Animal and plant hosts may still seem to be worlds apart, but it was not long ago that the importance of *C. elegans* to parasitologists of any persuasion was often dismissed.

Such is the transformation and scope of research on parasitic nematodes that a book representing the entire field, whilst simultaneously capturing the dynamism of the subject, would be almost impossible to produce. This book was therefore designed from the outset to be selective of subject matter, and we targeted authors who are authorities in their subjects, were active researchers in their own rights, and could be prospective in their thinking. We asked the authors to place their subjects in context before entering into a bold account of new findings and their implications, and to emphasize what future developments may now be possible. As a consequence of this strategy, there are many gaps in the coverage in terms of the species dealt with, but we make no apologies for this; it was, in fact, quite deliberate – the principles dealt with will apply broadly, and cross-system research is clearly a way ahead for new principles to be discovered in biological science. In this spirit we have included articles on areas in which *C. elegans* is playing a major role, and on novel areas of nematode parasitism illuminated by plant-parasitic nematodes. The chapters are also diverse in style and approach, which we encouraged.

There are so many scientists of high calibre in nematode parasitology that we found it extremely difficult to choose, but in order to produce a balanced approach we were unable to call on all of the authors whom we would otherwise wished to have included. If this book appears in another edition, however, then we may be able to recruit from amongst the authors who were not included on this occasion.

Shortly after completing his chapter, one of the authors, Alan Bird, died suddenly and unexpectedly from a heart attack. It is a fitting tribute to Alan that his last published work is in a book that unites plant- and animal-parasitic nematodes within a framework of basic biology encompassing the *C. elegans* literature. Alan was one of the first scientists to appreciate the power of the *C. elegans* model to study parasitic nematodes, and also to realize that the conceptual mechanisms of parasitism by nematodes are catholic, no matter what the host. And this central idea was the driving force behind this book.

Malcolm Kennedy
William Harnett

Glasgow, June 2000

Access to Colour Illustrations

In order to reduce the cost of the book, and thereby improve its accessibility, there are no colour reproductions. Many of the illustrations, however, can only properly be appreciated and understood in colour (particularly those in Chapter 20). The colour illustrations can therefore be viewed and downloaded from the following internet site:

http://www.cabi.org/Bookshop/book_detail.asp?isbn = 0851994237

See also http://www/gla.ac.uk/Acad/IBLS/II/mwk/res/images.html

If you have any suggestions or problems relating to the illustrations appearing in the book, then please feel free to contact malcolm. kennedy@bio.gla.ac.uk

Molecular Analysis of Nematode Evolution

Mark L. Blaxter

Institute of Cell, Animal and Population Biology, University of Edinburgh, Edinburgh EH9 3JT, UK

Introduction

The phylum Nematoda is both speciose and biologically and ecologically diverse (Chitwood and Chitwood, 1974; Andrássy, 1976; Anderson, 1992; Malakhov, 1994). Nematodes live at the bottoms of the deepest oceans (Ditlevsen, 1926), in the frozen deserts of the Antarctic and in terrestrial soils and inshore muds, often in incredible numerical abundance (Platonova and Gal'tsova, 1976). One of the best-known features of the Nematoda is that it includes a large number of parasitic species, many of which infect humans, domestic animals and food crops (Nickle, 1991; Anderson, 1992; Blaxter and Bird, 1997; Blaxter, 1998). Phylogenetic analysis can answer questions pertaining to the evolution of the parasitic lifestyle. How often has it arisen? Are there common features of the non-parasitic relatives of parasitic groups that point to 'preadaptations' necessary in the evolution of the trait? Can a directionality of evolution be inferred for contrasting parasitic characters, such as host utilization or mode of infection? Is parasitism an ancient phenotype or one that has developed recently, or, how closely related are parasites to each other?

The free-living nematode *Caenorhabditis elegans* is the subject of one of the most wide-ranging analyses of biology of a single species yet attempted (Riddle *et al.*, 1997; Bargmann, 1998; Blaxter, 1998; *C. elegans* Genome Sequencing Consortium, 1998; Chervitz *et al.*, 1998; Clark and Berg, 1998; Ruvkun and Hobert, 1998). In order to use this understanding to develop novel treatments for nematode-induced diseases, the relationship of *C. elegans* to parasitic species must be understood (Blaxter and Bird, 1997;

Blaxter, 1998). Once the evolutionary framework is in place, the search for genes and processes unique to or important in parasitism can be taken up in earnest.

This chapter discusses recent advances in molecular phylogenetic analysis of the phylum Nematoda. These advances offer for the first time a relatively unbiased view of the phylogenetic structure of the Nematoda, free of unquantifiable observer bias. In addition, they allow the testing of hypotheses of parasitic nematode evolution in a rigorous way. It is clear that molecular phylogenetic analysis promises insights into many corners of the phylum, and only a few of these are highlighted here; the main wave of molecular genetic insights will come soon from the accelerating genomic analysis of the Nematoda.

The Molecular Revolution

Before the advent of molecular phylogenetic analysis of nematodes, the systematic study of the phylum was hindered by: (i) the number of species and morphological diversity; (ii) the limitations of light microscopical analysis of nematode morphology; and (iii) the inevitable specialization of nematode sytematists. When most species under study are less than a couple of millimetres in size, and the characters being observed are at best difficult to classify and define homology in, the problem seems intractable. In particular, the separation of 'free-living' and 'parasitic' *nematologists* has led to parasites being classified in separate taxa (usually at ordinal level) unrelated to other forms. Several authors have presented schemas for the evolutionary relationships of the phylum and, not surprisingly, these are often in deep disagreement (Chitwood and Chitwood, 1974; Sudhaus, 1976; Maggenti, 1981, 1983; Poinar, 1983; Lorenzen, 1994; Malakhov, 1994). The best-known analysis proposes that the Nematoda can be split into two great classes, the 'Secernentea' and the 'Adenophorea', with most terrestrial and parasitic species being 'Secernentea' and most marine species being 'Adenophorea'. Recently, alternative hypotheses have been proposed, involving a threefold split in the nematodes (Malakhov, 1994). At lower systematic levels, there are disagreements about the relationship of parasitic orders to each other and to free-living groups, and many families and genera are disputed entities.

In order to bring some clarity to the field, a universal metric is needed: a measure taken from any species that allows it to be compared, using verifiable methods, with other species, and to derive phylogenetic information from the comparison. Molecular phylogenetic markers offer such a solution (Hillis, 1987; Hillis *et al.*, 1996). By choosing a gene present in all species, and subject to the same (or at least similar) evolutionary constraint in all species, a database of comparable characters can be built and used to derive phylogenetic trees. There are several methods for building trees from aligned sequence datasets, and each has its own problems and benefits. However, when used together, and with correct application of statistical testing, these methods can yield testable phylogenetic hypotheses that appear to be robust and informative (Swofford *et al.*, 1996; Swofford, 1999).

Genomic DNA sequences evolve at different rates depending on the constraints under which they are held. Non-coding, non-transcribed sequences will evolve faster than sequences encoding an essential protein. Mitochondrial DNA also appears to evolve faster than nuclear DNA. Thus, when addressing a phylogenetic question, it is important to choose a segment of DNA that will have accumulated changes at a rate comparable to the phyletic events under study. For the analysis of populations or congeneric species, a rapidly evolving gene is chosen. For the analysis of deep phylogeny (the relationship of the Nematoda to other phyla, or the interrelationship of nematode orders), a slowly evolving gene is used. One problem is that if a set of phyletic events happened in quick succession a long time ago it may not be possible to resolve them. The slowly evolving gene used to resolve distant events might not retain a signal of the order of phylesis occurring over the short time span. Thus, adaptive radiations are often difficult to resolve, as a burst of phylesis takes place in a short space of deep time.

Two additional caveats, of a methodological sort, are important to note. One is that species identification is still of paramount importance. A taxon to be analysed needs to be placed into the known biology of nematodes in order for its inclusion to be meaningful. Misidentification will still lead to confounding errors. The second is that contamination of nematode sample DNA with DNA from other organisms (particularly of hosts or of food sources) can lead to isolation and sequencing of a gene from the wrong species (from a different phylum or even kingdom). Highly conserved genes are often conserved not only between species, but also between kingdoms, and PCR primers used to amplify nematode genes could accidentally amplify homologues from mammalian host or bacterial and fungal contaminants/food. This consideration is of particular importance in the analysis of museum specimens (Thomas *et al.*, 1997; Herniou *et al.*, 1998). While it is now possible to amplify genes from even long-term formalin-fixed nematodes, the presence of even small amounts of unfixed contaminant DNA can lead to mistakes. This is evident in the sequence dataset, as taxa that root erroneously have unexpected affinities or show remarkable identity to other organisms. Within the laboratory, contamination with fungal DNA from human commensals is a common and worrying problem.

Molecular Markers

In the short history of nematode molecular phylogenetics, a number of different genes have already been used for analysis: cytochrome *c* (Vanfleteren *et al.*, 1994), globin (Blaxter *et al.*, 1994a,b; Vanfleteren *et al.*, 1994), RNA polymerase II (Baldwin *et al.*, 1997), heat shock protein 70 (Snutch and Baillie, 1984; Beckenbach *et al.*, 1992), ribosomal RNAs and their spacer segments (Aleshin *et al.*, 1998; Blaxter *et al.*, 1998; Kampfer *et al.*, 1998; Dorris *et al.*, 1999) and mitochondrial genes (Hyman and Slater, 1990; Anderson *et al.*, 1993; Pelonquin *et al.*, 1993; Powers *et al.*, 1993; Grant, 1994; Hyman and Beck Azevedo, 1996; Blouin *et al.*, 1997; Hugall *et al.*, 1997; Keddie *et al.*, 1998). The slowly evolving genes (cytochrome *c*, globin, coding regions of ribosomal RNAs, RNA polymerase II, heat shock protein 70) are suitable for the analysis of deep events in nematode evolution, while the mitochondrial and ribosomal spacer genes are more suited to intra-species, intra-genus, and intra-family analyses. One of the issues clearly raised in these studies is the problem associated with multigene families where different family members perform different functions (i.e. are paralogues) and have evolved differently: the phylogenies constructed will thus reflect gene evolution (including gene duplication) rather than species evolution (Blaxter *et al.*, 1994a). As the nematode genome projects progress, additional genes may become available for analysis (Moore *et al.*, 1996; Blaxter *et al.*, 1997a, 1999; Daub *et al.*, 2000).

The ribosomal RNAs (rRNA) have been most extensively studied. The rRNA genes are favourite choices because they are present in multiple (and probably identical) tandemly arrayed copies (and thus provide a large molar excess of target in PCR reactions compared with single-copy genes) and there is an extensive literature on the mode and tempo of evolution of the genes from other studies. These studies reveal that the evolution of rRNA gene clusters is mosaic in that some regions – essentially the coding regions: small subunit or 18S (ssu), large subunit or 28S (lsu) and 5.8S genes – evolve relatively slowly; while others – the internal (ITS-1 and ITS-2) and external transcribed spacer regions, and the non-transcribed spacer – evolve relatively rapidly. Even within the coding regions there are regions with widely differing rates of molecular evolution. The choice of a single gene or set of genes for analysis promotes collaboration and synthesis of datasets from different laboratories, while the use of multiple different genes allows independent assessment/confirmation of hypotheses.

The Structure of the Nematoda and the Origins of Parasitism

Small subunit rRNA gene sequences have been determined for a large number of nematode taxa distributed across the phylum (Ellis *et al.*, 1986; Zarlenga *et al.*, 1994a,b; Fitch *et al.*, 1995; Fitch and Thomas, 1997; Aleshin *et al.*, 1998; Blaxter *et al.*, 1998; Kampfer *et al.*, 1998; Dorris *et al.*, 1999). These can thus be used to examine the relationships of the different nematode orders. The pattern of nematode evolution that emerges is radically different from most published schemata, but agrees with some of the conclusions reached in recent, cladistic analyses of morphology. The Nematoda has a tripartite structure, and the 'Adenophorea'/'Secernentea' division is not supported. In our original analysis (Blaxter *et al.*, 1998), we used 54 different nematode ssu rRNA genes. In the interim, additional sequences have been published (Nadler, 1992, 1995; Aleshin *et al.*, 1998; Kampfer *et al.*, 1998). Addition of these to our dataset confirms and extends the original findings (Dorris *et al.*, 1999). Importantly, we still do not find an 'Adenophorea'/'Secernentea' division, unlike one study using ssu rRNA sequences that may have been in error due to taxon sampling biases (Kampfer *et al.*, 1998).

The basic divisions of the Nematoda are between clades I, II and C&S (Fig. 1.1). Clade I, which corresponds to the Dorylaimida (with the addition of the free-living Mononchida, insect-parasitic Mermithida and vertebrate-parasitic Triplonchida) includes invertebrate, vertebrate and plant parasites, and both marine (benthic) species as well as terrestrial ones. The human parasites *Trichuris* and *Trichinella* are in this clade. Clade II corresponds to the Enoplida (with the addition of the Triplonchida) and includes marine and plant-parasitic species. Clade C&S (Chromadorida and Secernentea) is a novel combination of the marine Chromadorida and all the taxa previously grouped in the class 'Secernentea'. Chromadorids are free-living marine species, with a few terrestrial/freshwater representatives. A chromadorid radiation is evident, with the Plectidae being the sister taxon to the Secernentea. The Secernentea can be further divided into three clades that group animal-parasitic, plant-parasitic and free-living species in novel combinations.

Clade III comprises only animal parasites. This was entirely unexpected, and such associations had not been made before, but the sequence data are striking in their 100% bootstrap support for this clade. The clade includes four traditional orders: the ascarids (*Ascaris, Toxocara*), the spirurids (filaria such as *Brugia* and *Onchocerca*), the oxyurids (the pinworms such as *Enterobius*) and the rhigonematids (millipede parasites). The different orders are very close genetically. The Spirurida may be an invalid taxon: the gnathostomes lie basal to a compound clade of the other spirurids and the other orders. Resolution of species within each order is poor with the ssu rRNA sequence (Nadler, 1992, 1995), and no close association with intermediate or definitive host biology/systematics is readily discerned.

Clade IV is also a novel association of animal-parasitic, plant- and fungus-parasitic and free-living groups, unexpected from previous analyses. It includes the free-living Cephalobida, the plant-parasitic and insect-parasitic Tylenchida, fungivorous Aphelenchida and a group of insect and vertebrate parasites including *Strongyloides* and *Steinernema*. Clade IV is much more diverse genetically than clade III and can be split into two parts (Dorris *et al.*, 1999). The reality of this division is unclear as there are

(A)

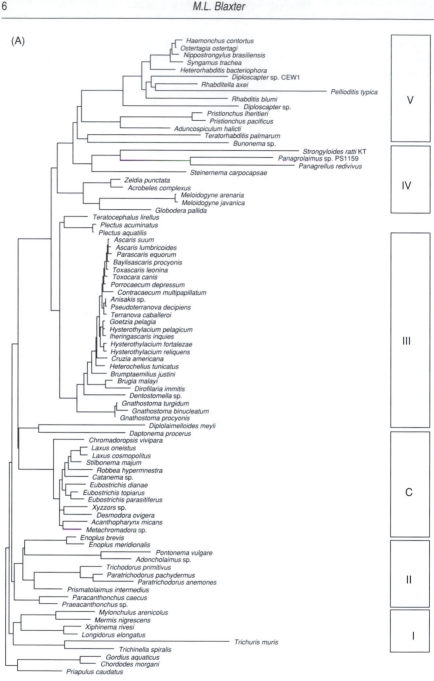

0.05 changes

Fig. 1.1. The phylogenetic structure of the Nematoda revealed by small subunit ribosomal RNA analysis. (A) Neighbour-joining (NJ) analysis of aligned ssu rRNA genes from nematodes. The alignment is based on that of Blaxter *et al.* (1998), with the addition of sequences from Aleshin *et al.* (1998), Nadler (1998)

(B)

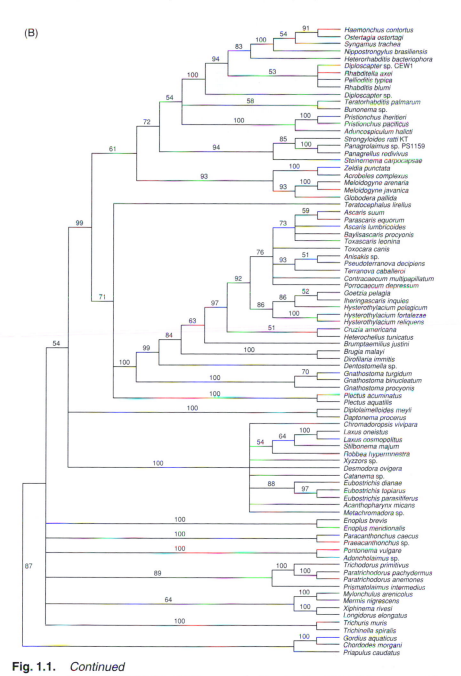

Fig. 1.1. *Continued*

and Kampfer *et al.* (1998). The NJ tree was built using the general time reversible model of nucleotide substitution with a gamma distribution with shape parameter 0.7. All analyses were performed using PAUP* 4.02 (Swofford, 1999). Beside the phylogram are the clades described in the text. C, chromadorida. (B) A bootstrap

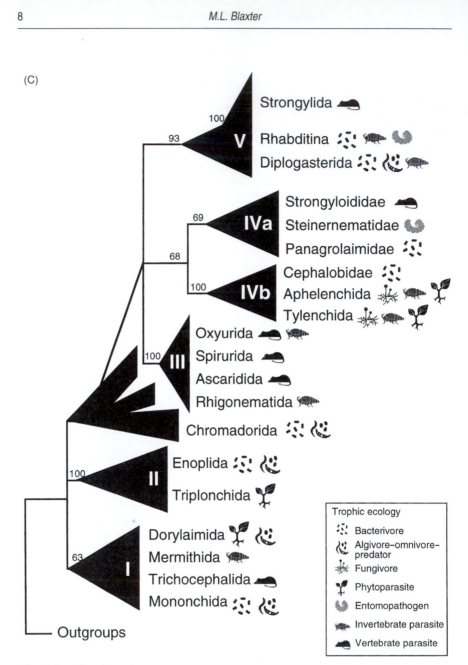

Fig. 1.1. *Continued*
analysis of the NJ tree, with the same parameters as in (A). Numbers correspond
to percentage bootstrap support for the given node. (C) Trophic ecologies of
each named nematode Order; above each node is maximal bootstrap support
(not necessarily as in (A) and (B) – the support is derived from more exhaustive
analyses; see Blaxter *et al.* (1998) for the numbered and named clades).
Adapted from Dorris *et al.* (1999).

possible methodological problems associated with very long predicted branch lengths in some taxa.

Clade V links the free-living, microbivorous Rhabditida and Diplogasterida with the vertebrate parasitic Strongylida. *C. elegans* is a rhabditid, and its closest major parasitic sister taxa are thus the gut- and lung-parasitic strongyles, such as the human hookworms (*Necator*, *Ancylostoma*). The nematodes in clade V are very diverse genetically, and analysis using ssu rDNA reveals significant structure (Fig. 1.2) (Fitch *et al.*, 1995). Within the Rhabditida are many examples of close association (phoretic association) and even near-parasitism of insects. In particular, *Heterorhabditis* is a pathogen of insect larvae in the soil, and invades and kills its host with a symbiotic bacterium.

Strongylid Evolution Revealed by Three Different Genes

The Strongylida (within clade V) are a diverse and important parasitic group. Their evolutionary relationships have been a subject of research and debate for many years (Beveridge and Durette-Desset, 1994; Durette-Desset *et al.*, 1994, 1999; Chilton *et al.*, 1997a,b; Sukhdeo *et al.*, 1997). In particular, the concepts of horizontal transmission between host species and vertical transmission within lineages of hosts have been examined and discussed: strongylids are found in hosts as diverse as birds and eutherian mammals. The abundance of 'morphology' in these relatively large nematodes, particularly the male bursa and its rays and the cuticularization of the buccal capsule/stoma, has permitted researchers to generate robust and testable hypotheses of strongylid evolution (Durette-Desset *et al.*, 1999).

Strongylid evolution has been traced using five genes. The globin genes are found in multiple isoforms in each species (a myoglobin and a cuticle globin, but in several strongyles there are multiple copies of the cuticle globin gene; Blaxter, 1993; Blaxter *et al.*, 1994a; Daub *et al.*, 2000; Blaxter, unpublished) making evolutionary reconstruction of species evolution problematic. The mitochondrial cytochrome oxidase I (COI) gene has been proposed as a marker and used to examine the mode of evolution of parasitic phenotypes (Sukhdeo *et al.*, 1997). However, as can be seen from Fig. 1.3, the COI gene does not yield a strong phylogenetic signal and leaves the strongylids unresolved. This is because there have been many multiple substitutions in the gene, leading to convergence. The 5.8 S rRNA gene has also been used, but this too yields unresolved phylogenies, as there is too little variation (Chilton *et al.*, 1997b). There is not such an extensive dataset for the ssu rRNA gene (Zarlenga *et al.*, 1994a,b; Blaxter *et al.*, 1998; Dorris *et al.*, 1999), but analysis of the available clade V sequences reveals some strongly supported structure in the Strongylida, with the two representatives of the Metastrongyloidea significantly linked as a monophyletic clade (Fig. 1.2). There is, again, too

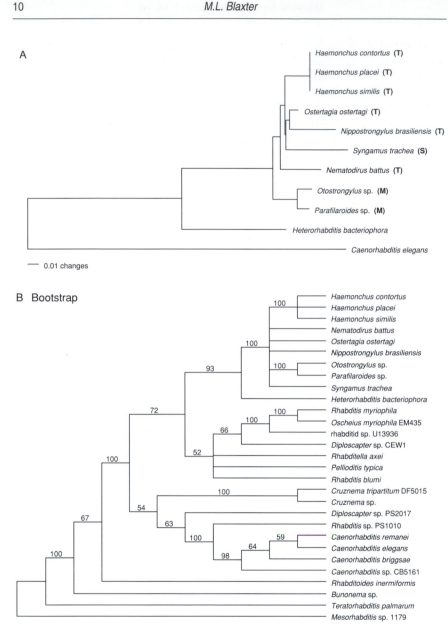

Fig. 1.2. Phylogenetic analysis of clade V: the Rhabditina–Strongylida–Diplogasterida. (A) NJ analysis of ssu rRNA sequences of strongylid taxa using *C. elegans* as an outgroup. The three *Haemonchus* species have identical sequences, while the other taxa differ at only a very few positions. Data from Blaxter *et al.* (1998) and Zarlenga *et al.* (1994a,b) and Blaxter, unpublished (*Syngamus*). (T), Trichostrongyloidea; (S), Strongyloidea; (M), Metastrongyloidea. (B) Bootstrap analysis of the nematode taxa in clade V. Within the Rhabditina there are several strongly supported groups, but within Strongylida only the *Haemonchus* spp. and Metastrongyloidea are supported as distinct clades.

little variation in the ssu gene to resolve the phylogeny further. There is now a large number of ITS sequences, particularly ITS-2 sequences, from strongylida in the databases (Campbell *et al.*, 1995; Chilton *et al.*, 1995, 1997a,b, 1998; Hoste *et al.*, 1995; Stevenson *et al.*, 1995, 1996; Gasser *et al.*, 1996, 1998; Hung *et al.*, 1997; Romstad *et al.*, 1997a,b, 1998; Newton *et al.*, 1998). Analysis of these (Fig. 1.4) confirms, for the taxa analysed, the division into superfamilies (Ancylostomatoidea, Trichostrongyloidea and Strongyloidea; no members of the Metastrongyloidea were analysed). Within the Strongyloidea there is significant structure that, interestingly, intermixes marsupial (metatherian) and eutherian mammal parasites, suggesting multiple transfers of parasites between these two host groups, and separates taxa currently placed in the same genus (*Oesophagostomum* is no longer monophyletic as the Oesophagostominae clade includes *Chabertia* nested within *Oesophagostomum* species) or family (the Phasco-strongylinae and Cloacininae are split) (Dorris *et al.*, 1999). Within the Trichostrongyloidea, the structure revealed is not statistically supported by bootstrap analysis, and no direction of evolutionary change between the different families can be proposed. In re-analysis of a dataset comprising just the Trichostrongyloidea, with *Strongylus edentatus* as an outgroup, the same result is achieved (not shown). Thus, the ITS-2 dataset is sufficient to resolve the Strongyloidea but not the sampled taxa of the Trichostrongyloidea (where there is insufficient variation in the ITS-2 sequence, and variable portions are difficult to align). This problem could be resolved by analysis of additional genera within the Trichostrongyloidea (only five out of more than 50 proposed are represented). Comparison with published analyses of morphological characters using cladistic methods (Durette-Desset *et al.*, 1999) is also compromised by the lack of generic representation.

Resolution of Species Complexes by Molecular Phylogenetics

Molecular phylogenetics can also be used to look at the population structure of species and the relationships between/within species complexes. Chapter 4 discusses the use of ITS sequence and ITS restriction fragment length polymorphism (RFLP) in distinguishing between different closely related species for diagnostic purposes. Differences in ITS sequence within species of human hookworms have also been demonstrated. Using highly variable sequences, questions of the relationships of isolates/subspecies can be resolved. For example, the closely related nematodes *Bursaphelenchus mucronatus* and *B. xylophilus* have very different disease phenotypes, with *B. xylophilus* associated with the economically important pine wilt disease while *B. mucronatus* is only associated with pines already damaged (Wingfield *et al.*, 1984; Rutherford *et al.*, 1990). The global

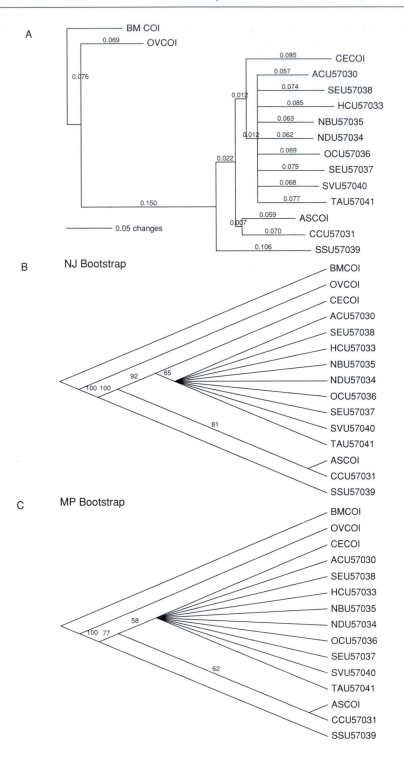

distribution of pathogenic *Bursaphelenchus* suggested that different 'xylophilus' isolates might be different species with different pathogenic profiles, and even that some 'mucronatus' nematodes might cause disease (De Guiran and Boulbria, 1986; De Guiran and Bruguier, 1989; Webster *et al.*, 1990; Abad *et al.*, 1991; Beckenbach *et al.*, 1992; Riga *et al.*, 1992; Iwahori *et al.*, 1998). The relationships between a panel of isolates from around the world was resolved using ITS phylogenetics and clearly demonstrates that the two species, while diverse, are distinct genetic entities (Fig. 1.5) (Beckenbach *et al.*, 1999). Isolates from different regions of the globe (Japan versus Europe) did not cluster. Similar analyses have been carried out to distinguish human and pig *Ascaris* using mitochondrial sequencing or haplotypes derived from PCR-RFLP analysis (Anderson *et al.*, 1993, 1995a,b; Anderson, 1995; Anderson and Jaenike, 1997; Peng *et al.*, 1998).

The Evolution of Parasitic Phenotypes

The evolution of parasitic phenotypes in nematodes is a topic of active practical and theoretical research (Skorping *et al.*, 1991; Read and Skorping, 1995a,b). Understanding the mode and tempo of acquisition of particular phenotypes associated with succesful parasitism will permit fuller appreciation of the evolutionary constraints experienced by organisms adapting to new hosts.

Within the Nematoda, parasitism has arisen multiple times (Blaxter *et al.*, 1998). The plant-parasitic Tylenchida, Dorylaimida and Triplonchida have acquired the phenotype independently. The insect/invertebrate parasitism of the species in clades V, IV, III and I have similarly arisen independently. Vertebrate parasitism has arisen at least four times, in the Trichocephalida of clade I, the three orders in clade III, the Strongyloididae of clade IV and the Strongylida of clade V. As there are still additional animal parasitic groups that have not been analysed (importantly including *Dioctophyme* and *Capillaria*) the number of independent acquisitions of vertebrate parasitism predicted may still be an underestimate. The import

---

Fig. 1.3. (Opposite) COI genes of strongylid and other nematodes. Aligned COI sequences (from Sukhdeo *et al.*, 1997 (where each is designated by the GenBank accession number), except for *Ascaris suum* ASCOI and *C. elegans* CECOI which are from Okimoto *et al.* (1992), *O. volvulus* COI which is from Keddie *et al.* (1998) and *B. malayi* COI sequence which is unpublished data of the author) were analysed using maximum parsimony and neighbour joining. (A) Phylogram of the bootstrap consensus tree derived by neighbour joining analysis of the aligned COI sequences. Branch lengths (in proportion of predicted changes) are given for each branch. (B) Percent bootstrap support for the resolved nodes mapped on a cladogram of the NJ analysis. (C) Bootstrap analysis using maximum parsimony of the same dataset. Nodes supported at less than 50% are collapsed to form a polytomy.

of this is that we must search for not a unifying single 'reason' for parasitism, but multiple and possibly complex adaptations that may be specific to each clade.

An oft-quoted preadaptation of free-living nematodes to parasitism is the existence of a dauer or resting stage (Riddle *et al.*, 1981; Riddle, 1988;

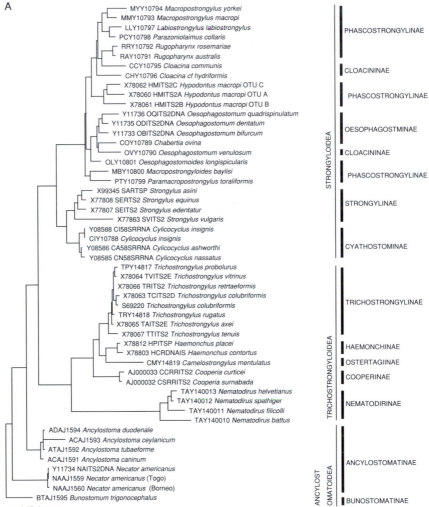

Fig. 1.4. Phylogenetic analysis of aligned ITS2 regions of strongylid nematodes. (A) NJ phylogram derived using logdet/paralinear distance correction. Branches supported at > 65% by bootstrap are darker. (B) Bootstrap cladogram showing the support for each node. The Strongyloidea are well resolved, while the Trichostrongyloidea are not. Data from Genbank (accession numbers given) as described by Chilton *et al.* (1995, 1997a,b, 1998) and Gasser *et al.* (1996, 1998).

B Bootstrap

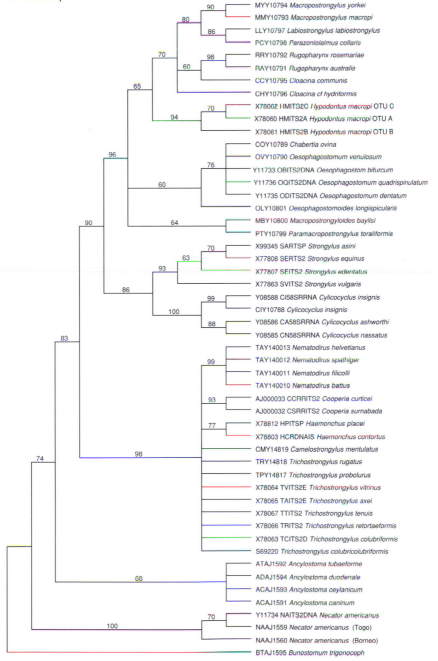

Fig. 1.4. *Continued.*

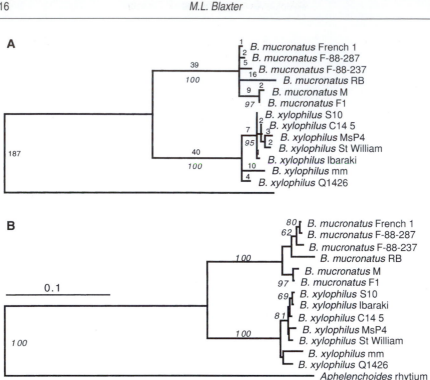

Fig. 1.5. Delineation of *Bursaphelenchus* species. (A) Maximum parsimony phylogram of an analysis of the aligned ITS1 and ITS2 regions from *Bursaphelenchus* isolates. Branch lengths are given above each branch, and bootstrap support below. (B) As for (A) but using NJ analysis (logdet/paralinear distance correction). Adapted from Beckenbach *et al.* (1999).

Viney, 1999). The dauer in most rhabditids (clade V free-living nematodes) is a modified or alternate third stage larva (L3). In *C. elegans* a genetic pathway for dauer induction, maintenance and exit has been elucidated that involves environmental sensing and intra-organismal hormone-like signalling (Riddle *et al.*, 1981; Riddle, 1988). The ability to generate dauer-constitutive mutants of *C. elegans* by changes at single genetic loci (Riddle *et al.*, 1981) offers a clear theoretical route to the evolution of constitutive infective L3 (dauer homologue) formation in parasitic groups. In this context, the genus *Strongyloides* has been held up as a group that has gone 'half-way' to parasitism. The *Strongyloides* life cycle includes a facultative free-living phase, where emergent L1 females can choose between proceeding to the infective L3, or becoming fecund, sexually reproducing adults and producing many infective L3 progeny, depending on their environment (Viney, 1994, 1996, 1999). Thus, analysis of the dauer-L3 decision can be taken as a model for the infective L3 versus free-living L3 decision in *Strongyloides*, and the *Strongyloides* pattern/process can be taken as a model for all infective L3 generations (and particularly those that, like *Strongyloides*, have an invasive mechanism).

While this argument is persuasive, it must be tempered by the phylogenetic relationships of the species. *Strongyloides* has traditionally been classified as a rhabditid (and thus a member of clade V). The ssu rRNA analysis clearly and robustly places *Strongyloides* in clade IV, close to free-living cephalobes and the plant-, fungus- and insect-parasitic aphelenchs and tylenchs. This novel association also unites *Strongyloides* with a number of other species with 'alternating' life cycles (Nickle, 1991), suggesting that *Strongyloides* is the highly evolved, vertebrate parasitic member of a clade of organisms with a propensity for complex life cycles. Unfortunately, not many of these other species have been studied molecularly, but two are now available. *Rhabdias bufonis* is a frog parasite that has an alternating lifecycle – in this case, an obligatory alternation between parasitic and free-living phases. The ssu rRNA data place *Rhabdias* as a sister taxon to *Strongyloides* (M. Dorris and M. Blaxter, unpublished). *Rhabdias* has morphological characters uniting it with the Alloionematida, a group of insect parasites with complex life cycles (Nickle, 1991). One non-parasitic member of the Alloionematidae has been sequenced and is placed in the *Rhabdias–Strongyloides* clade (M. Dorris and M. Blaxter, unpublished). Within clade IV, many taxa have the ability to arrest at multiple stages of their life cycles, not just the L3. Thus *Bursaphelenchus* arrests as an L4, and many cephalobes can arrest at several larval stages. Plant-parasitic tylenchs also arrest at stages other than L3. Analysis of the mechanisms underlying the *Strongyloides* L3 decision may thus reveal biology peculiar to the species in clade IV, rather than universal mechanisms behind infective L3 biology in all parasites.

One of the longest-standing puzzles in nematode parasitism is the presence of an obligatory tissue migration phase in the life cycles of some gut-parasitic species (Read and Skorping, 1995b). As this tissue migration results in an appreciable attrition of individuals, there must be a strong selective force maintaining the phenotype. Explanations of the phenotype take two forms. One suggests that it is an evolutionary relic revealing that the nematodes' ancestors were originally skin penetrators, which had to migrate through the tissues to reach the gut. The other suggests that it is an adaptation with current relevance to the nematodes' biology, and is a mechanism whereby they can grow larger as larvae and thus produce larger numbers of offspring as adults. The second view allows for independent gain and loss of the phenotype, depending on environmental and epidemiologial parameters, while the first would suggest that such a complex phenotype can only be lost.

Sukhdeo *et al.* (1997) presented an analysis of the phylogeny of strongylid nematodes which they claim demonstrated that the ancestral nematode was a skin-penetrating tissue migrator, and that one clade of parasites (the Trichostrongylidae and Heligmosomidae) has secondarily reduced or lost this tissue migration. The claim was based on a molecular phylogenetic analysis of mitochondrial COI genes from 12 species, including three from nematodes outside the Strongylida. As described above, the COI sequences do not yield a robust phylogeny (Fig. 1.3) and it is therefore not possible to make any inference about the direction of evolutionary change in terms of tissue migration phenotypes, or to derive an ancestral condition. The resolution of this question awaits further sequence analysis. However, the placement of *Heterorhabditis* as a sister taxon to the Strongylida suggests that invasion may have been the original phenotype, as *Heterorhabditis* invades the larvae of its insect vicitms by crossing the cuticle. Indeed, all the vertebrate-parasitic species of clades II, IV and V have closely related (and in IV and V closest sister) taxa that are insect parasites or pathogens. This association strongly suggests that insect parasitism may have been a repeated prelude to vertebrate parasitism.

Concluding Remarks

If we are to gain insight into the evolution of complex parasitic phenotypes such as those dealt with above, there must be rigorous application of both molecular phylogenetic inference and the comparative method (Harvey and Pagel, 1991; Skorping *et al.*, 1991; Read and Skorping, 1995a,b). The generation of datasets from large numbers of species will yield robust answers to questions of this kind in the near future.

References

Abad, P.S., Tares, S., Bruguier, N. and De Gurian, G. (1991) Characterization of the relationships in the pinewood nematode species complex (PWNSC) (*Bursaphelenchus* spp.) using a heterologous *unc-22* DNA probe from *Caenorhabditis elegans*. *Parasitology* 102, 303–308.

Aleshin, V.V., Kedrova, O.S., Milyutina, I.A., Vladychenskaya, N.S. and Petrov, N.B. (1998) Relationships among nematodes based on the analysis of 18S rRNA gene sequences: molecular evidence for a monophyly of chromadorian and secernentean nematodes. *Russian Journal of Nematology* 6, 175–184.

Anderson, R.C. (1992) *Nematode Parasites of Vertebrates. Their Development and Transmission.* CAB International, Wallingford, UK.

Anderson, T.J. (1995) *Ascaris* infections in humans from North America: molecular evidence for cross-infection. *Parasitology* 110, 215–219.

Anderson, T.J. and Jaenike, J. (1997) Host specificity, evolutionary relationships and macrogeographic differentiation among *Ascaris* populations from humans and pigs. *Parasitology* 115, 325–342.

Anderson, T.J.C., Romero-Abal, M.E. and Jaenike, J. (1993) Genetic structure and epidemiology of *Ascaris* populations: patterns of host affiliation in Guatemala. *Parasitology* 107, 319–334.

Anderson, T.J., Komuniecki, R., Komuniecki, P.R. and Jaenike, J. (1995a) Are mitochondria inherited paternally in *Ascaris? International Journal for Parasitology* 25, 1001–1004.

Anderson, T.J., Romero-Abal, M.E. and Jaenike, J. (1995b) Mitochondrial DNA and *Ascaris* microepidemiology: the composition of parasite populations from individual hosts, families and villages. *Parasitology* 110, 221–229.

Andrássy, I. (1976) *Evolution as a Basis for the Systematisation of Nematodes.* Akademiai Kiado, Budapest.

Baldwin, J.G., Frisse, L.M., Vida, J.T., Eddleman, C.D. and Thomas, W.K. (1997) An evolutionary framework for the study of developmental evolution in a set of nematodes related to *Caenorhabditis elegans. Molecular Phylogenetics and Evolution* 8, 249–259.

Bargmann, C.I. (1998) Neurobiology of the *Caenorhabditis elegans* genome. *Science* 282, 2028–2033.

Beckenbach, K., Smith, M.J. and Webster, J.M. (1992) Taxonomic affinities and intra- and inter-specific vartiation in *Bursaphelenchus* spp. as determined by the polymerase chain reaction. *Journal of Nematology* 24, 140–147.

Beckenbach, K., Blaxter, M.L. and Webster, J.M. (1999) Phylogeny of *Bursaphelenchus* species derived from analysis of ribosomal internal transcribed spacer DNA sequences. *Nematology* 1, 539–548.

Beveridge, I. and Durette-Desset, M.C. (1994) Comparative ultrastructure of the cuticle of trichostrongyle nematodes. *International Journal for Parasitology* 24, 887–898.

Blaxter, M.L. (1993) Nemoglobins: divergent nematode globins. *Parasitology Today* 9, 353–360.

Blaxter, M.L. (1998) *Caenorhabditis elegans* is a nematode. *Science* 282, 2041–2046.

Blaxter, M.L. and Bird, D.M. (1997) Parasites. In: Riddle, D., Blumenthal, T., Meyer, B. and Priess, J. (eds) *C. elegans II.* Cold Spring Harbor Laboratory Press, Cold Spring Harbor, New York.

Blaxter, M.L., Ingram, L. and Tweedie, S. (1994a) Sequence, expression and evolution of the globins of the nematode *Nippostrongylus brasiliensis. Molecular and Biochemical Parasitology* 68, 1–14.

Blaxter, M.L., Vanfleteren, J., Xia, J. and Moens, L. (1994b) Structural characterisation of an *Ascaris* myoglobin. *Journal of Biological Chemistry* 269, 30181–30186.

Blaxter, M.L., Daub, J., Waterfall, M., Guiliano, D., Williams, S., Jayaraman, K., Ramzy, R., Slatko, B. and Scott, A. (1997) The Filarial Genome Project. COST 819, The Commission of the European Community, Brussels.

Blaxter, M.L., De Ley, P., Garey, J., Liu, L.X., Scheldeman, P., Vierstraete, A., Vanfleteren, J., Mackey, L.Y., Dorris, M., Frisse, L.M., Vida, J.T. and Thomas, W.K. (1998) A molecular evolutionary framework for the phylum Nematoda. *Nature* 392, 71–75.

Blaxter, M.L., Aslett, M., Daub, J., Guiliano, D. and The Filarial Genome Project (1999) Parasitic helminth genomics. *Parasitology* 118, S39–S51.

Blouin, M.S., Yowell, C.A., Courtney, C.H. and Dame, J.B. (1997) *Haemonchus placei* and *Haemonchus contortus* are distinct species based on mtDNA evidence. *International Journal for Parasitology* 27, 1383–1387.

C. elegans Genome Sequencing Consortium (1998) Genome sequence of *Caenorhabditis elegans:* a platform for investigating biology. *Science* 282, 2012–2018.

Campbell, A.J.D., Gasser, R.B. and Chilton, N.B. (1995) Differences in a ribosomal DNA sequence of *Strongylus* species allows identification of single eggs. *International Journal for Parasitology* 25, 359–365.

Chervitz, S.A., Koonin, E.V., Aravind, L., Ball, C.A., Dwight, S.S., Harris, M.A., Dolinski, K., Sherlock, G., Mohr, S., Smith, T., Weng, S., Cherry, J.M. and Botstein, D. (1998) Comparison of the complete protein sets of worm and yeast: orthology and divergence. *Science* 282, 2022–2028.

Chilton, N.B., Gasser, R.B. and Beveridge, I. (1995) Differences in ribosomal DNA sequence of morphologically indistinguishable species within the *Hypodontus macropi* complex (Nematoda: Strongyloidea). *International Journal for Parasitology* 25, 647–651.

Chilton, N.B., Gasser, R.B. and Beveridge, I. (1997a) Phylogenetic relationships of Australian strongyloid nematodes inferred from ribosomal DNA sequence data. *International Journal for Parasitology* 27, 1481–1494.

Chilton, N.L.B., Hoste, H., Hung, G.-C., Beveridge, I. and Gasser, R.B. (1997b) The 5.8S rDNA sequences of 18 species of bursate nematodes (order Strongylida): comparison with Rhabditid and Tylenchid nematodes. *International Journal for Parasitology* 27, 119–124.

Chilton, N.B., Hoste, H., Newton, L.A., Beveridge, I. and Gasser, R.B. (1998) Common secondary structures for the second internal transcribed spacer pre-rRNA of two subfamilies of trichostrongylid nematodes. *International Journal of Parasitology* 28, 1765–1773.

Chitwood, B.G. and Chitwood, M.B. (1974) *Introduction to Nematology*. University Park Press, Baltimore.

Clark, N.D. and Berg, J.M. (1998) Zinc fingers in *C. elegans*: finding families and probing pathways. *Science* 282, 2018–2022.

Daub, J., Loukas, A., Pritchard, D.I. and Blaxter, M.L. (2000) A survey of genes expressed in adults of the human hookworm *Necator americanus*. *Parasitology* 120, 171–184.

De Guiran, G. and Boulbria, A. (1986) Le nématode des pins. Charactéristiques de la souche française et risque d'introduction et d'extension de *Bursaphelenchus xylophilus* en Europe. *OEPP/EPPO Bulletin* 16, 445–452.

De Guiran, G. and Bruguier, N. (1989) Hybridisation and phylogeny of pinewood nematode (*Bursaphelenchus* spp.). *Nematologica* 34, 321–330.

Ditlevsen, H. (1926) *Free-living Nematodes*. H. Hagerup, Copenhagen.

Dorris, M., De Ley, P. and Blaxter, M. (1999) Molecular analysis of nematode diversity. *Parasitology Today* 15, 188–193.

Durette-Desset, M., Gasser, R.B. and Beveridge, I. (1994) The origins and evolutionary expansion of the Strongylida (Nematoda). *International Journal for Parasitology* 24, 1139–1166.

Durette-Desset, M.C., Hugot, J.P., Darlu, P. and Chabaud, A.G. (1999) A cladistic analysis of the Trichostrongyloidea (Nematoda). *International Journal for Parasitology* 29, 1065–1086.

Ellis, R.E., Sulston, J.E. and Coulson, A.R. (1986) The rDNA of *C. elegans*: sequence and structure. *Nucleic Acids Research* 14, 2345–2364.

Fitch, D.H.A. and Thomas, W.K. (1997) Evolution. In: Riddle, D., Blumenthal, T., Meyer, D. and Priess, J. (eds) *C. elegans II*. Cold Spring Harbor Laboratory Press, Cold Spring Harbor, New York, pp. 815–850.

Fitch, D.H.A., Bugaj-gaweda, B. and Emmons, S.W. (1995) 18S ribosomal gene phylogeny for some rhabditidae related to *Caenorhabditis elegans*. *Molecular Biology and Evolution* 12, 346–358.

Gasser, R.B., Stewart, L.E. and Speare, R. (1996) Genetic markers in ribosomal DNA for hookworm identification. *Acta Tropica* 62, 15–21.

Gasser, R.B., Monti, J.R., Bao-Zhen, Q., Polderman, A.M., Nansen, P. and Chilton, N.B. (1998) A mutation scanning approach for the identification of hookworm species and analysis of population variation. *Molecular and Biochemical Parasitology* 92, 303–312.

Grant, W.N. (1994) Genetic variation in parasitic nematodes and its implications. *International Journal for Parasitology* 24, 821–830.

Harvey, P.H. and Pagel, M.D. (1991) *The Comparative Method in Evolutionary Biology*. Oxford University Press, Oxford.

Herniou, E.A., Pearce, A.C. and Littlewood, D.T.J. (1998) Vintage helminths yield valuable molecules. *Parasitology Today* 14, 289–292.

Hillis, D.M. (1987) Molecular versus morphological approaches to systematics. *Annual Review of Ecological Systems* 18, 23–42.

Hillis, D.M., Moritz, C. and Mable, B.K. (1996) *Molecular Systematics*. Sinauer Associates, Sunderland, Massachusetts.

Hoste, H., Chilton, N.B., Gasser, R.B. and Beveridge, I. (1995) Differences in the second internal transcribed spacer (ribosomal DNA) between five species of *Trichostrongylus* (Nematoda: Trichostrongylidae). *International Journal for Parasitology* 25, 75–80.

Hugall, A., Stanton, J. and Moritz, C. (1997) Evolution of the AT-rich mitochondrial DNA of the root-knot nematode, *Meloidogyne hapla*. *Molecular Biology and Evolution* 14, 40–48.

Hung, G.C., Chilton, N.B., Beveridge, I., McDonnell, A., Lichtenfels, J.R. and Gasser, R.B. (1997) Molecular delineation of *Cylicocyclus nassatus* and *C. ashworthi* (Nematoda: Strongylidae). *International Journal for Parasitology* 27, 601–605.

Hyman, B.C. and Beck Azevedo, J.L. (1996) Similar evolutionary patterning among repeated and single copy nematode mitochondrial genes. *Molecular Biology and Evolution* 13, 221–232.

Hyman, B.C. and Slater, T.M. (1990) Recent appearance and molecular characterization of mitochondrial DNA deletions within a defined nematode pedigree. *Genetics* 124, 845–853.

Iwahori, H., Tsuda, K., Kanzaki, N., Izui, K. and Futai, K. (1998) PCR-RFLP and sequencing analysis of ribosomal DNA of *Bursaphelenchus* nematodes related to pine wilt disease. *Fundamental and Applied Nematology* 21, 655–666.

Kampfer, S., Sturmbauer, C. and Ott, C.J. (1998) Phylogenetic analysis of rDNA sequences from adenophorean nematodes and implications for the Adenophorea–Secernetea controversy. *Invertebrate Biology* 117, 29–36.

Keddie, E.M., Higazi, T. and Unnasch, T.R. (1998) The mitochondrial genome of *Onchocerca volvulus*: sequence, structure and phylogenetic analysis. *Molecular and Biochemical Parasitology* 95, 111–127.

Lorenzen, S. (1994) *The Phylogenetic Systematics of Free-living Nematodes*. The Ray Society, London.

Maggenti, A. (1981) *General Nematology*. Springer-Verlag, New York.

Maggenti, A.R. (1983) Nematode higher classification as influenced by the species and family concepts. In: Stone, A.R., Platt, H.M. and Khalil, L.F. (eds) *Concepts in Nematode Systematics*. Academic Press, London, pp. 25–40.

Malakhov, V.V. (1994) *Nematodes. Structure, Development, Classification and Phylogeny*. Smithsonian Institution Press, Washington, DC.

Moore, T.A., Ramachandran, S., Gam, A.A., Neva, F.A., Lu, W., Saunders, L., Williams, S.A. and Nutman, T.B. (1996) Identification of novel sequences and codon usage in *Strongyloides stercoralis*. *Molecular and Biochemical Parasitology* 79, 243–248.

Nadler, S.A. (1992) Phylogeny of some ascaridoid nematodes, inferred from comparison of 18S and 28S rRNA sequences. *Molecular Biology and Evolution* 9, 932–944.

Nadler, S.A. (1995) Advantages and disadvantages of molecular phylogenetics: a case study of ascaridoid nematodes. *Journal of Nematology* 27, 423–432.

Nadler, S.A. and Hudspeth, D.S.S. (1998) Ribosomal DNA and phylogeny of the Ascaridoidea (Nemata: Secernentea): implications for morphological evolution and classification. *Molecular Phylogenetics and Evolution* 10, 221–236.

Newton, L.A., Chilton, N.B., Beveridge, I. and Gasser, R.B. (1998) Systematic relationships of some members of the genera *Oesophagostomum* and *Chabertia* (Nematoda: Chabertiidae) based on ribosomal DNA sequence data. *International Journal for Parasitology* 28, 1781–1789.

Nickle, W.R. (ed.) (1991) *Manual of Agricultural Nematology*. Marcel Dekker, New York.

Okimoto, R., MacFarlane, J.L., Clary, D.O. and Wolstenholme, D.R. (1992) The mitochondrial genomes of two nematodes, *Caenorhabditis elegans* and *Ascaris suum*. *Genetics* 130, 471–498.

Pelonquin, J.J., Bird, D.M., Kaloshian, I. and Matthews, W.C. (1993) Isolates of *Meloidogyne hapla* with distinct mitochondrial genomes. *Journal of Nematology* 25, 239–243.

Peng, W., Anderson, T.J., Zhou, X. and Kennedy, M.W. (1998) Genetic variation in sympatric *Ascaris* populations from humans and pigs in China. *Parasitology* 117, 355–361.

Platonova, T.A. and Gal'tsova, V.V. (1976) *Nematodes and their Role in the Meiobenthos*. Nakua, Leningrad.

Poinar, G.O. (1983) *The Natural History of Nematodes*. Prentice-Hall, Englewood Cliffs, New Jersey.

Powers, T.O., Harris, T.S. and Hyman, B.C. (1993) Mitochondrial DNA sequence divergence among *Meloidogyne incognita, Romanomermis culcivorax, Ascaris suum* and *Caenorhabditis elegans*. *Journal of Nematology* 25, 564–572.

Read, A.F. and Skorping, A. (1995a) Causes and consequences of life history variation in parasitic nematodes. In: Abad, P., Burnell, A., Laumond, C., Boemare, N. and Coudert, F. (eds) *Ecology and Transmission Strategies of Entomopathogenic Nematodes (COST 819)*. European Commission, Brussels, pp. 58–68.

Read, A.F. and Skorping, A. (1995b) The evolution of tissue migration by parasitic nematode larvae. *Parasitology* 111, 359–371.

Riddle, D.L. (1988) The dauer larva. In: Wood, W.B. (ed.) *The Nematode Caenorhabditis elegans*. Cold Spring Harbor Laboratory Press, Cold Spring Harbor, New York, pp. 393–412.

Riddle, D.L., Swanson, M.M. and Albert, P.S. (1981) Interacting genes in nematode dauer larva formation. *Nature* 290, 668–671.

Riddle, D., Blumenthal, T., Meyer, B. and Priess, J. (eds) (1997) *C. elegans II*. Cold Spring Harbor Laboratory Press, Cold Spring Harbor, New York.

Riga, E., Beckenbach, K. and Webster, J.M. (1992) Taxonomic relationships of *Bursaphelenchus xylophilus* and *B. mucronatus* based on interspecific and intra-specific cross-hybridisation and DNA analysis. *Fundamental and Applied Nematology* 15, 391–395.

Romstad, A., Gasser, R.B., Monti, J.R., Polderman, A.M., Nansen, P., Pit, D.S. and Chilton, N.B. (1997a) Differentiation of *Oesophagostomum bifurcum* from *Necator americanus* by PCR using genetic markers in spacer ribosomal DNA. *Molecular Cell Probes* 11, 169–176.

Romstad, A., Gasser, R.B., Nansen, P., Polderman, A.M., Monti, J.R. and Chilton, N.B. (1997b) Characterization of *Oesophagostomum bifurcum* and *Necator americanus* by PCR-RFLP of rDNA. *Journal of Parasitology* 83, 963–966.

Romstad, A., Gasser, R.B., Nansen, P., Polderman, A.M. and Chilton, N.B. (1998) *Necator americanus* (Nematoda: Ancylostomatidae) from Africa and Malaysia have different ITS-2 rDNA sequences. *International Journal for Parasitology* 28, 611–615.

Rutherford, T.A., Mamiya, Y. and Webster, J.M. (1990) Nematode-induced pine wilt disease: factors influencing its occurrence and distribution. *Forest Science* 36, 145–155.

Ruvkun, G. and Hobert, O. (1998) The genomic topography of *C. elegans* developmental control. *Science* 282, 2033–2041.

Skorping, A., Read, A.F. and Keymer, A.E. (1991) Life history covariation in intestinal nematodes of mammals. *Oikos* 60, 365–372.

Snutch, T.P. and Baillie, D.L. (1984) A high degree of DNA strain polymorphism associated with the major heat shock gene in *Caenorhabditis elegans*. *Molecular and General Genetics* 195, 329–335.

Stevenson, L.A., Chilton, N.B. and Gasser, R.B. (1995) Differentiation of *Haemonchus placei* from *H. contortus* (Nematoda: Trichostrongylidae) by the ribosomal DNA second internal transcribed spacer. *International Journal for Parasitology* 25, 483–488.

Stevenson, L.A., Gasser, R.B. and Chilton, N.B. (1996) The ITS-2 rDNA of *Teladorsagia circumcincta*, *T. trifurcata* and *T. davtiani* (Nematoda: Tricho-strongylidae) indicates these taxa are one species. *International Journal for Parasitology* 26, 1123–1126.

Sudhaus, W. (1976) Vergleichende Untersuchungen zur Phylogenie, Systematik, Ökologie, Biologie und Ethologie der Rhabditidae (Nematoda). *Zoologica* 43, 1–229.

Sukhdeo, S.C., Sukhdeo, M.V.K., Black, M.B. and Vrijenhoek, R.C. (1997) The evolution of tissue migration in parasitic nematodes (Nematoda: Strongylida) inferred from a protein-coding mitochondrial gene. *Biological Journal of the Linnean Society* 61, 281–298.

Swofford, D.L. (1999) PAUP* version 4.02b. Sinauer Associates, Sunderland, Massachusetts.

Swofford, D.L., Olsen, G.J., Waddell, P.J. and Hillis, D.M. (1996) Phylogenetic infer-ence. In: Hillis, D.M., Moritz, C. and Mable, B.K. (eds) *Molecular Systematics*. Sinauer Associates, Sunderland, Massachusetts, pp. 407–514.

Thomas, W.K., Vida, J.T., Frisse, L.M., Mundo, M. and Baldwin, J. (1997) DNA sequences from formalin-fixed nematodes: integrating molecular and morphological approaches to taxonomy. *Journal of Nematology* 29, 248–252.

Vanfleteren, J.R., Blaxter, M.L., Tweedie, S.A.R., Trotman, C., Lu, L., Van Heuwaert, M.-L. and Moens, L. (1994) Molecular genealogy of some nematode taxa as based on cytochrome *c* and globin amino acid sequences. *Molecular Phylogenetics and Evolution* 3, 92–101.

Viney, M. (1994) A genetic analysis of reproduction in *Strongyloides ratti. Parasitology* 109, 511–515.

Viney, M.E. (1996) Developmental switching in the parasitic nematode *Strongyloides ratti. Proceedings of the Royal Society of London, Series B* 263, 201–208.

Viney, M.E. (1999) Exploiting the life cycle of *Strongyloides ratti. Parasitology Today* 15, 231–235.

Webster, J.M., Anderson, R.V., Baillie, D.L., Beckenbach, K., Curran, J. and Rutherford, T.A. (1990) DNA probes for differentiating isolates of the pinewood nematode species complex. *Revue de Nematologie* 13, 255–263.

Wingfield, M.J., Blanchette, R.A. and Nicholls, T.H. (1984) Is the pinewood nematode an important pathogen in the United States? *Journal of Forestry* 82, 232–235.

Zarlenga, D.S., Lichtenfels, J.R. and Stringfellow, F. (1994a) Cloning and sequence analysis of the small subunit ribosomal RNA gene from *Nematodirus battus. Journal of Parasitology* 80, 342–344.

Zarlenga, D.S., Stringfellow, F., Nobary, M. and Lichtenfels, J.R. (1994b) Cloning and characterisation of ribosomal RNA genes from three species of *Haemonchus* (Nematoda: Trichostrongyloidea) and identification of PCR primers for rapid differentiation. *Experimental Parasitology* 78, 28–36.

The *Wolbachia* Endosymbionts of Filarial Nematodes

Claudio Bandi,[1] Timothy J.C. Anderson,[1]* Claudio Genchi[1] and Mark L. Blaxter[2]

[1]Istituto di Patologia Generale Veterinaria, Università di Milano, Milano, Italy; [2]Institute of Cell, Animal and Population Biology, University of Edinburgh, Edinburgh, UK

Introduction

Wolbachia endosymbionts are abundant in arthropods, where they promote a variety of reproductive manipulations, including feminization of genetic males, parthenogenesis and cytoplasmic incompatibility. *Wolbachia* is also present in filarial nematodes and has recently attracted a great deal of attention. This chapter reviews the studies so far published and discusses potential implications and future research prospects. Since this is a relatively young field, the chapter will also refer to unpublished studies and will include some speculation. The aim is to stimulate further work on the subject.

The Discovery and Rediscovery of Intracellular Symbiosis in Filarial Nematodes

At the beginning of the 1970s, ultrastructural investigations on the embryogenesis and fertilization of nematodes led to the observation of bacteria-like bodies in the oogonia, oocytes and embryos of the filarial nematode *Dirofilaria immitis* (Harada *et al.*, 1970; Lee, 1975). The bacterial nature of these bodies was fully recognized in 1975 and the presence of similar bacteria was reported for other filarial species, including *Brugia malayi* and

* Present address: Department of Genetics, Southwest Medical Foundation, PO Box 760549, San Antonio, TX 78245-0549, USA.

B. pahangi (McLaren *et al.*, 1975; Vincent *et al.*, 1975). Detailed studies on the tissue distribution and transovarial transmission of the bacteria of *B. malayi* and *Onchocerca volvulus* were then published (Kozek, 1977; Kozek and Figueroa, 1977). The possible implications of the presence of intracellular bacteria in filarial worms were fully recognized by the authors of these pioneering studies. The suggestion by Kozek (1977) that 'the possibility that the organisms within *B. malayi* cause some of the clinical symptoms and signs currently attributed to the worm warrant further detailed investigations' is particularly remarkable. Intracellular bacteria have subsequently been observed during studies on the ultrastructure of filarial nematodes (e.g. Franz and Andrews, 1986; Franz and Copeman, 1988) but until recently no investigations have been focused on these bacteria. The bacteria within *D. immitis* have been identified as members of the alpha proteobacteria (Bandi *et al.*, 1994) and have been shown to be closely related to the arthropod endosymbiont *Wolbachia* (Sironi *et al.*, 1995). *Wolbachia* endosymbionts are now known to be widespread among filarial nematodes, with ten species showing the infection out of the 11 so far examined (Bandi *et al.*, 1998; Genchi *et al.*, 1998; Henkle-Duhrsen *et al.*, 1998).

Background Information on Bacterial Symbiosis in Invertebrates

Intracellular symbiosis is extremely widespread in invertebrates. For example, mutualistic symbioses with intracellular bacteria can be found in almost all animal phyla, including sponges, cnidaria, nematodes, anellids, mollusca and arthropoda. Buchner (1965) thoroughly reviews most information published on bacterial symbiosis in animals up to 1964. After this monumental work, various reviews on more specific subjects have been published (e.g. Baumann, 1998, and references therein) including some recent reviews on *Wolbachia* (O'Neill *et al.*, 1997; Werren, 1997). In most of these papers, the term symbiosis is apparently used with a broad meaning: the intracellular bacterium is usually referred to as an endosymbiont even in the absence of data on effects on host fitness. Here only key points on intracellular symbiosis and *Wolbachia* will be summarized, so as to put the information available on symbiosis in filarial nematodes into a broader context.

Phylogenetic positionings of Wolbachia *and other intracellular* bacteria

Based on 16S rDNA analysis (Olsen and Woese, 1993), a dozen main lineages have been described for the eubacteria, and at least four of these lineages have been shown to include intracellular bacteria: the proteobacteria, the chlamydiae, the Gram-positives, and the flavobacteria-bacteroides (Werren and O'Neill, 1997). The proteobacterial group encompasses various lineages of intracellular symbionts, which are thought to have acquired their respective intracellular niches independently. For example, the beneficial symbionts of aphids have been assigned to the gamma proteobacteria (Baumann *et al.*, 1998), while *Wolbachia* has been placed into the alpha 2 subclass of the proteobacteria as a member of the Rickettsiales, one of the most typical families of intracellular bacteria (O'Neill *et al.*, 1992). Within this group, *Wolbachia* has been shown to be closely related to the genera *Anaplasma, Cowdria* and *Ehrlichia.*

Wolbachia, *vertical transmission, mutualistic symbiosis and reproductive parasitism*

Wolbachia is maternally transmitted to offspring in both arthropods and filarial nematodes. Paternal transmission of *Wolbachia* is thought to be rare in insect populations. Further investigations are required to confirm whether this is also the case in filarial nematodes (see later). The survival and reproduction of vertically transmitted endosymbionts is tightly linked to the survival and reproduction of its host. It is thus generally thought that vertically transmitted symbionts will increase their fitness by evolving mutualistic interactions with the hosts (Yamamura, 1993). A classic example of mutualistic symbiosis is found in aphids, where bacterial endosymbionts provide the host insect with essential amino acids (Baumann *et al.*, 1998). Inherited bacteria may also increase their fitness by manipulating host reproduction (reproductive parasitism: Hurst and Majerus, 1993; Werren and O'Neill, 1997). Reproductive parasites include those maternally transmitted microorganisms that are able to induce sex-ratio distortions toward females. For a maternally transmitted endosymbiont, increasing the proportion of females in the host population is a fairly obvious way to increase its fitness. In arthropods, *Wolbachia* determines sex-ratio distortions through parthenogenesis, feminization of genetic males or death of male embryos (Stouthamer *et al.*, 1990; Bouchon *et al.*, 1998; Hurst *et al.*, 1999).

A further kind of reproductive alteration determined by *Wolbachia* in arthropods is cytoplasmic incompatibility (CI). In CI, reduced fecundity is observed when infected males mate with uninfected females, or after matings between individuals harbouring different compatibility types of *Wolbachia* (bi-directional CI). Bi-directional CI may create reproductive barriers, which would prevent population fusion and possibly promote speciation (Werren, 1997). CI could derive from some kind of *Wolbachia*-induced modification of spermatozoans (e.g. through a 'sterilizing toxin') which is rescued in eggs harbouring *Wolbachia* (e.g. through an 'antitoxin'). In other words, *Wolbachia* is able to sterilize those females that do not carry

Wolbachia. In populations harbouring CI-inducing *Wolbachia,* the fitness of uninfected females is thus lower than that of infected ones. Maternally transmitted bacteria with the ability to reduce the fitness of uninfected females should spread in the host populations (Fine, 1978). Field experimental evidence from populations of *Drosophila simulans* confirms this prediction (Turelli and Hoffmann, 1991). *Wolbachia* is also present in various insects where its effects (if any) are not known. In addition, a *Wolbachia* has been described that is pathogenic to fruit flies (Min and Benzer, 1997).

Distribution of Wolbachia *in arthropods*

Wolbachia is estimated to occur in around 20% of insect species, as well as in mites and isopod crustaceans (Werren, 1997). Despite its widespread distribution among arthropod families, *Wolbachia* is patchily distributed within families and populations. Phylogenetic analyses of arthropod *Wolbachia* using 16S rRNA, *ftsZ, GroEL* and *Wolbachia* surface protein (*wsp*) genes have provided further insights into the evolutionary dynamics of *Wolbachia* infection in arthropods: host and bacterial phylogenies are discordant, suggesting horizontal transmission (O'Neill *et al.,* 1992; Werren *et al.,* 1995; Masui *et al.,* 1997; Zhou *et al.,* 1998). Evidence for horizontal transmission also comes from studies on host–parasitoid systems and from the fact that individual insects have been found infected by distantly related wolbachiae (Werren, 1997). These distribution and phylogenetic patterns are consistent with theoretical models on the population biology of a bacterium inducing CI: both sweeps of new compatibility types and losses of infection are expected to occur in the host populations. New compatibility types could thus invade already infected populations promoting some inconsistency of the host–symbiont phylogenies. On the other hand, once *Wolbachia* has become fixed within a population, selective pressures for maintaining the ability to modify sperm should become lower. This could be followed by the loss of the sterilizing trait and, finally, by the loss of the infection (Hurst and McVean, 1996).

Distribution of *Wolbachia* in Filarial Worms

Within the body of filarial nematodes, intracellular bacteria have been observed by electron microscopy in the lateral cords of both males and females. Within the cell cytoplasm, the bacteria are in membrane-bound vacuoles. In some cases, the cytoplasm of lateral cord cells is filled by bacteria: these bacteria-filled cells resemble in some ways insect bacteriocytes (Baumann *et al.,* 1998). In female worms, bacteria are present also in the oogonia, oocytes, developing embryos and in the cell layer surrounding the oviduct (McLaren *et al.*, 1975; Kozek, 1977; Kozek and Figueroa, 1977; Taylor *et al.*, 1999). Bacteria have also been observed in microfilariae and in second-, third- and fourth-stage larvae (Kozek, 1977). The presence of bacteria in the male reproductive system has not yet been recorded. More recently, antibodies directed against bacterial catalase and *GroEL* have been used for immunohistochemical staining of bacteria in filarial nematodes (Henkle-Duhrsen *et al.*, 1998; Hoerauf *et al.*, 1999). The staining of the lateral cords and of the female reproductive tract supports previous EM observations on the tissue distribution of filarial bacteria. It is not yet clear whether other bacteria in addition to *Wolbachia* are present in the body of filarial worms. However, direct sequencing of 16S rDNAs amplified using universal eubacterial primers produced unambiguous *Wolbachia* sequences from both *D. immitis* and *B. pahangi* (Sironi *et al.*, 1995; Bandi *et al.*, 1999). In addition, cloned 16S rDNA PCR products obtained from *B. pahangi* were all found to be *Wolbachia* sequences (Bandi *et al.*, 1999). These results suggest that *Wolbachia* is the only (or at least the dominant) bacterium in the body of these nematodes.

Out of the 11 species of filarial nematode so far examined, *Wolbachia* has been detected in ten species (Table 2.1). The presence of *Wolbachia* in these ten species has been revealed by PCR followed by sequencing of the amplified products (Sironi *et al.*, 1995; Bandi *et al.*, 1998; Genchi *et al.*, 1998; Henkle-Duhrsen *et al.*, 1998). That all the *Wolbachia* sequences obtained from the ten different species were different rules out the possibility that contaminating PCR products were sequenced. In addition, the results of PCR analyses on the presence/absence of *Wolbachia* in filarial nematodes are consistent with the EM data on the presence/absence of intracellular bacteria (Table 2.1). Indeed, the species in which *Wolbachia* has not so far been detected – *Acanthocheilonema viteae* – was also recorded as not harbouring intracellular bacteria in previous EM studies (McLaren *et al.*, 1975). It is uncertain whether *A. viteae* is naturally *Wolbachia*-free or whether symbionts were lost during laboratory maintenance of the strains. The phylogenetic position of *A. viteae* is also uncertain and other species of the genus *Acanthocheilonema* have not yet been examined for the presence of *Wolbachia*. If the lineage leading to *Acanthocheilonema* is ancestral to the *Wolbachia*-infected lineages, this would help to establish when *Wolbachia* first infected filarial worms (Casiraghi *et al.*, 2001). On the other hand, if *A. viteae* has infected ancestors, this would suggest that loss of *Wolbachia* has occurred in the course of filarial evolution. These questions illustrate the need for robust phylogenies of filarial nematodes for understanding the evolution of *Wolbachia* symbiosis.

Exhaustive surveys have not been carried out to determine the prevalence of infection within a single species of filarial nematode. However, infection was found in all eight specimens of *D. immitis* collected from worldwide locations, indicating that infection prevalence is likely to be close to 100% (Sironi *et al.*, 1995). Sequences of *Wolbachia ftsZ* genes from

Table 2.1.   Filarial nematodes infected with intracellular bacteria. The method of detection is shown. Neither electron microscopy nor the immuno-histochemical staining techniques used are to be regarded as *Wolbachia* specific (see note). Positive identification of intracellular bacteria as *Wolbachia* is shown only where PCR amplified products of rRNA or *ftsZ* genes have been sequenced.

Filaria species	PCR and sequencing	Electron microscopy	Immuno-histochemistry
Dirofilaria immitis	Yes[1]	Yes[1,7]	Yes[3]
D. repens	Yes[2]	Yes[7]	nd
Onchocerca fasciata	nd	Yes[8]	Yes[3]
O. flexuosa	nd	No[9]	No[3]
O. jakutensis	nd	Yes[9]	nd
O. gibsoni	Yes[2]	Yes[10]	Yes[3]
O. gutturosa	Yes[2]	nd	nd
O. ochengi	Yes[2]	Yes[8]	Yes[3]
O. volvulus	Yes[3]	Yes[11,12]	Yes[3]
Loa loa	nd	No[7,13]	nd
Mansonella ozzardi	nd	Yes[14]	nd
Brugia malayi	Yes[2,4]	Yes[4,15,16]	nd
B. pahangi	Yes[2,4,5]	Yes[5,7,16]	nd
Wuchereria bancrofti	Yes[2,4]	Yes[4]	nd
Litomosoides sigmodontis	Yes[6]	No/Yes[1,6]	Yes[6]
Dipetalonema setariosum	nd	nd	Yes[1]
Acanthocheilonema viteae	No[2]	No[7]	nd

[1]Sironi *et al.* (1995); [2]Bandi *et al.* (1998); [3]Henkle-Duhrsen *et al.* (1998); [4]Taylor *et al.* (1999); [5]Bandi *et al.* (1999); [6]Hoerauf *et al.* (1999); [7]McLaren *et al.* (1975); [8]Determann *et al.* (1997); [9]Plenge-Bonig *et al.* (1995); [10]Franz and Copeman (1988); [11]Kozek and Figueroa (1977); [12]Franz and Buttner (1983); [13]Franz *et al.* (1984); [14]Kozek and Raccurt (1983); [15]Kozek (1977); [16]Vincent *et al.* (1975).
Immuno-histochemical staining of intracellular bacteria in filarial nematodes has been obtained using antibodies against *GroEL* and catalase (Henkle-Duhrsen *et al.*, 1998; Hoerauf *et al.*, 1999); the specificity of these antibodies is unknown, but it is expected to be low because both *GroEL* and catalase show high level of amino acid conservation throughout the proteobacteria.
nd = not done.

these *D. immitis* specimens were extremely conserved, with only two synonymous changes in the *ftsZ* sequence from a nematode specimen from Cuba relative to the specimens from the other locations (Bandi *et al.*, 1998). More recently, further specimens of *D. immitis* and *Dirofilaria repens* from different Italian locations have been shown to harbour *Wolbachia*, and their *ftsZ* and *wsp* sequences have been shown to be identical (Bazzocchi *et al.*, 2000a; T.J.C. Anderson, C. Bandi, C. Bazzocchi and G. Favia, 2000, unpublished results). Similarly, in the cattle filaria *Onchocerca ochengi*, *Wolbachia* prevalence has been shown to be 100% (Langworthy *et al.*, 2000). In *D. immitis*, all males and females so far examined have been found to be infected

(Sironi *et al.*, 1995). In surveys of the human parasite *B. malayi*, *Wolbachia* was detected by PCR in all females and in 25% of males (Taylor *et al.*, 1999). Re-analysis of male worms using a nested PCR revealed that all individual male worms were infected (Taylor *et al.*, 2000).

Phylogeny of *Wolbachia* in Filarial Worms

Phylogenetic relationships among *Wolbachia* from filarial nematodes and from arthropods have been examined through the analysis of *ftsZ*, *wsp* and 16S rRNA genes. Four main *Wolbachia* lineages have been described: A and B (from arthropods: Werren *et al.*, 1995) and C and D (from nematodes: Bandi *et al.*, 1998; Bazzocchi *et al.*, 2000a). Figure 2.1 shows the relationship between nematode wolbachiae and representative arthropod

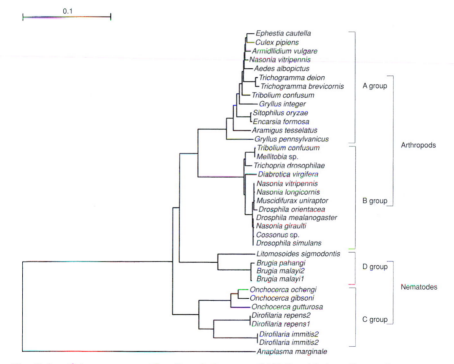

Fig. 2.1. A tree representing the phylogeny of *Wolbachia* in arthropods (groups A and B) and filarial nematodes (groups C and D). Group designations correspond to those proposed by Werren *et al.* (1995) and by Bandi *et al.* (1998). The names at the terminal nodes are those of the host species. The tree is based on the *ftsZ* gene sequence alignment used by Bandi *et al.* (1998). The tree was obtained using a distance matrix method (Jukes and Cantor correction; neighbour-joining method).

wolbachiae. The main conclusions of phylogenetic analysis of filarial wolbachiae (Bandi *et al.*, 1998) are set out below.

Interphylum transmission

The phylogenies based on *ftsZ, wsp* and 16S rDNA are star-like, with deep splits between the four main lineages. *Wolbachia* from *Dirofilaria* spp. and *Onchocerca* spp. form the C lineage, while bacteria from *B. malayi, Wuchereria bancrofti* and *Litomosoides sigmodontis* form the D lineage. The evolutionary split between C and D lineages was estimated to have occurred about 100 million years ago (Bandi *et al.*, 1998). Similarly, the level of nucleotide divergence between both C and D clusters with A and B clusters suggests a divergence date approximately 100 million years ago. These estimations are very approximate, but strongly suggest that recent horizontal transfer has not occurred between arthropods and nematodes. However, the date of the split between arthopods and nematodes (> 600 million years ago) clearly predates the split between the lineages of *Wolbachia* infecting nematodes and arthropods. As such, it is clear either that interphylum transmission occurred at some point in the past or, alternatively, that both arthropods and nematodes were infected from a third source.

Is Wolbachia *monophyletic?*

One aim of the phylogenetic work was to resolve whether arthropod or nematode wolbachiae are ancestral. It has proved difficult to resolve the sister group relationships between the four major lineages of *Wolbachia* using *ftsZ, wsp* and rDNA data sets, although a recent study has provided weak evidence for the monophyly of nematode wolbachiae (Casiraghi *et al.*, 2000). Two factors may hamper the clear resolution of the relationship between the four groups. Firstly, the outgroup taxa used (*Anaplasma marginale, Ehrlichia* spp., *Cowdria* spp.) are relatively distant from *Wolbachia* (unfortunately, no more suitable outgroup taxa are known). Secondly, the sequences may be evolving too rapidly for analysis of ancient divergence events. Analysis of more slowly evolving sequences will be required to resolve this question.

Horizontal transmission of Wolbachia *between nematodes*

Comparison of host and symbiont phylogenies is a powerful method for assessing the occurrence and frequency of horizontal transmission (Moran and Baumann, 1994). There are limited sequence data available for assessing the phylogeny of filarial nematodes. Comparison of *Wolbachia* phylogeny with the available molecular phylogenies (Xie *et al.*, 1994; Casiraghi *et al.*, 2001) and a morphology-based classification (Anderson, 1992) of filarial nematodes does not provide evidence for discordance: all phylogenetic relationships which are well established for the host nematodes hold for the symbionts. Thus, the phylogenetic data suggest long-term association between filarial nematode and *Wolbachia* and strict vertical transmission. However, the host nematode phylogeny is currently poorly resolved and the number of *Wolbachia*/nematode associations examined is as yet quite small, so this conclusion should be viewed with caution. In particular, further data are needed to assess critically the correct placement of the genera *Acanthocheilonema* and *Litomosoides* (Bandi *et al.*, 1998; Casiraghi *et al.*, 2001).

Should Nematode *Wolbachia* Behave Differently from Arthropod *Wolbachia*?

The reproductive manipulations used by *Wolbachia* in arthropods are thought to be related to the uniparental mode of transmission. Unlike insects, which have flagellated sperm containing relatively little cytoplasm, many nematodes have large amoeboid sperm, which contain multiple mitochondria and may transmit significant numbers of *Wolbachia* (Scott, 1996). If there is a significant paternal component to transmission, then CI, feminization, parthenogenesis induction and male killing are rather unlikely to have evolved in nematode *Wolbachia*. Experimental crosses between *B. pahangi* and *B. malayi* have demonstrated *Wolbachia* transmission through female worms only (Taylor *et al.*, 1999). In addition, electron microscopy studies have not provided evidence for the presence of *Wolbachia* in sperm (Lee, 1975; Kozek, 1977). However, further studies are necessary to exclude the possiblity of significant paternal transmission.

In view of the diversity of reproductive effects induced by arthropod *Wolbachia*, it would not be surprising if *Wolbachia* behaves in different ways in different filarial nematodes. The C and D *Wolbachia* groups are estimated to have diverged approximately 100 million years ago (see earlier). Given this fact, it is quite possible that *Wolbachia* in different lineages have followed different evolutionary trajectories.

The *Wolbachia*–Filaria Relationship: an Obligate Mutualism?

Antibiotic 'curing' experiments provide a powerful approach in investigating endosymbiont effects on host biology and have been widely used in work on arthropod *Wolbachia*. A number of such experiments have been inadvertently conducted on filarial worms, in the course of testing the effects of antibiotic treatments on these nematodes. As early as 1973, the bacteriostatic drug tetracycline was shown to have prophylactic effects against the infection of the experimental mammalian host (the gerbil) by two filarial species, *B. malayi* and *L. sigmodontis* (for a retrospective discussion of this study, see McCall *et al.*, 1999). This study also showed that tetracycline is ineffective against *A. viteae*, which is now known not to harbour *Wolbachia*. Subsequently it has been shown that tetracycline inhibits larval development in mosquitoes (Sucharit *et al.*, 1978) and production of microfilariae in gerbils (Bosshardt *et al.*, 1993). More recently, tetracycline has been shown to inhibit embryogenesis in *D. immitis* (Genchi *et al.*, 1998). Treatment with tetracycline is known to be immunosuppressive in mammals, and inhibition of nematode development and reproduction in gerbils was regarded as an unexpected effect of this drug. The observed nematode attrition could thus result from direct effects of tetracycline on nematodes or could be a secondary effect of the action of the drug on *Wolbachia*.

Two studies have attempted to determine whether tetracycline treatment on experimental hosts infected by filariae had any effect on the wolbachiae harboured by these filariae. It has been shown that tetracycline treatment causes degeneration of *Wolbachia* and inhibition of *Wolbachia* transmission in three filarial species, *B. pahangi*, *D. immitis* and *L. sigmodontis* (Bandi *et al.*, 1999; Hoerauf *et al.*, 1999). In agreement with the results of previous studies, inhibitory effects of tetracycline on the reproduction and development of these nematodes have been recorded. Hoerauf *et al.* (1999) have also confirmed that tetracycline treatment is ineffective against the *Wolbachia*-free filaria *A. viteae*. More recently, intermittent tetracycline treatments have been shown to lead to *Wolbachia* degeneration and to the death of adult *O. ochengi* worms in cattle (Langworthy *et al.*, 2000). There is thus an overall consistency of data supporting the possibility that the effects of tetracycline on filarial nematodes are mediated by some effects on their *Wolbachia* endosymbionts.

However, the interpretation of these data is not obvious. One simple interpretation is that the *Wolbachia*/nematode system is an obligate symbiosis. If *A. viteae* lost *Wolbachia* during laboratory maintenance or recent evolutionary history, this would cast doubt about whether the symbiosis is truly obligate. A second interpretation of the experimental data is that the effects observed on filarial reproduction result from CI (see discussion in Genchi *et al.*, 1998, and in Bandi *et al.*, 1999) while the effects on development and long-term survival of the worms derive from some interference with the mutualistic interactions between the nematode and *Wolbachia*. For example, embryo degeneration following tetracycline treatment could derive from the fusion of sperm produced before the start of the treatment and stored in the spermatheca (and thus modified by *Wolbachia*) with eggs produced after the start of treatment (and thus unable to rescue the modified sperm). If this is the case, continuing observations for a long period after the end of treatment might reveal resumption of embryogenesis

due to generation of unmodified sperm. However, in humans harbouring
O. volvulus, embryogenesis did not appear to resume even 4 months after
the end of the treatment (Hoerauf *et al.*, 2000a). It is also possible that tetracycline could act directly on the nematode. Indeed, tetracycline is known to
accumulate in the body of *D. immitis* (Tobie and Beye, 1960). Whether this
has a direct detrimental effect on the nematode is unknown, thus it would
be interesting to test whether antibiotics different from tetracycline, but
effective on *Wolbachia*, have any effect on filarial nematodes. A study using
rifampicin seems to suggest that this is the case (Townson *et al.*, 1999). For
the use of antibiotics which were ineffective against both *Wolbachia* and the
filarial nematode *L. sigmodontis*, see Hoerauf *et al.* (2000b).

The *Wolbachia* Catalase

The effectors of the mammalian host immune attack against filaria include
reactive oxygen intermediates. Filarial nematodes express glutathione
peroxidase, thioredoxin peroxidase and superoxide dismutase at their
surface – enzymes believed to protect the nematode from this attack
(Selkirk *et al.*, 1998). A bacterial catalase gene has been identified that most
probably derives from the endosymbiont genome (Henkle-Duhrsen *et al.*,
1998); this enzyme may contribute with other enzymes to the protection of
both *Wolbachia* and its nematode host from oxygen radicals.

Genomics and Proteomics

Sequencing the genome of filarial *Wolbachia* would possibly allow
identification of: (i) genes implicated in the biology of the *Wolbachia*–filaria
symbiosis; (ii) prokaryotic targets for the control of filariases; and (iii)
proteins implicated in the pathology of the diseases (see later) or useful in
serological diagnosis.

Genome sequencing and genome comparative studies would provide a
complementary approach for investigating the *Wolbachia*–filaria relationship (Kalman *et al.*, 1999). Identifying missing genes in the endosymbionts
(or the host) will reveal whether the symbiont (or the host) relies
metabolically upon the host cell (or upon the symbiont). In other obligate
intracellular bacteria, genome sequencing has shown that the bacteria have
discarded almost all of their biosynthetic machinery in favour of scavenging
nutrients from the host (Andersson *et al.*, 1998; Stephens *et al.*, 1998). For
example, there are only two genes that can be implicated in amino acid
biosynthesis in the genome of *Rickettsia prowazekii* (Andersson *et al.*, 1998).
Finding genetic evidence for upregulated biosynthetic pathways in the
endosymbiont genome would aid understanding of what *Wolbachia* provide
for the host cell. It is well known, for example, that aphid bacterial

endosymbionts provide the host insect with some essential amino acids: in this symbiotic system, evidence for overexpression of the relevant proteins has been obtained through molecular genetic studies (Bäumann *et al.*, 1998). A further approach to finding evidence for overexpression of *Wolbachia* proteins would be 2D-electrophoresis-proteomics. In addition to metabolic genes, proteins of the endosymbiont cell wall are a focus of interest due to their potential role in nutrient uptake, waste excretion, interaction with the host cytoskeleton (particularly during mitosis), cell wall synthesis and extracellular activities.

An initiative to investigate the genome of *B. malayi* was established in 1994 (Blaxter *et al.*, 1999). A number of expressed sequence tags (ESTs) generated by the project appeared to derive from *Wolbachia* rather than the host genome. These ESTs are presumed to be *Wolbachia*-derived owing to their lack of similarity to other nematode (or eukaryotic) genes and high identity to genes from *Rickettsia* or other alpha proteobacteria (D. Guiliano and M.L. Blaxter, unpublished results). There may still be unidentified *Wolbachia* ESTs in the Filarial Genome Project dataset if the genes from which they derive are highly diverged or unique to *Wolbachia* of filaria. In addition, during the construction of a physical map of the *B. malayi* genome using bacterial artificial chromosomes (BAC), several BAC inserts appeared to derive from the endosymbiont chromosome (D. Guiliano, B. Slatko, J. Foster and M.L. Blaxter, unpublished results). These BACs have been end sequenced and also screened for hybridization to the presumed *Wolbachia* ESTs and other PCR-amplified *Wolbachia* genes (*ftsZ*, 16S rRNA, *wsp*). Contig analysis suggests that about 700 kb of the *Wolbachia* genome has been isolated in this first screen. Further contig analysis and chromosome walking from the current BACs is in progress. Of note is the discovery of a repetitive element in the *B. malayi Wolbachia* genome present in more than four copies and with sequence similarity to the IS200-like elements of an insect *Wolbachia* (Masui *et al.*, 1999; D. Guiliano and M.L. Blaxter, unpublished results).

It is expected that different *Wolbachia* genomes (from both arthropods and filarial nematodes) and genomes of other intracellular bacteria will be sequenced within the next few years (Pennisi, 1999). Comparisons of *Wolbachia* genomes and those of other intracellular bacteria would also provide a more general understanding of the evolution of 'resident genomes' (Andersson and Kurland, 1998; Kalman *et al.*, 1999). Finally, phylogenetics on the *Wolbachia* genes would allow investigation of whether host genes have been transferred into the symbiont genome.

Implications for the Pathogenesis of Filarial Diseases

Immunopathological phenomena play an important role in the pathology of filariases (such as blindness caused by *O. volvulus* and elephantiasis caused by *W. bancrofti* and *B. malayi*) (e.g. Ottesen, 1992; Cooper *et al.*, 1997; Freedman, 1998). Both nematode and *Wolbachia* antigens may be involved in disease pathogenesis. A major advantage of focusing research on *Wolbachia* antigens derives from the feasibility of identifying candidate immunomodulatory/proinflammatory/antigenic molecules from bacteria. The integration of proteomics, Western blotting and genomics could provide a straightforward approach to identifying candidate protein antigens and could allow rapid designing of primers for amplifying, cloning and expressing their genes. One productive strategy might be to target molecules that have already been shown in other bacteria to be antigenic or implicated in determining innate immune responses. Initial investigations in this area have revealed that cats infected with *D. immitis* generate a specific antibody response against the *Wolbachia* suface protein (Bazzocchi *et al.*, 2000b) and that *Wolbachia* GroEl (HSP-60) might be implicated in the activation of inflammatory responses (Ford *et al.*, 2000). It is notable that the gene coding for *GroEl* had been identified previously from an *O. volvulus* expression library using human infection sera (A.C. Koszarski and M. Gallin, 1997; unpublished GenBank accession Y09416).

Being a member of the proteobacteria, a group that encompasses the most typical Gram-negative bacteria, *Wolbachia* is expected to have lipopolysaccharide (LPS) in the outer envelope of the cell wall. Indeed, LPS (also known as endotoxin) has already been found in several members of the Rickettsiales (e.g. Amano and Williams, 1984). LPS typically induces the release of TNF alpha and IL-1 from macrophages, and this promotes a sequela of phenomena involving cells of the immune system and endothelial cells (Parillo, 1990). The outcomes can be both local and generalized, with dramatic consequences such as the septic shock syndrome. Does *Wolbachia* have LPS? LPS have not yet been purified from *Wolbachia*, but Taylor *et al.* (2000) have provided convincing evidence for the presence of LPS in crude extracts of *B. malayi* and for their role in the pathogenesis of lymphatic filariasis. Evidence has also been provided for the involvement of LPS in the pathogenesis and immunoregulation of human onchocerciasis (Brattig *et al.*, 2000).

Implications for Treatment

If *Wolbachia* is necessary to the host worm, the bacterium could become a target for the control of filariases. The prophylactic effect of tetracycline against filarial infection is particularly encouraging. In addition, tetracycline treatment inhibits the embryogenesis of filarial worms (see earlier). However, in view of the role of microfilariae in the pathogenesis of some filarial diseases (e.g. river blindness, caused by *O. volvulus*), treatments that reduce microfilaria production are worth testing in clinical trials, in particular when based on inexpensive drugs such as tetracycline. Indeed, a recent study has shown that tetracycline treatment results in long-term sterilization of *O. volvulus* females in humans (Hoerauf *et al.*, 2000a). It has also been shown that tetracycline can kill adult worms of *O. ochengi*, a close relative of *O. volvulus* (Langworthy *et al.*, 2000). It is notable that ivermectin, the drug currently used for controlling the microfilaraemia in *O. volvulus*, does not cause death or even sterilization of adult worms.

Treatments targeted toward the symbionts could also be used to reduce adverse reactions to chemotherapy, such as the systemic reaction which follows larvicidal treatment of dog and human patients with a high level of microfilariae, or adulticide heartworm treatment in dogs (e.g. Boreham and Atwell, 1983; Turner *et al.*, 1994). These side effects are thought to derive partially from the release of antigenic, inflammatory and toxic substances that follows worm damage and death. In addition to the recent evidence for the presence of LPS in the body of filarial worms, it is known that the injection of crude extracts of *D. immitis* triggers pathological effects which resemble those caused by LPS (Kitoh *et al.*, 1994). If endosymbiont-derived substances are shown to play some role in pathogenesis, antibiotic treatment prior to administration of microfilaricidal or adulticide drugs could reduce the density of *Wolbachia* present and thereby reduce side effects.

Concluding Remarks

This chapter focuses on *Wolbachia* in filarial nematodes. EM studies have also revealed intracellular bacteria in other nematodes (e.g. Shepperd *et al.*, 1973; Marti *et al.*, 1995) and the bacterial endosymbionts of plant pathogenic nematodes belonging to the genus *Xiphinema* have recently been identified as belonging to the verrucomicrobia group (Vandekerckhove *et al.*, 2000). However, most nematode bacteria are still to be identified. These may also play important roles in nematode biology.

Notes Added in Proof

1. *Implications for pathogenesis.* A recent study showed that the plasma levels of IL-6 and LPS-binding protein are associated with the severity of adverse reactions after diethylcarbamazine treatment of microfilaraemic patients (Haarbrink *et al.*, 2000, *Journal of Infectious Diseases* 182, 564–569). The authors discuss the possibility that *Wolbachia* LPS are involved in the side effects of therapy.

2. *Implications for treatment.* It has recently been shown that tetracycline inhibits the development of filarial nematodes from L3 to L4 *in vitro* (Smith and Rajan, 2000, *Experimental Parasitology* 95, 265–270). However, chloramphenicol, erythromycin and ciprofloxacin failed to inhibit the

L3–L4 moulting. The authors discuss the possibility that the effects of tetracycline on the nematode are direct (and not mediated by an effect on *Wolbachia*).

References

Amano, K. and Williams, J.C. (1984) Chemical and immunological characterisation of lipopolysaccharides from phase I and phase II *Coxiella burnetii*. *Journal of Bacteriology* 160, 994–1002.

Anderson, R.C. (1992) *Nematode Parasites of Vertebrates: their Development and Transmission*. CAB International, Wallingford, UK.

Andersson, S.G. and Kurland, C.G (1998) Reductive evolution of resident genomes. *Trends in Microbiology* 6, 263–268.

Andersson, S.G., Zomorodipour, A., Andersson, J.O., Sicheritz-Ponten, T., Alsmark, U.C., Podowski, R.M., Naslund, A.K., Eriksson, A.S., Winkler, H.H. and Kurland, C.G. (1998) The genome sequence of *Rickettsia prowazekii* and the origin of mitochondria. *Nature* 396, 133–140.

Bandi, C., Damiani, G., Sacchi, L., Bardin, M.G., Sironi, M., Fani, R. and Magrassi, L. (1994) Caratterizzazione molecolare e identificazione di simbionti intracellulari. *Fondazione Iniziative Zooprofilattiche e Zootecniche Brescia* 37, 53–57.

Bandi, C., Anderson, T.J.C., Genchi, C. and Blaxter, M. (1998) Phylogeny of *Wolbachia* in filarial nematodes. *Proceedings of the Royal Society of London B* 265, 2407–2413.

Bandi, C., McCall, J.W., Genchi, C., Corona, S., Venco, L. and Sacchi, L. (1999) Effects of tetracycline on the filarial worms *Brugia pahangi* and *Dirofilaria immitis* and their bacterial endosymbionts *Wolbachia*. *International Journal for Parasitology* 29, 357–364.

Baumann, P. (1998) Symbiotic associations involving microorganisms. *BioScience* 48, 254–255.

Baumann, P., Baumann, L., Clark, M.A. and Thao, M.L. (1998) *Buchnera aphidicola*: the endosymbiont of aphids. *ASM News* 64, 203–209.

Bazzocchi, C., Jamnongluk, W., O'Neill, S.L., Anderson, T.J.C., Genchi, C. and Bandi C. (2000a) *wsp* gene sequences from the *Wolbachia* of filarial nematodes. *Current Microbiology* 41, 96–100.

Bazzocchi, C., Ceciliani, F., McCall, J.W., Ricci, I., Genchi, C. and Bandi, C. (2000b) Antigenic role of the endosymbionts of filarial nematodes: IgG response against the *Wolbachia* surface protein in cats infected with *Dirofilaria immitis*. *Proceedings of the Royal Society of London B* (in press).

Blaxter, M.L., Aslett, M., Daub, J., Guiliano, D. and The Filarial Genome Project (1999) Parasitic helminth genomics. *Parasitology* 118, S39–S51.

Boreham, P.F.L. and Atwell, R.B. (1983) Adverse drug reactions in the treatment of filarial parasites: haematological, biochemical, immunological and pharmacological changes in *Dirofilaria immitis* infected dogs treated with diethylcarbamazine. *International Journal for Parasitology* 13, 547–556.

Bosshardt, S.C., McCall, J.W., Coleman, S.H., Jones, K.L., Petit, T.A. and Klei, T.L. (1993) Prophylactic activity of tetracycline against *Brugia pahangi* infection in Jirds (*Meriones unguiculatus*). *Journal of Parasitology* 79, 775–777.

Bouchon, D., Rigaud, T. and Juchault, P. (1998) Evidence for widespread *Wolbachia* infection in isopod crustaceans: molecular identification and host feminization. *Proceedings of the Royal Society of London B* 265, 1081–1090.

Brattig, N.W., Rathjens, U., Ernst, M., Geisinger, F., Tischendorf, F.W. and Renz, A. (2000) Lipopolysaccharide-like molecules from *Wolbachia* endobacteria of the filaria *Onchocerca volvulus* are candidate mediators in the sequence of inflammatory and anti-inflammatory responses of human monocytes. *Microbes and Infection* 2, 1–11.

Buchner, P. (1965) *Endosymbiosis of Animals with Plant Microorganisms.* John Wiley & Sons, New York.

Casiraghi, M., Anderson, T.J.C., Bandi, C., Bazzocchi, C. and Genchi, C. (2001) A phylogenetic analysis of filarial nematodes: comparison with the phylogeny of *Wolbachia* endosymbionts. *Parasitology* (in press).

Cooper, P.J., Guderian, R.H., Proano, R. and Taylor, D.W. (1997) The pathogenesis of chorioretinal disease in onchocerciasis. *Parasitology Today* 13, 94–98.

Determann, A., Mehlhorn, H. and Ghaffar, F.A. (1997) Electron microscope observations on *Onchocerca ochengi* and *O. fasciata* (Nematoda: Filarioidea). *Parasitology Research* 83, 591–603.

Fine, P.E.M. (1978) On the dynamics of symbiote-dependent cytoplasmic incompatibility in culicine mosquitoes. *Journal of Invertebrate Pathology* 30, 10–18.

Ford, L., Prasad, G.B.K.S., Wu, Y., Cross, H. and Taylor, M.J. (2000) Filarial *Wolbachia* HSP60 and LPS share common pathways in the activation of inflammatory responses. In: *First International Wolbachia Conference* (Kolymbari, Crete, Greece, 7–12 June 2000), abstract book, pp. 125–126.

Franz, M. and Andrews, P. (1986) Fine structure of adult *Litomosoides carinii* (Nematoda: Filarioidea). *Zeitschrift für Parasitenkunde* 72, 537–547.

Franz, M. and Buttner, D.W. (1983) The fine structure of adult *Onchocerca volvulus*. IV. The hypodermal chords of the female worm. *Tropenmedizin und Parasitologie* 34, 122–128.

Franz, M. and Copeman, D.B. (1988) The fine structure of male and female *Onchocerca gibsoni. Tropical Medicine and Parasitology* 39 (Supplement 4), 466–468.

Franz, M., Melles, J. and Buttner, D.W. (1984) Electron microscope study of the body wall and the gut of adult *Loa loa. Zeitschrift für Parasitenkunde* 70, 525–536.

Freedman, D.O. (1998) Immune dynamics in the pathogenesis of human lymphatic filariasis. *Parasitology Today* 14, 229–234.

Genchi, C., Sacchi, L., Bandi, C. and Venco, L. (1998) Preliminary results on the effect of tetracycline on the embryogenesis and symbiotic bacteria of *Dirofilaria immitis*. An update and discussion. *Parassitologia* 40, 247–249.

Harada, R., Maeda, T., Nakashima, A., Sadakata, M., Ando, M., Yonomine, K., Otsuji, Y. and Sato, H. (1970) Electron-microscopical studies on the mechanism of oogenesis in *Dirofilaria immitis*. In: Sasa, M. (ed.) *Recent Advances in Researches on Filariasis and Schistosomiasis in Japan.* Baltimore University Press, Baltimore, Maryland, pp. 99–121.

Henkle-Duhrsen, K., Eckelt, V.H., Wildenburg, G., Blaxter, M. and Walter, R.D. (1998) Gene structure, activity and localization of a catalase from intracellular bacteria in *Onchocerca volvulus. Molecular and Biochemical Parasitology* 96, 69–81.

Hoerauf, A., Nissen-Pahle, K., Schmetz, C., Henkle-Duhrsen, K., Blaxter, M.L., Buttner, D.W., Gallin, M., Al-Qaoud, K.M., Lucius, M. and Fleischer, B. (1999) Tetracycline therapy targets intracellular bacteria in the filarial nematode

Litomosoides sigmodontis and results in filarial infertility. *Journal of Clinical Investigation* 103, 11–18.

Hoerauf, A., Volkmann, L., Hamelmann, C., Adjei, O., Autenrieth, I.B., Fleischer, B. and Buttner, D.W. (2000a) Endosymbiotic bacteria in worms as targets for a novel chemotherapy in filariasis. *Lancet* 355, 1242–1243.

Hoerauf, A., Volkmann, L., Nissen-Pachle, K., Schmetz, C., Autenrieth, I., Buttner, D.W. and Fleischer, B. (2000b) Targeting of *Wolbachia* endobacteria in *Litomosoides sigmodontis*: comparison of tetracyclines with chloramphenicol, macrolides and ciprofloxacin. *Tropical Medicine and International Health* 5, 275–279.

Hurst, G.D.D. and Majerus, M.E.N. (1993) Why do maternally inherited micro-organisms kill males? *Heredity* 71, 81–95.

Hurst, G.D.D., Jiggins, F.M., von der Schulenburg, J.H.G., Bertrand, D., West, S.A., Goriacheva, I.I., Zakharov, I.A., Werren, J.H., Stouthamer, R. and Majerus, M.E.N. (1999) Male-killing *Wolbachia* in two insect species. *Proceedings of the Royal Society of London B* 266, 735–740.

Hurst, L.D. and McVean, G.T. (1996) Clade selection, reversible evolution and the persistence of selfish elements: the evolutionary dynamics of cytoplasmic incompatibility. *Proceedings of the Royal Society of London B* 263, 97–104.

Kalman, S., Mitchell, W., Marathe, R., Lammel, C., Fan, J., Hyman, R.W., Olinger, L., Grimwood, J., Davis, R.W. and Stephens, R.S. (1999) Comparative genomes of *Chlamydia pneumoniae* and *C. trachomatis*. *Nature Genetics* 21, 385–389.

Kitoh, K., Watoh, K., Chaya, K., Kitagawa, H. and Sasaki, Y. (1994) Clinical, hemato-logic, and biochemical findings in dogs after induction of shock by injection of heartworm extracts. *American Journal of Veterinary Research* 55, 1535–1541.

Kozek, W.J. (1977) Transovarially-transmitted intracellular microorganisms in adult and larval stages of *Brugia malayi*. *Journal of Parasitology* 63, 992–1000.

Kozek, W.J. and Figueroa, M. (1977) Intracytoplasmic bacteria in *Onchocerca volvulus*. *American Journal of Tropical Medicine and Hygiene* 26, 663–678.

Kozek, W.J. and Raccurt, C. (1983) Ultrastructure of *Mansonella ozzardi* microfilaria, with a comparison of the South American (simuliid-transmitted) and the Carib-bean (culicoid-transmitted) forms. *Tropenmedizin und Parasitologie* 34, 38–53.

Langworthy, N.G., Renz, A., Mackenstedt, U., Henkle-Duhrsen, K., de Bronsvoort, M.B., Tanya, V.N., Donncly, M.J. and Trees, A.J. (2000) Macrofilaricidal activity of tetracycline against the filarial nematode *Onchocerca ochengi*: elimination of *Wolbachia* precedes worm death and suggests a dependent relationship. *Proceedings of the Royal Society of London B* 267, 1063–1069.

Lee, C.-C. (1975) *Dirofilaria immitis*: ultrastructural aspects of oocyte development and zygote formation. *Experimental Parasitology* 37, 449–468.

Marti, O.G., Rogers, C.E. and Styer, E.L. (1995) Report on intracellular bacterial symbiont in *Noctuidonema guayanense*, an ectoparasitic nematode of *Spodoptera frugiperda*. *Journal of Invertebrate Pathology* 66, 94–96.

Masui, S., Sasaki, T. and Ishikawa, H. (1997) groE-homologous operon of *Wolbachia*, an intracellular symbiont of arthropod: a new approach to their phylogeny. *Zoological Science* 14, 701–706.

Masui, S., Kamoda, S., Sasaki, T. and Ishikawa, H. (1999) The first detection of the insertion sequence ISW1 in the intracellular reproductive parasite *Wolbachia*. *Plasmid* 42, 13–19.

McCall, J.W., Jun, J.J. and Bandi, C. (1999) *Wolbachia* and the antifilarial properties of tetracycline. An untold story. *Italian Journal of Zoology* 66, 7–10.

McLaren, D.J., Worms, M.J., Laurence, B.R. and Simpson, M.G. (1975) Micro-organisms in filarial larvae (Nematoda). *Transactions of the Royal Society of Tropical Medicine and Hygiene* 69, 509–514.

Min, K.T. and Benzer, S. (1997) *Wolbachia*, normally a symbiont of *Drosophila*, can be virulent, causing degeneration and early death. *Proceedings of the National Academy of Sciences USA* 94, 10792–10796.

Moran, N. and Baumann, P. (1994) Phylogenetics of cytoplasmically inherited microorganisms of arthropods. *Trends in Ecology and Evolution* 9, 15–20.

Olsen, G.J. and Woese, C.R. (1993) Ribosomal RNA: a key to phylogeny. *The FASEB Journal* 7, 113–123.

O'Neill, S.L., Giordano, R., Colbert, A.M.E., Karr, T.L. and Robertson, H.M. (1992) 16S rRNA phylogenetic analysis of the bacterial endosymbionts associated with cytoplasmic incompatibility in insects. *Proceedings of the National Academy of Sciences USA* 89, 2699–2702.

O'Neill, S.L., Hoffmann, A.A. and Werren, J.H. (eds) (1997) *Influential Passengers*. Oxford University Press, Oxford.

Ottesen, E.A. (1992) Infection and disease in lymphatic filariais: an immunological perspective. *Parasitology* 104, S71-S79.

Parillo, J.E. (1990) Septic shock in humans: advances in the understanding of pathogenesis, cardiovascular dysfunction, and therapy. *Annals of Internal Medicine* 113, 227–242.

Pennisi, E. (1999) Bacterial partners for filaria. *Science* 283, 1105.

Plenge-Bonig, A., Kromer, M. and Buttner, D.W. (1995) Light and electron micros-copy studies on *Onchocerca jakutensis* and *O. flexuosa* of red deer show different host–parasite interactions. *Parasitology Research* 81, 66–73.

Scott, A.L. (1996) Nematode sperm. *Parasitology Today* 12, 425–430.

Selkirk, M.E., Smith, V.P., Thomas, G.R. and Gounaris, K. (1998) Resistance of filarial nematode parasites to oxidative stress. *International Journal for Parasitology* 28, 1315–1332.

Shepherd, A.M., Clark, S.A. and Kempton, A. (1973) An intracellular micro-organism associated with tissues of *Heterodera* spp. *Nematologica* 19, 31–34.

Sironi, M., Bandi, C., Sacchi, L., Di Sacco, B., Damiani, G. and Genchi, C. (1995) A close relative of the arthropod endosymbiont *Wolbachia* in a filarial worm. *Molecular and Biochemical Parasitology* 74, 223–227.

Stephens, R.S., Kalman, S., Lammel, C., Fan, J., Marathe, R., Aravind, L., Mitchell, W., Olinger, L., Tatusov, R.L., Zhao, Q., Koonin, E.V. and Davis, R.W. (1998) Genome sequence of an obligate intracellular pathogen of humans: *Chlamydia trachomatis. Science* 282, 754–759.

Stouthamer, R., Luck, R.F. and Hamilton, W.D. (1990) Antibiotics cause partheno-genetic *Trichogramma* to revert to sex. *Proceedings of the National Academy of Sciences USA* 87, 2424–2427.

Sucharit, S., Viraboonchai, S., Panavut, N. and Harinasuta, C. (1978) Studies on the effects of tetracycline on *Brugia pahangi* infection in *Aedes togoi. Southeast Asian Journal of Tropical Medicine and Public Health* 9, 55–59.

Taylor, M.J., Bilo, K., Cross, H.F., Archer, J.P. and Underwood, A.P. (1999) 16S rDNA phylogeny and ultrastructural characterization of *Wolbachia* intracellular bacteria of the filarial nematodes *Brugia malayi, B. pahangi,* and *Wuchereria bancrofti. Experimental Parasitology* 91, 356–361.

Taylor, M.J., Cross, H.F. and Bilo, K. (2000) Inflammatory responses induced by the filarial nematode *Brugia malayi* are mediated by lipopolysaccharide-like activity from endosymbiotic *Wolbachia* bacteria. *Journal of Experimental Medicine* 191, 1429–1436.

Tobie, J.E. and Beye, H.K. (1960) Fluorescence of tetracycline in filarial worms. *Proceedings of Society of Experimental Biology and Medicine* 104, 137–140.

Townson, S., Siemienska, J., Hollik, L. and Hutton, D. (1999) The activity of rifampicin, oxytetracycline and chloramphenicol against *Onchocerca linealis* and *O. gutturosa. Transactions of the Royal Society of Tropical Medicine and Hygiene* 93, 123–124.

Turelli, M. and Hoffmann, A.A. (1991) Rapid spread of an inherited incompatibility factor in California *Drosophila. Nature* 353, 440–442.

Turner, P.F., Rockett, K.A., Ottesen, E.A., Francis, H., Awadzi, K. and Clark, I.A. (1994) Interleukin-6 and tumor necrosis factor in the pathogenesis of adverse reactions after treatment of lymphatic filariasis and onchocerciasis. *Journal of Infectious Diseases* 169, 1071–1075.

Vandekerckhove, T.M., Willems, A., Gillis, M. and Coomans, A. (2000) Novel verrucomicrobua in *Xiphinema americanum*-group species (Nematoda, Longidoridae): an obligate symbiosis with a history of over 150 million years of mathernal transmission. In: *First International Wolbachia Conference* (Kolymbari, Crete, Greece, 7–12 June 2000), abstract book, pp. 140–142.

Vincent, A.L., Portaro, J.K. and Ash, L.R. (1975) A comparison of the body wall ultrastructure of *Brugia pahangi* with that of *Brugia malayi. Journal of Parasitology* 63, 567–570.

Werren, J.H. (1997) Biology of *Wolbachia. Annual Review of Entomology* 42, 587–609.

Werren, J.H., Zhang, W. and Guo, L.R. (1995) Evolution and phylogeny of *Wolbachia*: reproductive parasites of arthropods. *Proceedings of the Royal Society of London B* 261, 55–63.

Werren, J.H. and O'Neill, S.L. (1997) The evolution of heritable symbionts. In: O'Neill, S.L., Hoffmann, A.A. and Werren, J.H. (eds) *Influential Passengers: Inherited Microorganisms and Arthropod Reproduction.* Oxford University Press, Oxford, pp. 1–41.

Xie, X., Bain, O. and Williams, S.A. (1994) Molecular phylogenetic studies on filarial parasites based on 5S ribosomal spacer sequences. *Parasite* 1, 141–151.

Yamamura, N. (1993) Vertical transmission and evolution of mutualism from parasitism. *Theoretical and Population Biology* 44, 95–109.

Zhou, W., Rousset, F. and O'Neill, S. (1998) Phylogeny and PCR-based classification of *Wolbachia* strains using *wsp* gene sequences. *Proceedings of the Royal Society of London B* 265, 509–515.

Forward Genetic Analysis of Plant-parasitic Nematode–Host Interactions

Charles H. Opperman

Plant Nematode Genetics Group, Box 7616, Departments of Plant Pathology and Genetics, North Carolina State University, Raleigh, NC 27695-7616, USA

Introduction

This is, by necessity, a short contribution. Forward genetic analysis of any obligate parasitic species is extremely difficult. When one considers that most of the economically important parasitic nematode species are endoparasites, the task becomes nigh impossible. Nevertheless, there are several endoparasitic plant nematodes that lend themselves to classical genetic strategies, and these are their stories. Although the dataset is small, it offers an intriguing look into the genetic world of parasitic nematodes.

An understanding of the genetics of nematode–plant interaction is essential to development of novel management strategies. Unfortunately, the study of this interaction has been very one-sided, focusing primarily on the genetics of plant resistance and almost not at all on the genetics of nematode parasitism. The small size and obligately parasitic life habit of phytophagous nematodes have hindered genetic analysis of nematode parasitism. In addition, many of the most important sedentary endoparasitic forms exhibit modified reproductive strategies (e.g. mitotic or meiotic parthenogenesis in *Meloidogyne* spp.) that preclude classical genetic approaches to analysis. The cyst nematodes however, are primarily amphimictic and are amenable to genetic analyses.

Nematode–host interactions are complex and poorly understood. The relationship between cyst nematodes and their hosts appears to have co-evolved, and as a result there are numerous genes for host resistance that are complemented by nematode parasitism genes (Triantaphyllou, 1987). Different alleles of these genes may interact in various combinations to give a range of nematode–host interactions. Because there may be numerous genes for resistance in a given host species, the interpretation of these interactions is complicated. A lack of knowledge regarding the functions of either resistance or parasitism genes further confuses this picture.

The genetic basis of the nematode–host interaction is poorly characterized. In the case of the potato cyst nematode (*Globodera rostochiensis*)–potato interaction, a gene-for-gene relationship appears to be in operation (Janssen *et al.*, 1991). In this system, nematode genes for parasitism are recessive. Potatoes carrying the dominant *H1* gene are resistant to certain pathotypes of *G. rostochiensis*, but those nematodes carrying recessive parasitism genes can reproduce. Pure parasitic and non-parasitic lines of *G. rostochiensis* have been selected, and crosses using these lines have revealed that parasitism is inherited at a single locus in a recessive manner (Janssen *et al.*, 1990, 1991). Results from reciprocal crosses suggested that there is no evidence for sex-linked inheritance of parasitism. The expected segregation patterns of 3 : 1 non-parasitic to parasitic combined with the dominant nature of the *H1* resistance gene suggest that this interaction functions in a classical gene-for-gene type of mechanism (Janssen *et al.*, 1991).

Ten pathotypes of the cereal cyst nematode (*Heterodera avenae*) have been reported, based on the response of resistant barley cultivars (Andersen and Andersen, 1982). Extensive variations in pathogenicity have been reported within these pathotypes, however, and it is clear that there is more genetic variation within this nematode species than is represented by the current schemes. The use of an alternative host plant, such as oat or wheat, further complicates the picture (Cook and York, 1982). *H. avenae* populations from Europe and Australia exhibit extensive variation in their relative abilities to parasitize resistant cultivars, indicating the presence of multiple genes conferring parasitic ability on a given host genotype (Triantaphyllou, 1987). Limited data from crossing experiments indicates that at least two dominant and one or more recessive genes are involved in *H. avenae* pathogenicity on cereal crops (Andersen, 1965; Person and Rivoal, 1979).

The interaction of soybean cyst nematode (SCN) with soybean (*Glycine max*) has been extensively studied in the United States. This system has been used as a model to dissect the genetics of nematode parasitism, mainly due to the tractability of classical genetic manipulation of the nematode. The remainder of this brief review will focus on the SCN side of the interaction with its host and the progress that has been made in unravelling the complex genetic systems controlling parasitic behaviour.

Soybean Cyst Nematode Biology

The soybean cyst nematode, *Heterodera glycines*, exists throughout all major soybean-growing regions of the world (Riggs, 1977; Doupnik, 1993).

Planting of resistant soybean cultivars is the most widely used method for limiting yield losses caused by this nematode. For example, it is estimated that the resistant cultivar 'Forrest' prevented crop losses worth US$405 million between 1975 and 1980 (Bradley and Duffy, 1982). *H. glycines* populations are dynamic with respect to their ability to parasitize resistant cultivars; thus, resistance-breaking *H. glycines* genotypes may be selected over time in soybean-production fields, resulting in non-durable resistance (Young, 1994).

H. glycines is an obligate cross-fertile species and a sedentary endo-parasitic plant nematode. The life cycle consists of six stages: the egg, four juvenile stages and the sexually dimorphic adults. The second-stage juvenile (J2) is the infective form. After the J2 penetrates the root, it migrates to an area near the vascular cylinder, where it establishes a complex feeding site (Jones, 1981; Endo, 1992). SCN males migrate out of the root for mating within 15–20 days after infection. The adult female nematode produces 200–400 eggs, which remain primarily in her swollen, hardened body, forming a cyst (Triantaphyllou and Hirschmann, 1962). Each life cycle takes 25–30 days and there may be several generations per growing season. The nematode egg is able to survive in the cyst for a number of years under very harsh environmental conditions (Alston and Schmitt, 1988; Young, 1992).

Previous studies have shown that SCN is a diploid nematode with nine chromosomes (Triantaphyllou and Hirschmann, 1962). The genome size is approximatley 92.5 Mb, as determined by flow cytometry (Opperman and Bird, 1998). Genome complexity is estimated to be approximately 82% unique sequence. In addition, a tetraploid form with 18 bivalents has been isolated (Triantaphyllou and Riggs, 1979). The tetraploid carries approximately 1.5 times the amount of DNA per nucleus than the diploid (Goldstein and Triantaphyllou, 1979; Triantaphyllou and Riggs, 1979). The juveniles and adults have significantly larger body size than those of the diploid form as well, which is typical of tetraploid forms (Triantaphyllou and Riggs, 1979). Crosses between the diploid and tetraploid forms have yielded viable aneuploid ($n = 14$) hybrids (Goldstein and Triantaphyllou, 1979).

Genetic Analysis of Parasitism

A parasite must reproduce to complete its life cycle successfully. In this sense, the ability of an *H. glycines* individual to parasitize a soybean plant is measured by reproduction. In general, resistant hosts do not permit the female nematode to develop to reproductive maturity. Parasitism is a qualitative trait that the individual nematode either does or does not possess. In addition, nematode populations may be described quantitatively by their level of reproduction on a given host plant. Field populations of *H. glycines* are mixtures of many genotypes, some of which may confer the ability to overcome host resistance genes, because selection pressure from growing resistant cultivars can alter the frequency of alleles in the population for reproducing in a resistant host.

Research on the genetic basis of parasitism in *H. glycines* is complicated. Results from population measurements are usually biased by genetic variability among and within *H. glycines* populations, and the frequency of a certain gene for parasitism (nematode genes necessary to overcome host resistance) may affect phenotypic designation of either parasitism or the levels of reproduction (Luedders, 1983; Opperman *et al.*, 1995; Dong and Opperman, 1997). Single pair mating and F1 hybrid host range tests of *H. glycines* populations suggested that the parasitism gene(s) in these populations of races 2 and 4 were partially dominant to the parasitism gene(s) in races 1 and 3 (Price *et al.*, 1978). Single cyst selection and inbreeding on a resistant host for many generations indicated that this nematode would tolerate concurrent selection and inbreeding (Dropkin and Halbrendt, 1986). Secondary selection of these inbred lines on a different resistant host resulted in suppressed cyst development on the previous selection host, suggesting that alleles of parasitism genes exist for some hosts (Luedders and Dropkin, 1983; Luedders, 1985). However, other studies demonstrated that continuous selection of *H. glycines* on a resistant soybean increased frequency of parasitism genes in that group, but the frequencies of parasitism genes in other groups were not affected, which suggested that the parasitism genes in the PI88788 and PI90763 groups are not allelic, but are independent loci (Triantaphyllou, 1975). Reciprocal crosses between field populations indicated that the parasitism genes were not sex linked in the progeny (Triantaphyllou, 1975).

In soybean, it is believed that both major and minor genes, dominant, partially dominant and recessive, are all involved to some degree in conferring resistance to *H. glycines* (Triantaphyllou, 1987). However, it is not clear which genes are essential and which are specific to certain nematode genotypes, if any. Interpretation is further complicated by the previous use of *H. glycines* field populations to evaluate resistant soybeans. Recently, we developed pure lines of SCN that carry single genes for parasitic ability on soybeans, and used them to demonstrate that SCN contains unlinked dominant and recessive genes for parasitism of various host genotypes (Dong and Opperman, 1997). A non-parasitic SCN line, which fails to reproduce on the resistant soybean lines PI88788 and PI90763, was used as the female and recurrent parent and was crossed to a parasitic line that does reproduce on these resistant hosts. The segregation ratio of the progeny lines developed by single female inoculation revealed that parasitism to these soybean lines is controlled by independent, single genes in the nematode. These loci were named *ror*, for reproduction on a resistant host

(Dong and Opperman, 1997). In the inbred lines, *Ror-1(kr1)* confers the ability to reproduce on PI88788 and is dominant. The recessive gene *ror-2(kr2)* controls reproduction on PI90763. A second recessive gene, *ror-3(kr5)*, controls the ability to parasitize 'Peking'. Although not verified, it is an intriguing possibility that some genes controlling parasitism may be acting additively. Examination of F_1 data from controlled crosses (Opperman, unpublished) reveals that the presence of two genes results in twice as many females being formed on PI88788 as when only one gene is present. This may explain varying levels of aggressiveness between different nematode populations on the same host genotype. It is particularly significant to note that these loci are entirely independent and do not appear to interact, i.e. no novel host ranges are detected when combinations of *ror* genes are present in a particular nematode line. In addition to alleles for parasitism of resistant soybeans, there are SCN lines that have been selected to reproduce on tomato. The genes controlling this host acquisition remain to be characterized, at either the genetic or molecular level.

The mating technique we developed was a practical necessity because individually inoculated infective nematodes do not always penetrate soybean roots (Dong and Opperman, 1997). This, in turn, makes direct analysis of segregating F_2 populations highly variable and inaccurate. Crosses and backcrosses were conducted using bulked individuals from each inbred line as parents, and single female descent populations were developed for testing the major and independent gene numbers. Using this strategy, dominance or recessiveness did not affect the analysis of gene numbers. For example, non-parasitism can only arise from individually derived lines that were homozygous for the non-parasitism allele in the F_2 generation. Parasitism would always be detected from lines that descended from heterozygous F_2 females due to segregation and random mating during the amplification process. Possible linkage between the parasitism genes can also be detected among the progeny lines, although the amplification process on the susceptible host may cause linkage equilibrium within each line.

Using these types of crossing strategies, linkage maps have been developed from cyst nematodes. The SCN map has nine linkage groups which correspond to $n = 9$ of *H. glycines*. We have placed both parasitism (*ror-2*) and other genes (*gcy-1*) on distinct linkage groups on this map (Heer, Sosinski, Burgwynn and Opperman, unpublished observations). This map is significant because it contains the first mapped parasitism loci from a plant nematode. It is also a merged map, combining RAPD, AFLP, microsatellite and phenotypically mapped markers. A map from the potato cyst nematode (*G. rostochiensis*) has also been developed (van der Voort, 1999). This map has nine linkage groups, which correspond to the nine chromosomes of *G. rostochiensis* (van der Voort *et al.*, 1996), and consists of AFLP markers.

Concluding Remarks

Although it seems like there should be more, the data presented herein is the result of years of painstaking work. This chapter has deliberately avoided discussing genomics projects or EST sequencing, mainly to illustrate how little is actually known about classical genetics of nematode parasitism. In the coming years, many of the questions about nematode parasitism will be answered, or at least addressed, by the considerable power of genomic analysis. Obviously, nematode parasitic ability is not a single-gene trait and much will be learned about pathways involved in many different parasite functions. Yet, one cannot help but long for the simple phenotype: does it infect or does it not? And what is the gene that controls that event?

References

Alston, D.G. and Schmitt, D.P. (1988) Development of *Heterodera glycines* life stages as influenced by temperature. *Journal of Nematology* 20, 366.

Andersen, S. (1965) Heredity of race 1 or race 2 in *Heterodera avenae*. *Nematologica* 11, 121.

Andersen, S. and Andersen, K. (1982) Suggestions for determination and terminology of pathotypes and genes for resistance in cyst-forming nematodes, especially *Heterodera avenae*. *OEPP/EPPO Bulletin* 12, 379.

Bradley, E.B. and Duffy, M. (1982) *The Value of Plant Resistance to the Soybean Cyst Nematode: a Case Study of Forrest Soybean*. USDA-ARS Staff Report AGES820929, Washington, DC.

Cook, R. and Evans, K. (1987) Resistance and tolerance. In: Brown, R.H. and Kerry, B.R. (eds) *Principles and Practice of Nematode Control in Crops*. Academic Press, Sydney.

Cook, R. and York, P.A. (1982) Resistance of cereals to *Heterodera avenae*: methods of investigation, sources, and inheritance of resistance. *OEPP/EPPO Bulletin* 12, 423–434.

Dong, K. and Opperman, C.H. (1997) Genetic analysis of parasitism in the soybean cyst nematode *Heterodera glycines*. *Genetics* 146, 1311–1318.

Doupnik, B., Jr (1993) Soybean production and disease estimate for north central United States from 1989 to 1991. *Plant Disease* 77, 1170–1171.

Dropkin, V.H. and Halbrendt, J.M. (1986) Inbreeding and hybridizing soybean cyst nematodes on pruned soybeans in petri plates. *Journal of Nematology* 18, 200.

Endo, B.Y. (1992) Cellular responses to infection. In: Riggs, R.D. and Wrather, J.A. (eds) *Biology and Management of the Soybean Cyst Nematode*. APS Press, St Paul, Minnesota.

Goldstein, P. and Triantaphyllou, A.C. (1979) Karyotype analysis of the plant-parasitic nematode *Heterodera glycines* by electron microscopy. II. The tetraploid and an aneuploid hybrid. *Journal of Cell Science* 43, 225.

Janssen, R., Bakker, J. and Gommers, F.J. (1990) Selection of virulent and avirulent lines of *Globodera rostochiensis* for the *H1* resistance gene in *Solanum tuberosum* ssp. *andigena* CPC 1673. *Review of Nematology* 13, 265.

Janssen, R., Bakker, J. and Gommers, F.J. (1991) Mendelian proof for a gene-for-gene relationship between *Globodera rostochiensis* and the *H1* resistance gene from *Solanum tuberosum* ssp. *andigena* CPC 1673. *Review of Nematology* 14, 213.

Jones, M.G.K. (1981) The development and function of plant cells modified by endoparasitic nematodes. In: Zuckerman, B.M. and Rohde, R.A. (eds) *Plant Parasitic Nematodes*, Vol. 3. Academic Press, New York.

Luedders, V.D. (1983) Genetics of the cyst nematode–soybean symbiosis. *Phytopathology* 73, 944–948.

Luedders, V.D. (1985) Selection and inbreeding of *Heterodera glycines* on *Glycine max*. *Journal of Nematology* 17, 400.

Luedders, V.D. and Dropkin, V.D. (1983) Effects of secondary selection on cyst nematode reproduction on soybean. *Crop Science* 23, 263.

Opperman, C.H., Dong, K. and Chang, S. (1995) Genetic analysis of the soybean–*Heterodera glycines* interaction. In: Lamberti, F., De Giorgi, C. and McK. Bird, D. (eds) *Advances in Molecular Plant Nematology*. Plenum Press, New York, pp. 65–75.

Person, F. and Rivoal, R. (1979) Hybridation entre les races Fr1 et Fr4 d'*Heterodera avenae* Wollenweber en France et étude du comportement d'agressivité des descendants F1. *Revue de Nematologie* 2, 177.

Price, M., Caviness, C.E. and Riggs, R.D. (1978) Hybridization of races of *Heterodera glycines*. *Journal of Nematology* 10, 114.

Riggs, R.D. (1977) Worldwide distribution of soybean cyst nematode and its economic importance. *Journal of Nematology* 9, 34–38.

Triantaphyllou, A.C. (1975) Genetic structure of races of *Heterodera glycines* and inheritance of ability to reproduce on resistant soybean. *Journal of Nematology* 7, 356.

Triantaphyllou, A.C. (1987) Genetics of nematode parasitism on plants. In: Veech, J.A. and Dickson, D.W. (eds) *Vistas on Nematology*. Society of Nematologists, Hyattsville, Maryland.

Triantaphyllou, A.C. and Hirschmann, H. (1962) Oogenesis and mode of reproduction in the soybean cyst nematode, *Heterodera glycines*. *Nematologica* 7, 235.

Triantaphyllou, A.C. and Riggs, R.D. (1979) Polploidy in an amphimictic population of *Heterodera glycines*. *Journal of Nematology* 11, 371.

van der Voort, J.N.A.M.R., Van Enckevort, L.J.G., Pijnacker, L.P., Helder, P., Gommers, F.J. and Bakker, J. (1996) Chromosome number of the potato cyst nematode *Globodera rostochiensis*. *Fundamental and Applied Nematology* 19, 369–374.

van der Voort, J.N.A.M.R., Eck, H. van, Zandvoort, P. van, Overmars, H., Helder, J. and Bakker, J. (1999) Linkage analysis by genotyping sibling populations: a genetic map of the potato cyst nematode using a 'pseudo-F2' mapping strategy. *Molecular and General Genetics* 261, 1021–1031.

Young, L.D. (1992) Epiphytology and life cycle. In: Riggs, R.D. and Wrather, J.A. (eds) *Biology and Management of the Soybean Cyst Nematode*. APS Press, St Paul, Minnesota.

Young, L.D. (1994) Changes in the *Heterodera glycines* female index as affected by ten year cropping sequences. *Journal of Nematology* 26, 505–510.

Identification of Parasitic Nematodes and Study of Genetic Variability Using PCR Approaches

Robin B. Gasser

Department of Veterinary Science, The University of Melbourne, 250 Princes Highway, Werribee, Victoria 3030, Australia

Introduction

The accurate identification of nematodes (irrespective of developmental stage) has important implications for many areas, including systematics (taxonomy and phylogeny), population genetics, ecology and epidemiology, and is also central to diagnosis, treatment and control of the diseases they cause. Individual nematodes are frequently identified and distinguished on the basis of morphological features, the host they infect, their transmission patterns, their pathological effect(s) on the host or their geographical origin. However, these criteria are often insufficient for specific identification (Lichtenfels *et al.*, 1997; Andrews and Chilton, 1999). Immunological, biochemical and nucleic acid techniques provide powerful tools to overcome this limitation (reviewed by McManus and Bowles, 1996; Andrews and Chilton, 1999). In particular, the advent of the PCR method (Saiki *et al.*, 1985; Mullis *et al.*, 1986) has revolutionized nematode taxonomy and genetics, mainly because its sensitivity permits the amplification of genes or gene fragments from minute amounts of genomic DNA. This is of particular relevance because it is frequently impossible to obtain or isolate sufficient amounts of material from some nematodes at their different life-cycle stages (i.e. eggs or larvae) for conventional analyses.

The method of PCR allows selective amplification from a complex genome by enzymatic amplification *in vitro*. The double-stranded genomic DNA template is denatured by heating, and the temperature is then decreased to allow oligonucleotide primers to hybridize (anneal) to their complementary sequences on opposite strands of the template. The template-directed DNA synthesis (extension) then proceeds in both orientations from the primer sites by enzymatic catalysis with a thermostable DNA polymerase and results in double-stranded products. This synthesis is repeated approximately 30 times in an automated thermocycler. During each cycle, the template is replicated by a factor of two, so that upon completion of the cycling, millions of copies of the original template are available for subsequent analyses.

The intent of this chapter is to present some examples of recent applications of the PCR to parasites of socio-economic significance and to highlight approaches that should find broad applicability to nematodes. The focus is on methods for the identification of nematode species and strains (and/or diagnosis of infections) and on mutation detection methods for the analysis of genetic variation. The second section considers some important technical aspects of the PCR and choice of a DNA target region for systematic or population genetic studies. The third section describes various PCR approaches that have been applied effectively to parasites or have potential. The final section highlights the attributes of mutation detection methods for the high-resolution analysis of PCR-amplified DNA fragments and describes their applicability to parasites. As it was not possible to cover all of the current literature in this short chapter, the interested reader may wish to consult additional review articles relating to parasite systematics and/or genetic variation (e.g. LeJambre, 1993; Grant, 1994; Comes *et al.*, 1996; McManus and Bowles, 1996; Gasser, 1997; Lichtenfels *et al.*, 1997; Prichard, 1997; Anderson *et al.*, 1998).

Technical Aspects and Choice of the DNA Target for PCR

Template preparation

The isolation and purification of the nucleic acid template is an important first step toward achieving high amplification efficiency and specificity in PCR, particularly when low-stringency amplification (Pena *et al.*, 1994; Gomes *et al.*, 1997) is used. It can sometimes be difficult to obtain adequately pure amounts of genomic DNA template from some nematodes, both because of their tough cuticle (Dawkins and Spencer, 1989) and because of flocculate substance(s) found to co-precipitate with nucleic acids during isolation (Gasser *et al.*, 1993) which inhibit subsequent enzymatic amplification. Small-scale sodium dodecyl-sulphate(SDS)-proteinase K treatment, followed by phenol/chloroform extraction and ethanol precipitation, can be used (Gasser *et al.*, 1993). Also, a range of mini-columns for template purification is commercially available, and it is now possible (for example) to isolate genomic DNA by direct purification from SDS-proteinase K digests without the need for phenol/chloroform extraction or ethanol precipitation (see later). Using such mini-columns, DNA can be purified effectively from single eggs, larvae and adults of tiny parasitic nematodes. A recent evaluation has shown that 0.1–1 pg of DNA of *Oesophagostomum* spp. is adequate for efficient PCR amplification of ribosomal DNA (rDNA) (Gasser *et al.*, 1998b).

PCR: precautions, optimization and errors

Important in the establishment of any PCR procedure is the implementation of stringent precautions in the laboratory to prevent contamination (Yap *et al.*, 1994) and the optimization of the amplification protocol(s) (Innis and Gelfand, 1990). If a PCR is not optimized, there may be problems due to PCR errors or artefacts in the amplification products (Abrams and Stanton, 1992), particularly when relatively long regions of DNA are amplified. Specificity and fidelity of the PCR represent two important parameters to be considered when establishing a protocol. PCR with low specificity amplifies one or more sequences in addition to the intended target sequence. PCR with low fidelity amplifies sequence(s) with a high frequency of nucleotide errors. Specificity can be optimized by modifying buffer conditions, primers and cycling conditions (Innis and Gelfand, 1990), while fidelity is predominantly dependent on the amount of starting template and the DNA polymerase used (Krawczak *et al.*, 1989; Cha and Thilly, 1995). DNA polymerases (such as *Taq* from the bacterium *Thermus aquaticus*) can introduce nucleotide errors during amplification to a rate of approximately $1–2 \times 10^{-4}$ per base (e.g. Scharf *et al.*, 1986; Saiki *et al.*, 1988; Keohavong and Thilly, 1989). The use of high-fidelity polymerases should minimize problems associated with nucleotide misincorporations (Cline *et al.*, 1996).

Whether PCR-induced errors are significant is determined by the method of analysis of the amplification product. If the population of molecules within a product is examined simultaneously (for example, by direct sequencing), artefactual misincorporations may be represented only by a weak background against the sequence and will thus remain undetected. Similarly, when PCR products are analysed directly by mutation detection methods (see later), errors should remain undetected. However, when amplicons are cloned and sequenced, a significant number of individual clones may contain PCR-induced errors. 'Jumping PCR' (Pääbo, 1990; Abrams and Stanton, 1992) can also occur when a (multicopy) region containing different sequence polymorphisms (or loci) is amplified. It appears to be associated with a failure of the polymerase to complete replication of all template strands during each round of PCR. Ends of some incomplete strands will lie between the two polymorphic sites, and, in subsequent PCR rounds, incomplete strands will act as long primers, anneal to heterologous templates and be extended. This results in the amplification of recombinant molecules which can include all possible combinations of the original polymorphic positions. The number of the different species of molecule is affected by the PCR conditions and the distance between the polymorphisms.

Target region

The choice of a target region or regions for amplification by PCR depends on the questions to be addressed and the purpose it should serve. All regions of the nuclear and mitochondrial genomes of parasites accumulate mutations over time, and some regions are more accessible to nucleotide changes than others. For example, non-coding regions and introns usually evolve more rapidly than coding regions as they are unlikely to be constrained by function, whereas genes associated with a particular function or functions are less likely to accumulate spontaneous mutations, as the particular function may be related to the survival of the organism. Generally, if the target region should provide genetic markers for the identification of species, then the level of within-species variation in the sequence should be substantially lower than the degree of variation between or among species. If the region should provide markers for the identification of strains, then a significant level of sequence variation should exist within the species under investigation.

Nuclear rDNA is a useful target for the definition of species or strain markers. The rDNA of eukaryotic organisms is a large multigene family consisting of tandemly arrayed sequence repeats (frequently several hundred), usually found in clusters in specific chromosomes. The molecular process involved in the evolution of rDNA is mutational change (Elder and Turner, 1995). rDNA sequences exhibit patterns of 'concerted evolution', which results in sequence similarity (i.e. homogeneity) tending to be greater within a species than between species (Arnheim, 1983; Gerbi, 1986; Schlötterer and Tautz, 1994). Consequently, rDNA can provide useful genetic markers for parasite species. For example, recent studies have demonstrated that internal transcribed spacers (ITS-1 and ITS-2) of rDNA provide accurate species markers for a range of bursate nematodes (e.g. Hoste *et al.*, 1993, 1995, 1998; Campbell *et al.*, 1995; Chilton *et al.*, 1995; Gasser and Hoste, 1995; Stevenson *et al.*, 1995, 1996; Gasser *et al.*, 1996b,c,d; Hung *et al.*, 1996, 1997; Epe *et al.*, 1997; Newton *et al.*, 1997; Romstad *et al.*, 1997; Samson-Himmelstjerna *et al.*, 1997; Höglund *et al.*, 1999) which can be utilized to develop PCR-based diagnostic systems. Intraspecific variation in the internal transcribed spacers is usually low, but other parasite groups (e.g. platyhelminths and arthropods) exhibit significant sequence or length heterogeneity therein (Wesson *et al.*, 1992, 1993; Anderson *et al.*, 1993; Bowles and McManus, 1993a; Kane and Rollinson, 1994; Bowles *et al.*,

1995; Gasser and Chilton, 1995; Tang *et al.*, 1996; van Herwerden *et al.*, 1998), reflecting population variation.

The mitochondrial genome is small, usually circular and maternally inherited, and evolves independently from and more rapidly than the nuclear genome (Brown, 1985; Anderson *et al.*, 1995b). Mitochondrial DNA (mtDNA) is therefore useful for analysing genetic variation within and among parasite populations (e.g. Bowles *et al.*, 1992; Bowles and McManus, 1993b; Anderson *et al.*, 1995a, 1998; Okamoto *et al.*, 1995; Blouin *et al.*, 1997; Hashimoto *et al.*, 1997; Scott *et al.*, 1997; Peng *et al.*, 1998; Zhang *et al.*, 1998; Bøgh *et al.*, 1999), although some species can display low levels of intraspecific variation (Anderson *et al.*, 1998). Establishing the population genetic structure of nematode species has important implications for understanding epidemiology, evolutionary processes such as host-race formation, adaptation to host defences and the development of drug resistance (Blouin, 1998; Viney, 1998). In a detailed study, Blouin *et al.* (1995) used a PCR-based sequencing approach to analyse the population genetic structure of parasitic trichostrongyloid species (*Ostertagia ostertagi* and *Haemonchus placei* from cattle, *H. contortus* and *Teladorsagia circumcincta* from sheep, and *Mazamastrongylus odocoilei* from white-tailed deer) from North America, utilizing a long non-coding region of mtDNA. The sequence data allowed a comparison to be made among the structures of the five species. The parasites of sheep and cattle displayed a pattern consistent with high gene flow among populations, whilst the parasite of deer had a pattern of substantial population subdivision and isolation by distance. The results suggested that host movement represents an important factor determining the population genetic structure of these nematodes. High gene flow was considered to provide a unique opportunity for the spread of rare alleles that confer anthelmintic resistance. All species, including the parasite of deer, had high within-population diversities that appeared to be related to large population sizes and a relatively rapid rate of mutations. In another study, Anderson *et al.* (1995a) employed a PCR-linked restriction fragment length polymorphism (RFLP) analysis of mtDNA to investigate the genetic subdivision of *Ascaris* populations in Guatemala. These workers studied sequence variation in 265 individual *Ascaris* worms collected from human and porcine hosts from three different geographical areas. Restriction mapping of individual worms revealed 42 different genotypes. *Ascaris* populations were strongly structured at the level of the individual host in both humans and pigs. Also, significant heterogeneity was detected among populations from different villages, but not from different families within a village. The clustering of related parasites within hosts suggested a similar clustering of related infective stages in the environment and may also explain the female-biased sex ratios found in *Ascaris* populations.

Other regions of the genome(s), such as repetitive elements (see later), also evolve rapidly and can be exploited as polymorphic markers in random or selective PCR assays. Characterization of such markers is of significance for the construction of genetic linkage maps for nematode species (Grant, 1994; Roos *et al.*, 1998).

Approaches to the Identification of Nematode Species and Strains

Amplification from specific DNA regions

For establishing a PCR approach, appropriate target sequences can be selected based on knowledge from previous studies of similar organisms. If no prior DNA sequence data are available, oligonucleotide primers are usually designed to 5′ and 3′ (complementary) regions that are conserved for related parasites and assumed to hybridize to the complementary DNA strand of the same region in the parasite(s) under study. Preliminary PCR amplification determines whether the region can be amplified effectively, and DNA sequencing or hybridization confirms the identity of the amplicon. The PCR conditions are then modified to achieve optimum amplification efficiency, specificity and fidelity. The detection of amplicons then usually relies on size separation by agarose or polyacrylamide gel electrophoresis.

Primers conserved in sequence across a range of organisms can be employed in PCR for various applications. For instance, this approach can be used for the characterization of ribosomal or mitochondrial DNAs of particular nematodes, or for the development of diagnostic assays (McManus and Bowles, 1996). For example, Zarlenga *et al.* (1994) developed a specific PCR system for the differentiation of different species of *Haemonchus*, economically important parasites of ruminants. Enzymatic amplification of external transcribed rDNA spacer using primers complementary and proximal to the 3′-end of the large subunit and the 5′-end of the small subunit rDNAs enabled rapid differentiation of individual adults of *H. contortus* from *H. placei* by utilizing size variability in this region of the repeat. In another study, Zarlenga *et al.* (1998) reported the development of a semi-quantitative PCR assay for the diagnosis of patent *Ostertagia ostertagi* infection in cattle. Conserved oligonucleotide primers were used in PCR to amplify an approximately 1 kb rDNA fragment, spanning the ITS-1 and part of the 5.8S rRNA gene, from *O. ostertagi*, while fragments of approximately 600 bp were amplified from *H. contortus*, *Cooperia oncophora* and *Oesophagostomum radiatum* DNA. When DNA samples derived from adult nematodes of the different genera were mixed and amplified simultaneously, no inhibition was apparent in PCR and *O. ostertagi* amplicons were readily detected on agarose gels. There was a correlation between the intensity of the 1 kb and 600 bp PCR products on gels and the percentage of *O. ostertagi* DNA within the DNA mix of heterologous species. There was also a high correlation between the percentage of *O. ostertagi* DNA and percentage of *O. ostertagi* eggs in faeces. Effective amplification was achieved from 1/20th of a single egg of *O. ostertagi*. Hence, the establishment of this PCR assay has major implications for diagnosis of ostertagiasis in cattle as well as for studying the epidemiology and population biology of the parasite.

Oligonucleotide primers designed to specific regions of DNA can also be used for diagnostic applications (e.g. Roos and Grant, 1993; Favia *et al.*, 1996; Jacobs *et al.*, 1997; Newton *et al.*, 1997; Romstad *et al.*, 1997; Samson-Himmelstjerna *et al.*, 1997; Monti *et al.*, 1998). Using rDNA targets, this strategy has also been employed for the development of PCR assays for the species-specific identification of different developmental stages of strongyloid nematodes representing the subfamilies Strongylinae and Cyathostominae. For instance, Hung *et al.* (1999a) characterized the ITS-1 and ITS-2 sequences of 28 species of horse strongyles and designed specific oligonucleotide primers for five important species (*Strongylus vulgaris, Cyathosomum catinatum, Cylicocyclus nassatus, Cylicostephanus longibursatus* and *Cylicostephanus goldi*) based on the nucleotide differences among all species examined. Utilizing these primers, a PCR approach was developed for the specific amplification of picogram to nanogram amounts of rDNA. Effective amplification was achieved from egg and larval DNA isolated from faeces and copro-cultures, respectively. These results have major implications for studying the prevalence and biology of equine strongyles and for investigating the distribution of anthelmintic resistance (in conjunction with faecal egg-count reduction testing).

PCR-linked restriction fragment length polymorphism (PCR-RFLP) analysis

PCR products amplified from a particular region of the genome can be analysed for sequence variation by PCR-RFLP. Fragments amplified using a specific primer set are digested with one or more restriction endonucleases and separated (usually) by agarose gel electrophoresis. Then, the digestion profiles are detected by ultraviolet transillumination of ethidium-bromide stained gels and recorded by photography (Saperstein and Nickerson, 1991; Bowles and McManus, 1993a). Improved resolution of restriction profiles can be achieved by denaturation of digests and separation on denaturing polyacrylamide gels (Cupolillo *et al.*, 1995). PCR-RFLP has been used effectively for the definition of strain and species-specific genetic markers in the rDNA of nematodes of socio-economic importance (e.g. Gasser *et al.*, 1996b,c,d; Newton *et al.*, 1998a; Zhu *et al.*, 1998b,c, 1999a). For instance, Newton *et al.* (1998a) characterized 24 species of strongylid nematodes (representing the families Trichostrongylidae, Molineidae and Chabertidae) from livestock (sheep, goats, cattle and pigs) by this approach. The rDNA region spanning the ITS-1, 5.8S rRNA gene and the

ITS-2 was amplified from genomic DNA by PCR, digested separately with endonucleases *Rsa*I, *Hin*fI, *Dra*I or *Nla*III, and the fragments separated by agarose gel electrophoresis. The PCR products amplified from individual species appeared as single bands of approximately 870 bp in size, except for *O. ostertagi*, whose product was approximately 1250 bp. The analysis revealed characteristic restriction profiles for all species, except for *Cooperia surnabada* and *C. oncophora*, which are now considered to represent one species based on sequence data (Newton *et al.*, 1998c). Although applied to strongylid nematodes of livestock, the PCR-RFLP approach has broad applicability to other nematode groups for species and/or strain identification.

Arbitrarily primed PCR

Arbitrarily primed PCR (AP-PCR), or random amplification of polymorphic DNA (RAPD), is based on the enzymatic amplification of random fragments of genomic DNA with (usually) single primers of arbitrary sequence (Welsh and McClelland, 1990; Williams *et al.*, 1990). Advantages of RAPD over other DNA techniques are speed, simplicity, ability to amplify from small amounts of genomic DNA and the capacity to screen the entire genome using a simple approach without requiring prior DNA sequence information. However, there can be problems with both reproducibility of results and specificity as a consequence of low stringency in PCR (Ellsworth *et al.*, 1993; MacPherson *et al.*, 1993).

RAPD analysis has proved useful for the definition of genetic markers for a broad range of species and strains of protozoan and metazoan parasites (e.g. Dias-Neto *et al.*, 1993; Siles-Lucas *et al.*, 1994; Epe *et al.*, 1995; Guo and Johnson, 1995; Humbert and Cabaret, 1995; Jacquiet *et al.*, 1995; Nadler *et al.*, 1995; Felleisen and Gottstein, 1996; Gasser *et al.*, 1996e; Joachim *et al.*, 1997; Leignel *et al.*, 1997; Mathis *et al.*, 1997; Felleisen, 1998). Numerous research groups have exploited the approach to characterize nematodes (first-stage muscle larvae) of the genus *Trichinella* (e.g. Bandi *et al.*, 1993, 1995; Pozio *et al.*, 1995; Rodriguez *et al.*, 1996; Wu *et al.*, 1998). For instance, Bandi *et al.* (1995) used RAPD to genetically type *Trichinella* isolates. Five 10-mer or 20-mer primers were employed under different PCR conditions to produce multibanded fingerprints from muscle larvae representing 40 different isolates. Of the five primers evaluated, one of them could be used effectively to genotype individual larvae. The resultant data were then analysed following a numerical, taxonomic approach. The classification of the isolates into five species and three phenotypes of uncertain taxonomic status by RAPD was in agreement with allozyme electrophoretic data and supported the polyspecific structure of the genus *Trichinella* (Pozio *et al.*, 1992). These results demonstrated this approach to be complementary to the *in vivo* characterization of *Trichinella* in mice and allozyme electrophoresis.

RAPD can also be used to search for 'hidden variation' within bands that appear to be monomorphic. Bands common to a range of geographical isolates of a particular organism can be scanned for genetic variation (e.g. Carter *et al.*, 1995). This approach was recently employed to analyse genetic variability in the mycopathogen, *Histoplasma capsulatum* (Carter *et al.*, 1996). The strategy involves excising specific bands and reamplifying them for subsequent analysis by mutation scanning (see later) and DNA sequencing. Once sequence variants are defined within the bands, primers can be designed for the direct enzymatic amplification of specific loci. These modifications increase reproducibility and resolution of the procedure for the analysis of sequence variation within and among populations. Although used to strain-type a microorganism, this strategy could also be applied to nematodes.

Analysis of mini- and microsatellites, and other repetitive elements

Minisatellites and microsatellites have been described as being both abundant and ubiquitous in the genomes of all eukaryotes (Tautz and Renz, 1984; Tautz, 1993). These sequences consist of tandem repeats of short motifs which are randomly dispersed throughout the genome. They are usually non-transcribed and maintain polymorphism as a consequence of the accumulation of mutations. The variation in repeat number allows the alleles present at a locus to be scored by size on electrophoretic gels. The satellites are characterized by allelic ('hyper-') variability in repeat length, and consequently have been used to study the genetic structure of populations as well as for linkage analysis and genetic mapping (e.g. Tautz, 1989; Love *et al.*, 1990; Bell and Ecker, 1994; Goldstein and Clark, 1995). By utilizing primers designed to unique sequences flanking a microsatellite, the PCR can be employed to amplify the repeat region, which can then be analysed by denaturing gel electrophoresis and displayed, for example, by autoradiography or staining. The analysis is relatively simple technically and allows the simultaneous analysis of multiple genetic markers.

Attention has focused on the use of satellite DNA as taxonomic markers and for studying genetic variation in nematode populations (e.g. Arnot *et al.*, 1994; Fisher and Viney, 1996; Grenier *et al.*, 1996, 1997; Zarlenga *et al.*, 1996; Hoekstra *et al.*, 1997; Gasser *et al.*, 1998f). For instance, Hoekstra *et al.* (1997) isolated and characterized microsatellite markers for the ovine parasite, *H. contortus*. A library of 2–2.3 kb *Hind*III-*Eco*RI fragments of *H. contortus* DNA constructed in pBluescript SK+ was screened by hybridization with radioactively labelled $(CA)_{25}$ and $(GA)_{15}$ GG oligonucleotides, and 12 CA/GT and one CT/GA (imperfect) microsatellites were isolated. The majority of the isolated CA/GT dinucleotide repeats were located in a conserved region and linked with the repetitive element, HcREP1. Using the microsatellites in a PCR-based system, extensive genetic diversity was detected within and among four different geographical populations of *H. contortus*. Similar approaches have been applied to study the malarial parasite, *Plasmodium falciparum* (Arnot *et al.*, 1994), the mite, *Sarcoptes scabiei* (Walton *et al.*, 1997) and *Strongyloides ratti* (Fisher and Viney, 1996) as well as entomopathogenic nematodes (Grenier *et al.*, 1996), and are likely to be applicable to a wide range of parasites of socio-economic significance.

Other studies (Callaghan and Beh, 1994a,b; Christensen *et al.*, 1994) have characterized repetitive DNA elements of parasitic nematodes of livestock, such as species of *Trichostrongylus, Ostertagia, Haemonchus, Cooperia* and *Oesophagostomum*. These sequence elements used as probes in hybridization assays have proved to be specific for a particular parasite genus, but were shown to cross-hybridize among closely related members within a genus, preventing their use for species or strain identification. However, a recent study demonstrated that such elements can be used effectively for species/genotypic identification by PCR. Using parasite species of the genus *Trichostrongylus* as a model, Gasser *et al.* (1995) developed an amplification fingerprinting system for species identification using single primers to the repetitive element, TcREP (Callaghan and Beh, 1994a). The fingerprinting technique was exquisitely sequence dependent and appeared to be a complex reaction, involving multiple interactions of the primer with the template as well as interactions among the amplicons themselves, to generate a multibanded profile. Advantages of this finger-printing approach over RAPD are that the thermocycling takes place at high stringency (60°C annealing temperature), that the protocol is shorter and that primer annealing is parasite directed, and thus specific for certain members within a group or genus of parasites. The technique is less time consuming and less labour intensive compared with hybrid-ization, PCR-RFLP and sequencing techniques. Although established for *Trichostrongylus* spp., the approach (using a polyacrylamide gel system) could be applied to other nematode groups where highly repetitive and cross-hybridizing sequence elements have been defined. The sensitivity of such an assay should allow species determination of individual develop-mental stages of parasites (i.e. larvae or eggs), which cannot be identified morphologically. Given its ability to detect intraspecific variation in fingerprint profiles (Gasser *et al.*, 1995), it could also be used to study population variation.

DNA sequencing approaches

Sequencing of parasite genes provides a powerful tool for the accurate identification of parasites and for systematic studies (Johnson and Baver-stock, 1989; Reddy, 1995; McManus and Bowles, 1996), and is based on either of the two original protocols (Sanger *et al.*, 1977; Maxam and Gilbert, 1980). Cycle-sequencing (Murray, 1989), a PCR-based modification of the dideoxy method (Sanger *et al.*, 1977), is an attractive approach as it is rapid to perform and can be used for sequence analysis from minute amounts of starting template. This approach also allows the DNA template to be sequenced at high temperatures, thus reducing technical problems due to secondary structure formations. Manual cycle-sequencing protocols have been established for the sequence analysis of rDNA of nematodes (e.g. Gasser *et al.*, 1993) and automated systems are now widely used. Datasets generated using such protocols have been exploited to define strain and species-specific genetic markers for parasites (McManus and Bowles, 1996) and identify cryptic (morphologically similar but genetically distinct) species (Chilton *et al.*, 1995; Zhu *et al.*, 1998b; Hung *et al.*, 1999b), and have allowed the phylogenetic reconstruction of parasitic groups (e.g. Chilton *et al.*, 1997; Blaxter *et al.*, 1998; Nadler and Hudspeth, 1998; Newton *et al.*, 1998b).

Direct sequencing of PCR products has major advantages over methods that rely on a cloning step prior to sequencing, in that variation within a product is displayed directly. If multiple different sequence types (alleles) exist within an amplicon, nucleotide variation is represented by polymorphism (one or more nucleotides present at one or more sequence positions) on the sequencing gel. However, when high levels of sequence heterogeneity exist within an amplicon (e.g. for multicopy genes), it may be difficult or impossible to read the sequencing gel (Gasser and Chilton, 1995). To circumvent this, amplicons can be cloned and then a number of clones isolated for sequencing. However, once the amplicon (representing a multicopy gene) is cloned, an estimation of the degree of sequence variation may be difficult. If significant sequence variation exists within an amplicon of a particular size, it is possible to precede cycle sequencing with 'preparative' mutation detection approaches (i.e. excising sequence variants from gels and subjecting those to direct sequence analysis) (see later). This approach has been exploited recently for the detailed analysis of microsatellite variation in the rDNA of porcine lungworms of the genus *Metastrongylus* (Conole *et al.*, 2001).

Analysis of Genetic Variability by PCR-coupled Mutation Detection Methods

The accurate analysis of genetic variation in nematodes has major implications for parasite identification and for investigating population genetic structures (LeJambre, 1993; Grant, 1994). Conventional DNA methods have been valuable and their application has provided a wealth of molecular data (reviewed by McManus and Bowles, 1996), but little attention has been paid to their capacity to resolve sequence variation (reviewed by Gasser, 1997). For instance, PCR-RFLP analysis using (multiple) restriction endonucleases is useful, but sequence variation may go undetected as each enzyme used scans only a subset of potentially variable nucleotide positions. Mutation detection methods provide powerful alternatives for the direct analysis of sequence variation (Lessa and Applebaum, 1993). They are widely used in biomedical research areas for analytical and diagnostic purposes (Cotton, 1997) and rely on the separation of DNA molecules differing by one or more nucleotides by physical properties rather than by size. Given the capacity of these methods to resolve low-level nucleotide variation, they are now finding increased application to parasites (e.g. Craig and Kain, 1996; Gasser *et al.*, 1996a; Kain *et al.*, 1996; Mathis *et al.*, 1996; Gasser and Monti, 1997; Haag *et al.*, 1997; Stothard *et al.*, 1997, 1998; Tang and Unnasch, 1997; Zhu and Gasser, 1998). Recently, the techniques of denaturing gradient gel electrophoresis (DGGE), single-strand conformation polymorphism (SSCP) and their modifications have been applied to nematodes (reviewed by Gasser, 1997).

Denaturing gradient gel electrophoresis (DGGE)

This method was first described by Fischer and Lerman (1983) and entails electrophoresis of DNA fragments at high temperature (50–60°C) in a polyacrylamide gel containing a gradient of denaturant (such as urea or formamide) (Myers *et al.*, 1985a,b,c, 1987). The principle of the method is that nucleotide differences cause a change in the melting behaviour of DNA fragments (Fig. 4.1). Fragments are separated electrophoretically in a

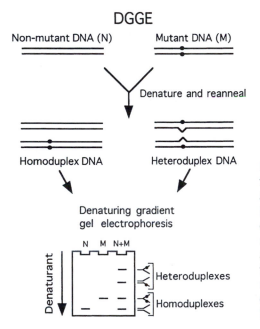

Fig. 4.1. Schematic representation of heteroduplex formation and principle of DGGE. Mutant homoduplexes (M) melt at a lower denaturant concentration than non-mutant homoduplexes (N) and are consequently retarded in the gel. Heteroduplexes (N + M) melt at even lower denaturant concentrations (modified from Børresen, 1996).

gradient of denaturant at high temperature. As a DNA fragment migrates through the gel, it encounters increasing concentrations of denaturant. At some point in the gel, the fragment becomes partially or fully denatured. Partial denaturation causes a marked decrease in the electrophoretic mobility of the DNA molecule. The position in the gel at which the DNA melts is determined by its nucleotide sequence and composition. Therefore, DNA fragments of the same size but differing in sequence (e.g. by a point mutation) will melt at different positions within the gradient of the gel, and hence will be separated. DGGE was originally used to analyse genomic DNA, but the technique has been adapted for the analysis of PCR products. Incorporation of a GC-rich sequence (GC-clamp) into one of the primers used for PCR alters the melting behaviour of the fragment and permits the separation of almost 100% of variant molecules (Sheffield *et al.*, 1989, 1992).

DGGE has been used to display sequence variation within rDNA of trichostrongyloid species of veterinary importance (Gasser *et al.*, 1996a). The results of that study showed that DGGE provides a powerful analytical technique for visually displaying different sequence types of ITS-2 within and among different geographical isolates of a species. In another study (Gasser *et al.*, 1998e), a DGGE-sequencing approach was also exploited to define nucleotide variations in the ITS-2 of the parasitic nematode, *H. contortus*. For 94 individuals of *H. contortus*, representing nine different populations, 13 different profiles were displayed. Eighteen bands representing those profiles were excised from gels and subjected to cycle-sequencing. Thirteen different types of ITS-2 sequence representing 12 nucleotide variations (four transitions, five transversions, one insertion and two deletions) could be defined and were associated with particular positions of the predicted secondary structure for the ITS-2 pre-rRNA. The results showed that the approach could achieve the separation of clonal types of ITS-2 sequence for amplicons of 350 bp (including the GC-clamp) and could accurately quantify the frequency of ITS-2 sequence types ('alleles') within and among different populations without the need for conventional Southern blotting or cloning. Therefore, this approach should be useful for studying the mechanisms of sequence homogenization in rDNA, for elucidating speciation events and population differentiation at the molecular level (Elder and Turner, 1995), as well as for studying aspects of pre-rRNA structure and processing (cf. van der Sande *et al.*, 1992; van Nues *et al.*, 1995; Chilton *et al.*, 1998).

Single-strand conformation polymorphism (SSCP) analysis

PCR-based SSCP is technically relatively simple and can be used effectively to display major sequence types over short sequence lengths (usually approximately 100–400 bp). SSCP relies on the principle that the electrophoretic mobility of a single-stranded DNA molecule in a non-denaturing gel is dependent on its structure and size (Orita *et al.*, 1989; Hayashi, 1991) (Fig. 4.2). In solution, the single-stranded molecules take on secondary and tertiary conformations as a result of base pairing between nucleotides within individual strands. These conformations are dependent on the length of the strand, its sequence and the location and number of regions of base pairing. This means that a mutation at a particular nucleotide position in the primary sequence can change the conformation of the molecule. When separated on a non-denaturing polyacrylamide gel, molecules differing by a single nucleotide can be separated because of changes in their mobility as a consequence of differing conformations.

The mutation detection rate of SSCP can vary depending on parameters such as size of the fragment, base composition of the sequence, electrophoresis temperature and gel composition (i.e. pore size and cross-linking) (Hayashi and Yandell, 1993; Sheffield *et al.*, 1993; Teschauer *et al.*, 1996). It appears that fragment size and electrophoresis time are the most important factors determining the mutation detection rate. Usually, 50–100% of mutations are detected for amplicons of 100–200 bp. This rate can decrease for amplicons of greater than 200 bp (Cotton, 1997), although we have shown that a single nucleotide difference can be detected between rDNA amplicons of approximately 530 bp (Zhu and Gasser, 1998). Using appropriate DNA targets, PCR-based SSCP methods have proved useful for the identification of nematode species or strains where

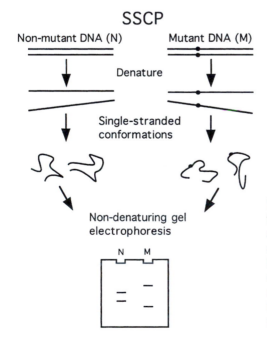

Fig. 4.2. Schematic representation of the principle of SSCP analysis. A point mutation (represented by a dot on a DNA strand) leads to different single-strand conformations of the mutant DNA (M) as compared with non-mutant DNA (N), resulting in differential mobilities in a non-denaturing gel.

morphological characters are unreliable (Gasser, 1997; Gasser and Monti, 1997; Gasser *et al.*, 1997, 1998a–d,f; Zhu and Gasser, 1998; Zhu *et al.*, 1998a,b,c, 1999a). For instance, the high resolution capacity of SSCP was exploited to overcome limitations in the morphological identification of different developmental stages of hookworms of human and veterinary health importance (Gasser *et al.*, 1998a). Single-strand rDNA profiles enabled the identification of seven different species and also reflected population variation within some species. More recently, similar approaches have been applied to a range of helminth groups, demonstrating their broad applicability (e.g. Bøgh *et al.*, 1999; Gasser *et al.*, 1999; Zhang *et al.*, 1999; Zhu *et al.*, 1999b).

SSCP protocol

Given its technical simplicity and relatively high mutation detection capacity (Zhu and Gasser, 1998), PCR-SSCP provides a useful tool to identify nematodes and to screen for genetic variability. This section describes a protocol that is routinely employed in our laboratory to address taxonomic problems in nematodology. It includes all steps from the isolation of DNA from individual nematodes, through to the SSCP and sequence analysis of polymorphic bands. The use of radioactively labelled PCR products allows high resolution analysis and has the advantage (over staining procedures) of being able to detect sequence types that constitute as little as a few per cent of the total number of copies.

Isolation/purification of nematode DNA

Individual nematodes (adult, larval or egg stages; packed volume approximately 50 μl) are placed in Eppendorf tubes containing 100–200 μl of extraction buffer (20 mM Tris-HCl, pH 8, 50 mM EDTA, 1% w/v sodium dodecyl-sulphate plus 0.5 μg μl^{-1} proteinase K, Boehringer-Mannheim) and incubated at 37°C for 12–14 h. The tube is vortexed and centrifuged at 12,000 × *g*, and the DNA is purified directly over a spin column (Wizard™ DNA Clean-Up, Promega). To do this, the supernatant is mixed with 900 μl of the resin provided, vortexed, incubated at room temperature (22–24°C) for 1 min, then passed through the column using a syringe. The column is then washed (using the same syringe) with 2 ml 80% isopropanol and centrifuged to remove residual propanol. The column is then transferred to a fresh tube, and the DNA is eluted from the column by adding 25–50 μl of hot (70°C) H$_2$O, followed by incubation for 1 min at room temperature and centrifugation (12,000 × *g*) for 20 s. For long-term storage at –20°C or –70°C, the DNA is transferred to 0.6 ml double-lock PCR tubes. Using these spin columns, DNA is purified effectively from single eggs, larvae and adults of tiny parasitic nematodes, and can be used directly for enzymatic amplification.

Enzymatic amplification

DNA regions of approximately 100–500 bp are amplified by PCR (Saiki *et al.*, 1988) using appropriate primer sets end-labelled with γ-^{33}P-ATP (NEN, DuPont, cat. no. NEG206H) by polynucleotide kinase T4 (Promega) (Table 4.1). PCR reactions (25–50 μl) are performed in 10 mM Tris-HCl, pH 8.4; 50 mM KCl; 3–3.5 mM MgCl$_2$; 250 μM of each dNTP; 12.5 pmol of each primer and 1–2 U Ampli*Taq* (Perkin Elmer) polymerase (Table 4.2). For effective amplification it is important to optimize, by serial titration the MgCl$_2$ concentration, and we have found that 3–3.5 mM is best for Ampli*Taq* and sequences of strongylid and ascaridoid nematodes. For PCR, 1 μl (usually approximately 1–20 ng) of genomic DNA is added to each reaction, and appropriate controls are included in each amplification run. For rDNA or mtDNA regions, we cycle as follows: 94°C, 5 min, followed by 30–35 cycles of 94°C, 30 s (denaturation); 50–60°C, 30 s (annealing); 72°C,

Table 4.1.   Endlabelling of primers.

Component	Volume (μl)
Primer 1 (20-mer; 100 pmol μl^{-1})	5
Primer 2 (20-mer; 100 pmol μl^{-1})	5
H$_2$O	4
10× T4 kinase buffer (Promega)	2
T4 polynucleotide kinase (Promega, cat. no. M4101)	2
γ 33p-[ATP] (NEN, DuPont; cat. no. NEG302H)	2
Total	20

Incubate tube for 1–2 h at 37°C.
Then, inactivate polynucleotide kinase at 70°C for 3 min.
Add 20 μl H$_2$O (gives 40 μl end-labelled primer mix = total of 500 pmol of each primer).
Use 10 μl of each radiolabelled primer for the PCR reaction mix (see Table 4.2).

Table 4.2.   PCR reaction mix.

Component	Amount (μl)
H$_2$O	346
10× buffer (with 1.5 mM MgCl$_2$)	50
dNTPs (= 25 mM of each dATP, dCTP, dGTP and dTTP)	50
MgCl$_2$ (the final concentration of MgCl$_2$ is 3 mM)	30
γ ^{33}P-end-labelled primer mix (see Table 4.1) (250 pmol of each primer)	20
Ampli*Taq* (Perkin Elmer)	4
Total (enough for 19 reactions @ 24 μl each)	500

Make up reaction mix in 0.6 ml double-lock tubes (Robbins Scientific, cat. no. 1048-01-0).
For storage, freeze reaction mix without the polymerase at –20°C; add enzyme just prior to use.

30 s (extension); followed by 72°C for 5 min (final extension) (in a Perkin Elmer Cetus thermocycler). PCR products are run on 2–2.5% agarose-TBE (65 mM Tris-HCl, 22.5 mM boric acid, 1.25 mM EDTA, pH 9) gels, stained with ethidium bromide and photographed with Polaroid 667 film (Kodak). For radiolabelled PCR products, agarose gels are dried on to two layers of blotting paper and exposed to autoradiographic film (RP1, Agfa) for 12 h. This can be carried out to confirm the specificity of products and PCR conditions.

Non-denaturing gel electrophoresis

First, an aliquot (5 μl) of each PCR product is examined by agarose gel electrophoresis prior to SSCP analysis and should represent approximately 50–100 ng. Then, 10 μl of PCR product is mixed directly with 10 μl of loading buffer (10 mM NaOH, 95% formamide, 0.05% bromophenol blue and 0.05% xylene cyanole). After denaturation at 94°C for 2 min on a heat block and subsequent snap-cooling on a freeze block (−20°C) (in both cases, sample tubes must fit tightly in the block), 2–4 μl of each sample (same amounts loaded for individual samples) are subjected to electrophoresis in a 0.4 mm thick, 0.5–0.6 × mutation detection enhancement (MDE™, FMC BioProducts) gel (Table 4.3). The electrophoresis conditions depend on the nucleotide sequence to be analysed and need to be determined in order to achieve optimal band resolution. We run our gels at 7 W and 20°C (constant) for 10–20 h. The most critical point to achieving optimal resolution is a constant temperature (< 1°C fluctuation) throughout electrophoresis. We run our gels in a temperature-controlled room (20°C) in a conventional sequencing rig (S2, Life Technologies) fitted with

Table 4.3. MDE™ (FMC BioProducts) gels.

Components	Volume (ml)				
	60[b]	100	120	600	(Total)
0.5× MDE gel:					
2× MDE (stock)	15	25	30	150	
10× TBE[a]	3.6	6	7.2	36	
H$_2$O	41.4	69	82.8	414	
0.6× MDE gel:					
2× MDE (stock)	18	30	36	180	
10× TBE	3.6	6	7.2	36	
H$_2$O	38.4	64	76.8	384	

[a]Bio-Rad, cat. no. 161–0741.
[b]Add 310 μl of 10% w/v ammonium persulphate (Promega, cat. no.V3131) and 31 μl TMED (Promega, cat. no. V3161) to 60 ml.

a mantle connected to a temperature-control system (MultiTemp® III, Pharmacia Biotech) which circulates 10% v/v ethylene glycol. The ΦX174-*Hae* III marker (Promega) end-labelled with γ33p can be used as a reference (not size!) marker (this marker is not denatured prior to loading on the gel). After electrophoresis, gels are dried on to blotting paper and directly subjected (at room temperature) to autoradiography (RP1 X-ray film, Agfa) for 12–24 h.

Sequencing of SSCP bands

The sequences of individual SSCP bands can be determined. Bands are excised from dried gels with a scalpel. Positioning of the autoradiograph over the gel can be achieved using Glogos™ II (Stratagene) as reference markers, and the accuracy of the procedure can be confirmed by re-exposing the gel after band excision. Each gel/paper slice is suspended in 20 μl H_2O and incubated at 4°C overnight (this incubation period can be shorter). Then, 1 μl of this suspension is diluted 10^{-2}–10^{-3} in H_2O and re-amplified by PCR with unlabelled oligonucleotide primers (same conditions as primary amplification). After purification of the PCR product over a spin column (Wizard™ PCR-Prep, Promega) and elution into 30–40 μl H_2O, an aliquot (3 μl) is examined on a 2% agarose-TBE gel, and 1 μl of the purified product is subjected directly to cycle-sequencing with the *f*-mol kit (Promega) using the same primers and cycling conditions as for primary amplification. Sequences are read (recording polymorphisms), aligned manually and compared with appropriate sequences.

Concluding Remarks

The accurate identification of nematodes is central to many fundamental and applied areas in nematodology, but is frequently not possible using traditional diagnostic approaches. Biochemical and molecular methods have provided alternative, taxonomic tools. In particular, PCR has had a major impact in these areas in recent years, allowing the analysis of minute amounts of nematode DNA (from tiny adults, eggs or larvae). Using selected examples, this chapter has highlighted some applications of PCR to the identification of nematodes and diagnosis of nematodiases of socio-economic importance, and indicated the value of mutation detection methods for the high resolution analysis of genetic variation in nematode populations. Although important advances have been made, a multitude of basic questions relating to the population biology, population genetic structures, epidemiology (transmission dynamics and zoonotic potential) as well as inheritance and evolution of genes (to name just a few) remain to be addressed using this advanced technology.

Acknowledgements

All colleagues and students who have made valuable contributions to the research in our laboratory and who are authors or co-authors on previously published papers are gratefully acknowledged. Thanks to Xingquan Zhu for preparing the figures and comments on the manuscript. The author's research has been supported largely through grants from the Australian Research Council, the Department of Industry, Science and Tourism, the Melbourne University Equine Research Fund, the Rural Industries Research and Development Corporation, the Collaborative Research Program of the University of Melbourne, the Canine Research Foundation and the Australian Companion Animal Health Foundation.

References

Abrams, E.S. and Stanton, V.P. Jr (1992) Use of denaturing gradient gel electrophoresis to study conformational transitions in nucleic acids. *Methods in Enzymology* 212, 71–104.

Anderson, T.J.C., Romero-Abal, M.E. and Jaenike, J. (1993) Genetic structure and epidemiology of *Ascaris* populations: patterns of host affiliation in Guatemala. *Parasitology* 107, 319–334.

Anderson, T.J.C., Romero-Abal, M.E. and Jaenike, J. (1995a) Mitochondrial DNA and *Ascaris* microepidemiology: the composition of parasite populations from individual hosts, families and villages. *Parasitology* 110, 221–229.

Anderson, T.J.C., Komuniecki, R., Komuniecki, P. and Jaenike, J. (1995b) Are mitochondria inherited maternally in *Ascaris*? *International Journal for Parasitology* 25, 1001–1004.

Anderson, T.J.C., Blouin, M.S. and Beech, R.N. (1998) Population biology of parasitic nematodes: application of genetic markers. *Advances in Parasitology* 41, 220–283.

Andrews, R.H. and Chilton, N.B. (1999) Multilocus enzyme electrophoresis: a valuable technique for providing answers to problems in parasite systematics. *International Journal for Parasitology* 29, 213–253.

Arnheim, N. (1983) Concerted evolution of multigene families. In: Nei, M. and Koehn, R.K. (eds) *Evolution of Genes and Proteins*. Sinauer, Sunderland, Massachusetts, pp. 38–61.

Arnot, D.E., Roper, C. and Sultan, A.A. (1994) MVR-PCR analysis of hypervariable DNA sequence variation. *Parasitology Today* 10, 324–327.

Bandi, C., La Rosa, G., Comincini, S., Damiani, G. and Pozio, E. (1993) Random amplified polymorphic DNA technique for the identification of *Trichinella* species. *Parasitology* 107, 419–424.

Bandi, C., La Rosa, G., Bardin, M.G., Damiani, G., Comincini, S., Tasciotti, L. and Pozio, E. (1995) Random amplified polymorphic DNA fingerprints of the eight taxa of *Trichinella* and their comparison with allozyme analysis. *Parasitology* 110, 401–407.

Bell, C.J. and Ecker, J.R. (1994) Assignment of 30 microsatellite loci to the linkage map of *Arabidopsis*. *Genomics* 19, 137–144.

Blaxter, M.L., De-Ley, P., Garey, J.R., Liu, L.X., Scheldeman, P., Vierstraete, A., Vanfleteren, J.R., Mackey, L.Y., Dorris, M., Frisse, L.M., Vida, J.T. and Thomas, W.K. (1998) A molecular evolutionary framework for the phylum Nematoda. *Nature* 392, 71–75.

Blouin, M.S. (1998) Mitochondrial DNA diversity in nematodes. *Journal of Helminthology* 72, 285–289.

Blouin, M.S., Yowell, C.A., Courtney, C.H. and Dame, J.B. (1995) Host movement and the genetic structure of populations of parasitic nematodes. *Genetics* 141, 1007–1014.

Blouin, M.S., Yowell, C.A., Courtney, C.H. and Dame, J.B. (1997) *Haemonchus placei* and *Haemonchus contortus* are distinct species based on mtDNA evidence. *International Journal for Parasitology* 27, 1383–1387.

Bøgh, H.O., Zhu X.Q., Qian, B.-Z. and Gasser, R.B. (1999) Scanning for nucleotide variations in mitochondrial DNA fragments of *Schistosoma japonicum* by single-strand conformation polymorphism. *Parasitology* 118, 73–82.

Børresen, A.-L. (1996) Constant denaturant gel electrophoresis (CDGE) in mutation screening. In: Pfeifer, G.P. (ed.) *Technologies for Detection of DNA Damage and Mutations.* Plenum Press, New York.

Bowles, J. and McManus, D.P. (1993a) Rapid discrimination of *Echinococcus* species and strains using a polymerase chain reaction-based RFLP method. *Molecular and Biochemical Parasitology* 57, 231–240.

Bowles, J. and McManus, D.P. (1993b) NADH dehydrogenase 1 gene sequences compared for species and strains of the genus *Echinococcus. International Journal for Parasitology* 23, 969–972.

Bowles, J., Blair, D. and McManus, D.P. (1992) Genetic variants within the genus *Echinococcus* identified by mitochondrial DNA sequencing. *Molecular and Biochemical Parasitology* 54, 165–174.

Bowles J., Blair, D. and McManus, D.P. (1995) A molecular phylogeny of the genus *Echinococcus. Parasitology* 110, 317–328.

Brown, W.M. (1985) The mitochondrial genome of animals. In: MacIntyre, R.J. (ed.) *Molecular Evolutionary Genetics.* Plenum Press, New York, pp. 95–130.

Callaghan, M.J. and Beh, K.J. (1994a) A middle-repetitive DNA sequence element in the sheep parasitic nematode, *Trichostrongylus colubriformis. Parasitology* 109, 345–350.

Callaghan, M.J. and Beh, K.J. (1994b) Characterization of a tandemly repetitive DNA sequence from *Haemonchus contortus. International Journal for Parasitology* 24, 137–141.

Campbell, A.J.D., Gasser, R.B. and Chilton, N.B. (1995) Differences in a ribosomal sequence of *Strongylus* species allows identification of single eggs. *International Journal for Parasitology* 25, 359–365.

Carter, D.A., Burt, A. and Taylor, J.W. (1995) Direct analysis of specific bands from arbitrarily primed PCR reactions. In: Innis, M.A., Gelfand, D.H. and Sninsky, J.J. (eds) *PCR Strategies.* Academic Press, New York, pp. 325–332.

Carter, D.A., Burt, A., Taylor, J.W., Koenig, G.L. and White, T.J. (1996) Clinical isolates of *Histoplasma capsulatum* from Indianapolis have a recombining population structure. *Journal of Clinical Microbiology* 34, 2577–2584.

Cha, R.S. and Thilly, W.G. (1995) Specificity, efficiency and fidelity of PCR. In: Dieffenbach, C.W. and Dveksler, G.S. (eds) *PCR Primer. A Laboratory Manual.*

Cold Spring Harbor Laboratory Press, Cold Spring Harbor, New York, pp. 35–51.

Chilton, N.B., Gasser, R.B. and Beveridge, I. (1995) Differences in a ribosomal DNA sequence of morphologically indistinguishable species within the *Hypodontus macropi* complex (Nematoda: Strongyloidea). *International Journal for Parasitology* 25, 647–651.

Chilton, N.B., Gasser, R.B. and Beveridge, I. (1997) Phylogenetic relationships of Australian strongyloid nematodes inferred from ribosomal DNA sequence data. *International Journal for Parasitology* 27, 1481–1494.

Chilton, N.B., Hoste, H., Newton, L.A., Beveridge, I. and Gasser, R.B. (1998) Common secondary structures for the second internal transcribed spacer pre-rRNA of two subfamilies of trichostrongylid nematodes. *International Journal for Parasitology* 28, 1765–1773.

Christensen, C.M., Zarlenga, D.S. and Gasbarre, L.C. (1994) *Ostertagia, Haemonchus, Cooperia* and *Oesophagostomum*: construction and characterization of genus-specific DNA probes to differentiate important parasites of cattle. *Experimental Parasitology* 78, 93–100.

Cline, J., Braman, J.C. and Hogrefe, H.H. (1996) PCR fidelity of *Pfu* DNA polymerase and other thermostabile DNA polymerases. *Nucleic Acids Research* 24, 3546–3551.

Comes, A.M., Humbert, J.F., Cabaret, J. and Elard, L. (1996) Using molecular tools for diagnosis in veterinary parasitology. *Veterinary Research* 27, 333–342.

Conole, J.C., Chilton, N.B., Jarvis, T. and Gasser, R.B. (2001) Mutation scanning analysis of microsatellite variability in the ITS-2 (precursor rRNA) for three species of *Metastrongylus* (Nematoda: Metastrongyloidea). *Parasitology* (in press).

Cotton, R.G.H. (1997) *Mutation Detection*. Oxford University Press, Oxford.

Craig, A.A. and Kain, K.C. (1996) Molecular analysis of strains of *Plasmodium vivax* from paired primary and relapse infections. *Journal of Infectious Diseases* 174, 373–379.

Cupolillo, E., Grimaldi, G. Jr, Momen, H. and Beverley, S.M. (1995) Intergenic region typing (IRT): a rapid molecular approach to the characterization and evolution of *Leishmania. Molecular and Biochemical Parasitology* 73, 145–155.

Dawkins, H.J.S. and Spencer, T.L. (1989) The isolation of nucleic acid from nematodes requires an understanding of the parasite and its cuticular structure. *Parasitology Today* 5, 73–76.

Dias-Neto, E., Pereira de Souza, C., Rollinson, D., Katz, N., Pena, S.D.J. and Simpson, A.J.G. (1993) The random amplification of polymorphic DNA allows the identification of strains and species of schistosome. *Molecular and Biochemical Parasitology* 57, 83–88.

Elder, J.F. and Turner, B.J. (1995) Concerted evolution of repetitive DNA sequences in eukaryotes. *Quarterly Review of Biology* 70, 297–320.

Ellsworth, D.L., Rittenhouse, K.D. and Honeycutt, R.L. (1993) Artifactual variation in randomly amplified polymorphic DNA banding patterns. *BioTechniques* 14, 214–217.

Epe, C., Bienioschek, S., Rehbein, S. and Schnieder, T. (1995) Comparative RAPD-PCR analysis of lungworms (Dictyocaulidae) from fallow deer, cattle, sheep and horses. *Journal of Veterinary Medicine* B 42, 187–191.

Epe C., von Samson-Himmelstjerna, G. and Schnieder, T. (1997) Differences in a ribosomal DNA sequence of lungworm species (Nematoda: Dictyocaulidae) from fallow deer, cattle, sheep and donkeys. *Research in Veterinary Science* 62, 17–21.

Favia, G., Lanfrancotti, A., Della Torre, A., Gancrini, G. and Coluzzi, M. (1996) Polymerase chain reaction – identification of *Dirofilaria repens* and *Dirofilaria immitis*. *Parasitology* 113, 567–571.

Felleisen, R.S.J. (1998) Comparative genetic analysis of trichomonadid protozoa by the random amplified polymorphic DNA technique. *Parasitology Research* 84, 153–156.

Felleisen, R. and Gottstein, B. (1996) 'Arbitrarily primed-PCR'. Oder: Der Zufall im Dienste der Forschung. *Schweizer Archiv für Tierheilkunde* 138, 139–143.

Fischer, S.G. and Lerman, L.S. (1983) DNA fragments differing by single base pair substitutions are separated in denaturing gradient gels: correspondence with melting theory. *Proceedings of the National Academy of Sciences USA* 80, 1579–1583.

Fisher, M.C. and Viney, M.E. (1996) Microsatellites of the parasitic nematode *Strongyloides ratti*. *Molecular and Biochemical Parasitology* 80, 221–224.

Gasser, R.B. (1997) Mutation scanning methods for the analysis of parasite genes. *International Journal for Parasitology* 27, 1449–1463.

Gasser, R.B. and Chilton, N.B. (1995) Characterisation of taeniid cestode species by PCR-RFLP of ITS2 ribosomal DNA. *Acta Tropica* 59, 31–40.

Gasser, R.B. and Hoste, H. (1995) Genetic markers for closely-related parasitic nematodes. *Molecular and Cellular Probes* 9, 315–320.

Gasser, R.B. and Monti, J.R. (1997) Identification of parasitic nematodes by PCR-SSCP of ITS-2 rDNA. *Molecular and Cellular Probes* 11, 201–209.

Gasser, R.B., Chilton, N.B., Hoste, H. and Beveridge, I. (1993) Rapid sequencing of rDNA from single worms and eggs of parasitic helminths. *Nucleic Acids Research* 21, 2525–2526.

Gasser, R.B., Nansen, P. and Bøgh, H.O. (1995) Specific fingerprinting of parasites by PCR with single primers to defined repetitive elements. *Acta Tropica* 60, 127–131.

Gasser, R., Nansen, P. and Guldberg, P. (1996a) Fingerprinting sequence variation in ribosomal DNA of parasites by DGGE. *Molecular and Cellular Probes* 10, 99–105.

Gasser, R.B., Stewart, L.E. and Speare, R. (1996b) Genetic markers in ribosomal DNA for hookworm identification. *Acta Tropica* 62, 15–21.

Gasser, R.B., Stevenson, L.A., Chilton, N.B., Nansen, P., Bucknell, D.G. and Beveridge, I. (1996c) Species markers for equine strongyles detected in intergenic spacer rDNA by PCR-RFLP. *Molecular and Cellular Probes* 10, 371–378.

Gasser, R.B., LeGoff, L., Petit, G. and Bain, O. (1996d) Rapid delineation of closely-related filarial parasites using genetic markers in spacer rDNA. *Acta Tropica* 62, 143–150.

Gasser, R.B., Bao-Zhen, Q., Nansen, P., Johansen, M.V. and Bøgh, H.O. (1996e) Use of RAPD for the detection of genetic variation in the human blood fluke, *Schistosoma japonicum*, from mainland China. *Molecular and Cellular Probes* 10, 353–358.

Gasser, R.B., Monti, J.R., Zhu, X.Q., Chilton, N.B., Hung, G.-C. and Guldberg, P. (1997) PCR-SSCP of ribosomal DNA to fingerprint parasites. *Electrophoresis* 18, 1564–1566.

Gasser, R.B., Monti, J.R., Bao-Zhen, Q., Polderman, A.M., Nansen, P. and Chilton, N.B. (1998a) A mutation scanning approach for the identification of hookworm species and analysis of population variation. *Molecular and Biochemical Parasitology* 92, 303–312.

Gasser, R.B., Woods, W.G. and Bjørn, H. (1998b) PCR-based SSCP to distinguish *Oesophagostomum dentatum* from *O. quadrispinulatum* developmental stages. *International Journal for Parasitology* 28, 1903–1909.

Gasser, R.B., Zhu, X.Q. and McManus, D.P. (1998c) Display of sequence variation in PCR-amplified mitochondrial DNA regions of *Echinococcus* by single-strand conformation polymorphism. *Acta Tropica* 71, 107–115.

Gasser, R.B., Zhu, X.Q. and McManus, D.P. (1998d) Dideoxy fingerprinting: application to the genotyping of *Echinococcus. International Journal for Parasitology* 28, 1775–1779.

Gasser, R.B., Zhu, X.Q., Chilton, N.B., Newton, L.A., Nedergaard, T. and Guldberg, P. (1998e) Analysis of sequence homogenisation in rDNA arrays of *Haemonchus contortus* by DGGE. *Electrophoresis* 19, 2391–2395.

Gasser, R.B., Zhu, X.Q., Monti, J.R., Dou, L., Cai, X. and Pozio, E. (1998f) PCR-SSCP of rDNA for the identification of *Trichinella* isolates from mainland China. *Molecular and Cellular Probes* 12, 27–34.

Gasser, R.B., Zhu, X.Q. and Woods, W.G. (1999) Genotyping *Taenia* tapeworms by single-strand conformation polymorphism of mitochondrial DNA. *Electrophoresis* 20, 2834–2837.

Gerbi, S.A. (1986) The evolution of eukaryotic ribosomal DNA. *BioSystems* 19, 247–258.

Goldstein, D.B. and Clark, A.G. (1995) Microsatellite variation in North American populations of *Drosophila melanogaster. Nucleic Acids Research* 23, 3882–3886.

Gomes, M.A., Silva, E.F., Macedo, A.M., Vago, A.R. and Melo, M.N. (1997) LSSP-PCR for characterization of strains of *Entamoeba histolytica* isolated in Brazil. *Parasitology* 114, 517–520.

Grant, W.N. (1994) Genetic variation in parasitic nematodes and its implications. *International Journal for Parasitology* 24, 821–830.

Grenier, E., Bonifassi, E., Abad, P. and Laumond, C. (1996) Use of species-specific satellite DNAs as diagnostic probes in the identification of Steinernematidae and Heterorhabditidae entomopathogenic nematodes. *Parasitology* 113, 483–489.

Grenier, E., Castagnone-Sereno, P. and Abad, P. (1997) Satellite DNA sequences as taxonomimarkers in nematodes of agronomic interest. *Parasitology Today* 13, 398–401.

Guo, Z.G. and Johnson, A.M. (1995) Genetic characterization of *Toxoplasma gondii* strains by random amplified polymorphic DNA polymerase chain reaction. *Parasitology* 111, 127–132.

Haag, K.L., Zaha, A., Araujo, A.M. and Gottstein, B. (1997) Reduced genetic variability within coding and non-coding regions of the *Echinococcus multilocularis* genome. *Parasitology* 115, 521–529.

Hashimoto, K., Watanobe, T., Liu, C.X., Init, I., Blair, D., Ohnishi, S. and Agatsuma, T. (1997) Mitochondrial DNA and nuclear DNA indicate that the Japanese *Fasciola* species is *F. gigantica. Parasitology Research* 83, 220–225.

Hayashi, K. (1991) PCR-SSCP: a simple and sensitive method for detection of mutations in the genomic DNA. *PCR Methods and Applications* 1, 34–38.

Hayashi, K. and Yandell, D.W. (1993) How sensitive is PCR-SSCP? *Human Mutation* 2, 338–346.

Hoekstra, R., Criado-Fornelio, A., Fakkeldij, J., Bergman, J. and Roos, M.H. (1997) Microsatellites of the parasitic nematode *Haemonchus contortus*: polymorphism and linkage with a direct repeat. *Molecular and Biochemical Parasitology* 89, 97–107.

Höglund, J., Wilhelmsson, E., Christensson, D., Mörner, T., Waller, P. and Mattsson, J.G. (1999) ITS2 sequences of *Dictyocaulus* species from cattle, roe deer and moose in Sweden: molecular evidence for a new species. *International Journal for Parasitology* 29, 607–611.

Hoste, H., Gasser, R.B., Chilton, N.B., Mallet, S. and Beveridge, I. (1993) Lack of intraspecific variation in the second internal transcribed spacer (ITS-2) of *Trichostrongylus colubriformis* ribosomal DNA. *International Journal for Parasitology* 23, 1069–1071.

Hoste, H., Chilton, N.B., Gasser, R.B. and Beveridge, I. (1995) Differences in the second internal transcribed spacer (ribosomal DNA) between five species of *Trichostrongylus* (Nematoda: Trichostrongylidae). *International Journal for Parasitology* 25, 75–80.

Hoste, H., Chilton, N.B., Beveridge, I. and Gasser, R.B. (1998) Differences in the first internal transcribed spacer of ribosomal DNA among five species of *Trichostrongylus*. *International Journal for Parasitology* 28, 1251–1260.

Humbert, J.F. and Cabaret, J. (1995) Use of random amplified polymorphic DNA for identification of ruminant trichostrongylid nematodes. *Parasitology Research* 81, 1–5.

Hung, G.-C., Jacobs, D.E., Krecek, R.S., Gasser, R.B. and Chilton, N.B. (1996) *Strongylus asini* (Nematoda, Strongyloidea): genetic relationships with other *Strongylus* species determined by ribosomal DNA. *International Journal for Parasitology* 26, 1407–1411.

Hung, G.-C., Chilton, N.B., Beveridge, I., McDonnell, A., Lichtenfels, J.R. and Gasser, R.B. (1997) Molecular delineation of *Cylicocyclus nassatus* and *C. ashworthi* (Nematoda: Strongylidae). *International Journal for Parasitology* 27, 601–605.

Hung, G.-C., Gasser, R.B., Beveridge, I. and Chilton, N.B. (1999a) Species-specific amplification of ribosomal DNA from some species of equine strongyles by PCR. *Parasitology* 119, 69–80.

Hung, G.-C., Chilton, N.B., Beveridge, I. and Gasser, R.B. (1999b) Molecular evidence for cryptic species within *Cylicostephanus minutus* (Nematoda: Strongylidae). *International Journal for Parasitology* 29, 285–291.

Innis, M.A. and Gelfand, D.H. (1990) Optimization of PCR. In: Innis, M.A., Gelfand, D.H., Sninsky, J.J. and White, T.J. (eds) *PCR Protocols – a Guide to Methods and Applications*. Academic Press, New York, pp. 3–12.

Jacobs, D.E., Zhu, X.Q., Gasser, R.B. and Chilton, N.B. (1997) PCR-based methods for identification of potentially zoonotic ascaridoid parasites of the dog, fox and cat. *Acta Tropica* 68, 191–200.

Jacquiet, P., Humbert, J.F., Comes, A.M., Cabaret, J., Thiam, A. and Cheikh, D. (1995) Ecological, morphological and genetic characterization of sympatric *Haemonchus* spp. parasites of domestic ruminants in Mauritania. *Parasitology* 110, 483–492.

Joachim, A., Daugschies, A., Christensen, C.M., Bjørn, H. and Nansen, P. (1997) Use of random amplified polymorphic DNA-polymerase chain reaction for the definition of genetic markers for species and strains of porcine *Oesophagostomum. Parasitology Research* 83, 646–654.

Johnson, A.M. and Baverstock, B.P. (1989) Rapid ribosomal RNA sequencing and the phylogenetic analysis of protists. *Parasitology Today* 5, 102–105.

Kain, K.C., Craig, A.A. and Ohrt, C. (1996) Single-strand conformational polymorphism analysis differentiates *Plasmodium falciparum* treatment failures from re-infections. *Molecular and Biochemical Parasitology* 79, 167–175.

Kane, R.A. and Rollinson, D. (1994) Repetitive sequences in the ribosomal DNA internal transcribed spacer of *Schistosoma haematobium, Schistosoma intercalatum* and *Schistosoma matthei. Molecular and Biochemical Parasitology* 63, 153–156.

Keohavong, P. and Thilly, W.G. (1989) Fidelity of DNA polymerases in DNA amplification. *Proceedings of the National Academy of Sciences USA* 86, 9253–9257.

Krawczak, M., Reiss, J., Schmidtke, J. and Rosler, U. (1989) Polymerase chain reaction: replication errors and reliability of gene diagnosis. *Nucleic Acids Research* 17, 2197–2201.

Leignel, V., Humbert, J.F. and Elard, L. (1997) Study by ribosomal DNA ITS2 sequencing and RAPD analysis on the systematics of four *Metastrongylus* species (Nematoda: Metastrongyloidea). *Journal of Parasitology* 83, 606–611.

LeJambre, L.F. (1993) Molecular variation in trichostrongylid nematodes from sheep and cattle. *Acta Tropica* 53, 331–343.

Lessa, E.P. and Applebaum, G. (1993) Screening techniques for detecting allelic variation in DNA sequences. *Molecular Ecology* 2, 119–129.

Lichtenfels, J.R., Hoberg, E.P. and Zarlenga, D.S. (1997) Systematics of gastrointestinal nematodes of domestic ruminants: advances between 1992 and 1995 and proposals for future research. *Veterinary Parasitology* 72, 225–245.

Love, J.M., Knight, A.M., McAleer, M.A. and Todd, J.A. (1990) Towards construction of a high resolution map of the mouse genome using PCR-analysed microsatellites. *Nucleic Acids Research* 18, 4123–4130.

MacPherson, J.M., Eckstein, P.E., Scoles, G.J. and Gajadar, A.A. (1993) Variability of the random amplified polymorphic DNA assay among thermal cyclers, and effects of primer and DNA concentration. *Molecular and Cellular Probes* 7, 293–299.

Mathis, A., Weber, R., Kuster, H. and Speich, R. (1996) Reliable one-tube nested PCR for detection and SSCP-typing of *Pneumocystis carinii. Journal of Eukaryotic Microbiology* 43, 7S.

Mathis, A., Michel, M., Kuster, H., Müller, C., Weber, R. and Deplazes, P. (1997) Two *Encephalitozoon cuniculi* strains of human origin are infectious to rabbits. *Parasitology* 114, 29–35.

Maxam, A.M. and Gilbert, W. (1980) Sequencing end-labeled DNA with base-specific chemical cleavages. *Methods in Enzymology* 65, 499–560.

McManus, D.P. and Bowles, J. (1996) Molecular genetic approaches to parasite identification: their value in diagnostic parasitology and systematics. *International Journal for Parasitology* 26, 687–704.

Monti, J.R., Chilton, N.B., Qian, B.-Z. and Gasser, R.B. (1998) PCR-based differentiation of *Necator americanus* from *Ancylostoma duodenale* using specific markers in ITS-1 rDNA. *Molecular and Cellular Probes* 12, 71–78.

Mullis, K.B., Faloona, F., Scharf, S., Saiki, R., Horn, G. and Erlich, H. (1986) Specific enzymatic amplification of DNA *in vitro*: the polymerase chain reaction. *Cold Spring Harbor Symposium of Quantitative Biology* 51, 263–273.

Murray, V. (1989) Improved double-stranded DNA sequencing using the linear polymerase chain reaction. *Nucleic Acids Research* 17, 8889.

Myers, R.M., Fischer, S.G., Lerman, L.S. and Maniatis, T. (1985a) Nearly all base substitutions in DNA fragments joined to a GC-clamp can be detected by denaturing gradient gel electrophoresis. *Nucleic Acids Research* 13, 3131–3145.

Myers, R.M., Fischer, S.G., Maniatis, T. and Lerman, L.S. (1985b) Modification of the melting properties of duplex DNA by attachment of a GC rich DNA sequence as determined by denaturing gradient gel electrophoresis. *Nucleic Acids Research* 13, 3111–3129.

Myers, R.M., Lumelsky, N., Lerman, L.S. and Maniatis, T. (1985c) Detection of single base substitutions in total genomic DNA. *Nature* 313, 495–498.

Myers, R.M., Maniatis, T. and Lerman, L.S. (1987) Detection and localization of single base changes by denaturing gel electrophoresis. *Methods in Enzymology* 155, 501–527.

Nadler, S.A. and Hudspeth, D.S.S. (1998) Ribosomal DNA and phylogeny of the Ascaridoidea (Nemata: Secernentea): implications for morphological evolution and classification. *Molecular Phylogenetics and Evolution* 10, 221–236.

Nadler, S.A., Lindquist, R.L. and Near, T.J. (1995) Genetic structure of midwestern *Ascaris suum* populations: a comparison of isoenzyme and RAPD markers. *Journal of Parasitology* 81, 385–394.

Newton, L.A., Chilton, N.B., Monti, J.R., Bjørn, H., Varady, M., Christensen, C.M. and Gasser, R.B. (1997) Rapid PCR-based delineation of the porcine nodular worms, *Oesophagostomum dentatum* and *O. quadrispinulatum*. *Molecular and Cellular Probes* 11, 149–153.

Newton, L.A., Chilton, N.B., Beveridge, I., Hoste, H., Nansen, P. and Gasser, R.B. (1998a) Genetic markers for strongylid nematodes of livestock defined by PCR-based restriction analysis of spacer rDNA. *Acta Tropica* 69, 1–15.

Newton, L.A., Chilton, N.B., Beveridge, I. and Gasser, R.B. (1998b) Systematic relationships of some members of the genera *Oesophagostomum* and *Chabertia* (Nematoda: Chabertiidae) based on ribosomal DNA sequence data. *International Journal for Parasitology* 28, 1781–1789.

Newton, L.A., Chilton, N.B., Beveridge, I. and Gasser, R.B. (1998c) Genetic evidence indicating that *Cooperia surnabada* and *Cooperia oncophora* are one species. *International Journal for Parasitology* 28, 331–336.

Okamoto, M., Bessho, Y., Kamiya, M., Kurosawa, T. and Horii, T. (1995) Phylogenetic relationships within *Taenia taeniaeformis* variants and other taeniid cestodes inferred from the nucleotide sequence of the cytochrome c oxidase subunit I gene. *Parasitology Research* 81, 451–458.

Orita, M., Suzuki, Y., Sekiya, T. and Hayashi, K. (1989) Rapid and sensitive detection of point mutations and DNA polymorphisms using the polymerase chain reaction. *Genomics* 5, 874–879.

Pääbo, S. (1990) Amplifying ancient DNA. In: Innis, M.A., Gelfand, D.H., Sninsky, J.J. and White, T.J. (eds) *PCR Protocols – a Guide to Methods and Applications.* Academic Press, New York, pp. 159–166.

Pena, S.D.J., Barreto, G., Vago, A.R., de Marco, L., Reinach, F.C., Dias-Neto, E. and Simpson, A.J.G. (1994) Sequence-specific 'gene signatures' can be obtained by PCR with single specific primers at low stringency. *Proceedings of the National Academy of Sciences USA* 91, 1946–1949.

Peng, W., Anderson, T.J.C., Zhou, X. and Kennedy, M.W. (1998) Genetic variation in sympatric *Ascaris* populations from humans and pigs in China. *Parasitology* 117, 355–361.

Pozio, E., La Rosa, G., Murrell, D. and Lichtenfels, J.R. (1992) Taxonomic revision of the genus *Trichinella. Journal of Parasitology* 78, 654–659.

Pozio, E., Bandi, C., La Rosa, G., Jarvis, T., Miller, I. and Kapel, C.M. (1995) Concurrent infection with sibling *Trichinella* species in a natural host. *International Journal for Parasitology* 25, 1247–1250.

Prichard, R. (1997) Application of molecular biology in veterinary parasitology. *Veterinary Parasitology* 71, 155–175.

Reddy, G.R. (1995) Determining the sequence of parasite DNA. *Parasitology Today* 11, 37–42.

Rodriguez, E., Nieto, J., Castillo, J.A. and Garate, T. (1996) Characterization of Spanish *Trichinella* isolates by random amplified polymorphic DNA (RAPD). *Journal of Helminthology* 70, 335–343.

Romstad, A., Gasser, R.B., Polderman, A.M., Nansen, P., Pit, D.S.S. and Chilton, N.B. (1997) Differentiation of *Oesophagostomum bifurcum* from *Necator americanus* by PCR using genetic markers in spacer ribosomal DNA. *Molecular and Cellular Probes* 11, 169–176.

Roos, M.H. and Grant, W.N. (1993) Species-specific PCR for the parasitic nematodes *Haemonchus contortus* and *Trichostrongylus colubriformis. International Journal for Parasitology* 23, 419–421.

Roos, M.H., Hoekstra, R., Plas, M.E., Otsen, M. and Lenstra, J.A. (1998) Polymorphic DNA markers in the genome of parasitic nematodes. *Journal of Helminthology* 72, 291–294.

Saiki, R.K., Scharf, S.J., Faloona, F., Mullis, K.B., Horn, G.T., Erlich, H.A. and Arnheim, N. (1985) Enzymatic amplification of β-globin genomic sequences and restriction site analysis for diagnosis of sickle cell anemia. *Science* 230, 1350–1354.

Saiki, R.K., Gelfand, D.H., Stoffel, S., Scharf, S.J., Higuchi, R., Horn, G.T., Mullis, K.B. and Erlich, H.A. (1988) Primer-directed enzymatic amplification of DNA with a thermostable DNA polymerase. *Science* 239, 487–491.

Samson-Himmelstjerna, G. von, Woidtke, S., Epe, C. and Schnieder, T. (1997) Species-specific polymerase chain reaction for the differentiation of larvae from *Dictyocaulus viviparus* and *Dictyocaulus eckerti. Veterinary Parasitology* 68, 119–126.

Sanger, F., Nicklen, S. and Coulson, A.R. (1977) DNA sequencing with chain terminating inhibitors. *Proceedings of the National Academy of Sciences USA* 88, 2815–2819.

Saperstein, D. and Nickerson, J.M. (1991) Restriction fragment length polymorphism analysis using PCR coupled to restriction digests. *BioTechniques* 10, 488–489.

Scharf, S.J., Horn, G.T. and Erlich, H.A. (1986) Direct cloning and sequence analysis of enzymatically amplified genomic sequences. *Science* 233, 1076–1078.

Schlötterer, C. and Tautz, D. (1994) Chromosomal homogeneity of *Drosophila* ribosomal DNA arrays suggests intrachromosomal exchanges drive concerted evolution. *Current Biology* 4, 777–783.

Scott, J.C., Stefaniak, J., Pawlowski, Z.S. and McManus, D.P. (1997) Molecular genetic analysis of human cystic hydatid cases from Poland: identification of a new genotypic group (G9) of *Echinococcus granulosus*. *Parasitology* 114, 37–43.

Sheffield, V.C., Cox, D.R., Lerman, L.S. and Myers, R.M. (1989) Attachment of a 40-base pair G+C-rich sequence (GC-clamp) to genomic DNA fragments by polymerase chain reaction results in improved detection of single-base changes. *Proceedings of the National Academy of Sciences USA* 86, 232–236.

Sheffield, V.C., Beck, J.S., Nichols, B., Cousineau, A., Lidral, A. and Stone, E.M. (1992) Detection of multiallele polymorphisms within gene sequences by GC-clamped denaturing gradient gel electrophoresis. *American Journal of Human Genetics* 50, 567–575.

Sheffield, V.C., Beck, J.S., Kwitek, A.E., Sandstrom, D.W. and Stone, E.M. (1993) The sensitivity of single strand conformation polymorphism analysis for the detection of single base substitutions. *Genomics* 16, 325–332.

Siles-Lucas, M., Felleisen, R., Cuesta-Bandera, C., Gottstein, B. and Eckert, J. (1994) Comparative genetic analysis of Swiss and Spanish isolates of *Echinococcus granulosus* by Southern blot hybridisation and random amplified polymorphic DNA technique. *Applied Parasitology* 35, 107–117.

Stevenson, L.A., Chilton, N.B. and Gasser, R.B. (1995) Differentiation of *Haemonchus placei* from *H. contortus* (Nematoda: Trichostrongylidae) by the ribosomal second internal transcribed spacer. *International Journal for Parasitology* 25, 483–488.

Stevenson, L.A., Gasser, R.B. and Chilton, N.B. (1996) The ITS-2 rDNA of *Teladorsagia circumcincta*, *T. trifurcata* and *T. davtiani* indicates that these taxa are one species. *International Journal for Parasitology* 26, 1123–1126.

Stothard, J.R., Frame, I.A. and Miles, M.A. (1997) Use of polymerase chain reaction-based single strand conformational polymorphism and denaturing gradient gel electrophoresis methods for detection of sequence variation of ribosomal DNA of *Trypanosoma cruzi*. *International Journal for Parasitology* 27, 339–343.

Stothard, J.R., Frame, I.A., Carrasco, H.J. and Miles, M.A. (1998) Temperature gradient gel electrophoresis (TGGE) analysis of riboprints from *Trypanosoma cruzi*. *Parasitology* 117, 249–253.

Tang, J. and Unnasch, T.R. (1997) Heteroduplex analysis in medical entomology: a rapid and sequence-based tool for population and phylogenetic studies. *Parasitology Today* 13, 271–274.

Tang, J., Laurent, T., Back, C. and Unnasch, T. (1996) Intra-specific heterogeneity of the rDNA internal transcribed spacer in the *Simulium damnosum* (Diptera: Simulidae) complex. *Molecular Biology and Evolution* 13, 244–252.

Tautz, D. (1989) Hypervariability of simple sequences as a general source for polymorphic DNA markers. *Nucleic Acids Research* 17, 6463–6471.

Tautz, D. (1993) Notes on the definition and nomenclature of tandemly repetitive DNA sequences. In: Pena, S.D.J., Chakraborty, R., Epplen, J.T. and Jeffreys, A.J. (eds) *DNA Fingerprint: State of Science*. Birkhäuser Verlag, Basel, Switzerland, 21 pp.

Tautz, D. and Renz, M. (1984) Simple sequences are ubiquitous repetitive components of eukaryotic genomes. *Nucleic Acids Research* 12, 4127–4138.

Teschauer, W., Mussack, T., Braun, A., Waldner, H. and Fink, E. (1996) Conditions for single strand conformation polymorphism (SSCP) analysis with broad applicability: a study on the effects of acrylamide, buffer and glycerol concentrations in SSCP analysis of exons of the p53 gene. *European Journal of Clinical Chemistry and Clinical Biochemistry* 34, 125–131.

van der Sande, C.A.F.M., Kwa, M., van Nues, R.W., van Heerikhuizen, H., Raué, H.A. and Planta, R.J. (1992) Functional analysis of internal transcribed spacer 2 of *Saccharomyces cerevisiae* ribosomal DNA. *Journal of Molecular Biology* 223, 899–910.

van Herwerden, L., Blair, D. and Agatsuma, T. (1998) Intra- and inter-specific variation in nuclear ribosomal internal transcribed spacer 1 of the *Schistosoma japonicum* species complex. *Parasitology* 116, 311–317.

van Nues, R.W., Rientjes, J.M.J., Morre, S.A., Mollee, E., Planta, R.J., Venema, J. and Raué, H.A. (1995) Evolutionary conserved structural elements are critical for processing of internal transcribed spacer 2 from *Saccharomyces cerevisiae* precursor ribosomal RNA. *Journal of Molecular Biology* 250, 24–36.

Viney, M.E. (1998) Nematode population genetics. *Journal of Helminthology* 72, 281–283.

Walton, S.F., Currie, B.J. and Kemp, D.J. (1997) A DNA fingerprinting system for the ectoparasite *Sarcoptes scabiei*. *Molecular and Biochemical Parasitology* 85, 187–196.

Welsh, J. and McClelland, M. (1990) Fingerprinting genomes using PCR with arbitrary primers. *Nucleic Acids Research* 18, 7213–7218.

Wesson, D.M., Porter, C. and Collins, F.H. (1992) Sequence and secondary structure comparisons of ITS rDNA in mosquitoes (Diptera: Culicidae). *Molecular Phylogenetics and Evolution* 1, 253–269.

Wesson, D.M., McLain, D.K., Oliver, J.H., Piesman, J. and Collins, F.H. (1993) Investigation of the validity of species status of *Ixodes dammini* (Acari: Ixodidae) using rDNA. *Proceedings of the National Academy of Sciences USA* 90, 10221–10225.

Williams, J.G.K., Kubelik, A.R., Livak, K.J., Rafalski, J.A. and Tingey, S.V. (1990) DNA polymorphisms amplified by arbitrary primers are useful as genetic markers. *Nucleic Acids Research* 18, 6531–6535.

Wu, Z., Nagano, I. and Takahashi, Y. (1998) The detection of *Trichinella* with polymerase chain reaction (PCR) primers constructed using sequences of random amplified polymorphic DNA (RAPD) or sequences of complementary DNA encoding excretory-secretory (E-S) glycoproteins. *Parasitology* 117, 173–183.

Yap, E.P.H., Lo, Y.-M.D., Fleming, K.D. and McGee, J.O'D. (1994) False positives and contamination in PCR. In: Griffin, H.G. and Griffin, A.M. (eds) *PCR Technology. Current Innovations.* CRC Press, Boca Raton, Florida, pp. 249–258.

Zarlenga, D.S., Stringfellow, F., Nobary, M. and Lichtenfels, J.R. (1994) Cloning and characterization of ribosomal RNA genes from three species of *Haemonchus* (Nematoda: Trichostrongyloidea) and identification of PCR primers for rapid differentiation. *Experimental Parasitology* 78, 28–36.

Zarlenga, D.S., Aschenbrenner, R.A. and Lichtenfels, J.R. (1996) Variations in microsatellite sequences provide evidence for population differences and multiple ribosomal gene repeats within *Trichinella pseudospiralis*. *Journal of Parasitology* 82, 534–538.

Zarlenga, D.S., Gasbarre, L.C., Boyd, P., Leighton, E. and Lichtenfels, J.R. (1998) Identification and semi-quantitation of *Ostertagia ostertagi* eggs by enzymatic amplification of ITS-1 sequences. *Veterinary Parasitology* 77, 245–257.

Zhang, L.H., Chai, J.J., Jiao, W., Osman, Y. and McManus, D.P. (1998) Mitochondrial genomic markers confirm the presence of the camel strain (G6 genotype) of *Echinococcus granulosus* in north-western China. *Parasitology* 116, 29–33.

Zhang, L.H., Gasser, R.B., Zhu, X.Q. and McManus, D.P. (1999) Screening for different genotypes of *Echinococcus granulosus* within China and Argentina by SSCP-sequencing. *Transactions of the Royal Society of Tropical Medicine and Hygiene* 93, 1–6.

Zhu, X.Q. and Gasser, R.B. (1998) SSCP-based mutation scanning approaches to fingerprint sequence variation in ribosomal DNA of ascaridoid nematodes. *Electrophoresis* 19, 1366–1373.

Zhu, X.Q., Chilton, N.B., Beveridge, I. and Gasser, R.B. (1998a) Detection of sequence variation in parasite ribosomal DNA by electrophoresis in agarose gels supplemented with a DNA-intercalating agent. *Electrophoresis* 19, 671–674.

Zhu, X.Q., Jacobs, D.E., Chilton, N.B., Sani, R.A., Cheng, N.A.B.Y. and Gasser, R.B. (1998b) Molecular characterization of a *Toxocara* variant from cats in Kuala Lumpur, Malaysia. *Parasitology* 117, 155–164.

Zhu, X.Q., Gasser, R.B., Podolska, M. and Chilton, N.B. (1998c) Characterisation of anisakid nematodes of potential zoonotic significance by ribosomal DNA. *International Journal for Parasitology* 28, 1911–1921.

Zhu, X.Q, Chilton, N.B., Jacobs, D.E., Boes, J. and Gasser, R.B. (1999a) Characterisation of *Ascaris* from human and pig hosts by nuclear ribosomal DNA sequences. *International Journal for Parasitology* 29, 469–478.

Zhu, X.Q., Bøgh, H.O. and Gasser, R.B. (1999b) Dideoxy fingerprinting of low-level nucleotide variation in mitochondrial DNA of the human blood fluke, *Schistosoma japonicum*. *Electrophoresis* 20, 2830–2833.

Diversity in Populations of Parasitic Nematodes and its Significance

Mark E. Viney

School of Biological Sciences, University of Bristol, Woodland Road, Bristol BS8 1UG, UK

Introduction

Parasitic nematodes, as all other organisms, are genetically and phenotypically diverse. The genetic diversity of animal parasitic nematodes has begun to be investigated in some detail and this information has been used to explore their population genetics. All of this work has recently been thoroughly reviewed by Anderson *et al.* (1998). However, the phenotypic diversity of animal parasitic nematodes has been investigated much less thoroughly. This chapter will begin to redress the imbalance and will consider three examples of phenotypic diversity of animal parasitic nematodes. It will then extend this theme by speculating on the existence of diversity in other phenotypes that have yet to be investigated. In doing so it is hoped to show why investigating and understanding this aspect of the biology of parasitic nematodes is important to the proper understanding of this important group of pathogens, but it is also an exciting research challenge for the future.

A Developmental Choice for *Strongyloides ratti*

Strongyloides ratti has two developmental routes in its life cycle (Fig. 5.1) and this developmental choice shows phenotypic diversity. The parasitic phase of *S. ratti* is a parthenogenetic female (Viney, 1994). Eggs that pass out of a host can complete the free-living phase of the life cycle by two alternative developmental routes, termed heterogonic and homogonic. In homogonic

©CAB *International* 2001. *Parasitic Nematodes*
(eds M.W. Kennedy and W. Harnett)

development, eggs passed in the faeces moult through two larval stages into infective third-stage larvae (iL3s), which infect new hosts. In heterogonic development, larvae moult through four larval stages, finally maturing as free-living adult males and females. These reproduce by conventional sexual reproduction and the female then lays eggs (Viney *et al.*, 1993). The larvae that hatch from these eggs moult to iL3s, as in homogonic development. For *S. ratti*, there is only one free-living generation and the progeny of this generation always moult into iL3s. Both the heterogonic and homogonic developmental pathways can be undergone by the progeny of a single parasitic female (Graham, 1938; Viney *et al.*, 1992). Thus, young larvae have a developmental choice between homogonic and heterogonic development.

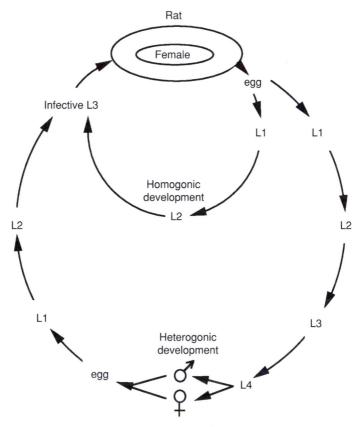

Fig. 5.1. The life cycle of *Strongyloides ratti*. The progeny of the parasitic female can develop by two different routes, termed heterogonic and homogonic. L, larval stage.

Diversity in Developmental Route

Different populations of *S. ratti* vary in their developmental propensity. This can be seen by measuring the relative proportion of heterogonic and homogonic development that occurs during an infection. This is measured by the homogonic index (Hi), which is the proportion of worms that develop by the homogonic route, such that if all worms develop via this route, Hi = 1 (Viney *et al.*, 1992). Isofemale lines (a line of parasites derived from a single iL3) from different geographical origins differ in their developmental propensity during an infection (Fig. 5.2). Isofemale lines of UK isolates (e.g. ED40) develop almost exclusively by the homogonic route throughout an infection (Fig. 5.2) (Viney *et al.*, 1992; Fisher and Viney, 1998). Isofemale lines of isolates from Japan (ED53 and ED54) develop homogonically at the start of an infection. As the infection progresses, more heterogonic development occurs. However, the magnitude and rate of this increase in heterogonic development is reproducibly characteristic of each line but different between the two lines. An isofemale line from an American isolate (ED5) shows the same increasing heterogonic

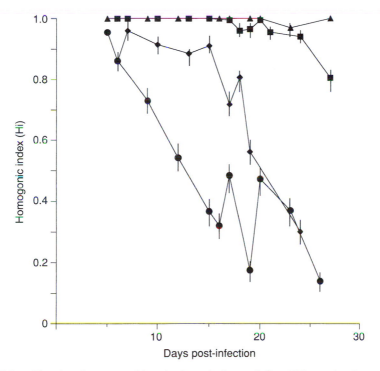

Fig. 5.2. The development of four isofemale lines of *S. ratti* throughout an infection. Error bars are 95% confidence intervals. The origins of the lines are ED5 (●), USA; ED53 (■) and ED54 (♦), Japan; ED40 (▲), UK.

development during an infection but at a rate greater than in any other isofemale line examined (Viney *et al.*, 1992). These observations show that different parasite lines have different inherent developmental propensities. Thus, there is phenotypic diversity in this trait. These differences are probably adaptations to the environments found in the different geographical locations from which these lines were isolated.

Artificial Selection for Developmental Route

Developmental propensity can also be selected for. Isofemale lines of *S. ratti* that underwent mixed heterogonic and homogonic development have been artificially selected for each developmental route. This was done by the separate passage of iL3s that had developed by the homogonic route of development or by the heterogonic route of development. Isofemale line ED5 was selected in this way and responded to the selection (Fig. 5.3). The per generation effect of selection, measured by comparing

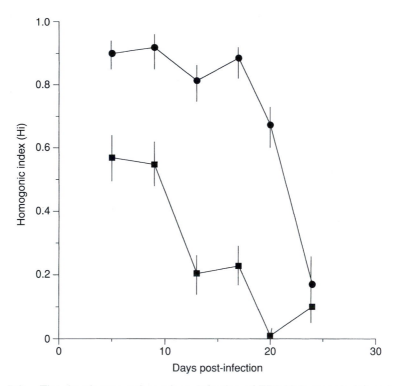

Fig. 5.3. The development through an infection of ED5 Heterogonic (■) and ED5 Homogonic (●) after selection for heterogonic and homogonic development, respectively, for 71 and 78 generations, respectively. Error bars are 95% confidence intervals.

the developmental route of larvae whose parents were either parasitic females or free-living females, was significant (Viney, 1996). A similar selection regime on a different isofemale line (ED132) showed that a significant response to the selection regime occurred after just 14 generations (Viney, 1996). These results show that the trait under selection, 'developmental route', is heritable and therefore that it has a genetic basis. It also shows that the isofemale lines subjected to the selection regime contained variation for this trait.

The Environment and Developmental Route

Observations on many species of *Strongyloides* have shown that the conditions to which young larvae are exposed affect their developmental route (Schad, 1989). For *S. ratti*, one such factor is the temperature at which L1s and L2s are maintained, with homogonic development favoured at lower temperatures and heterogonic development at higher temperatures (Viney, 1996). Lines selected for heterogonic or homogonic development (above) both show the same direction of response to temperature, but the magnitude of the effect and the range over which it occurs is different for the selected lines (Viney, 1996).

Closer inspection of the developmental response to temperature shows that it is actually due to a developmental choice between alternative female morphs, namely free-living females and homogonic iL3s (Fig. 5.1) (Harvey *et al.*, 2000). Comparison of the proportion of larvae that develop into each of three morphs (free-living males, free-living females and iL3s which developed by the homogonic route of development; Fig. 5.1; Viney *et al.*, 1992) at 19 and at 25°C shows that the proportion of larvae that develop into free-living males is constant between the two temperatures, but that the proportion of larvae that develop into free-living females or iL3s which developed by the homogonic route of development changes between the two incubation temperatures. This, combined with other data (Harvey *et al.*, unpublished), has shown that the parasitic female produces offspring of two chromosomal sexes. Genetically male larvae develop into free-living males only. Genetically female larvae develop into free-living females or homogonic iL3s. The developmental choice of female larvae is affected by the temperature experienced by early-stage larvae, with development into free-living females favoured at higher temperatures and homogonic development into iL3s favoured at lower temperatures.

Phenotyplc Plasticity in Response to Temperature

The development of isofemale lines selected for different developmental routes is affected by temperature, but the magnitude of the effect differs between different selected lines. To explore why this difference occurs, the concept of phenotypic plasticity can be useful. Phenotypic plasticity is when a particular genotype has the capacity to produce a different phenotype in response to a change in the environment. The discrete change between alternative morphs, as seen with *S. ratti*, is known as developmental conversion (Smith-Gill, 1983). The difference between the phenotypes in the different environments is known as the sensitivity or the reaction norm of the response. In this example with *S. ratti*, the sensitivity is the difference in Hi at two temperatures. The sensitivity can most easily be thought of as the slope of a line describing the developmental response in the two environments. These data for two lines selected for heterogonic development (ED5 Heterogonic and ED248 Heterogonic) and for two lines selected for homogonic development (ED5 Homogonic and ED132 Homogonic) at 19 and 25°C temperature treatments are shown in Fig. 5.4. These data show that ED248 Heterogonic has the greater sensitivity (i.e. the line is steeper) compared with the other lines. More generally, this suggests that lines

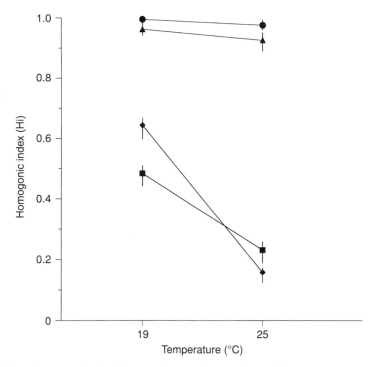

Fig. 5.4.   The sensitivity of four isofemale lines of *S. ratti* in response to temperature. ED248 Heterogonic (♦) and ED132 Homogonic (▲) are two lines selected for heterogonic and homogonic development, respectively, from an ancestral line (ED132). ED5 Heterogonic (■) and ED5 Homogonic (●), as for Fig. 5.3. Error bars are 95% confidence intervals.

selected for heterogonic development may have greater temperature sensitivity than lines selected for homogonic development.

How has this difference in sensitivity come about? These isofemale lines were selected for propensity of developmental route and the selection clearly affected this trait (Fig. 5.3). However, these lines also vary in their temperature sensitivities, a trait for which they were not directly selected. This difference may be a direct consequence of the response to the selection to which the lines were subjected. That is, lines that pre-dominantly undergo homogonic development have, for some reason, a low temperature sensitivity and, conversely, lines that predominantly undergo heterogonic development lines have greater temperature sensitivity. An alternative reason for the difference in sensitivity of the selected lines may be that the selection regime that was applied selected on the target trait (developmental propensity) but, in addition, it inadvertently or cryptically selected on the trait of 'temperature sensitivity'. From these observations, the reason for the difference in temperature sensitivity is not known and indeed separating genetic correlation of traits from correlation of cryptic selection needs to be approached with caution (Hill and Caballero, 1992).

The Molecular Basis of Variation in Sensitivity

The basis of the difference in these sensitivities is not known but, in princi-ple, it can be assumed that the temperature cue is, in some way, sensed by larvae and this is transduced through a molecular pathway or cascade. The consequence of this process is different (quantitative, qualitative or both) gene expression and this is the basis of the switch between the heterogonic and homogonic routes of development. For individual worms, this process is a binary choice. However, this process applied at a population level will result in a change in the proportion of worms that develop by each route.

Variation in the temperature sensitivity of development could be brought about in many different ways. For example, different isofemale lines may have different thermosensory resolution. That is, at an extreme, a parasite line whose greatest temperature resolution was 6° would simply not 'recognize' the different temperatures to which it was exposed. Thus, variation in thermosensory resolution could be the basis of variation in temperature sensitivity. Alternatively, if thermosensory resolution was the same between different isofemale lines, the molecular mechanism that transduces such a signal could (positively or negatively) modulate that sig-nal, such that the magnitude of the developmental output varied between lines in response to the same temperature input. This would imply a quanti-tative difference in the molecular mechanism that transduces the primary thermosensory signal. Differences in sensitivity could occur by either, or a combination, of these or other mechanisms. Determining and separating which mechanisms are operating will be difficult. However, implicit in these ideas is the concept that variation in sensitivity requires variation in the molecular processes underlying the development of the phenotype.

Some isofemale lines of *S. ratti*, apparently, show no heterogonic development (Viney *et al.*, 1992). This could mean one of two things. Firstly, these lines may, in effect, be mutants that have lost the ability to undergo heterogonic development. Alternatively, the sensitivity of their temperature response may be so low (i.e. the slope of the lines, as in Fig. 5.4, is essentially zero) that no difference in developmental route is seen over the temperature range examined. This difference may appear semantic, i.e. there may be no response over a physiological temperature range. However, an important biological difference between these two possibilities is that parasite lines with a very low, or zero, sensitivity still have the genetic and developmental capability to develop by both developmental routes; thus future selection could alter their sensitivity and hence the developmental propensity of the line. Lines that have lost the genetic and developmental mechanisms to undergo one type of development will never recover from that position. The genetic basis of, and the way in which selection acts on, phenotypic plasticity and sensitivity are not understood despite much theoretical and some empirical attention (Schlichting and Pigliucci, 1998). The diversity in phenotypic plasticity of *S. ratti* and in its temperature sensitivity demonstrates the deeper level at which phenotypic diversity occurs and suggests the multiple levels at which natural selection can act. It also highlights the obvious, but important, fact that such phenotypic diversity has an underlying genetic and molecular basis.

Developmental Route and Host Immunity

In addition to the effect of temperature on phenotypic plasticity, the host immune response is also an effector. A normal infection of *S. ratti* in rats lasts about a month. During this time, rats become increasingly immune and progressively lose their infection. This change coincides with an increase in the proportion of heterogonic development that occurs (Fig. 5.3). When immunocompromised, such as hypothymic (nude) or γ-irradiated, hosts are infected with *S. ratti*, development of the free-living generation is more homogonic compared with control infections (Gemmill *et al.*, 1997). Moreover, the predominantly heterogonic development of an established infection is reversed to predominantly homogonic development by the administration of corticosteroids (Gemmill *et al.*, 1997). Together, the results of these three methods of host immunosuppression show that host immune status affects the subsequent development of the free-living generation, with heterogonic development favoured in immune hosts. How this effect of host immunity on future development occurs is not understood. It may occur via the parasitic female as a consequence of the physiological stress imposed by the immune response. The host immune response presumably also changes the microenvironment and physiology of the gut and this may directly affect the eggs/larvae as they pass through the gut.

Interaction of Host Immunity and Temperature on Developmental Route

It has already been noted that the temperature sensitivity of developmental route is different for different isofemale lines when compared at one time point in an infection (Fig. 5.4). If this sensitivity is considered through an infection, it can be seen that the sensitivity increases during an infection (Fig. 5.5) (Harvey *et al.*, 2000); here the difference in the proportion of larvae that develop into free-living females and iL3s at 19 and at 25°C becomes greater as an infection progresses. However, if the temperature sensitivity of larvae passed from immunologically intact animals is compared with that of larvae passed from hypothymic (nude) rats, it can be

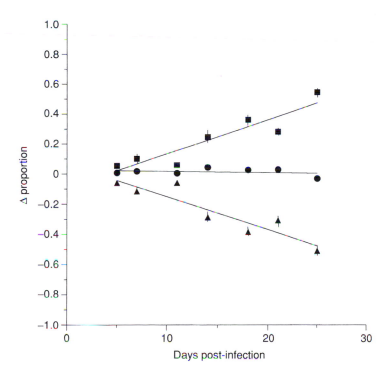

Fig. 5.5. Best-fit regression lines of Δ proportion (the difference in the proportion of larvae developing at 25 and 19°C) into free-living males (●), free-living females (■) or iL3s that developed by the homogonic route of development (▲) through an infection. Error bars are 95% confidence intervals. Some error bars are smaller than the symbol.

also means that the base population contained diversity in the developmental arrest trait. *Obeliscoides cuniculi*, a parasite of rabbits, was artificially selected for increased propensity to arrest. This was highly successful after just five generations (Watkins and Fernando, 1984). Together, these observations have clear parallels with the developmental switch of *S. ratti*.

The interaction of the factors that affect the propensity for arrest are not completely clear. It is probable that many of the observations of changes in propensity for arrest (above) are consequences of exposure to the timing, length and extent of arrest-inducing conditions, particularly temperature. Thus, similarly to *S. ratti*, developmental arrest is a phenotypically plastic trait that responds to environmental conditions. There is no information available about how different parasite isolates respond to arrest-inducing conditions. This would be worthy of investigation. However, it is not unreasonable to suppose that there are differences in the response of different parasite isolates to environmental conditions, i.e. there are different sensitivities to factors that affect the propensity for arrest.

Phenotypic Diversity and Immunology

For much of its life cycle a parasite's environment is its host. A principal, and potentially lethal, feature of this environment is the host immune response. Diversity in parasites' interactions with this aspect of their environment may therefore be expected. The host immune response varies between different hosts and so, if this is considered as a selection pressure, it may produce the ideal conditions for the generation of diversity in immune-mediated parasite–host interactions.

There is good evidence, for a number of helminth species, that different parasite lines vary in their infection characteristics in hosts, much of which is reviewed by Read and Viney (1996). For example, different isolates of *Trichinella spiralis* vary in the kinetics of their primary infection in the same mouse strain. Crucially, these differences are removed when mice are immunosuppressed (Bolas-Fernandez and Wakelin, 1989). Analogous observations have been made for *Trichuris muris* in mice. Different isolates differed in the kinetics of infection and expulsion. However, in immunosuppressed mice, all isolates had similar fecundity (Bellaby *et al.*, 1995). Combined, these observations show immune-dependent variation between parasite lines in their infection kinetics.

A number of attempts have been made to select on nematode immune-related traits. Lines of *Heligmosomoides polygyrus* have been selected in hosts with different histories of previous exposure (naïve, once previously exposed, or multiply previously exposed). This resulted in parasite lines that differed in fitness when tested in semi-immune animals, with the lines selected in the hosts with greatest previous exposure surviving best and being most fecund (Dobson and Tang, 1991). A further comparison of these selected parasite lines, in mice selected for high or low immune response to *H. polygyrus* or in unselected mice, also showed that parasites selected in previously exposed hosts survived better and were more fecund in all lines of mice (Su and Dobson, 1996). An analysis of various immunological parameters during an infection showed that the parasite lines differed in the immune response that they induced in the mouse strains. For example, it was found that the parasite line selected in previously exposed hosts induced a lower level of IgG in all mice strains compared with the other parasite line. The parasite line selected in naïve hosts, which in general survived less well in all the mouse strains, usually induced a greater peripheral eosinophilia compared with the parasite line selected in previously exposed mice (Su and Dobson, 1996). Thus, the functional measures of the host immune response (e.g. parasite survival and fecundity) are correlated (and, presumably, causally related) to the degree and nature of the host's immune response that was measured.

Taken as a whole, these observations show that parasite lines differ in an immune-dependent manner in their infection/expulsion kinetics. Furthermore, there is heritable variation in survival and fecundity in previously exposed hosts and quantitative variation in the immune response that selected parasite lines elicit. Again, taken as a whole, these observations have the necessary corollary that variation in these traits exists not only in laboratory-maintained isolates but also in helminth species in nature. The phenotypes under consideration here (infection/expulsion kinetics, survival, fecundity) are multifactorial life-history traits. Understanding the basis of variation in the components and interplay of these complex, immune-responsive phenotypes must be of crucial relevance to understanding the immunology of infections of parasitic nematodes. This is of particular relevance in view of current attempts to develop immunological methods of nematode control.

The interaction between helminths and their hosts' immune systems occurs at a molecular level, and thus the functional variation that has been seen in these experiments will ultimately have a molecular basis. Despite the widespread interest in the antigenic and surface molecule variation of parasitic protozoa in relation to host immune responses, there has been little analogous investigation in nematodes. There are, though, some isolated and tantalizing reports of molecular diversity of putatively immune-significant molecules. Individual infective larvae of *Ascaris lumbricoides* vary in the surface binding of antibody from different donors (Fraser and Kennedy, 1990). With some donor serum, all larvae responded similarly, whereas with other donors there was considerable heterogeneity between individual infective larvae. Although the basis of this heterogeneity is not known, one possibility is that it is due to antigenic diversity within the parasite population (Fraser and Kennedy, 1990). A similar observation has been made for *T. trichiura*. A comparison of human immune responses against individual *T. trichiura*, measured by Western blots, appears to show antigenic differences between individual worms (Currie *et al.*, 1998). Both of these studies found variation between individual worms in their recognition by serum from individual donors. This finding implies that there was also heterogeneity in the worms that infected the hosts that eventually became the serum donors. Individual infective larvae of *Onchocerca lienalis* have been observed to have minor molecular weight heterogeneity of a 23 kDa molecule (Bianco *et al.*, 1990). These observations show that there is qualitative molecular diversity of putatively important antigens. Differences such as these, or quantitative differences in certain molecules, or combinations of both, may be the basis of the different immune responses elicited by different parasite lines observed by Su and Dobson (1996).

Investigation of the diversity in the interaction of parasitic nematodes with host immune responses and its molecular basis is an under-investigated area. It is clear that, when looked for, there is diversity in immune-dependent interactions and immune-significant molecular heterogeneity. There is a need to extend these studies, especially given the continued attention towards the development of nematode vaccines. The host immune response is a strong selection pressure on parasitic nematodes. The use of a vaccine based on an individual antigen will act as a strong selection pressure on the target parasite population and there should perhaps be concern that this will select for strains that are 'immune' to the effects of such a vaccine.

Summary So Far

Of the examples considered above, two are of phenotypic diversity in a life-history trait where the life-history trait under consideration is clearly a facultative phenomenon. That is, for developmental route in *S. ratti* and for arrested development, there are distinct, mutually exclusive developmental routes. Thus, diversity in these traits between different parasite lines is relatively easy to observe, as is the response to selection. Both these traits are, in part, affected by environmental conditions and so are phenotypically plastic. For *S. ratti*, variation in the sensitivity of this plasticity can also be seen. Although environmental sensitivity of arrested development is as yet uninvestigated, by analogy with *S. ratti* it is likely to vary.

The third example considered the interaction of life-history traits (survival rates, fecundity, immunogenicity) with an environmental factor specific to parasites, namely the host immune system. Here phenotypic diversity in response to environmental conditions (host immunity) is not so readily apparent. To observe phenotypic diversity, different parasite lines need to be compared in their kinetics of infection and, to show immune-dependence, these must be complemented by control experiments in immunosuppressed hosts. Experiments seeking to select on this diversity require a very significant effort (see Su and Dobson, 1996) and, not surprisingly, are rather rare. However, on the few occasions when such endeavours have been undertaken a response to such selection is readily apparent. Such studies are very much in their infancy and much work needs to be done, but one conclusion from these observations is that phenotypic diversity and plasticity in parasitic nematodes exist. This should come as no surprise, since diversity in life-history traits and diversity in their interaction with the environment are likely to be the rule rather than the exception. Given this, where else may we expect to find other interesting and important diversity?

Parasite Modulation of Host Immune Responses

Many parasitic nematodes survive in their hosts by actively altering the host's immune response. This is particularly well known for human filariasis, in which the induction of specific T-cell subsets is inhibited by an active infection (Maizels *et al.*, 1993, 1995). Also, for *H. polygyrus* (*Nematospiroides dubius*), the presence of adult worms in the intestine causes suppression of the immune system, such that challenge infections are successful (Behnke *et al.*, 1983). The molecular details by which such parasite-dependent immunomodulation occurs are not yet understood. Presumably, the parasites produce and release various effector molecules that interact, and thereby alter, the host's immune response. There is obviously a strong positive selection pressure acting on parasites for any mechanism that will allow parasite survival in the face of a host's immune system. More specifically for immunomodulatory methods, *a priori*, the selection is equally strong – the consequence of not doing so is death. However, these mechanisms are also, presumably, energetically expensive. Therefore an advantageous strategy for an individual is to cheat. In this way, an individual parasite would not induce host immunomodulation but, rather, would rely on the immunomodulatory effects of coinfecting parasites. It is not known whether this happens, but such a strategy may be predicted from evolutionary considerations; indeed there are many precedents for such 'cheating' survival strategies with many other animals. It is not known whether immunomodulatory cheating occurs for parasitic nematodes, but then it has never been looked for. There is also the intriguing possibility that the prevalence of 'cheating' in a host population is related to the pattern of distribution of an infection within a host population, since this is related to the probability of being with non-'cheating', coinfecting parasites. Immunomodulatory 'cheating' may be seen in a parasite population as individual parasites (or sub-isolates) that do not cause immunomodulation. Perhaps this is happening all the time, but in the laboratory is seen as 'experimental noise'.

Consequences of Laboratory Maintenance

Many species of parasitic nematodes are maintained in the laboratory in host species in which they are not found in nature. This has the potential consequence that the laboratory population is, in some way, different from the natural population. Transfer and adaptation of a parasite from a natural host into a different species in the laboratory entails a process of selection. The selection will act on the trait 'ability to survive in a non-natural host'. Most of the parasite population may have had little, or indeed no, ability to survive in the non-natural host. Thus, at its most extreme form, this selection will have been for the very small proportion of the parasite population with the ability to survive in a non-natural host. A consequence of this is that the parasite population will have gone through a genetic bottleneck.

Despite the potential importance, not to mention intrinsic interest, of the genetics of host specificity, this process has received only limited experimental attention. The most complete study is of *Nippostrongylus brasiliensis*, a natural parasite of rats. This was selected for 40 generations in the hamster (*Mesocricetus auratus*) and was found to have significantly increased its establishment rate in hamsters after eight generations without a reduction in establishment rate in rats (Haley, 1966). Similarly, two studies passaged *N. brasiliensis* through mice and both found a significant increase in its establishment rate after relatively few generations (Solomon and Haley, 1966; Westcott and Todd, 1966). When these mouse-adapted lines were then tested in rats, different results were obtained. Westcott and Todd (1966) found that the adapted lines did less well in rats, whereas Solomon and Haley (1966) found no difference. These studies show that *N. brasiliensis* does adapt to new host species and we can therefore infer that base populations contain variation for the trait 'ability to survive in a non-natural host'. As discussed above, the consequence of such as process is that a laboratory population under study may be a small, unrepresentative sample of the natural parasite population both in terms of adaptation to a host species and in its genetic diversity. It is probable that the immunological interaction between a parasite and its host is a major determinant of the success of the transfer of a parasite to a non-natural host. If this is so, subsequent immunological studies of that parasite may be little short of irrelevant to the natural host–parasite relationship (Gemmill *et al.*, 2000). However, it is also clear that the nature of adaptation of parasites to new host species is an under-investigated area. This is despite the fact that understanding the nature of the laboratory populations that we work with would seem to be rather basic to being able to understand the significance and relevance of laboratory data to field situations.

A regular, routine passage of laboratory infections (in natural or non-natural hosts) may well act as a selection process in itself. Thus, in natural infections the stages that are produced to allow transmission to new hosts will be sampled in an essentially stochastic way. For example, *Ascaris* eggs pass into the environment and remain there until a new potential host is contaminated. *A priori*, the particular eggs that contaminate this new host are a random sample of the available eggs. The maintenance of a similar life cycle in the laboratory based on a fixed protocol would, typically, collect infective stages on day *N* post-infection and use this material to infect new animals. Iteration of this would act as a strong selection pressure on the trait 'produce infective stages on day *N*'. The response to the selection will depend on the variation in this trait present in the population. Indeed, selecting for this trait may not matter very much for molecular biology studies. However, inadvertent selection on a life-history trait may make subsequent work on life-history traits in that parasite line little more than meaningless.

Concluding Remarks

Other contributions to this book have taken a molecular view of parasitic nematodes, yet molecules make only a rather brief appearance here. This chapter has tried to show that parasitic nematodes are fascinatingly and tantalizingly diverse at a phenotypic level. It has focused particularly on diversity in phenotypes that are apparent in response to environmental conditions within or outside a host. The interaction of parasites with within-host factors is a major current research effort. However, helminth immunology is particularly notable for its inattention to diversity, especially when compared with the immunology of parasitic protozoa (Read and Viney, 1996). Observations of the interaction of host immunity with subsequent development in *S. ratti* show the potential power of such interactions. It is also clear that a principal mechanism of the action of host immune responses is against nematode fecundity (Stear *et al.*, 1997). This is likely to be a molecularly complex interaction. Understanding this interaction, as well as variation in the interaction is interesting, but could also form the basis of control by transmission-reduction rather than eradication *per se*.

Much of the work described elsewhere in this book is trying to understand the molecular basis of various aspects of the biology of parasitic nematodes. The phenotypic diversity considered here, such as the different temperature sensitivities of *S. ratti*, will have a molecular basis. Understanding this for any organism is a major research challenge, but understanding immune-relevant analogues has special and applied relevance for the study of parasitic nematodes. Apparently complex traits need not have a complex molecular basis. For example, natural isolates of the free-living nematode *Caenorhabditis elegans* vary in their feeding behaviour (solitary versus social). This difference has been found to be due to different isoforms of a putative neuropeptide receptor (De Bono and Bargman, 1998).

Investigating phenotypic diversity is not easy. A basic requirement is to have different lines of parasite available and in natural host species. However, being aware of the possibility of variation between individual worms would be a start. The few studies that have molecularly considered individual worms (Bianco *et al.*, 1990; Fraser and Kennedy, 1990; Currie *et al.*, 1998) have found variation between individual worms. Such variation may be the basis of some experimental 'noise'. Perhaps efforts should be focused on this noise? The phenotypic diversity that exists in natural, and even laboratory, populations of nematodes is maintained there by natural selection. This tells us, anthropomorphically, that such diversity matters to parasitic nematodes. It is hoped that this chapter has shown that it should also matter to us.

References

Anderson, T.J.C., Blouin, M.S. and Beech, R.N. (1998) Population biology of parasitic nematodes: applications of genetic markers. *Advances in Parasitology* 41, 219–283.

Armour, J. and Duncan, M. (1987) Arrested larval development in cattle nematodes. *Parasitology Today* 3, 171–176.

Behnke, J.M., Hannah, J. and Pritchard, D.I. (1983) *Nematospiroides dubius* in the mouse: evidence that adult worms depress the expression of homologous immunity. *Parasite Immunology* 5, 397–408.

Bellaby, T., Robinson, K., Wakelin, D. and Behnke, J.M. (1995) Isolates of *Trichuris muris* vary in their ability to elicit protective immune responses to infection in mice. *Parasitology* 111, 353–357.

Bianco, A.E., Robertson, B.D., Kuo, Y.-M., Townson, S. and Ham, P.J. (1990) Developmentally regulated expression and secretion of a polymorphic antigen by *Onchocerca* infective-stage larvae. *Molecular and Biochemical Parasitology* 39, 203–211.

Bolas-Fernandez, F. and Wakelin, D. (1989) Infectivity of *Trichinella* isolates in mice is determined by host immune responsiveness. *Parasitology* 99, 83–88.

Currie, R.M., Needham, C.S., Drake, L.S., Cooper, E.S. and Bundy, D.A.P. (1998) Antigenic variability in *Trichuris trichiura* populations. *Parasitology* 117, 347–353.

De Bono, M. and Bargmann, C.I. (1998) Natural variation in a neuropeptide Y receptor homolog modifies social behaviour and food response in *C. elegans*. *Cell* 94, 679–689.

Dobson, C. and Tang, J. (1991) Genetic variation and host–parasite relations: *Nematospiroides dubius* in mice. *Journal of Parasitology* 77, 884–889.

Fisher, M.C. and Viney, M.E. (1998) The population genetic structure of the facultatively sexual parasitic nematode *Strongyloides ratti* in wild rats. *Proceedings of the Royal Society of London B* 265, 703–709.

Fraser, E.M. and Kennedy, M.W. (1990) Heterogeneity in the expression of surface exposed epitopes among larvae of *Ascaris lumbricoides*. *Parasite Immunology* 13, 219–225.

Gemmil, A.W., Viney, M.E. and Read, A.F. (1997) Host immune status determines sexuality in a parasitic nematode. *Evolution* 51, 393–401.

Gemmil, A.W., Viney, M.E. and Read, A.F. (2000) The evolutionary ecology of host-specificity: experimental studies with *Strongyloides ratti. Parasitology* 120, 429–437.

Gibbs, H.C. (1986) Hypobiosis in parasitic nematodes – an update. *Advances in Parasitology* 25, 129–174.

Graham, G.L. (1938) Studies on *Strongyloides*. II. Homogonic and heterogonic progeny of the single homogonically derived *S. ratti* parasite. *The American Journal of Hygiene* 27, 221–234.

Haley, A.J. (1966) Biology of the rat nematode *Nippostrongylus brasiliensis* (Travassos, 1914). IV. Characteristics of *N. brasiliensis* after 1 to 40 serial passages in the Syrian hamster. *Journal of Parasitology* 52, 109–116.

Harvey, S.C., Gemmill, A.W., Read, A.F. and Viney, M.E. (2000) The control of morph development in the parasitic nematode *Strongyloides ratti. Proceedings of the Royal Society of London B* 267, 2057–2063.

Hill, W.G. and Caballero, A. (1992) Artificial selection experiments. *Annual Review of Ecology and Systematics* 23, 287–310.

Maizels, R.M., Bundy, D.P., Selkirk, M.E., Smith, D.F. and Anderson, R.M. (1993) Immunological modulation and evasion by helminth parasites in human populations. *Nature* 365, 797–805.

Maizels, R.M., Sartono, E., Kurniawan, A., Partono, F., Selkirk, M.E. and Yazdanbash, M. (1995) T-cell activation and the balance of antibody isotypes in human lymphatic filariasis. *Parasitology Today* 11, 50–56.

Michel, J.F., Lancaster, M.B. and Hong, C. (1973) Inhibition of development: variation within a population of *Ostertagia ostertagi. Journal of Comparative Pathology* 83, 351–356.

Read, A.F. and Viney, M.E. (1996) Helminth immunogenetics: why bother? *Parasitology Today* 12, 337–343.

Schad, G.A. (1989) Morphology and life history of *Strongyloides stercoralis*. In: Grove, D.I. (ed.) *Strongyloidiasis: a Major Roundworm Infection of Man*. Taylor and Francis, London, pp. 85–104.

Schlichting, C.D. and Pigliucci, M. (1998) *Phenotypic Evolution: a Reaction Norm Perspective*. Sinauer Associates, Sunderland, Massachusetts.

Smeal, M.G. and Donald, A.D. (1982) Inhibited development of *Ostertagia ostertagi* in relation to production systems for cattle. *Parasitology* 85, 21–25.

Smith-Gill, J.S. (1983) Developmental plasticity: developmental conversion versus phenotypic modulation. *American Zoologist* 23, 47–55.

Solomon, M.S. and Haley, J.A. (1966) Biology of the rat nematode *Nippostrongylus brasiliensis* (Travassos, 1914). V. Characteristics of *N. brasiliensis* after serial passage in the laboratory mouse. *Journal of Parasitology* 52, 237–241.

Stear, M.J., Bairden, K., Duncan, J.L., Holmes, P.H., McKelar, Q.A., Park, M., Strain, S., Murray, M., Bishop, S.C. and Gettinby, G. (1997) How hosts control worms. *Nature* 389, 27.

Su, Z. and Dobson, C. (1996) Genetic and immunological adaptation of *Heligmosomoides polygyrus* in mice. *International Journal for Parasitology* 27, 653–663.

Viney, M.E. (1994) A genetic analysis of reproduction in *Strongyloides ratti. Parasitology* 109, 511–515.

Viney, M.E. (1996) Developmental switching in the parasitic nematode *Strongyloides ratti. Proceedings of the Royal Society of London B* 263, 201–208.

Viney, M.E., Matthews, B.E. and Walliker, D. (1992) On the biological and biochemical nature of cloned populations of *Strongyloides ratti. Journal of Helminthology* 66, 45–52.

Viney, M.E., Matthews, B.E. and Walliker, D. (1993) Mating in the parasitic nematode *Strongyloides ratti*: proof of genetic exchange. *Proceedings of the Royal Society of London B* 254, 213–219.

Watkins, A.R.J. and Fernando, M.A. (1984) Arrested development of the rabbit stomach worm *Obeliscoides cuniculi*: manipulation of the ability to arrest through processes of selection. *International Journal of Parasitology* 14, 559–570.

Westcott, R.B. and Todd, A.C. (1966) Adaptation of *Nippostrongylus brasiliensis* to the mouse. *Journal of Parasitology* 52, 233–236.

New Insights into the Intestinal Niche of *Trichinella spiralis*

Judith A. Appleton

James A. Baker Institute for Animal Health, and Department of Microbiology and Immunology, College of Veterinary Medicine, Cornell University, Ithaca, NY 14853, USA

Introduction

In the textbook entitled *Biological Science*, Gould and Keeton (1996) presented the following definition: 'The ways in which an organism uses its environment to make a living define its niche.' The L1 larva of *Trichinella spiralis* makes its living using two distinct environments of its vertebrate host: intestine and muscle. Larvae initiate infection by invading columnar epithelial cells in the small intestine. Because the larva is 1 mm in length, and epithelial cells are approximately 10–20 µm in diameter, the larva creates a syncytium by occupying a large number of cells at any one time (Wright, 1979). The worm moves through the epithelium, most commonly at the crypt–villus junction, leaving trails of dead cells behind (Gardiner, 1976; Wright *et al.*, 1987). The nematode is believed to moult (Capo *et al.*, 1984), mature and reproduce in this site (Gardiner, 1976), although uncertainties exist because these aspects of the parasite's life cycle have been difficult to study. Newborn larvae are released by female worms, migrate into the lamina propria and then travel via blood and lymph to the striated muscle, where they invade muscle cells and mature to become infective L1 larvae. The occupying larva modifies both the muscle cell and its immediate environment in ways that would appear to promote the development and long-term survival of the parasite (reviewed by Despommier, 1998). The cell biology of the muscle niche of *T. spiralis* is discussed in Chapter 7.

This review discusses the activities of *T. spiralis* L1 larvae in the intestine. Understanding of these activities derives, in part, from studies of immunity. Antibodies specific for larvae have been shown to afford protection against intestinal infection (Appleton *et al.*, 1988). Recent studies of antibody-mediated disruption of the intestinal niche have provided insight into the interactions between host epithelial cell and nematode (McVay *et al.*, 1998, 2000). The earliest events in niche establishment by *T. spiralis* are likely to involve recognition of the host cell by the parasite. It is no surprise that results of experiments conducted *in vitro* suggest that this recognition requires the active participation of both the larva and the enterocyte (ManWarren *et al.*, 1997; Butcher *et al.*, 2000). Although details of the molecular events in the process have not yet been elucidated, some of the parameters of invasion have been established. Furthermore, novel experimental approaches and reagents have been developed that allow for the necessary studies to be undertaken.

Excretory/Secretory and Surface Glycoproteins of *T. spiralis* L1 Larvae

L1 larvae of *T. spiralis* synthesize highly immunogenic glycoproteins. The glycoproteins originate in the stichosome and are found in granules of α, β and γ stichocytes (Ellis *et al.*, 1994; Ortega-Pierres *et al.*, 1996). When stichocytes degranulate, the granule contents are released into the lumen of the nematode's intestinal tract and are then excreted or disgorged; these products are commonly referred to as excretory/secretory (ES) products. Glycoproteins that share epitopes with ES products also occur on the surface of the larva (Appleton and Usack, 1993), though the cellular origin of surface glycoproteins is not known. The family of antigenically cross-reactive glycoproteins has been referred to as TSL-1 (*T. spiralis* larva-1) antigens (Appleton *et al.*, 1991). Every host species that becomes infected with *T. spiralis* makes antibodies against TSL-1 antigens (Denkers *et al.*, 1991). In rodents, antibodies specific for these antigens are induced during the muscle stage of infection (Appleton and McGregor, 1987; Denkers *et al.*, 1990b). It is not known whether secreted stichocyte or surface glyco-proteins escape from the encapsulated nurse cell, or are exported by them, or whether these antigens are released from larvae that for some reason fail and die in the muscle. Regardless, the antigens reach lymphoid tissue and induce high specific antibody titres.

Tyvelose and *Trichinella* Glycoprotein Function

The immunodominant epitope in TSL-1 antigens is found on glycans (Denkers *et al.*, 1990a). The epitope is formed by an unusual sugar, tyvelose (3,6-dideoxy-D-*arabino*hexose; Tyv) (Wisnewski *et al.*, 1993). Tyv-bearing glycans are large, tri- and tetra-antennary, N-linked structures (Reason *et al.*,

1994). Each antenna is capped by Tyv in the β-anomeric conformation (Ellis *et al.*, 1997). The more abundant tetra-antennary glycans comprise three tetrasaccharide antennae: β-D-Tyv*p*(1→3)-β-D-GalNAc*p*(1→4)-[α-L-Fuc*p*(1→3)]-β-D-GlcNAc*p*; and one non-fucosylated trisaccharide antenna: β-D-Tyv*p*(1→3)-β-D-GalNAc*p*(1→4)-β-D-GlcNAc*p* (Ellis *et al.*, 1997). Synthesis of Tyv appears to be restricted to the stichocytes of L1 larvae (Ellis *et al.*, 1994; Ortega-Pierres *et al.*, 1996).

Three general functions may apply to glycan moieties in glycoproteins: (i) to protect polypeptides against proteolytic degradation; (ii) to promote and preserve proper folding of polypeptides; and (iii) to mediate cellular adhesion by binding to protein receptors (Dwek, 1995). It is easy to argue that Tyv-capped glycans of *T. spiralis* would protect the surface and secreted products of the larva from destruction by glycosidases and proteases in the intestine. The tetra-antennary glycan has a mass of approximately 5 kDa and should cover a large area on the protein surface. For example, a major ES glycoprotein (the so-called 43 kDa species) bears two glycans which comprise 25% of the mass of the molecule (Gold *et al.*, 1990; Su *et al.*, 1991; Vassilatis *et al.*, 1992). Tyv-bearing glycoproteins that cover the larval body surface appear to be densely packed (McVay *et al.*, 1998). The exposed glycans would protect the surface and cuticle from proteolytic attack in the stomach and intestine. Thus, the Tyv-capped glycan is likely to protect proteins from degradation and to influence protein folding; however, a role for the glycan in adhesion has yet to be demonstrated. The only other organism known to synthesize Tyv is *Salmonella*, which incorporates α-Tyv into the *O*-polysaccharide (Stacey and Barker, 1960). The fact that *Salmonella* is also an enteric pathogen supports the idea that Tyv may contribute to the survival and success of these agents in the gut.

Function of ES Products

It is certain that additional activities of larval glycoproteins reside in the polypeptides. An ES glycoprotein may perform a function in the muscle, or the intestine, or both, so that the significance of an activity or a sequence identity may be difficult to assign to a particular niche. cDNAs encoding four distinct ES glycoproteins have been analysed; no function for any has been proven (Sugane and Matsuura, 1990; Zarlenga and Gamble, 1990; Su *et al.*, 1991; Vassilatis *et al.*, 1992; Arasu *et al.*, 1994). A serine/threonine kinase activity (Arden *et al.*, 1997) and an endonuclease activity (Mak and Ko, 1999) have been described in larval ES products but proteins or cDNAs for these activities have not been characterized.

It has been known for some time that *T. spiralis* antigens localize to Nurse cell nuclei (Ooi and Kamiya, 1986; Despommier *et al.*, 1990; Sanmartin *et al.*, 1991) and recently it has been shown that some of these antigens bear Tyv (Yao *et al.*, 1998). It has been proposed that these molecules influence gene expression in muscle cells. The major glycoprotein species that have been detected in muscle nuclei appear to be, at best, minor constituents of ES products; the majority of Tyv-bearing ES proteins have not been detected in the nucleus (Jasmer *et al.*, 1994; Yao *et al.*, 1998).

The ways in which ES proteins may function in the intestine are not known. These molecules have been detected in intestinal epithelia by immunohistochemistry (Capo *et al.*, 1986). The storage of Tyv-bearing glycoproteins in stichocyte granules and their discharge during entry into the intestine suggests that they participate in the process of invasion and niche establishment. The L1 larva of *T. spiralis* has no oral appendages and does not possess a stylet. Larvae invade epithelial cells 'head first' (personal observation) so that disgorged Tyv-bearing glycoproteins are positioned to play a role in invasion and transit of larvae through epithelial cells. In addition, glycoproteins that cover the body surfaces of larvae are in intimate contact with the cytoplasm of epithelial cells (Wright, 1979; ManWarren *et al.*, 1997), where they may play an active role in the niche. The observation that antibodies specific for Tyv cause larvae to be expelled from the intestine further supports the notion that glycoproteins play an important role at the interface between parasite and host.

Current understanding of parasitism by *T. spiralis* is compartmentalized, and so glycoprotein function has been considered in the context of one compartment or another. However, L1 larvae of *T. spiralis* have evolved under selective pressure to parasitize both intestine and muscle. Biological economy may require a duality of function in larval glycoproteins such that they are able to perform distinct roles in each of the two niches. Dualism is common in proteins, and elucidation of such properties in parasitic nematode products would provide unique insights into the basis of host adaptation.

Protective Activity of Anti-tyvelose IgG

Rats infected with muscle-stage *T. spiralis* protect themselves against reinfection using an immune defence that expels larvae from the intestine (McCoy, 1940; Castro *et al.*, 1976; Love *et al.*, 1976; Bell *et al.*, 1979; Alizadeh and Wakelin, 1982). This phenomenon has been called rapid expulsion (Bell *et al.*, 1979). Rat pups that suckle a dam bearing muscle-stage larvae also rapidly expel L1 larvae from the intestine following infection (Appleton and McGregor, 1984). Immunity in neonatal rats was first described by Culbertson (1943), who showed that protection could be transferred to pups with serum from infected rats. More recent studies showed that this protection targets the L1 larva and not later stages of the life cycle (Appleton and McGregor, 1985b). Furthermore, specific rat monoclonal as well as polyclonal antibodies confer a high level of protection (as much as 100%) in passive transfer experiments (Appleton *et al.*, 1988). These antibodies are of various IgG isotypes. All protective monoclonal antibodies studied to date are specific for Tyv (Ellis *et al.*, 1994, 1997).

Neonatal rats are valuable subjects for the study of antibody-mediated intestinal immunity, because they are free from complicating variables associated with immune and inflammatory responses that are induced in animals during prior infection with the parasite. This simplifies both experimental design and interpretation of passive immunization studies. It must be remembered that neonatal rats differ significantly from adults in terms of diet and intestinal physiology. These differences do not seem to influence parasitism by *T. spiralis*: the distribution of larvae in the intestine as well as the completion of the life cycle proceeds normally in neonatal rats (Appleton and McGregor, 1985; Otubu *et al.*, 1993). The expression of antibody-mediated intestinal immunity may be quite different, however, due to the Fc receptor (nFcR)-driven transport of IgG across enterocytes from the lumen to the mucosa in neonates (Borthistle *et al.*, 1977). The role of the nFcR in rapid expulsion was evaluated by testing F(ab')$_2$ fragments of otherwise protective monoclonal antibodies. Fragments afford passive immunity to pups when delivered orally or systemically. This observation argues against a requirement for the nFcR in neonatal rapid expulsion. It may be more important that the neonatal gut is hospitable to IgG, allowing it to remain intact in the lumen.

Mucus Trapping and Expulsion

In passively immunized neonatal rats, Tyv-specific antibodies exclude larvae from the epithelium (Appleton *et al.*, 1988), where large numbers of excluded larvae become entrapped in mucus (Carlisle *et al.*, 1991a). Similarly, when immune adult rats are challenged with larvae, many luminal parasites are observed entrapped in mucus (Lee and Ogilvie, 1982; Bell *et al.*, 1984). Larvae are neither injured nor killed by mucus entrapment, which is reversible and is not a requirement for expulsion (Carlisle *et al.*, 1990). Rather, mucus appears to participate in expulsion by temporarily confining larvae to the lumen, thus facilitating their elimination from the intestine by normal physiological processes.

Administration of Tyv-specific monoclonal antibodies to rat pups already infected with intestinal larvae causes larvae in the epithelium to be expelled (Carlisle *et al.*, 1990). Only the L1 stage is susceptible to expulsion; once the larva has moulted to L2 it resists the effects of the antibodies (Carlisle *et al.*, 1990). Expulsive immunity is transferred by three IgG isotypes, F(ab')$_2$ fragments, as well as IgM (Carlisle *et al.*, 1991a). These findings argue against a role for Fc-mediated effector functions and imply that antibodies against Tyv can disturb the larva's niche in a direct fashion.

The large number of Tyv-bearing glycoproteins and their distribution in ES products and on the surfaces of larvae make it difficult to identify the relevant targets of protective antibodies. Coating the surfaces of larvae with anti-Tyv IgG prior to infection of rat pups promotes mucus entrapment and affords moderate protection against infection (41% reduction in worm burden after 24 h) (Carlisle *et al.*, 1990). It is not clear whether protection is the result of mucus entrapment or some other effect of surface-bound antibodies. An inhibitory role for surface-bound antibodies has been evaluated in experiments performed *in vitro*. When living *T. spiralis* larvae were surface labelled with biotin, neither avidin nor polyclonal anti-biotin antibodies prevented labelled larvae from entering epithelial cells. Nevertheless, biotinylation of the larva reduced binding of Tyv-specific antibodies to the body surface and coincidentally reduced the efficacy of those antibodies in excluding larvae from epithelial cells. Mucus is absent from this *in vitro* system, making it more likely that antibody binding to the larval body surface has an indirect, inhibitory effect on invasion of the epithelial cell (McVay *et al.*, 1998) (see below). The mechanism of inhibition is unknown.

In contrast to the protection afforded to suckling rats by Tyv-specific antibodies, passive immunization of weaned rats fails to cause expulsion of *T. spiralis* (Otubu *et al.*, 1993). Nevertheless, Tyv-specific antibodies do affect the behaviour of larvae in the intestines of weaned rats in the early hours following infection in that larvae are immobile in the intestinal tissue of such rats, though immobility is reversed when the larvae moult. These findings provide further evidence that antibodies specific for Tyv interfere with the L1 larva's niche.

Simple passive immunization with Tyv-specific antibodies does not protect adult rats against *T. spiralis*; however, it has been shown that prior infection with an unrelated intestinal nematode (*Heligmosomoides polygyrus*) in combination with passive immunization with Tyv-specific antibodies promotes expulsion of *T. spiralis* larvae (Bell *et al.*, 1992). The way(s) that *H. polygyrus* infection synergizes with antibodies is not known.

Invasion of Epithelia *in vitro*

The experimental results described above provide indirect evidence that the processes of invasion and intercellular transit by *T. spiralis* are facilitated by Tyv-bearing glycoproteins. More detailed investigation of the molecular events in invasion and niche establishment has been greatly facilitated by the use of an *in vitro* model of the intestinal epithelium (ManWarren *et al.*, 1997). In this system, epithelial cells are grown to confluence in plastic culture dishes or on glass chamber slides, coverglasses or filter inserts. Cultures are inoculated with 'activated' larvae, that is, larvae recovered from rat muscle by pepsin digestion and subsequently treated with rat intestinal contents or with bile from rats (ManWarren *et al.*, 1997) or pigs (Li *et al.*, 1998). Activated larvae are suspended in semi-solid agarose and then overlaid on the cell monolayer. Inoculated monolayers can be incubated for 30 min to several days, depending on the experiment. Under these conditions, larvae invade and migrate through epithelial cells in the monolayer. Invasive behaviour of larvae is enhanced dramatically by exposure to the intestinal milieu and depends upon the semi-solid medium. Cells that are invaded eventually die, and by applying the method that Wright *et al.* (1987) used on mouse intestinal tissues, dead cells can be detected in monolayers using propidium iodide (Fig. 6.1) or by staining with trypan blue (Fig. 6.2A). Damage to a culture is quantifiable using computer-assisted image capture and analysis of stained monolayers (ManWarren *et al.*, 1997).

Several parameters of the larval niche have been evaluated in this *in vitro* model. Results indicate that *T. spiralis* establishes its niche in cell cultures. First, the broad host range of *T. spiralis* has been reproduced in cultured cells. The nematode is known to infect a diversity of vertebrate

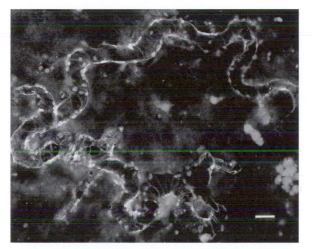

Fig. 6.1. MDCK-AA7 cells were grown on coverglasses, inoculated with *T. spiralis* larvae, labelled with propidium iodide, fixed, and incubated with rat anti-tyvelose monoclonal antibody and then FITC-conjugated goat anti-rat IgG (H and L chain) as described in ManWarren *et al.* (1997). Photomicrograph is a double exposure, taken firstly with 546 nm excitation and 580 nm barrier filters for imaging propidium iodide (red) and secondly with 450–490 nm excitation and 520–560 nm barrier filters for imaging FITC (green) using a Nikon Diaphot inverted microscope fitted for fluorescence (Opti-Quip, Highland Mills, New York). Tyvelose stains green and is limited to the serpentine path travelled by a larva. There are no parasites visible in this field. Nuclei of the dead cells stain intensely and uniformly red or, where they overlap with FITC, yellow. Nuclei of the live cells in the surrounding monolayer are very lightly fluorescent. Bar = 50 μm. Photograph prepared by L.F. Gagliardo, Cornell University.

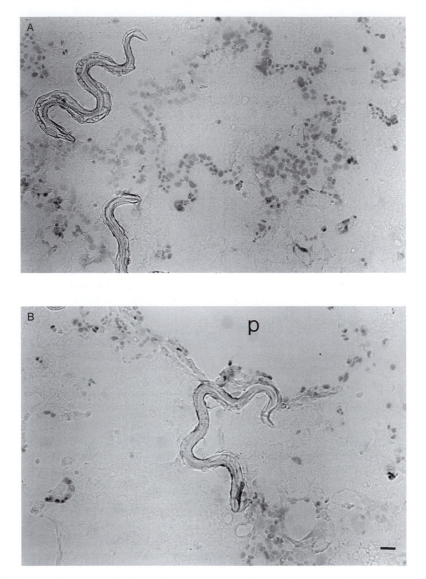

Fig. 6.2. Caco-2 epithelial cell monolayers cultured with *T. spiralis* L1 larvae in (A) the absence or (B) presence of 1 mg ml⁻¹ rat monoclonal, tyvelose-specific antibody 9D4 (McVay *et al.*, 2000). Monolayers were fixed and stained with trypan blue as described in ManWarren *et al.* (1997). (A) Serpentine trails of nuclei in dead cells are evident, revealing the paths travelled by larvae. (B) Tyvelose-specific antibody has inhibited the migration of the larva such that it is encumbered in cell debris and has pulled up a large area of the monolayer, creating a plaque (P). Bar = 50 μm. Photomicrograph prepared by C. McVay, TTUHSC, Lubbock, Texas.

hosts, and rat, human, monkey, dog and pig epithelial cell lines are susceptible to invasion *in vitro*. Several intestinal epithelial cell lines are susceptible, as are kidney epithelial cells from three different species (ManWarren
et al., 1997). The process of invasion is not simply one of physical penetration, as some cell lines resist entry by the larva. Fibroblast (WI-38), muscle
(C2C12) (ManWarren *et al.*, 1997) and two rat intestinal epithelial cell lines
(IEC-6 and IEC-18) (Butcher *et al.*, 2000) are resistant. Although the basis
for resistance in these cell lines is not known, they serve as tools for further
investigation of the cellular requirements for invasion.

Microscopic observation of larvae during invasion is a powerful
advantage of this system, revealing that larvae first browse the monolayer
by probing cells with their heads. A larva may browse but otherwise ignore
a number of cells before invading one. Once in the monolayer, the behaviour of the larva depends upon the cell type. Larvae that enter Madin-Darby
canine kidney (MDCK) epithelial cells often reverse direction or emerge
after travelling a short distance in the monolayer. Emergent larvae may
keep a portion of the body in the monolayer while the anterior browses the
surfaces of nearby cells and then re-enters the cell layer. In contrast, larvae
invade and migrate comparatively long distances in Caco-2 cells without
reversing direction or emerging from the monolayer. Perhaps for this
reason, Caco-2 cells support the development of *T. spiralis* larvae to
the adult stage (McVay *et al.*, 2000; L. Gagliardo and J.A. Appleton,
unpublished observations).

Approximately 25% of L1 larvae will moult to L2 or later stages when
cultured for 24 h in Caco-2 cells grown on plastic surfaces (McVay *et al.*,
2000). If one considers that the primary goal of intestinal *T. spiralis* is
reproduction, then if the epithelial culture accurately models the intestine,
it should support development of the reproductive tract of the worm. This
occurs to a limited extent, as adult males have been observed routinely,
albeit at low numbers, in prolonged co-culture experiments. Development
and growth of female worms also occurs at low frequency (L. Gagliardo and
J.A. Appleton, unpublished results), suggesting that epithelial cells may
meet all of the needs of the developing parasite. Other types of cells or
host factors that may support development have not yet been identified.
Lymphocytes can be ruled out as crucial to *T. spiralis* development, as the
parasite reproduces in nude mice (Perrudet-Badoux *et al.*, 1983; Vos *et al.*,
1983) and XID mice (Lim *et al.*, 1994).

Electron microscopic evaluation of infected Caco-2 monolayers grown
on membranes, under conditions that induced cellular polarization,
prompted the conclusion that the larvae occupy the cytoplasm of cells they
invade (ManWarren *et al.*, 1997). Apical and basal plasma membranes
appeared to be preserved in infected cells (Fig. 6.3). These findings reproduced the observations of Wright (1979) in his examination of intestinal
tissues from infected mice. In contrast, when Li *et al.* (1998) performed
similar experiments in HT29 monolayers grown on plastic, they concluded

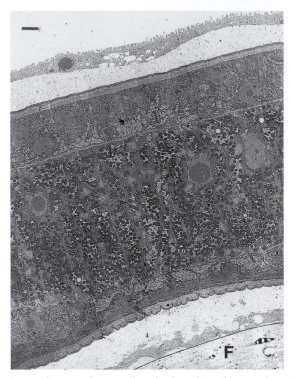

Fig. 6.3. Electron micrograph revealing the location of larvae in polarized Caco-2 monolayers grown on filter inserts (ManWarren *et al.*, 1997). Apical microvilli provide evidence of epithelial cell polarization. Epithelial cell cytoplasm is evident above and below the larva. Bar = 2 μm. The position of the filter substrate is marked (F). Photomicrograph prepared by S. Pearce-Kelling and J. Alling, Cornell University.

that penetration of cells by larvae caused morphological changes compatible with the loss of osmotic regulation and cell bursting. The differences in conclusions drawn from the two studies may reflect differences in growth conditions. Cells grown on plastic may be less able to expand in order to accommodate the larva, in comparison with cells growing in the intestine or grown on membranes *in vitro*.

Tyvelose and Invasion

Larvae deposit Tyv-bearing glycoproteins in cells they invade (ManWarren *et al.*, 1997) (Fig. 6.1). Most of the glycoproteins detected in infected monolayers are ES products (Butcher *et al.*, 2000); Tyv-bearing surface glycoproteins do not seem to be shed into cells during migration (McVay *et al.*, 1997). Despite this evidence of contact between host cells and ES products, direct evidence that parasite glycoproteins actually mediate invasion has not been forthcoming. Challenges that arise in the investigation of the earliest events in invasion include the speed with which larvae enter susceptible cells together with lack of synchrony and low multiplicity of infection in the cultures. Furthermore, infected cells are so quickly and dramatically altered by the invading larva that they are not easily studied. Finally, methods for the genetic manipulation of *T. spiralis* that could be used to knock out genes of interest are not yet available.

Use of resistant cells, such as the rat IEC-6 line, circumvents some of these problems and allows for further investigation of cellular invasion. Larvae exhibit browsing behaviour on resistant cell monolayers but fail to invade them. During browsing, larvae deliver ES products to cells. These glycoproteins are detected in Western blots of cell lysates or by fluorescent antibody staining of monolayers using Tyv-specific antibodies (Butcher *et al.*, 2000). The route of entry into the cell is not known. Tyv-bearing glycoproteins are detected in the cytoplasm of some cells and in the nucleus in others, and often co-localize with mitochondria-specific dyes. Delivery of ES products to a cell in the absence of invasion suggests that glycoprotein delivery to resistant cells is somehow different from delivery to susceptible cells. Alternatively, the resistant cell may fail to respond to browsing and glycoprotein delivery in a way that allows or promotes larval entry into the cell. One can speculate that the cellular response may take any of several possible forms, including cell membrane modification, release of attractive stimuli for nematode sensory receptors, or release of repellent sensory stimuli. The nature of any of these events and their mediators remains to be elucidated.

Cell Wounding and Pore Formation

Attempts to study the entry of ES products into cells using markers of fluid phase endocytosis yielded unexpected results. When larvae browse resistant IEC-6 cells in the presence of extracellular fluorescent dextran, dextran enters the cytoplasm of a significant proportion of the cells in the monolayer (Butcher *et al.*, 2000). The parameters of dextran entry are most compatible with the conclusion that larvae wound the plasma membranes of IEC-6 cells; that is, they create transient breaches in the membrane that allow impermeant markers to enter the cell (McNeil and Ito, 1989). Wounding is considered to be a common occurrence in intestinal epithelia (McNeil and Ito, 1989). Injured cells are able to heal their wounds by recruiting vesicles to seal the breach (Steinhardt *et al.*, 1994). In an experimental system, healing allows the injured cell to retain cytoplasmic dextran. In epithelial cell cultures inoculated with *T. spiralis* larvae, the relationship between glycoprotein delivery and injury of plasma membranes is not clear, i.e. dextran-laden cells do not always stain with Tyv-specific antibodies and

Tyv can be detected in cells that lack cytoplasmic dextran (Butcher *et al.*, 2000). Furthermore, there is some evidence for size restriction associated with dextran entry. This would be compatible with parasite-induced membrane pores.

The observations that *T. spiralis* larvae wound or create pores in IEC-6 cells yet fail to enter them, and deposit ES glycoproteins in IEC-6 cells yet fail to enter them, implicate an inappropriate cellular response to the parasite rather than inappropriate parasite behaviour as the basis for host cell resistance. Susceptible epithelial cells in culture do respond to infection with *T. spiralis* larvae as shown by the detection of IL-1β, IL-8 and ENA-78 at 6 h post-inoculation (Li *et al.*, 1998). The type of response required for invasion would be more immediate and the evidence favours parasite recognition of cytoplasmic host factors over cell surface recognition as a requirement for invasion. Taken together, these findings offer some refinement of our understanding of invasion. *In vitro* studies aimed at elucidating the action of Tyv-specific antibodies against invading larvae have provided some additional insights.

Interference with the Epithelial Niche of *T. spiralis* by Anti-Tyv IgG *in vitro*

Evidence from passive immunization studies indicates that the specific interaction of antibodies with tyvelose moieties on secreted glycoproteins has three inhibitory effects on *T. spiralis* larvae. Larvae may be excluded from epithelia and entrapped in mucus, dislodged from host intestinal epithelium, or immobilized in that site (Carlisle *et al.*, 1990, 1991a,b; Otubu *et al.*, 1993). Recently, these phenomena have been further investigated *in vitro* using epithelial cell lines that vary in sensitivity to invasion by larvae (McVay *et al.*, 2000). In cultures of Caco-2 cells, which are highly sensitive to invasion, Tyv-specific monoclonal antibodies encumber larvae that have invaded monolayers. As the encumbered parasite struggles to migrate through the cells, it pulls the cells it occupies as well as adjacent cells off the substrate, thereby creating plaques in the monolayer (Fig. 6.2B). Thus, although the larva succeeds in invading cells, it fails to establish a niche. In an infected animal, such encumbered larvae would eventually be released into the lumen of the intestine. The large aggregates of cellular debris that adhere to the body of the worm would likely compromise motility as well as promote entrapment in mucus. Encumbrance *in vivo* may manifest as the 'immobility' in intestinal tissue observed in passively immunized rats (Carlisle *et al.*, 1991b; Otubu *et al.*, 1993).

In cultures of the more resistant cell line, MDCK, the predominant effect of Tyv-specific monoclonal antibodies is to exclude larvae from the monolayer. This exclusion is associated with the formation of cap-like structures that cover the stoma of the larva. Such caps were first described by Oliver-Gonzalez (1940) on larvae that were incubated at 37°C with sera from infected animals. Caps comprise immune complexes as revealed by staining with fluorescent anti-rat IgG antibodies (Fig. 6.4). Caps may prevent invasion of epithelial cells by forming a barrier between the larva and the host cell. Alternatively, they may block sensory receptors located in the amphids, positioned on either side of the stoma, thereby preventing the worm from receiving sensory input required for invasive behaviour.

Although caps of immune complexes interfere with invasion, this is not the sole means of preventing larvae from entering cells. Monovalent Fab fragments of Tyv-specific antibodies also exclude larvae from MDCK cells (McVay *et al.*, 2000). This finding indicates that simple blocking of tyvelose moieties interferes with invasion, and suggests that glycans may interact with surface receptors or other binding partners in host cells. Thus, tyvelose-bearing molecules are implicated further as active participants in the process of invasion.

Finally, addition of Tyv-specific antibodies to Caco-2 cultures significantly inhibits larval moulting (McVay *et al.*, 2000). It seems likely that the inhibitory effect is an indirect result of antibody-mediated encumbrance of larvae in the monolayer rather than a direct effect on the process of moulting. Nevertheless, the observation provides further support for the conclusion that antibodies specific for Tyv interfere with the niche of the parasite.

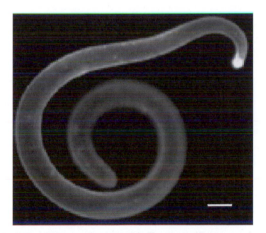

Fig. 6.4. *T. spiralis* L1 larva cultured with Tyv-specific monoclonal antibody 18H followed by FITC-conjugated goat antibody against rat IgG. Bright spot appears as a cap by phase contrast microscopy (McVay *et al.*, 1998) and is shown here to comprise immune complexes of Tyv-bearing ES products and Tyv-specific antibodies. Bar = 30 μm. Specimen was examined and FITC was imaged as described in Fig. 6.1. Image was captured using a charge coupled device camera (Hamamatsu Photonics K.K.) and NIH Image 1.58, as described in McVay *et al.* (1998). Photomicrograph prepared by L.F. Gagliardo, Cornell University.

Concluding Remarks

In summary, the *in vivo* protective effects of Tyv-specific antibodies, exclusion and immobility, can now be effectively studied using an *in vitro* model of the intestinal epithelium. Larvae are prevented from entering epithelial cells by caps of immune complexes or by binding of antibody to Tyv in the absence of immune complex formation. These effects would correlate with exclusion of larvae from epithelia observed in passively immunized rats. Larvae are encumbered as they migrate within epithelial monolayers, an effect that may correlate with immobility of larvae observed *in vivo*. It is reasonable to conclude that in the animal host the different effects work in combination, most likely in cooperation with innate host defences, to cause nematode expulsion from the intestine.

Parasitism by *T. spiralis* has been a subject of scientific interest for over 150 years. Recently, considerable attention has been paid to the parasite by immunologists interested in immunity to nematodes in general, and mucosal immunity in particular. It has been shown that glycan-specific antibodies are highly effective mediators of host defence against intestinal *T. spiralis* infection. Protective monoclonal antibodies have been used to elucidate mechanisms of worm expulsion, as well as to identify molecules that the parasite uses to create its niche. In the future, detailed characterization of these molecules and their functions should afford additional insights into parasitism by *Trichinella spiralis*, and possibly also by other types of pathogen.

Acknowledgements

I thank Catherine McVay, Lucille Gagliardo, Sue Pearce-Kelling and Julie Alling for preparing the photomicrographs included in this chapter, Fernanda Romaris and Daniel Beiting for helpful comments, and Anita Hesser for assistance with preparation of the manuscript. Research described in this chapter from the laboratory of J.A. Appleton is supported by the National Institute of Allergy and Infectious Diseases of the National Institutes of Health (AI-14490).

References

Alizadeh, H. and Wakelin, D. (1982) Comparison of rapid expulsion of *Trichinella spiralis* in mice and rats. *International Journal for Parasitology* 12, 65–73.

Appleton, J.A. and McGregor, D.D. (1984) Rapid expulsion of *Trichinella spiralis* in suckling rats. *Science* 226, 70–72.

Appleton, J.A. and McGregor, D.D. (1985a) Induction and expression of rapid expulsion in suckling rats. In: Kim, C.W. (ed.) *Trichinellosis* (Proceedings of the VIth International Conference on Trichinellosis), pp. 58–62.

Appleton, J.A. and McGregor, D.D. (1985b) Life-phase specific induction and expression of rapid expulsion in rats suckling *Trichinella spiralis*-infected dams. *Immunology* 55, 225–232.

Appleton, J.A. and McGregor, D.D. (1987) Characterization of the immune mediator of rapid expulsion of *Trichinella spiralis* in suckling rats. *Immunology* 62, 477–484.

Appleton, J.A. and Usack, L. (1993) Identification of potential antigenic targets for rapid expulsion of *Trichinella spiralis*. *Molecular and Biochemical Parasitology* 58, 53–62.

Appleton, J.A., Schain, L.R. and McGregor, D.D. (1988) Rapid expulsion of *Trichinella spiralis* in suckling rats: mediation by monoclonal antibodies. *Immunology* 65, 487–492.

Appleton, J.A., Bell, R.G., Homan, W. and van Knapen, F. (1991) Consensus on *Trichinella spiralis* antigens and antibodies. *Parasitology Today* 7, 190–192.

Arasu, P., Ellis, L.A., Iglesias, R., Ubeira, F.M. and Appleton, J.A. (1994) Molecular analysis of antigens targeted by protective antibodies in rapid expulsion of *Trichinella spiralis*. *Molecular and Biochemical Parasitology* 65, 201–211.

Arden, S.R., Smith, A.M., Booth, M.J., Tweedie, S., Gounaris, K. and Selkirk, M.E. (1997) Identification of serine/threonine protein kinases secreted by *Trichinella spiralis* infective larvae. *Molecular and Biochemical Parasitology* 90, 111–119.

Bell, R.G., McGregor, D.D. and Despommier, D. (1979) *Trichinella spiralis*: Mediation of the intestinal component of protective immunity in the rat by multiple, phase-specific, antiparasitic responses. *Experimental Parasitology* 47, 140–157.

Bell, R.G., Adams, L.S. and Ogden, R.W. (1984) Intestinal mucus trapping in the rapid expulsion of *Trichinella spiralis* by rats: induction and expression analyzed by quantitative worm recovery. *Infection and Immunity* 45, 267–272.

Bell, R.G., Appleton, J.A., Negrao-Correa, D.A. and Adams, L.S. (1992) Rapid expulsion of *Trichinella spiralis* in adult rats mediated by monoclonal antibodies of distinct IgG isotypes. *Immunology* 75, 520–527.

Borthistle, B.K., Kubo, R.T., Brown, W.R. and Grey, H.M. (1977) Studies on receptors for IgG on epithelial cells of the rat intestine. *Journal of Immunology* 119, 471–476.

Butcher, B.A., Gagliardo, L.F., ManWarren, T. and Appleton, J.A. (2000) Larvae-induced plasma membrane wounds and glycoprotein deposition are insufficient for *Trichinella spiralis* invasion of epithelial cells. *Molecular and Biochemical Parasitology* 107, 207–218.

Capo, V., Despommier, D.D. and Silberstein, D.S. (1984) The site of ecdysis of the L_1 larva of *Trichinella spiralis*. *Journal of Parasitology* 70, 992–994.

Capo, V., Silberstein, D. and Despommier, D.D. (1986) Immunocytolocalization of two protection-inducing antigens of *Trichinella spiralis* during its enteral phase in immune and non-immune mice. *Journal of Parasitology* 72, 931–938.

Carlisle, M.S., McGregor, D.D. and Appleton, J.A. (1990) The role of mucus in antibody-mediated rapid expulsion of *Trichinella spiralis* in suckling rats. *Immunology* 70, 126–132.

Carlisle, M.S., McGregor, D.D. and Appleton, J.A. (1991a) Intestinal mucus entrapment of *Trichinella spiralis* larvae induced by specific antibodies. *Immunology* 74, 546–551.

Carlisle, M.S., McGregor, D.D. and Appleton, J.A. (1991b) The role of the antibody Fc region in rapid expulsion of *Trichinella spiralis* in suckling rats. *Immunology* 74, 552–558.

Castro, G.A., Badial-Aceves, F., Adams, P.R., Copeland, E.M. and Dudrick, S.J. (1976) Response of immunized, parenterally nourished rats to challenge infection with the nematode, *Trichinella spiralis. Journal of Nutrition* 106, 1484–1491.

Culbertson, J.T. (1943) Natural transmission of immunity against *Trichinella spiralis* from mother rats to their offspring. *Journal of Parasitology* 29, 114–116.

Denkers, E.Y., Wassom, D.L. and Hayes, C.E. (1990a) Characterization of *Trichinella spiralis* antigens sharing an immunodominant, carbohydrate-associated determinant distinct from phosphorylcholine. *Molecular and Biochemical Parasitology* 41, 241–249.

Denkers, E.Y., Wassom, D.L., Krco, C.J. and Hayes, C.E. (1990b) The mouse antibody response to *Trichinella spiralis* defines a single, immunodominant epitope shared by multiple antigens. *Journal of Immunology* 144, 3152–3159.

Denkers, E.Y., Hayes, C.E. and Wassom, D.L. (1991) *Trichinella spiralis*: influence of an immunodominant, carbohydrate-associated determinant on the host antibody response repertoire. *Experimental Parasitology* 72, 403–410.

Despommier, D.D. (1998) How does *Trichinella spiralis* make itself at home? *Parasitology Today* 14, 318–323.

Despommier, D.D., Gold, A.M., Buck, S.W., Capo, V. and Silberstein, D. (1990) *Trichinella spiralis*: secreted antigen of the infective L_1 larva localizes to the cytoplasm and nucleoplasm of infected host cells. *Experimental Parasitology* 71, 27–38.

Dwek, R.A. (1995) Glycobiology: towards understanding the function of sugars. *Biochemical Society Transactions* 23, 1–25.

Ellis, L.A., Reason, A.J., Morris, H.R., Dell, A., Iglesias, R., Ubeira, F.M. and Appleton, J.A. (1994) Glycans as targets for monoclonal antibodies that protect rats against *Trichinella spiralis. Glycobiology* 4, 585–592.

Ellis, L.A., McVay, C.S., Probert, M.A., Zhang, J., Bundle, D.R. and Appleton, J.A. (1997) Terminal β-linked tyvelose creates unique epitopes in *Trichinella spiralis* glycan antigens. *Glycobiology* 7, 383–390.

Gardiner, C.H. (1976) Habit and reproductive behavior of *Trichinella spiralis. Journal of Parasitology* 62, 865–870.

Gold, A.M., Despommier, D.D. and Buck, S.W. (1990) Partial characterization of two antigens secreted by L_1 larvae of *Trichinella spiralis. Molecular and Biochemical Parasitology* 41, 187–196.

Gould, J.L. and Keeton, W.T. (1996) *Biological Science*, 6th edn. W.W. Norton & Company, New York.

Jasmer, D.P., Yao, S., Vassilatis, D., Despommier, D. and Neary, S.M. (1994) Failure to detect *Trichinella spiralis* p43 in isolated host nuclei and in irradiated larvae of infected muscle cells which express the infected cell phenotype. *Molecular and Biochemical Parasitology* 67, 225–234.

Lee, G.B. and Ogilvie, B.M. (1982) The intestinal mucus layer in *Trichinella spiralis* infected rats. In: Strober, W., Hanson, L.A. and Sell, K.W. (eds) *Recent Advances in Mucosal Immunity*. Raven Press, New York, p. 319.

Li, C.K., Seth, R., Gray, T., Bayston, R., Mahida, Y.R. and Wakelin, D. (1998) Production of proinflammatory cytokines and inflammatory mediators in human intestinal epithelial cells after invasion by *Trichinella spiralis. Infection and Immunity* 66, 2200–2206.

Lim, P.L., Choy, W.F., Chan, S.T.H., Leung, D.T. and Ng, S.S. (1994) Transgene-encoded antiphosphorylcholine (T15[+]) antibodies protect CBA/N (*xid*) mice against infection with *Streptococcus pneumoniae* but not *Trichinella spiralis. Infection and Immunity* 62, 1658–1661.

Love, R.J., Ogilvie, B.M. and McLaren, D.J. (1976) The immune mechanism which expels the intestinal stage of *Trichinella spiralis* from rats. *Immunology* 30, 7–15.

Mak, C.-H. and Ko, R.C. (1999) Characterization of endonuclease activity from excretory/secretory products of a parasitic nematode, *Trichinella spiralis. European Journal of Biochemistry* 260, 477–481.

ManWarren, T., Gagliardo, L., Geyer, J., McVay, C., Pearce-Kelling, S. and Appleton, J. (1997) Invasion of intestinal epithelia *in vitro* by the parasitic nematode *Trichinella spiralis. Infection and Immunity* 65, 4806–4812.

McCoy, O.R. (1940) Rapid loss of *Trichinella spiralis* larvae fed to immune rats and its bearing on the mechanisms of immunity. *American Journal of Hygiene* 32, 105–116.

McNeil, P.L. and Ito, S. (1989) Gastrointestinal cell plasma membrane wounding and resealing *in vivo. Gastroenterology* 96, 1238–1248.

McVay, C.S., Tsung, A. and Appleton, J.A. (1998) Participation of parasite surface glycoproteins in antibody-mediated protection of epithelial cells against *Trichinella spiralis. Infection and Immunity* 66, 1941–1945.

McVay, C.S., Bracken, P., Gagliardo, L.F. and Appleton, J.A. (2000) Antibodies to tyvelose exhibit multiple modes of interference with the epithelial niche of *Trichinella spiralis. Infection and Immunity* 68, 1912–1918.

Oliver-González, J. (1940) The *in vitro* action of immune serum on the larvae and adults of *Trichinella spiralis. Journal of Infectious Diseases* 67, 292–300.

Ooi, H.K. and Kamiya, M. (1986) Antibodies to nurse cell and various stages of *Trichinella spiralis* in patient sera. *Japanese Journal of Parasitology* 35, 109–113.

Ortega-Pierres, M.G., Yepez-Mulia, L., Homan, W., Gamble, H.R., Lim, P.L., Takahashi, Y.,. Wassom, D.L. and Appleton, J.A. (1996) Workshop on a detailed characterization of *Trichinella spiralis* antigens: a platform for future studies on antigens and antibodies to this parasite. *Parasite Immunology* 18, 273–284.

Otubu, O.E., Carlisle-Nowak, M.S., McGregor, D.D., Jacobson, R.H. and Appleton, J.A. (1993) *Trichinella spiralis*: the effect of specific antibody on muscle larvae in the small intestines of weaned rats. *Experimental Parasitology* 76, 394–400.

Perrudet-Badour, A., Boussac-Aron, Y., Ruitenberg, E.J., Kruizinga, W. and Elgersma, A. (1983) Protection to challenge with *Trichinella spiralis* after primary oral infection in congenitally athymic (nude) mice. *Journal of Parasitology* 69, 253–255.

Reason, A.J., Ellis, L.A., Appleton, J.A., Wisnewski, N., Grieve, R.B., McNeil, M., Wassom, D.L., Morris, H.R. and Dell, A. (1994) Novel tyvelose-containing tri- and tetra-antennary N-glycans in the immunodominant antigens of the intracellular parasite *Trichinella spiralis. Glycobiology* 4, 593–603.

Sanmartin, M.L., Iglesias, R., Santamarina, M.T., Leiro, J. and Ubeira, F.M. (1991) Anatomical location of phosphorylcholine and other antigens on encysted *Trichinella* using immunohistochemistry followed by Wheatley's trichrome stain. *Parasitology Research* 77, 301–306.

Stacey, M. and Barker, S.A. (1960) *Polysaccharides of Micro-organisms.* Oxford University Press, London.

Steinhardt, R.A., Bi, G. and Alderton, J.M. (1994) Cell membrane resealing by a vesicular mechanism similar to neurotransmitter release. *Science* 263, 390–393.

Su, X., Prestwood, A.K. and McGraw, R.A. (1991) Cloning and expression of complementary DNA encoding an antigen of *Trichinella spiralis. Molecular and Biochemical Parasitology* 45, 331–336.

Sugane, K. and Matsuura, T. (1990) Molecular analysis of the gene encoding an antigenic polypeptide of *Trichinella spiralis* infective larvae. *Journal of Helminthology* 64, 1–8.

Vassilatis, D.K., Despommier, D., Misek, D.E., Polvere, R.I., Gold, A.M. and Van der Ploeg, L.H.T. (1992) Analysis of a 43-kDa glycoprotein from the intracellular parasitic nematode *Trichinella spiralis. Journal of Biological Chemistry* 267, 18459–18465.

Vos, J.G., Ruitenberg, E.J., Van Basten, N., Buys, J., Elgersma, A. and Kruizinga, W. (1983) The athymic nude rat. IV. Immunocytochemical study to detect T-cells, and immunological and histopathological reactions against *Trichinella spiralis. Parasite Immunology* 5, 195–215.

Wisnewski, N., McNeil, M., Grieve, R.B. and Wassom, D.L. (1993) Characterization of novel fucosyl- and tyvelosyl-containing glycoconjugates from *Trichinella spiralis* muscle stage larvae. *Molecular and Biochemical Parasitology* 61, 25–36.

Wright, K.A. (1979) *Trichinella spiralis*: an intracellular parasite in the intestinal phase. *Journal of Parasitology* 65, 441–445.

Wright, K.A., Weidman, E. and Hong, H. (1987) The distribution of cells killed by *Trichinella spiralis* in the mucosal epithelium of two strains of mice. *Journal of Parasitology* 73, 935–939.

Yao, C., Bohnet, S. and Jasmer, D.P. (1998) Host nuclear abnormalities and depletion of nuclear antigens induced in *Trichinella spiralis*-infected muscle cells by the anthelmintic mebendazole. *Molecular and Biochemical Parasitology* 96, 1–13.

Zarlenga, D.S. and Gamble, H.R. (1990) Molecular cloning and expression of an immunodominant 53-kDa excretory-secretory antigen from *Trichinella spiralis* muscle larvae. *Molecular and Biochemical Parasitology* 42, 165–174.

Genetic Reprogramming of Mammalian Skeletal Muscle Cells by *Trichinella spiralis*

Douglas P. Jasmer

Department of Veterinary Microbiology and Pathology, Washington State University, Pullman, WA 99164-7040, USA

Introduction

Trichinellosis is caused by the parasitic nematode *Trichinella spiralis*. This parasite has a complex life cycle that alternates between intestinal and muscle cell compartments of the host. This nematode infection is unusual because *T. spiralis* is an intracellular parasite of mammalian cells. In addition, the broad host range of this parasite includes most mammals. The disease in humans has intrigued parasitologists, other biologists and public health workers for over a century (Cambell, 1983). The attraction to trichinellosis partly stems from the debilitating and sometimes fatal effects that characterize this disease.

Another fascinating aspect of trichinellosis is the muscle phase of the infection, which will be the focus of this chapter. From an epidemiological perspective, it is the muscle phase which serves as a reservoir for parasite transmission. Muscle larvae are infective to the next host upon ingestion, and the muscle phase is chronic, lasting months to years. From a biological perspective, the mechanisms by which the parasite establishes this reservoir are most intriguing.

The muscle phase of the infection is initiated by newborn larvae (immature L1 larvae). These larvae originate from female worms located in the small intestine of the host. On recognition of skeletal muscle cells, larvae invade and initiate the processes that culminate in long-term intracellular infection of these cells. Following infection, both the parasite and host muscle cell undergo significant changes, over the course of the first 15 days (referred to here as the initiation phase), which is followed by stabilization and maintenance of changes (chronic phase), which can last for years. Ingestion of muscle larvae (L1) leads to initiation of the intestinal phase in a subsequent mammalian host. Some basic questions that are being addressed in current research relate to clarifying cellular changes that are induced in the host muscle cell and larvae, how these changes are regulated, and how important these changes are for growth, development and survival of the parasite. Answers to these questions are significant relative to fundamental mechanisms of host/parasite interactions and intracellular muscle infections. The biological significance of anticipated answers is heightened by the remarkable changes that are induced in host skeletal muscle cells.

Numerous changes have been summarized for both the host muscle cell and parasite (Despommier *et al.*, 1975; Stewart, 1983; Jasmer, 1995; Despommier, 1998). Figure 7.1 presents some of these changes, which are elaborated on in this chapter. Relevant to this discussion are parasite changes that include an initial lag phase in growth for the first 5 days post-infection (dpi), followed by exponential growth which climaxes about 20 dpi (Despommier *et al.*, 1975). Prominent developmental changes include the elaboration of a set of oesophageal secretory cells called

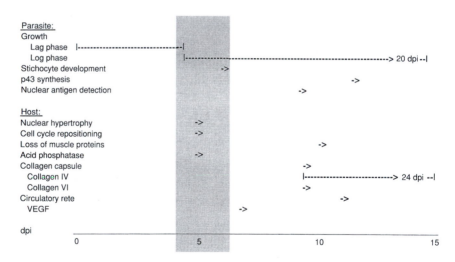

Fig. 7.1.   Timing of selected parasite and host characteristics acquired during infection of skeletal muscle cells with *Trichinella spiralis*. Characteristics indicated are discussed in the text. Arrow: (→) indicates time point at which expression of persistently maintained characteristic has been first detected; |----|, time frame of expression for transient characteristics; VEGF, vascular epithelial cell growth factor; dpi, day post infection. The grey box highlights: (i) early, basic changes in host cells that are expected to contribute to (e.g. cell cycle repositioning) or be markers of (e.g. acid phosphatase) genetic reprogramming; and (ii) parasite changes that may be linked to induced host cell changes.

stichocytes. These cells begin to develop about 6 dpi (Wu, 1965). At least four different classes of stichocytes, α, β, γ and δ, have been described, based on histochemical staining properties and cytoplasmic vesicle content (Sanmartin *et al.*, 1991). The functions of stichocyte secretory products are relevant to concepts on this host/parasite interaction, as discussed below. Infectivity of the parasite is acquired by about 15 dpi (mature L1). It is of some interest that major changes in the parasite are initiated concomitant with or following the occurrence of fundamental alterations in the infected cell. These correlations could reflect causal host/parasite interactions.

Host changes discussed in the following paragraphs indicate that striated skeletal muscle cells become genetically reprogrammed, as suggested previously (Despommier, 1975b; Stewart, 1983; Jasmer, 1995), during the initiation phase of infection. The discussion will initially concentrate on this genetic re-programming and how it is likely to be a focal point from which regulation of other phenotypic changes emanate in these cells. The role of the parasite in inducing and chronically regulating these changes will next be discussed. In addition, prospects for future research will be discussed. Guiding philosophies in this discussion are: (i) induced host cell changes that are critical to parasite growth, development and survival *might* have been achieved in a background of host phenotypic changes that are neutral to these parasite parameters; and (ii) identifiable parasite products expressed by muscle larvae, which conceivably modulate host processes, might function exclusively in the subsequent intestinal phase of the life cycle. These reminders aid in limiting assumptions about the significance of both host and parasite characteristics extant in the muscle phase.

Genetic Reprogramming of the Host Muscle Cell

Four fundamental changes induced in host cells will be discussed initially: (i) infection-induced cell cycle re-entry; (ii) suspension of host cells in apparent G_2/M; (iii) repression of host muscle gene expression; and (iv) further induction of the infected cell phenotype. A clear understanding of these fundamental changes induced by the infection is critical in elucidating the cellular mechanisms involved and, possibly, the host regulatory factors that are interfered with by parasite products.

Repositioning of the Host Cell in the Cell Cycle

Infection-induced cell cycle re-entry and suspension in G_2/M occur early in infection and are likely to influence regulation of some of the other host cell effects (Jasmer, 1993). The earliest indication that *T. spiralis* induces terminally differentiated skeletal muscle cells to re-enter the cell cycle came with the ability to label infected cell nuclei with tritiated thymidine (3HT) during asynchronous infections (Gabryel and Gustowska, 1968). However, one alternative state of skeletal myofibres involves damage of these cells followed by repair in a regeneration response. Nuclei of damaged myotubes that degenerate are replaced by nuclei from myoblasts (satellite cells) upon fusion of the latter with damaged myotubes. In this process, unfused satellite cells undergo cell division followed by myotube fusion and terminal differentiation. This process has been elucidated *in vivo* by pulse chase experiments using 3HT (Snow, 1997). However, muscle irradiation prevented regeneration and 3HT labelling of nuclei in damaged muscle, but not in *T. spiralis*-infected muscle cells (Jasmer, 1993). This result indicated that 3HT incorporation by infected cell nuclei did not result from a regeneration response. Additional evidence indicated that DNA synthesis in infected cell nuclei occurred within infected cells, and these nuclei were derived from terminally differentiated muscle nuclei present at the time of infection, not myoblast nuclei. Consequently, the infected skeletal muscle cell is apparently induced to re-enter the cell cycle and pass through S phase. More than 80% of nuclei became labelled between 4 and 5 dpi, indicating an early and relatively synchronized effect. In addition, nuclei labelled early in infection were stably maintained during chronic infection.

Despite close scrutiny of infected cell nuclei, there is no reported evidence of passage into mitosis. Isolated infected cell nuclei had an approximate 4N content of DNA (Jasmer, 1993). Therefore, *T. spiralis* infection evidently induces terminally differentiated skeletal muscle cells to re-enter the cell cycle and chronically arrest in G_2/M.

Since the gene regulatory environment can be determined by cell cycle position, as discussed below, this cell cycle repositioning during *T. spiralis* infection is expected to influence the infected cell phenotype. For instance, terminal differentiation and muscle gene expression are normally restricted to G_0/G_1 of the cell cycle (Lathrop *et al.*, 1985; Clegg *et al.*, 1987). Differentiation can be prevented by manipulating cell cycle factors that regulate the G_1 check point (Rao *et al.*, 1994; Skapek *et al.*, 1995). In addition, terminally differentiated skeletal muscle cells can be induced to re-enter the cell cycle in *in vitro* culture by DNA tumour viruses (Fogel and Defendi, 1967; Yaffe and Gershon, 1967). SV40 large T antigen is one viral factor responsible (Endo and Nidal-Ginard, 1988). Large T antigen binding to and interfering with the tumour suppressor protein, retino-blastoma protein, has been implicated in the mechanism (Gu *et al.*, 1993). Interactions of numerous cellular proteins are normally integrated to prevent promiscuous passage into S phase (Bartkova *et al.*, 1997). While the network of G_1 regulatory proteins is complex, this point in the cell cycle is obviously breached by *T. spiralis* infection and provides one focus for investigating host factors, parasite products and mechanisms involved. Complicating research on this topic is the fact that cell cycle re-entry occurs early in the infection when isolation of infected cells and immature muscle larvae is problematic.

There are other examples in infectious diseases in which transient interference with host G_2/M is evident (Jowett *et al.*, 1995; Planelles *et al.*, 1996). However, the long-term suspension (for months) in apparent G_2/M observed with *T. spiralis* is unusual. In contrast to interference with the G_1 checkpoint, it is less clear that suspension in G_2/M of the cell cycle involves disruption of regulatory pathways in host muscle cells. Dependence on centrioles for passage through mitosis has been established for some cells (Maniotis and Schliwa, 1991), and skeletal muscle cells apparently lose these structures following terminal differentiation (Tassin *et al.*, 1985). Consequently, once induced to re-enter the cell cycle, terminally differentiated skeletal muscle cells may be destined to arrest in G_2/M (Jasmer, 1993). For these reasons, discussion on possible parasite subversion of the G_2 checkpoint will be minimized. Nevertheless, there is a large body of information on eukaryotic cell factors that regulate the G_2 checkpoint (Jackman and Pines, 1997). These G_2 factors could become relevant to the *T. spiralis*/muscle cell interaction with additional information.

Repression of the Differentiated Skeletal Muscle Phenotype

Chronic repression of muscle gene expression clearly results from this infection. This effect is indicated by the dissolution of myofibrils that was observed by 5 dpi (Despommier, 1975), quantitative reduction of contractile proteins from these cells (Jasmer, 1990; Jasmer *et al.*, 1991), observed by 10 dpi and loss of transcripts for structural and regulatory muscle protein genes in chronically infected cells (Jasmer, 1993). The loss of transcripts for the muscle differentiation factors myogenin and MyoD (basic helix-loop-helix transcription factors) indicates that repression occurs at the level of muscle gene transcription (Jasmer, 1993).

Theoretical points at which muscle gene repression might be initiated are numerous and transcend cell cycle (cyclins and cyclin-dependent kinase inhibitors) (Rao *et al.*, 1994; Halevy *et al.*, 1995; Skapek *et al.*, 1995; Franklin and Xiong, 1996), retinoblastoma protein (RB) (Gu *et al.*, 1993), growth factor-regulated signal transduction (fibroblast growth factor, TGF-β, Ras) (Lassar *et al.*, 1989; Brennan *et al.*, 1991; Li *et al.*, 1992), and positive (Fos) and negative (Id) gene regulatory factors (Lassar *et al.*, 1989; Benezra *et al.*, 1990), as examples. The normal cell proliferation pathways are integrated and activation of one pathway can lead to activation of others, further masking the initial point(s) of disruption in the infected cell. One simple concept that aids in dealing with this complexity is that cell cycle progression presents a regulatory environment that is inhibitory for skeletal muscle gene expression, which is normally restricted to G_0 and early G_1. Hence, expression of regulatory proteins that promote cell proliferation tend to inhibit muscle differentiation.

Therefore, it is anticipated that chronic repression of muscle gene expression does not require chronic regulation by the parasite. Rather, this effect is expected to result from chronic suspension of infected cells in a non-G_0/G_1 gene regulatory environment. A simple hypothesis derived from known characteristics of skeletal muscle cells is that induction of cell cycle re-entry is sufficient to cause both chronic suspension in G_2/M and repression of muscle gene expression. This possibility has implications for the number of regulatory points at which the parasite might influence the host cell and possible methods to screen for those products.

Expanding on the preceding theme, tumour suppressor proteins, such as RB and to a lesser extent p107, appear to be situated at the dichotomous intersection through which the exclusive processes of proliferation and differentiation are regulated in skeletal muscle cells (Gu *et al.*, 1993; Schneider *et al.*, 1994). Underphosphorylated RB binds to and sequesters the E2F transcription factor, which when free from this complex activates genes essential for passage into S phase. Accordingly, RB appears to be required for normal muscle differentiation (Gu *et al.*, 1993; Schneider *et al.*, 1994). Additionally, there is evidence that muscle differentiation factors bind to RB (Gu *et al.*, 1993). This interaction might contribute to stabilization of the RB/E2F interaction or inhibit hyperphosphorylation of RB, which releases E2F from this complex. Multiple cellular pathways converge on the RB intersection to modulate RB functions. It is not surprising that many of the same regulatory factors and pathways that disrupt muscle differentiation normally lead to the hyperphosphorylation of RB and cell cycle progression. Therefore, products secreted by muscle larvae at about 5 dpi that are capable of promoting mammalian cell proliferation would represent candidate factors that may initiate repression of the muscle differentiation programme and expression of the infected cell phenotype.

Acquisition of the Infected Cell Phenotype

The aforementioned 'dressing-down' of the host muscle cell represents an early step in the infection-induced reprogramming of this cell. The acquisition of specialized infected cell characteristics is the next point to consider, and the cellular background on which this reprogramming occurs seems relevant to the mechanisms involved. The cell in question is a previously differentiated skeletal muscle cell that has been terminally repositioned in apparent G_2/M. Expectations of the global gene regulatory environment for this situation lack precedence. For instance, do gene methylation patterns in these cells reflect those of myoblasts, differentiated muscle cells, combinations of both, or neither? The general question can be extended to many other regulatory factors, both positive and negative.

For brevity, established infected cell characteristics that are easily related to concepts of genetic reprogramming will be discussed. Foremost are the unusual infected cell nuclei, some of which approach 17 μm in diameter. Enlargement of host cell nuclei and nucleoli is evident by at least 5 dpi (Despommier *et al.*, 1991), which coincides with cell cycle re-entry, and persists throughout the life of the infected cell. Nuclei of replicating mammalian cells do not approach the size of infected cell nuclei. Therefore, a 4N complement of DNA seems insufficient to account for this striking nuclear hypertrophy. Enhanced transcription in these nuclei has been suggested as one explanation (Gustowska *et al.*, 1989) and is supported by the abnormally large nucleoli, suggestive of elevated RNA Pol I activity. The relatively high levels of nuclear lamins in these nuclei (Yao *et al.*, 1998), which have implicated roles in transcription, might also support this idea. These collective observations lead to the suspicion that nuclear hypertrophy reflects fundamental changes with implications for the infected cell phenotype.

The collagen capsule is a hallmark of the encysted *T. spiralis* infected cell, the origin of which has been debated (Teppema *et al.*, 1973; Haehling *et al.*, 1995; Polvere *et al.*, 1997). However, collagen types IV and VI are both components of the capsule, and messenger RNAs for both genes were localized to infected cell cytoplasm at time points (9 dpi) coincident with initiation of capsule formation (Polvere *et al.*, 1997). Furthermore, collagen deposition around 10 dpi can be seen to demarcate unaffected (myofibril rich) and affected (myofibril depleted) cytoplasmic regions of infected myotubes (D.P. Jasmer, unpublished results). These unexpected observations stimulate many questions on how this process is orchestrated and what is regulating it. In addition, while type VI collagen expression persisted for 8 months, type IV expression was undetected after 24 dpi, indicating differential regulation of these infected cell genes during parasite habitation.

A circulatory rete develops to intimately surround individual encapsulated infected cells (Humes and Akers, 1952; Baruch and Despommier, 1991). While the rete occurs outside the infected cell, it represents another unusual feature of what can be considered an infected cell complex composed of the parasite, infected cell, collagen capsule and rete. This microcirculatory system might represent an adaptation that deals with unusual kinds or volume of metabolic waste derived from worms and the infected cell. Additionally, it might function to regulate specific aerobic/anaerobic needs of host cell, parasite or both (Capo *et al.*, 1998; Despommier, 1998). The source of angiogenesis factors that stimulate production of the rete is of interest. The first insight on this question came with detection of vascular endothelial cell growth factor (Capo *et al.*, 1998), raising the possibility of regulation by the infected cell. However, stimulation of angiogenesis by host infiltrating cells that surround infected cells, or even the parasite, cannot be excluded.

Many cytoplasmic changes have been documented at the morphological and histochemical levels for infected cells (summarized in Stewart, 1983). Two-dimensional protein profiles of chronically infected cells clearly support acquisition of a phenotype distinct from muscle cells (Jasmer, 1990). Elevated expression of host lysosomal acid phosphatase (Maier and Zaiman, 1966) is one of the earliest documented enzymatic changes in host cells and was detected by 5 dpi (Jasmer *et al.*, 1991). This timing correlates with cell cycle re-entry and nuclear hypertrophy, possibly reflecting initiation of genetic reprogramming.

What, then, is the importance of these changes to the growth, development and survival of the parasite and what are the relative roles of the parasite and host in regulating these dramatic host cell changes? While answers to these questions are relevant to concepts in cell biology and parasitology, no direct evidence addresses either of these issues. The general lack of reliable culture systems is a major obstacle for investigating these questions. In the absence of such systems, methods that inhibit expression of selected infected cell characteristics *in vivo* provide one possible approach. For instance, use of transgenic mice that contain null mutations for selected infected cell products is one possibility. Alternatively, pharmacological inhibition of components in the infected cell complex, such as the circulatory rete, provides another potential approach. Methods that inhibit host cell cycle re-entry, such as irradiation or antiproliferation drugs, are also conceivable. Such approaches also demand that the treatment has no direct effect on the parasite. This criterion can be met when inhibition is irreversibly achieved prior to infection, or when pharmacological effects are differentially restricted to the host or parasite (which is more problematic to ensure). Some applications of this kind are discussed below.

Parasite Regulation of the Infected Cell Phenotype

When considering the influence of *T. spiralis* on host muscle cells, it is sobering to consider that nematodes are at the apex of genetic complexity among intracellular parasites. Hence, the creative potential for designing parasite molecules that manipulate host functions seems boundless. In addition, intracellular release of a barrage of organic waste products by this intracellular parasite is likely. Adequate knowledge is lacking to predict confidently how the host cell will respond to the presumptive parasite effluent. However, available evidence supports a role for the parasite in manipulating the infected cell. Firstly, muscle cell changes induced by most other intracellular parasites of muscle do not parallel those of *T. spiralis* (Stewart and Gianini, 1982), supporting the theory of a specific effect by this parasite. In addition, use of newborn larvae that were irradiated prior to infection caused a delay in host cell cycle re-entry (Jasmer and Neary, 1994). Irradiation clearly interfered with intramuscular cell growth and development of these larvae, but did not prevent timely infection. This ability to modulate host cell cycle re-entry indirectly by an independent treatment of larvae implicates a role for larvae in host cell manipulation.

Secondly, detection of parasite products in the host cell provides additional support for parasite manipulation. Antibodies made against *T. spiralis* localized corresponding antigens to infected cell nuclei, referred to as nuclear antigens (NA) (Despommier *et al.*, 1990; Lee *et al.*, 1991; Jasmer *et al.*, 1994). In addition, drug treatments that deplete NA from host nuclei caused morphological and molecular alterations in these nuclei (Yao *et al.*, 1998). While not direct, collective evidence supports the hypothesis that parasite products interfere with normal host cell regulatory pathways.

Finally, several experiments indicate that excretory/secretory products (ESP), derived from newborn larvae or pepsin-HCl isolated mature muscle larvae, induce muscle cell basophilia (Blotna-Filipiak *et al.*, 1998; Wranicz *et al.*, 1998) or nuclear hypertrophy (Leung and Ko, 1997), respectively. Effects were observed *in vivo* and *in vitro*. Although correspondence of these general changes to biochemical or genomic characteristics of the infected cell and nuclei were not established, these observations may facilitate dissection of the parasite products that are responsible.

General Consideration of Parasite Products

Information on which parasite products might regulate infected muscle cell characteristics is unresolved. Parasite proteins will be the focus of this discussion. This focus results in part from general lack of information on other secreted products/metabolic wastes and their potential influences on the host cell. In addition, arguments for cell-permeable parasite products are less compelling, and no clear evidence exists for a bystander effect in which bona fide infected cell characteristics become established in neighbouring, uninfected host muscle cells.

The first compelling evidence that muscle larval products localize to host cell compartments came with monoclonal antibodies which bind to mature muscle larvae and to nuclei of the infected cell (Despommier *et al.*, 1990; Lee *et al.*, 1991). A carbohydrate determinant appears to be responsible for this result. The determinant is formed by an unusual dideoxy sugar, tyvelose, which is a constituent of a complex glycan added to many ESP products of muscle larvae maintained in *in vitro* culture (Wisnewski *et al.*, 1993; Reason *et al.*, 1994; see also Chapter 6). The determinant can be highly immunodominant in mice (Denkers *et al.*, 1990) and provides for specific detection of parasite proteins modified with the glycan. Proteins so modified predominate in α- and β-stichocytes, which suggested that proteins from these stichocytes are secreted into infected muscle cells. Multiple proteins from α-, β- or γ-stichocytes (p43 and p45–50, respectively) have been molecularly characterized. However, localization to host nuclei was not detected for any of these with antibodies against cognate recombinant proteins (Arasu *et al.*, 1994; Jasmer *et al.*, 1994; Vassilatis *et al.*, 1996). Antibodies against recombinant p43 did bind to infected cell cytoplasm around 10 dpi (Vassilatis *et al.*, 1996), suggesting transient release into host cells.

A major question, then, becomes: do α- or β-stichocyte products modulate the host muscle cell? Some evidence indicates that the answer is no. Stichocyte development is initiated in muscle larvae around 6 dpi (Wu, 1965) and expression of the p43 protein in developing muscle larvae was observed by 11 dpi (Vassilatis *et al.*, 1996). Irradiation of newborn larvae prior to infection inhibited both the development of stichocytes and detectable expression of p43 (Jasmer *et al.*, 1994) in muscle larvae evaluated at 15 dpi. Despite the lack of demonstrable stichocytes and proteins, infected cells re-entered the cell cycle (although there was a delay), acquired a collagen capsule, expressed increased acid phosphatase and had reduced levels of muscle proteins. These results suggest that products of α- and β-stichocytes are not required to induce these basic infected cell characteristics. Alternatively, products from these cells may be presynthesized for functions in the subsequent intestinal stage. This possibility is consistent with the release of glycan-modified proteins by isolated, mature muscle larvae during *in vitro* invasion of epithelial cells (ManWarren *et al.*, 1997).

Due to lack of specific reagents, products of δ- and γ-stichocytes were not evaluated in irradiated larval experiments and cannot be excluded from a role in the muscle infection. Interestingly, NA were detected in nuclei of muscle cells infected by irradiated larvae, which suggested that these products may influence the infected cell phenotype but are not derived from α- and β-stichocytes.

Nuclear Antigens

It is important to emphasize that detection of NA has depended on antibody determinants shared by other parasite products, such as the tyvelose containing glycan. The earliest detection of NA is about 9 dpi (Despommier *et al.*, 1990) and persists throughout the chronic infection. Therefore, current evidence supports a potential role for NA only after cell cycle repositioning.

Using isolated infected cell nuclei, prominent NA were identified that have predicted molecular weights of 79, 87 and 96 kDa (Jasmer *et al.*, 1994). However, a background smear between the major bands may indicate additional heterogeneity. A less prominent band at 71 kDa was also observed. NA co-localized with nucleoplasm and nuclear chromatin, but appeared to be excluded from nucleoli (Despommier *et al.*, 1990; Yao and Jasmer, 1998). Curiously, no proteins of a similar molecular weight have been detected in ESP of pepsin-HCl isolated muscle larvae. This apparent incongruity might result from repression of NA expression by conditions used for muscle larval isolation. Mebendazole (MBZ) treatment of chronically infected mice provided other evidence supporting the suggestion that NA come from the parasite (Yao *et al.*, 1998). MBZ was expected to inhibit parasite secretion without disrupting host cell functions. NA were demonstrably depleted from infected cell nuclei by 4 days of MBZ treatment, as measured with anti-glycan antibodies.

Associated with NA depletion from host nuclei was the diminution and deformation of these nuclei, significant depletion of host nucleoskeletal protein lamin b, and significant reductions in levels of host cell RNA, protein and acid phosphatase activity. In contrast, host infiltrating cells showed no such changes, supporting an infected cell-specific effect. These observations support the possibility that NA are synthesized by the parasite and modulate host nuclear functions. However, release of other parasite secretory products, not analysed, might have been inhibited in these experiments also.

In contrast, inactivity of muscle gene expression and elevated DNA content of infected cell nuclei did not change detectably with MBZ treatment, which may indicate that these host cell parameters are not chronically regulated by the parasite.

NA isolation and molecular characterization will be important to define the origin and functions of these proteins. At this time, infected cell nuclei offer the only source of these proteins, and NA have proved resistant to classic nuclear extraction methods (Yao and Jasmer, 1998). NA can be solubilized under conditions that co-extract nuclear lamins a/c and b (4 M urea, pH 8.0). Despite these similar physical properties, NA do not co-localize with lamins in the nucleoskeleton. However, both disulphide bonds and ionic interactions appear to contribute to nuclear complexes containing NA. In addition, NA can be cross-linked within host nuclei with protein cross-linking reagents. The foregoing properties represent current information available for the development of strategies to isolate and characterize these proteins and to investigate host proteins with which NA interact.

Other Possible *T. spiralis* Regulators of Host Muscle Cells

In contrast to NA which remain largely uncharacterized, better defined parasite proteins have been identified for which localization and interactions within the host cell remain ill-defined. Despite results from irradiated newborn larva experiments, α- and β-stichocyte proteins cannot be excluded from a regulatory role in host muscle cells. For instance, p43 was described as having a potential helix–loop–helix (HLH) motif (Vassilatis *et al.*, 1992). In a manner similar to the inhibitor of DNA binding (ID) (Benezra *et al.*, 1990), this motif in p43 might antagonize the heterodimerization of basic-HLH muscle differentiation factors, which is required for muscle gene activation. However, other findings suggest that p43 has a different function. Secretory DNase activity was recently described from *T. spiralis* muscle larvae isolated by pepsin-HCl digestion (Mak and Ko, 1999). Alignment of p43 with DNase II sequences from mammals demonstrated significant similarity (D.P. Jasmer, unpublished data). While possibly relevant to remodelling of host muscle cell DNA (Mak and Ko, 1999), a function of parasite DNase during the intestinal phase of infection should not be overlooked.

Two additional parasite proteins with apparent HLH binding properties were identified, based on the idea that direct antagonism of host HLH transcription factors might be responsible for muscle cell subversion. One is a MyoD-like protein (Connolly *et al.*, 1996) and another (110 kDa) is located in the vicinity of the hypodermis or somatic musculature of muscle larvae (Lindh *et al.*, 1998). Secretion of these proteins into the infected cell has not been reported.

Serine/threonine kinase activity has been reported in ESP of pepsin-HCl isolated muscle larvae (Arden *et al.*, 1997) and kinase activity was associated with proteins of 70 and 135 kDa. Phosphorylation status is functionally significant for multiple regulatory factors, including those involved in muscle differentiation (Li *et al.*, 1992). Therefore, kinase activity in parasite secretions may be significant in either the muscle or intestinal phases of infection.

Concluding Remarks

Based on current knowledge, several lines of inquiry will be mentioned here that seem approachable regarding mechanisms of this host/parasite interaction. The foremost of these is related to parasite proteins that might regulate infected cell characteristics during the initiation and chronic phases of infection. Methods exist for the extraction of nuclear antigens from host cell nuclei, which may facilitate the characterization of these proteins and their potential role in regulating host cell characteristics during the chronic infection. Currently, NA have not been associated with the early initiation phase involving cell cycle repositioning. If induced directly by the parasite, muscle larvae present at that time or newborn larvae are expected to produce the products responsible. While isolation of muscle cell larvae at the time of infection is problematic, newborn larvae can be obtained in relatively large quantities (Despommier, 1975b). It would be interesting to know if these larvae are pre-adapted for inducing cell cycle re-entry. In parallel, it will be important to assess critically whether muscle cell changes experimentally induced by newborn larval and muscle larval ESP represent bona fide infected cell changes, since degenerative/regenerative muscle changes exhibit superficial similarities to infected cell characteristics. The phenotype of the infected cell is without precedence. A more comprehensive understanding of the host cell phenotype should facilitate discovery of specific host cell features that contribute to growth, development and survival of the parasite. Given the ability to isolate RNA from this cell and the growing availability of microchip gene arrays for mammalian cells, detailed phenotypic analysis may be approachable. Finally, the ability to manipulate angiogenesis *in vivo* may allow the requirement of this host response for parasite survival to be investigated.

References

Arasu, P., Ellis, L.A., Iglesias, R., Ubeira, F.M. and Appleton, J.A. (1994) Molecular analysis of antigens targeted by protective antibodies in rapid expulsion of *Trichinella spiralis*. *Molecular and Biochemical Parasitology* 65, 201–211.

Arden, S.R., Smith, A.M., Booth, M.J., Tweedie, S., Gounaris, K. and Selkirk, M.E. (1997) Identification of serine/threonine protein kinases secreted by *Trichinella spiralis* infective larvae. *Molecular and Biochemical Parasitology* 90, 111–119.

Bartkova, J., Lukas, J. and Bartek, J. (1997) Aberrations of the G1- and G1/S-regulating genes in human cancer. *Progress in Cell Cycle Research* 3, 211–220.

Baruch, A.M. and Despommier, D.D. (1991) Blood vessels in *Trichinella spiralis* infections: a study using vascular casts. *Journal of Parasitology* 77, 99–103.

Benezra, R., Davis, R.L., Lockshon, D., Turner, D.L. and Weintraub, H. (1990) The protein Id: a negative regulator of helix–loop–helix DNA binding proteins. *Cell* 61, 49–59.

Blotna-Filipiak, M., Gabryel, P., Gustowska, L., Kucharska, E. and Wranicz, M.J. (1998) *Trichinella spiralis*: induction of the basophilic transformation of muscle cells by synchronous newborn larvae. II. Electron microscopy study. *Parasitology Research* 84, 823–827.

Brennan, T.J., Edmondson, D.G., Li, L. and Olson, E.N. (1991) Transforming growth factor β represses the actions of myogenin through a mechanism independent of DNA binding. *Proceedings of the National Academy of Sciences USA* 88, 3822–3826.

Cambell, W.C. (1983) Historical introduction. In: Campbell, W.C. (ed.) *Trichinella and Trichinosis*. Plenum Press, New York, pp. 1–30.

Capo, V.A., Despommier, D.D. and Polvere, R.I. (1998) *Trichinella spiralis*: vascular endothelial growth factor is up-regulated within the nurse cell during the early phase of its formation. *Journal of Parasitology* 84, 209–214.

Clegg, C.H., Linkhart, T.A., Olwin, B.B. and Hauschka, S.D. (1987) Growth factor control of skeletal muscle differentiation occurs in G_1-phase and is repressed by fibroblast growth factor. *Journal of Cell Biology* 105, 949–956.

Connolly, B., Trenholme, K. and Smith, D.F. (1996) Molecular cloning of a myoD-like gene from the parasitic nematode, *Trichinella spiralis*. *Molecular and Biochemical Parasitology* 81, 137–149.

Denkers, E.Y., Wassom, D.L., Krco, C.J. and Hayes, C.E. (1990) The mouse antibody response to *Trichinella spiralis* defines a single immunodominant epitope shared by multiple antigens. *Journal of Immunology* 144, 3152–3159.

Despommier, D.D. (1975) Adaptive changes in muscle fibers infected with *Trichinella spiralis*. *American Journal of Pathology* 78, 477–496.

Despommier, D.D. (1998) How does *Trichinella spiralis* make itself at home? *Parasitology Today* 14, 318–323.

Despommier, D.D., Aron, L. and Turgeon, L. (1975) *Trichinella spiralis*: growth of the intracellular (muscle) larva. *Experimental Parasitology* 37, 108–116.

Despommier, D.D., Gold, A.M., Buck, S.W., Capo, V. and Silberstein, D. (1990) *Trichinella spiralis*: Secreted antigen of the infective L_1 larva localizes to the cytoplasm and nucleoplasm of infected host cells. *Experimental Parasitology* 71, 27–38.

Despommier, D., Symmans, W.F. and Dell, R. (1991) Changes in nurse cell nuclei during synchronous infection with *Trichinella spiralis*. *Journal of Parasitology* 77, 290–295.

Endo, T. and Nidal-Ginard, B. (1988) SV40 large T antigen induces reentry of terminally differentiated myotubes into the cell cycle. In: Kedes, L.H. and Stockdale, F.E. (eds) *Cellular and Molecular Biology of Muscle Development*. Alan R. Liss, New York, pp. 95–104.

Fogel, M. and Defendi, V. (1967) Infections of muscle cultures from various species with oncogenic DNA viruses (SV40 and Polyoma). *Proceedings of the National Academy of Sciences USA* 58, 967–973.

Franklin, D.S. and Xiong, Y. (1996) Induction of p18INK4c and its predominant association with CDK4 and CDK6 during myogenic differentiaion. *Molecular Biology of the Cell* 7, 1587–1599.

Gabryel, P. and Gustowska, L. (1968) Veranderungen der quergestreiften muskelfasern im fruzen stadium einer *Trichinella spiralis*-infection. *Gegenbaurs Morphologishes Jahrbuch* 111, 174–180.

Gu, W., Schneider, J.W., Condorelli, G., Kaushal, S., Mahdavi, V. and Nadal-Ginard, B. (1993) Interaction of myogenic factors and the retinoblastoma protein mediates muscle cell commitment and differentiation. *Cell* 72, 309–324.

Gustowska, L., Gabryel, P., Blotna-Filipiak, M. and Rauhut, W. (1989) The role of the muscle cell nucleus in the mechanism of its transformation after infection by *Trichinella spiralis* larvae. II. Histochemical features of the functional transformation of the muscle cell nucleus in the course of infection. *Wiadomosci Parazytologiczne* 35, 1401–1411.

Haehling, E., Niederkorn, J.Y. and Stewart, G.L. (1995) *Trichinella spiralis* and *Trichinella pseudospiralis* induce collagen synthesis by host fibroblasts *in vitro* and *in vivo*. *International Journal of Parasitology* 25, 1393–1400.

Halevy, O., Novitch, B.G., Spicer, D.B., Skapek, S.X., Rhee, J., Hannon, G.J., Beach, D. and Lassar, A.B. (1995) Correlation of terminal cell cycle arrest of skeletal muscle with induction of p21 by MyoD. *Science* 267, 1018–1021.

Humes, A.G. and Akers, R.P. (1952) Vascular changes in the cheek pouch of the golden hamster during infection with *Trichinella spiralis* larvae. *Anatomical Record* 114, 103–113.

Jackman, M.R. and Pines, J.N. (1997) Cyclins and the G2/M transition. *Cancer Surveys* 29, 47–73.

Jasmer, D.P. (1990) *Trichinella spiralis*: altered expression of muscle proteins in trichinosis. *Experimental Parasitology* 70, 452–465.

Jasmer, D.P. (1993) *Trichinella spiralis* infected skeletal muscle cells: arrest in G_2/M is associated with the loss of muscle gene expression. *Journal of Cell Biology* 121, 785–793.

Jasmer, D.P. (1995) *Trichinella spiralis*: subversion of differentiated mammalian skeletal muscle cells. *Parasitology Today* 11, 185–188.

Jasmer, D.P. and Neary, S.M. (1994) *Trichinella spiralis*: inhibition of muscle larva growth and development is associated with a delay in expression of infected skeletal muscle characteristics. *Experimental Parasitology* 78, 317–325.

Jasmer, D.P., Bohnet, S. and Prieur, D.J. (1991) *Trichinella* spp.: differential expression of acid phosphatase and myofibrillar proteins in infected muscle cells. *Experimental Parasitology* 72, 321–331.

Jasmer, D.P., Yao, S., Vassilatis, D.K., Despommier, D. and Neary, S.M. (1994) Failure to detect *Trichinella spiralis* p43 in isolated host nuclei and in irradiated larvae of infected muscle cells which express the infected cell phenotype. *Molecular and Biochemical Parasitology* 67, 225–234.

Jowett, J.B., Planelles, V., Poon, B., Shah, N.P., Chen, M.L. and Chen, I.S. (1995) The human immunodeficiency virus type 1 *vpr* gene arrests infected T cells in the G_2/M phase of the cell cycle. *Journal of Virology* 69, 6304–6313.

Lasser, A.B., Thayer, M.J., Overell, R.W. and Weintraub, H. (1989) Transformation by activated ras or fos prevents myogenesis by inhibiting expression of MyoD1. *Cell* 58, 659–667.

Lathrop, B.K., Thomas, K. and Glaser, L. (1985) Control of myogenic differentiation by fibroblast growth factor is mediated by position of the G_1 phase of the cell cycle. *Journal of Cell Biology* 101, 2194–2198.

Lee, D.L., Ko, R.C., Yi, X.Y. and Yeung, M.H. (1991) *Trichinella spiralis*: antigenic epitopes from the stichocytes detected in the hypertrophic nuclei and cytoplasm of the parasitized muscle fibre (nurse cell) of the host. *Parasitology* 102, 117–123.

Leung, R.K. and Ko, R.C. (1997) *In vitro* effects of *Trichinella spiralis* on muscle cells. *Journal of Helminthology* 71,113–118.

Li, L., Zhou, J., James, G., Heller-Harrison, R., Czech, M.P. and Olson, E.N. (1992) FGF inactivates myogenic helix-loop-helix proteins through phosphorylation of a conserved protein kinase C site in their DNA-binding domains. *Cell* 71, 1182–1194.

Lindh, J.G., Connolly, B., McGhie, D.L. and Smith, D.F. (1998) Identification of a developmentally regulated *Trichinella spiralis* protein that inhibits MyoD-specific protein: DNA complexes *in vitro*. *Molecular and Biochemical Parasitology* 92, 163–175.

Mak, C.H. and Ko, R.C. (1999) Characterization of endonuclease activity from excretory/secretory products of a parasitic nematode, *Trichinella spiralis*. *European Journal of Biochemistry* 260, 477–481.

Maier, D.M. and Zaiman, H. (1966) The development of lysosomes in rat skeletal muscle in trichinous myositis. *Journal of Histochemistry and Cytochemistry* 14, 396–400.

Maniotis, M. and Schliwa, M. (1991) Microsurgical removal of centrosomes blocks cell reproduction and centriole generation in BSC-1 cells. *Cell* 67, 495–504.

ManWarren, T., Gagliardo, L., Geyer, J., McVay, C., Pearce-Kelling, S. and Appleton, J. (1997) Invasion of intestinal epithelia *in vitro* by the parasitic nematode *Trichinella spiralis*. *Infection and Immunity* 65, 4806–4812.

Planelles, V., Jowett, J.B., Li, Q.X., Xie, Y., Hahn, B. and Chen, I.S. (1996) Vpr-induced cell cycle arrest is conserved among primate lentiviruses. *Journal of Virology* 70, 2516–2524.

Polvere, R.I., Kabbash, C.A., Capo, V.A., Kadan, I. and Despommier, D.D. (1997) *Trichinella spiralis*: synthesis of type IV and type VI collagen during nurse cell formation. *Experimental Parasitology* 86, 191–199.

Rao, S.S., Chu, C. and Kohtz, D.S. (1994) Ectopic expression of cyclin D1 prevents activation of gene transcription by myogenic basic helix-loop-helix regulators. *Molecular and Cellular Biology* 14, 5259–5267.

Reason, A.J., Ellis, L.A., Appleton, J.A., Wisnewski, N., Grieve, R.B., McNeil, M., Wassom, D.L., Morris, H.R. and Dell, A. (1994) Novel tyvelose-containing tri- and tetra-antennary N-glycans in the immunodominant antigens of the intracellular parasite *Trichinella spiralis*. *Glycobiology* 4, 593–603.

Sanmartin, M.L., Iglesias, R., Sanmartin, M.T., Leiro, J. and Ubeira, F.M. (1991) Anatomical location of phosphorylcholine and other antigens on encysted *Trichinella* using immunohistochemistry followed by Wheatley's trichrome stain. *Parasitology Research* 77, 301–306.

Schneider, J.W., Gu, W., Zhu, L., Mahdavi, V. and Nadal-Ginard, B. (1994) Reversal of terminal differentiation mediated by p107 in Rb$^{-/-}$ muscle cells. *Science* 264, 1467–1471.

Skapek, S.X., Rhee, J., Spicer, D.B. and Lassar, A.B. (1995) Inhibition of myogenic differentiation in proliferating myoblasts by cyclin D1 dependent kinase. *Science* 267, 1022–1024.

Snow, M.H. (1997) Myogenic cell formation in regenerating rat skeletal muscle injured by mincing. *Anatomical Record* 188, 201–208.

Stewart, G.L. (1983) Pathophysiology of the muscle phase. In: Campbell, W.C. (ed.) *Trichinella and Trichinosis*. Plenum Press, New York, pp. 241–264.

Stewart, G.L. and Gianini, S.H. (1982) *Sarcocystis, Trypanosoma, Toxoplasma, Ancylostoma,* and *Trichinella* spp.: a review of the intracellular parasites of striated muscle. *Experimental Parasitology* 53, 406–447.

Tassin, A.-M., Maro, B. and Bornens, M. (1985) Fate of microtubule organizing centers during myogenesis *in vitro*. *Journal of Cell Biology* 100, 35–46.

Teppema, J.S., Robinson, J.E. and Ruitenberg, E.J. (1973) Ultrastructural aspects of capsule formation in *Trichinella spiralis* infection in the rat. *Parasitology* 66, 291–296.

Vassilatis, D.M., Despommier, D., Misek, D., Polvere, R., Gold, A.M. and Van der Ploeg, L.H.T. (1992) Analysis of a 43-kDa glycoprotein from the intracellular parasitic nematode *Trichinella spiralis*. *Journal of Biological Chemistry* 267, 18459–18465.

Vassilatis, D.K., Polvere, R.I., Despommier, D.D., Gold, A.M. and Van der Ploeg, L.H. (1996) Developmental expression of a 43-kDa secreted glycoprotein from *Trichinella spiralis*. *Molecular and Biochemical Parasitology* 78, 13–23.

Wisnewski, N., McNeil, M., Grieve, R.B. and Wassom, D.L. (1993) Characterization of novel fucosyl- and tyvelosyl-containing glycoconjugates from *Trichinella spiralis* muscle stage larvae. *Molecular and Biochemical Parasitology* 61, 25–35.

Wranicz, M.J., Gustowska, L., Gabryel, P., Kucharska, E. and Cabaj, W. (1998) *Trichinella spiralis*: induction of the basophilic transformation of muscle cells by synchronous newborn larvae. *Parasitology Research* 84, 403–407.

Wu, L.-Y. (1965) The development of the stichosome and associated structures in *Trichinella spiralis*. *Canadian Journal of Zoology* 33, 440–446.

Yaffe, D. and Gershon, D. (1967) Multinucleated muscle fibers: induction of DNA synthesis and mitosis by polyoma virus infection. *Nature* 215, 421–424.

Yao, C. and Jasmer, D.P. (1998) Nuclear antigens in *Trichinella spiralis* infected muscle cells: nuclear extraction, compartmentalization and complex formation. *Molecular and Biochemical Parasitology* 92, 207–218.

Yao, C., Bohnet, S. and Jasmer, D.P. (1998) Host nuclear abnormalities and depletion of nuclear antigens induced in *Trichinella spiralis*-infected muscle cells by the anthelmintic mebendazole. *Molecular and Biochemical Parasitology* 96, 1–13.

Plant-parasitic Nematodes

David McK. Bird[1] and Alan F. Bird[2]

[1]Department of Plant Pathology, Box 7616, North Carolina State University, Raleigh, NC 27695, USA; [2]CSIRO Land and Water, Private Bag No. 2, Glen Osmond, SA 5062, Australia

Introduction

It is becoming increasingly apparent to many researchers that both plant- and animal-parasitic nematodes face very similar biological challenges in interacting with their respective hosts. Although the precise nature of the molecules mediating key aspects of the host–parasite interaction will almost certainly be different in the different kingdoms, the underlying principles will be the same, and model systems based on plant-parasitic nematodes offer certain practical advantages over those involving animal hosts in elucidating these principles. In particular, the development of the soybean cyst nematode (SCN) as a genetic model (Dong and Opperman, 1997), in conjunction with the ability to manipulate host plants by forward and reverse genetics, permits these powerful techniques to be employed to dissect the host–parasite interaction. The burgeoning deployment of genomics in studies of parasitic nematode biology (Blaxter, 1998; Opperman and Bird, 1998; Bird *et al.*, 1999) will provide the tools to link SCN genetics to less tractable parasitic species, including animal parasites. Parasite genetics are discussed in Chapters 3, 4 and 5.

It is obviously not possible to provide a comprehensive review of plant nematology in one chapter, and readers are directed to descriptions of the taxonomy (Nickle, 1991), morphology (Bird and Bird, 1991), physiology and biochemistry (Perry and Wright, 1998) and cell biology (Fenoll *et al.*, 1997b) of plant-parasitic nematodes. Rather, examples will be provided from plant-parasitic nematodes that emphasize the catholic nature of nematode parasitism. Understanding how the host and parasite communicate, both mechanistically and also in an evolutionary context, is arguably the greatest current challenge in parasitic nematology. This chapter presents the groundwork upon which those experiments on plant parasites will be based, and emphasizes research areas likely to be fruitful in the near future.

Somewhat surprising has been the slow adoption of the free-living nematode *Caenorhabditis elegans* as a model for parasitic forms, with some notable exceptions (Riddle and Georgi, 1990; Bird and Bird, 1991). Although nematologists studying parasites have generally been aware of the ongoing development of *C. elegans* as a major metazoan model (Wood, 1988; Riddle *et al.*, 1997), it has not until very recently begun to act as a catalyst to unify nematology into a discipline that encompasses studies of plant and animal parasites (Blaxter and Bird, 1997; Bird and Opperman, 1998; Blaxter, 1998). The sheer volume of biological information obtained for *C. elegans* (Riddle *et al.*, 1997), along with its mature genome project (*C. elegans* Genome Sequencing Consortium, 1998) and suite of research tools (Epstein and Shakes, 1995), ensures that *C. elegans* will remain a unifying force in parasitic nematology.

Root-knot and Cyst Nematodes

Throughout this chapter a range of plant-parasitic nematode species will be mentioned, but many examples will be drawn from the root-knot (*Meloidogyne* spp.) and cyst (*Globodera* and *Heterodera* spp.) nematodes, as these have been the most extensively studied. Root-knot nematodes have a very broad host range, whereas the cyst nematodes are much more specific. These species hatch in the soil as a second-stage larva (L2) that penetrates and migrates within a host root to establish permanent feeding sites, which are characterized by extensive modifications to host cells (Fig. 8.1). The nematodes undergo dramatic developmental and morphological changes and adopt a sedentary life style. Eggs are either released in masses on to the surface of the root gall (root-knot) or encased in the body of the female, thus forming a cyst. Depending on the particular nematode species and host, and environmental conditions, there are typically between one and four generations per year.

The Impact of Plant-parasitic Nematodes

Nematodes are devastating parasites of crop plants in agricultural production and certainly contribute significantly to net reduction in yield, although assessing the true magnitude of the problem is difficult. Based on an extensive international survey (Sasser and Freckman, 1987) it has been estimated that overall yield loss averages 12.3% annually; this figure approaches 20% for some crops (e.g. banana). In monetary terms

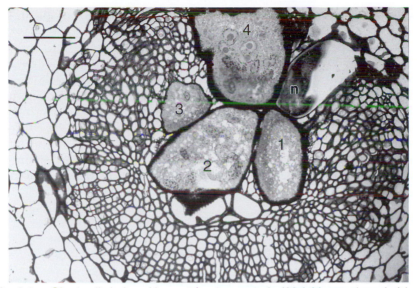

Fig. 8.1. Giant cells induced by root-knot nematode (*Meloidogyne javanica*) in tomato (*Lycopersicon esculentum*) roots. Four giant cells (1–4) are evident. The nematode (n) has contracted during fixation. Scale bar = 50 μm.

this figure certainly exceeds US$100 billion annually. Most of the damage is caused by a small number of nematode genera, principally the sedentary root-knot and cyst nematodes, and several migratory nematodes (including *Pratylenchus* and *Radopholus* spp.).

Another way to consider the impact of plant-parasitic nematodes is through the management strategies employed in their control. In 1982, 109 million pounds (approximately 49.4 Mt) of nematicide active ingredient were applied to crops in the USA, at a cost exceeding US$1 billion (Landels, 1989). Between 1986 and 1990 in The Netherlands, nematicide application was more than three times the combined total of chemicals needed to combat insects, fungi and weeds on experimental farms (Lewis *et al.*, 1997). However, in recent decades, issues such as groundwater contamination, toxicity to mammals and birds, and residues in food have caused much tighter restrictions on the use of agricultural chemicals, including suspension of use of nematicides in many countries (Thomason, 1987).

Host resistance is the most environmentally and economically sound approach for nematode management, and in those crops where resistance pertains it has proved to be an extremely valuable commodity. For example, the introduction of the *Heterodera glycines*-resistant cultivar 'Forrest' saved soybean growers in the southern USA over US$400 million during a 5-year period (Bradley and Duffy, 1982). Regrettably, nematode resistance is yet to be identified for many crop plants, although several naturally occurring resistance genes have recently been cloned (Williamson, 1999) and the potential use of these dominant loci to construct transgenic plants to circumvent breeding difficulties is an intriguing approach. Indeed, transfer of $Hs1^{pro-1}$, a gene from a wild relative of sugarbeet that confers resistance against *H. schachtii* (the beet cyst nematode), was shown to confer nematode resistance to susceptible sugarbeet roots (Cai *et al.*, 1997). However, experiments to transfer resistance from tomato into tobacco using the cloned *Mi* gene (which conditions resistance to *Meloidogyne incognita*) have so far been unsuccessful (Williamson, 1998). Other approaches to make transgenic, nematode-resistant crop plants based on an understanding of the host–parasite interaction have been proposed and were reviewed in detail by Atkinson *et al.* (1998).

Nematode Adaptations for Plant Parasitism

The nematode body plan has proved to be a remarkably adaptable platform upon which a wide range of modifications have evolved (Bird and Bird, 1991; Blaxter and Bird, 1997), permitting a diverse range of habitats to be exploited. Although various plant nematodes exhibit a range of modifications depending on their particular parasitic niche, several adaptations are widespread, including morphological and developmental specializations.

Mouthparts

The most obvious change related to parasitism of plants has been to their mouthparts. In particular, all plant parasitic nematodes have a mouth spear or buccal stylet for at least some of the stages in their life cycle; it may be absent in male forms. However, although a mouth spear is required for the parasitism of plants, its presence is not necessarily indicative of a plant host, as mouth spears are also found in entomopathogenic nematodes and in predatory dorylaims (Fig. 8.2A). Buccal stylets are clearly an adaptation towards penetrating the various structural polysaccharide barriers of potential hosts or prey (cellulose in plants and chitin in insects). In the endoparasites of plants, these structures have become more refined. Two types of spear have evolved and these exhibit different ontogeny (Bird and Bird, 1991), though in both cases they are formed by self-assembly of secreted components of the stomodeum. The dorylaim odontostyle is duplicated in larvae between moults, one odontostyle being functional and the other being stored in the pharyngeal wall until its deployment after the moult. Buccal stylets may be hollow and used as food channels as in the stomatostyle of tylenchids (Fig. 8.2B) and the odontostyle of longidorids, or solid and used as a pick as in the odontostyle of trichodorids (Yeates, 1998).

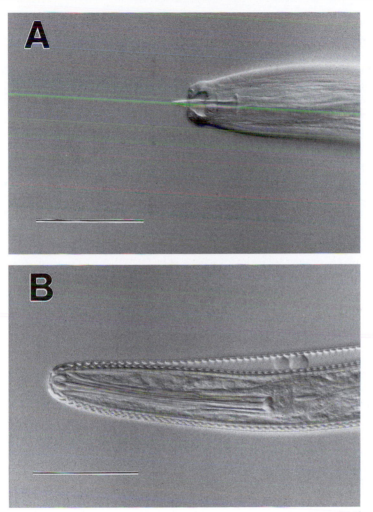

Fig. 8.2. Feeding stylets: (A) predatory dorylaim odontostyle; (B) parasitic tylenchid (*Hemicyclophora* spp.) stomatostyle. Scale bar = 50 μm.

Pharynx and Pharyngeal Glands

The extensible stylet of plant-parasitic nematodes is connected to a well-developed pharynx containing three or five gland cells. The most extensively studied are those of the tylenchid nematodes, where the pharynx is composed of a muscular metacorpus containing a triradiate pump chamber and three large and complex pharyngeal glands (Bird, 1967, 1968; Endo, 1984). Each gland is a single large secretory cell with a cytoplasmic extension that terminates in an ampulla, itself connected to the pharyngeal lumen via a valve. The valve of the single dorsal gland is located near the base of the stylet whereas the two subventral gland cells empty into the pharynx just posterior to the metacorporal pump chamber (Hussey and Mims, 1990).

Microscopy studies have revealed marked changes in the shape and volume of the pharyngeal glands that appeared to correlate with key events in establishment of the parasitic interaction, and the role played by gland contents has long been the subject of speculation (Linford, 1937; Bird, 1967, 1968, 1969; Hussey, 1989; Bird, 1992; Hussey *et al.*, 1994). In root-knot and cyst nematodes, the subventral glands seem to be more active prior to host penetration, with the reduction of secretory activity coordinated with the onset of parasitism (Endo, 1987, 1993; Endo and Wergin, 1988), at which time activity of the dorsal gland increases (Bird, 1983). Similarly, in non-feeding stages of *Nacobbus aberrans*, no evidence of secretory activity was observed in the pharyngeal glands, whereas upon onset of feeding a highly active dorsal gland was observed (Souza and Baldwin, 1998). Using video-enhancement techniques (Wyss *et al.*, 1992) it was demonstrated that migration through the root is accompanied by copious secretion of material from the head of the nematode. It was postulated that this material originated in the subventral glands and was presumed to be secreted via the stylet.

Various enzymatic functions for the secretions have been proposed, and convincing biochemical evidence obtained at least for the secretion of root-knot nematode-encoded cellulase (Bird *et al.*, 1975). However, it has not been until the recent cloning of genes encoding gland proteins that the nature of the secretion products have been discerned with confidence. The first to be isolated, *sec-1* from *M. incognita*, appears to encode a protein that is not secreted *per se*, but which plays a role in the secretion process (Ray *et al.*, 1994). Also demonstrated to be synthesized in the *Meloidogyne* pharyngeal glands (but not formally shown to be secreted) is a gene encoding chorismate mutase (Lambert *et al.*, 1999), an enzyme typically associated with the biosynthesis of the essential amino acid phenylalanine; a highly speculative role in the host–parasite interaction has been conjectured for this enzyme (Lambert *et al.*, 1999). It is interesting to note that the auxins (a family of phytohormones; see below) are synthesized from amino acid precursors.

To date, the most extensively studied secretion protein genes are from cyst nematodes. Using monoclonal antibodies directed to subventral gland antigens demonstrated to be truly secreted (de Boer *et al.*, 1996), sufficient protein for sequencing was affinity-purified, and the peptide sequences exploited to isolate genes defining a small family of endoglucanases from the potato cyst nematode (*Globodera rostochiensis*) and also from soybean cyst nematode (Smant *et al.*, 1998; Yan *et al.*, 1998). These *eng* genes, which most likely encode cellulases used during migration and perhaps also host penetration, appear to be widely present in plant-parasitic nematodes, having been isolated from root-knot nematodes (Rosso *et al.*, 1999) and detected in plant nematodes with diverse parasitic habits, including *Pratylenchus agilis, Paratrichodorus minor, Bursaphalenchus xylophilus, Rotylenchulus reniformis* and *Ditylenchus dipsaci* (Y. Yan and E.L. Davis, North Carolina, 1999, personal communication). Understanding the specific role of each family member in each of the various nematode–plant interactions will likely shed considerable light on the infection process.

Rectal Glands

In most parasitic nematodes the alimentary tract remains intact and functional and associated with food uptake and elimination of faeces. However, recent microscopy analyses using immunohistochemistry techniques (M.A. McClure, Arizona, 1999, personal communication) indicate that the anus of the *Meloidogyne* second-stage larva is in fact an orifice through which surface coat proteins are secreted. The source of these secretions is presumably the rectal glands, which become greatly enlarged in the adult stage and from which a copious gelatinous matrix is secreted (Dropkin and Bird, 1978) and which acts in various ways to protect the eggs. The anus apparently is never connected with the intestine. In highly specialized endoparasitic forms, such as in the genus *Meloidogyne*, where food is obtained as a sterile solution, there is presumably no need for the elimination of insoluble solid materials, such as bacterial cell walls.

Nothing is yet known of the molecular identity of the rectal gland secretions, although it is interesting to speculate that they might include enzymes able to degrade or remodel plant cell walls (e.g. endoglucanases, pectate lyases). Although the adult female root-knot nematode is clearly an endoparasite, the eggs are laid outside the root, presumably requiring breakdown of root cortical and epidermal cells.

Developmental Changes

Undoubtedly one of the key adaptations that has permitted nematodes to become such successful parasites is the ability to suspend development so as to couple their biology temporally with that of the host or other environmental cues. The canonical example of developmental arrest is provided by the dauer larva of *C. elegans* (Riddle and Albert, 1997) where it serves as an environmentally resistant, dispersal stage. The term 'dauer' refers to an enduring or lasting quality, and is used as a descriptor for the larvae of many different species of nematodes undergoing facultative or obligate dormancy. Dauer formation is an effective survival strategy in which stress is resisted and in which ageing and development are arrested.

The term 'dauer stage' is perhaps more appropriate since this type of dormancy is not restricted to the larval stages but has been observed in the adults of *Aphelenchoides ritzemabosi* (Wallace, 1963) and *Anguina australis* (I.T. Riley, South Australia, 1999, personal communication). Dormancy includes quiescence and diapause. The former can lead to a state of no measurable metabolism and a state of suspended animation (cryptobiosis or anabiosis). Diapause differs in that its type of dormancy is temporarily irreversible and requires other triggers to bring about reversal (Womersley *et al.*, 1998). Thus the dauer larvae of *Anguina agrostis* (*funesta*) exhibit a dormancy involving quiescence whereas the dauer of *C. elegans* exhibits a dormancy involving a diapause. Despite these differences, it is likely that the underlying mechanisms leading to the arrest involve the dauer pathway described in *C. elegans* (Riddle and Albert, 1997).

An extensive genetic analysis (Riddle and Albert, 1997) has revealed numerous genes controlling *dauer formation* (*daf* genes) in *C. elegans*. By testing for epistasis of various pairwise combinations of *daf* genes, Riddle's group has defined a pathway through which environmental signals are perceived and processed into developmental (e.g. dauer entry/exit) and behavioural (e.g. egg-laying) changes, and antigenic switching on the nematode surface (Grenache *et al.*, 1996). Microscopy of Daf mutants and cellular localization of *daf* gene expression (Riddle and Albert, 1997) have demonstrated that the dauer pathway is primarily a neuronal one, making it an ideal conduit for a rapid response to the environment. Thus, the dauer pathway plays a pivotal role in linking a wide range of developmental and behavioural responses of the nematode to changes in the environment, suggesting that rather than being a specialized adaptation to the *C. elegans* life style, the dauer pathway is a fundamental aspect of nematode biology.

Dauer formation is facultative in *C. elegans* and corresponds to an alternative third larval stage. For many parasitic nematodes, the dauer stage is obligate and is often (but not always) the infective stage. In fact, it is not always the case that the dauer stage corresponds to a distinct larval stage, and in *Anguina* the dauer larva is formed gradually during the development of the L2 (Riddle and Bird, 1985). In contrast with the L3 dauers of *C. elegans*, the L2 dauers of *Meloidogyne* and *Heterodera* spp. reflect precocious dauer development. Conversely, species such as *B. xylophilus*, which makes L4 dauers, are considered retarded in comparison with *C. elegans* (Riddle and Georgi, 1990).

Although the precise nature of the molecules involved remains elusive, it is well established that *C. elegans* integrates the environmental cues of nematode-produced pheromone, food signal and temperature to control entry to and exit from the dauer stage (Golden and Riddle, 1984). These cues permit individual nematodes to assess predictively whether or not sufficient resources are and will be available to complete the next life cycle. The obligate formation of a dauer stage by plant-parasitic nematodes implies either that the entry cue is constitutively provided, or that there is no *a priori* need for an entry cue *per se*, these alternatives have been discussed elsewhere (Bird and Opperman, 1998).

Within 50–60 min of exposure to a suitable ratio of food-signal to pheromone, *C. elegans* dauers commit to recovery, and feeding is initiated within 2–3 h. Dauer recovery is far less understood in plant-parasitic nematodes. Some plants, while not resistant, fail to be recognized as hosts (Reynolds *et al.*, 1970). Infective *Meloidogyne* L2 enter these non-hosts, and frequently leave, retaining the ability to reinfect a true host. This result is consistent with the failure of the L2 to reinitiate development and suggests that dauer recovery does not occur simply in response to entering a root. However, in a compatible host, changes can be discerned in the L2 following root penetration (Bird, 1967). For example, 99% of freshly hatched L2 will move through a 1 cm sand column within 6 h while only 12% of 1–2-day-old parasitic L2 dissected from the root can do this in the same time, though no morphological differences associated with movement have been observed. This might suggest that resumption of development by the parasite occurs prior to the establishment of a feeding site. Initiation of feeding has not been studied *in planta*, and conceivably might even begin whilst the L2 migrates through plant tissues. However, as is the case for *C. elegans*, feeding presumably supervenes the commitment to recover (i.e. to resume development). Perhaps the necessity to perceive a recovery cue prior to feeding is the reason that attempts to establish *in vitro* systems in which sedentary endoparasites are able to feed and develop have been unsuccessful (Bolla, 1987). It is possible that the reduced motility of post-penetration larvae reflects alterations to the musculature. One of the most obvious changes that occur with the onset of feeding by root-knot and cyst nematodes in plants is the loss of function of somatic musculature to the extent that movement becomes restricted to the head region.

A role of the dauer pathway in processing environmental cues upon which various developmental decisions are based has previously been suggested for animal-parasitic nematodes (Riddle and Albert, 1997) and the same is probably true for plant-parasitic nematodes. Both root-knot and cyst nematodes base developmental decisions on as yet unidentified host signals. Sex determination for parthenogenetic *Meloidogyne* species is based on perception of host status, a character that is conceptually equivalent to the 'food signal' perceived by *C. elegans*. Similarly, soybean cyst nematodes couple progeny diapause with host senescence. In both cases, it is likely that the dauer pathway mediates between the host cue and the developmental outcome. Obviously, the chemical nature of those cues will differ from species to species, and indeed such differences may play a central role in determining the host range of any given parasite or species.

Niches Occupied by Plant-parasitic Nematodes

Perusal of many recent reviews of plant nematology (Hussey, 1989; Williamson and Hussey, 1996; Hussey and Grundler, 1998) might suggest that roots are the only organs attacked by nematodes, but the stems, leaves and flowers of plants are parasitized by many species of nematodes. Indeed, the first plant-parasitic nematode to have been observed was the seed gall-forming species *Anguina tritici* (Needham, 1743). Economically important and scientifically interesting examples of aerial plant nematodes include those belonging to the genera *Ditylenchus* (Sturhan and Brzeski, 1991), *Aphelenchoides* and *Bursaphelenchus* (Nickle and Hooper, 1991) and *Anguina, Heteroanguina* and *Mesoanguina* (Krall, 1991).

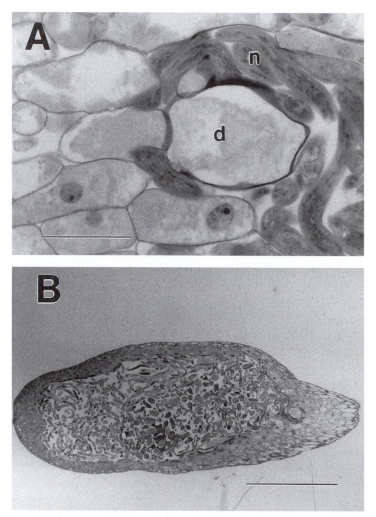

Fig. 8.3. *Anguina agrostis* (*funesta*) seed galls on annual ryegrass (*Lolium rigidum*). (A) Host cells with dense, granular cytoplasm and enlarged nuclei, adjacent to anhydrobiotic nematodes (n). Some host cells (d) are damaged and are anucleate. Scale bar = 50 μm. (B) Low magnification section through a galled seed showing the cavity filled with nematodes. Scale bar = 1 mm.

An Aerial Plant Parasite

Anguina agrostis (*funesta*) is able to survive essentially indefinitely in a state of anhydrobiosis in a gall established inside a grass seed. Most of the life cycle takes place in the developing inflorescence, but begins in the soil following rain (Stynes and Bird, 1982). Nematodes exit anhydrobiosis, emerge from the seed gall as infective larvae and make their way to the growing rye grass which initiates its floral structure close to the ground. Nematodes are carried up with the inflorescence, where galls are induced in tissues that would normally become the ovules (or sometimes stamens, glumes or the rachis). Changes take place in a large number of host cells. The cytoplasm becomes dense and granular, the nuclei enlarge (Fig. 8.3A) and gradually the cell contents become vacuolated and the cells become empty and collapse. This process ultimately leads to a gall with a cavity that becomes filled with nematodes (Fig. 8.3B). *A. agrostis* (*funesta*) is amphimictic and each gall usually has two or three of each sex, which undergo a single round of reproduction resulting in more L2 that subsequently become anhydrobiotic dauer larvae as the host senesces. Eggs have not been observed in galls with a single nematode (A.C. McKay, South Australia, 1999, personal communication) suggesting that neither parthenogenesis nor hermaphroditism occurs.

In certain grasses, including the important forage crop annual ryegrass (*Lolium rigidum*), these seed galls can become toxic to grazing animals when a bacterium (*Clavibacter toxicus*) is brought into the developing seed heads attached to the cuticle of the invading nematode. Intriguingly, the toxin is encoded by particles first observed under the transmission electron microscope (Bird *et al.*, 1980) and subsequently confirmed to be bacteriophage (Ophel *et al.*, 1993). This unique association of microorganisms is dependent on the nematode providing both transport and a niche in the seed gall that it initiates.

Root Parasites

Several authors have developed schemes to classify root-parasitic nematodes according to the site of feeding, and have conveniently presented these in cartoon format (Dropkin, 1969; Sijmons *et al.*, 1994; Wyss, 1997; Hussey and Grundler, 1998). In the simplest versions, nematodes are considered to be migratory or sedentary, and endo- or ectoparasites. Additional categories recognize classes such as 'ecto-endoparasite'. As part of these classification schemes, the nature of the host response is incorporated, such that the most elaborate versions recognize 14 'modes' of root parasitism (Hussey and Grundler, 1998). The morphology of the various feeding sites that nematodes induce in the epidermis, cortex and vascular cylinder of the roots of their host plants ranges from cytologically subtle changes that take place in single cells to the complex syncytia and giant cells induced by the sedentary endoparasites. A clear and detailed summation of the mostly descriptive work on the various interactions has been provided (Wyss, 1997) and will not be repeated here.

However, although a useful tool for describing individual host–parasite interactions, these classification schemes provide no information on mechanisms of the host–parasite interaction, nor do they give any clues to the evolution of parasitism. Despite assertions that parasitic interactions involving the sedentary endoparasites are evolutionarily more advanced than the others (e.g. Wyss, 1997), all extant plant-parasitic nematodes should be considered to be equally evolved. Differences between parasitic strategies reflect adaptations to exploit different ecological niches within the host, and this is particularly true of feeding behaviour. Because of this, great care should be exercised in extrapolating results obtained with one genus to interactions involving another. Further, much insight will be gained from careful examination of even the seemingly simple interactions. For example, an EM study revealed surprisingly complex host responses to the ecto-endoparasite *Scutellonema brachyurum* (Schuerger and McClure, 1983) and the ectoparasite *Criconemella xenoplax* (Hussey *et al.*, 1992). Nevertheless, the best-studied feeding sites remain those induced by root-knot and cyst nematodes, termed giant cells and syncytia, respectively. This is partly because of the great economic importance of these groups, but also because the large feeding sites induced by these sedentary parasites are amenable to microscopy and to biochemical and molecular analyses.

Feeding Sites of Sedentary Endoparasites

The subject of numerous studies under the light and electron microscope (e.g. Christie, 1936; Bird, 1961; Jones and Northcote, 1972; Jones and Payne, 1978), the anatomy and cytology of giant cells and syncytia is well established. The ontogeny of syncytia (Golinowski *et al.*, 1997) and giant cells (Bleve-Zacheo and Melillo, 1997) and the physiology of both types (Grundler and Böckenhoff, 1997) are the subject of recent comprehensive reviews and thus will only be briefly described.

Giant cells (Fig. 8.1) arise by expansion of individual parenchyma cells in the vascular cylinder. The developing cells undergo rounds of synchronous nuclear division uncoupled from cytokinesis, and individual nuclei become highly polyploid. The cell wall is extensively remodelled, including a marked reduction in plasmodesmatal connections with cells other than other neighbouring giant cells, and the development of finger-like projections into the cell. These events are tightly coupled to the developmental status of the nematode, and giant cells reach maximal size and activity at the onset of egg-laying (Bird, 1971). Interestingly, the transition from a parenchyma cell to a fully differentiated giant cell occurs early in the parasitic association. In other words, once the giant cells have been initiated, their characteristics do not change appreciably throughout the period of nematode feeding (apart from getting bigger, having more nuclei, etc.). In many hosts (but not all) cortical and pericycle cells around the giant cells expand and divide, resulting in the formation of a gall or knot.

In contrast to giant cells, the syncytia induced by cyst nematodes arise by coalescence of adjacent cells, resulting in a multinucleate cell in the absence of mitosis. Thus, despite resembling each other, giant cells and syncytia apparently have different ontogeny, although it is less clear that this is true at the molecular level as many genes specifically induced in giant cells also are specifically induced in syncytia (D. Bird, unpublished results).

Feeding Site Induction

Although detailed analysis of gene expression in feeding sites will ultimately define how these cells function, these approaches give no clue as to how these unique plant cells are initiated. Linford (1937), and numerous investigators since then, speculated that an inductive signal emanates from the parasite (Bird, 1962; Hussey, 1989; Bird, 1992) and, more specifically, from one or all of the pharyngeal glands (Linford, 1937; Bird, 1967, 1968, 1969; Hussey, 1989; Hussey *et al.*, 1994); the role of other secretory organs, such as the amphids, has also been formally discussed (Bird, 1992). Indeed, it has become dogma that proteinaceous secretory products are the inductive signal, with some models including direct interaction between pharyngeal gland proteins and host genes (Williamson and Hussey, 1996; Hussey and Grundler, 1998), presumably functioning as transcription factors. Formation of giant cells is perhaps the most studied, although is still far from understood.

Induction of Giant Cells

The genus *Meloidogyne* is very cosmopolitan, inducing stereotypical giant cells in a vast range of vascular plants, and any model for giant cell induction must be consistent with this fact. We believe this indicates that the process that leads to giant cell formation must involve some fundamental and widely conserved aspect(s) of plant biology. Phytohormones meet this requirement, and it is very appealing to speculate that giant cell induction is mediated via hormones (or hormone agonists/antagonists) that are directly nematode-produced, or through some hormone production or response pathway able to be directly manipulated by the parasite. *Meloidogyne* L2 have been shown to produce biologically active cytokinin (Bird and Loveys, 1980). Further, the level of cytokinin *in planta* is increased in nematode-infected roots, but the systemic cytokinin levels (as measured by the amount in xylem exudate) were not appreciably elevated (Bird and Loveys, 1980). Unfortunately, experiments in which the expression of phytohormone-responsive genes is measured are difficult to interpret as there is considerable functional interplay between plant hormone systems. Thus, until experimentally demonstrated, the role of nematode-produced cytokinin (or any other hormone, including peptides) remains equivocal.

A crucial feature of any nematode-specified inducer is the mode of delivery to the host cells. Molecules such as phytohormones or peptides might reasonably be expected to be diffusable in the host, and thus be free to act on any susceptible cell within a range of the nematode; this range would thus be concentration dependent. Alternatively, molecules might be actively transported to specific cells or cellular targets; the distribution of such proteins would thus depend on the nature and extent of the transport system. The third possibility is that the nematode selects cells to be induced and delivers an inducer directly to those cells. For genera such as *Heterodera*, where the stylet unequivocally crosses the cell wall, the stylet is an obvious candidate as the conduit for an inducer. However, we are unaware of any direct evidence showing the *Meloidogyne* stylet breaching the wall, although wall modifications have been attributed to damage caused by prior stylet penetration (Hussey and Mims, 1991). Thus it seems unlikely that these nematodes supply pharyngeal gland contents *into* host cells. Consistent with this is the failure (to date) to demonstrate any nematode proteins in giant cells, but rod-like structures called 'feeding tubes' have been observed in giant cells and, although the morphology of these structures varies widely from host to host (Hussey and Mims, 1991), they are widely suspected to be of nematode origin (Hussey and Grundler, 1998). An obvious ramification of the apparent failure of *Meloidogyne* to breach the host cell wall is that feeding would appear be extracellular, presumably requiring either active secretion or passive leakage from the host.

A further requirement of direct stylet injection is that the stylet must be able to reach the cells that are destined to become giant cells. Root-knot nematodes that have initiated feeding are sessile, having only a restricted amount of head movement, yet in many micrographs giant cells exist apparently beyond the reach of the stylet (e.g. giant cell number 3 in Fig. 8.1). The direct-injection model needs to reconcile this. Because giant cells are interconnected by plasmodesmata, solutes should flow freely, requiring the parasite to feed from only one giant cell. Might plasmodesmata also permit movement of the inductive signal?

The temporal requirement for a specific inductive signal is unknown. Temperature shift experiments have shown that the induction of resistance mediated by the *Mi* locus in tomato is restricted to the first 24–48 h after infection by *Meloidogyne* L2, showing that one aspect of the host–parasite interaction at least is temporally restricted (Dropkin, 1969). In the 'developmental switch' model (Bird, 1996) a transient induction is sufficient, but it is clear that some ongoing interaction between parasite and giant cells is required, as removal of the nematode leads to feeding site dissolution (Bird, 1962). Whether this constitutive stimulus is simply a physiological effect caused by the metabolic sink of feeding (Bird, 1996) or something more specific remains unknown.

It has been proposed that the question of feeding cell induction be considered in terms of regulation of host differentiation (Dropkin and Boone, 1966; Bird, 1974; Bird, 1996) and a number of groups have begun to identify genes expressed in feeding cells. It is likely that in the future the use of classical genetics, on both the host and the parasite side, will prove to be especially powerful in studying nematode–plant interactions, but such approaches are too new to have yet proved informative.

Genes Expressed in Feeding Cells

Nematode-induced feeding cells are unique cell types, and presumably have unique gene expression profiles. A wide range of strategies to identify genes up-regulated (or down-regulated) in feeding sites have been employed, and these have recently been extensively reviewed (Fenoll *et al.*, 1997a). Apart from genetics, strategies to identify constitutively expressed plant genes with crucial roles in the host–parasite interaction (such as the resistance genes) are less apparent.

The most productive approach in identifying giant cell-specific transcripts (i.e. not expressed in spatially or temporally equivalent healthy cells) has been a subtractive cDNA cloning which defined hundreds of genes (Bird and Wilson, 1994; Wilson *et al.*, 1994). These genes have been extensively characterized (Bird, 1996; A. Green and D. Bird, unpublished observations) and their sequences are available from GenBank. The great challenge with these (and, indeed, all the differentially expressed genes) is to relate their expression specifically to feeding site function, and ultimately this requires functional tests (e.g. inactivation in a transgenic plant). However, some inference can be made solely from sequence. For example, the gene defined by the clone DB#280 (Bird and Wilson, 1994) encodes the tomato orthologue of a *myb*-class of transcription factor named *Le-phan* (Thiery *et al.*, 1999). The *phan* gene has also been cloned from snapdragon (Waites *et al.*, 1998) and maize (Tsiantis *et al.*, 1999), where it is called *rough sheath2*. *Phan* appears to control the developmental transitions: division to determination, and determination to differentiation, by repression and de-repression, respectively, of the *knox* class of homeodomain genes (Timmermans *et al.*, 1999). Thus the presence of *Le-phan* transcripts in giant cells suggests that normal terminal differentiation (of what was initially a parenchyma cell) might be suppressed. A detailed model in which giant cell formation is initiated via an incompletely executed developmental programme has been presented (Bird, 1996), and it will be interesting to elucidate the specific role played by *Le-phan* in giant cells. Interestingly, mutations at the *knox* loci can be phenocopied and the penetrance modulated by application of exogenous hormones, and it has been suggested that knox proteins might be key regulators of the yet-to-be-discovered plant cytokinin biosynthesis genes (Frugis *et al.*, 1999).

Cell Cycle Regulation

A productive experimental approach has been to focus on the cell cycle perturbations evident in feeding sites. Based on their cytology, it can be surmised that giant cells exhibit defects in at least three points of the cell cycle: G_1 to S phase transition without prior completion of mitosis; failure to pass from metaphase to anaphase (resulting in endo-reduplicated nuclei); and failure to complete the anaphase to telophase transition (resulting in multinucleate cells). Recent work in yeast has shown that these three points are major sites of cell cycle control. Because the cell cycle has been intensively studied in *Arabidopsis*, it has proved possible to probe giant cells and syncytia by blocking various stages of the cycle using genetic and chemical inhibitors, and the results of these experiments have been recently reviewed (Gheysen *et al.*, 1997). Importantly, it was found that blocking the cell cycle also arrested development of giant cells (Gheysen *et al.*, 1997).

Genes with potential roles in the feeding-site cell cycle have also been found in more global screens. For example, the clone DB#103 (Bird and Wilson, 1994) defines *Le-ubc4*, which encodes a ubiquitin conjugating enzyme (UBC) that is strongly up-regulated in tomato giant cells. UBCs play a central role in cellular metabolism as part of the complex that regulates specific proteolysis. Significantly, cell cycle regulation is achieved to a large degree by ubiquitin-mediated degradation of specific proteins that block cell cycle progression at each of the three key points of the cycle. The best-studied cell cycle regulator is the anaphase-promoting complex (APC), which is a large ubiquitination complex that promotes the metaphase to anaphase and anaphase to telophase transitions via destruction of specific inhibitors (Townsley and Ruderman, 1998). Further, exit from mitosis is initiated by APC degradation of mitotic cyclins.

In healthy plants, *Le-ubc4* is expressed only in meristematic cells, where it encodes a nuclear localized protein (Bird, 1996). What, therefore, might its role be in giant cells, which are non-dividing? One possibility is that *Le-ubc4* has a role in controlling nuclear proliferation (i.e. cell cycle control) and it appears that *Le-ubc4* itself is cell cycle regulated. Le-UBC4 protein is first detectable in mid-G_2 phase. It remains present throughout mitosis, and disappears in late G_1 phase (A. Green and D. Bird, unpublished results). This suggests that Le-UBC4's role is not simply in synthesis of new chromosomal material *per se* (e.g. ubiquitination of histones). Might Le-UBC4 be a component of the APC, or some other specific regulatory complex? Alternatively, the influence exerted by Le-UBC4 might be much less subtle. Western blot data (D. Bird, unpublished results) indicate that Le-UBC4 levels are very high in giant cells, perhaps causing the fine balance of regulated cellular proteolysis to be non-specifically disrupted. Understanding the precise role of this gene in giant cells awaits specific function analysis.

Evolution of Parasitism: an Ancient Symbiosis?

Because of their soft bodies, the fossil record for nematodes is poor (Poinar, 1983) and both the manner in which they may have evolved and the relationship between different orders of nematodes has been the subject of much speculation (Coomans, 1983). In both cases the application of molecular phylogenetic methods has provided a framework from which to address these issues. Analysis of small subunit ribosomal DNA sequences led to the linking of groups into clades whose members appeared on morphological and prior phylogenetic grounds to be unrelated (Blaxter *et al.*, 1998; see also Chapter 1). Significantly, it appears that parasitism has evolved independently on numerous occasions; even the relatively small number of species examined (53) revealed four independent origins for animal parasitism, and three independent origins for plant parasitism (Blaxter *et al.*, 1998).

Parasitism is an acquired trait, and one unlikely to have evolved prior to evolution of the host, which for vascular plants is 400 million years ago. Presumably most of the genes in extant parasites share a common origin with most of the genes in extant free-living forms, and indeed an analysis in which the deduced proteins from a large number of randomly generated cDNA sequences from the filarial nematode *Brugia malayi* were searched against the entire suite of *C. elegans* predicted proteins revealed a match with 86% of the genes (Bird *et al.*, 1999). Although for technical reasons this number is probably an underestimate, it is nevertheless likely that for some genes in parasitic nematodes there will be no homologues in free-living species. The *eng* loci (Smant *et al.*, 1998) provide a good example. Using the deduced ENG-1 protein from *G. rostochiensis* as a query does identify two candidates from the entire *C. elegans* protein set (with BLAST scores of 55 and 46), but examination of these sequences indicates that neither encodes a cellulase, nor a related protein. Thus, these genes have either diverged from an ancient ancestor gene such that no homology can be detected, or they have evolved independently. Although the evidence is circumstantial, a compelling argument can be made that the latter is the case, and that an inter-kingdom gene transfer was involved (Keen and Roberts, 1998).

Acquisition of Parasitism Genes by Horizontal Gene Transfer

Enough genomic genes from tylenchid nematodes, including *Mi-sec-1* (Ray *et al.*, 1994), the *eng* family members (Yan *et al.*, 1998), *Mj-col-3* (Koltai *et al.*, 1997), *Ma-cut-1* (De Giorgi *et al.*, 1996) and *Mj-CM-1* (Lambert *et al.*, 1999), have been characterized in sufficient detail to confirm that these are all ordinary eukaryotic genes, with normal regulatory elements and with features typical of genes in *C. elegans* (including small introns and trans-spliced transcripts). It was therefore surprising to find that the deduced protein sequences of the cellulase genes (Smant *et al.*, 1998) and the chorismate mutase gene (Lambert *et al.*, 1999) were not similar to these enzymes from other eukaryotes, but rather have strong similarities to bacterial proteins. It seems likely that these pharyngeal gland genes were acquired from microbes via horizontal gene transfer (Keen and Roberts, 1998). Significantly, the homology between the ENG proteins and these enzymes from plant sources is low, suggesting that plants were not the source of the nematode genes.

It is likely that the concept of ancient nematodes acquiring microbial genes will be controversial, and ultimately might turn out not to be the case. It does, however, help to explain how plant parasitism has apparently evolved on multiple, independent occasions, and it will be especially interesting to examine in detail the candidate cellulase genes identified in other plant nematode species (Y. Yan and E.L. Davis, North Carolina, 1999, personal communication; see above). For example, *B. xylophilus*, another clade IV member (Blaxter *et al.*, 1998), is a fungal feeder; will its *eng* loci resemble bacterial or fungal genes? *P. minor* is a member of clade II; what will its *eng* loci most resemble? Placing these and other genes into the context of the ever-developing phylogenies will undoubtedly further the understanding of the origins of plant parasitism, and may provide clues to the mechanism of inter-kingdom gene transfer. Presumably one requirement for this is a physical interaction between the organisms involved. Interactions between bacteria and members of clade IV in particular (Blaxter *et al.*, 1998), many of which belong to the Tylenchida, including the genera *Anguina, Globodera* and *Meloidogyne*, have been examined in detail and are all pervading. Understanding these interactions may provide some clues to the origins of horizontally acquired genes.

Clade IV Nematode–Bacterial Associations

The simplest interaction is where the nematode, for example *Acrobeloides nanus* (a cephalobid), merely eats bacteria. *A. nanus* can feed on a wide range of bacteria, including *C. toxicus* (Bird and Ryder, 1993), an observation confirmed under the TEM (Fig. 8.4). By contrast, the plant parasite

A. agrostis (*funesta*) (a tylenchid) is unable to use *C. toxicus* as a food source, but this same bacterium has the capacity to adhere to the surface of the *Anguina* cuticle (Fig. 8.5) and, as mentioned above, be carried by the nematode into the plant, where it can continue its development at

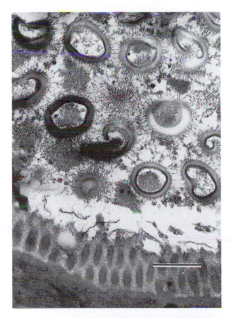

Fig. 8.4. Transmission electron micrograph of a section through the intestine of *Acrobeloides nanus*, which has been grown in a culture of *Clavibacter toxicus*. Note the crushed cells of the bacteria. Scale bar = 500 nm.

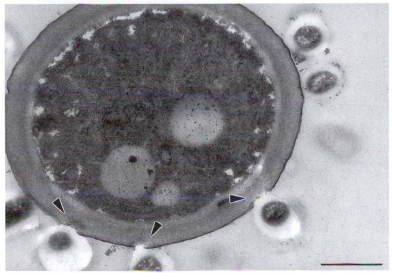

Fig. 8.5. Transmission electron micrograph of a transverse section through an infective dauer larva of *Anguina agrostis* (*funesta*) showing *Clavibacter toxicus* adhering to the cuticle and causing pathological changes to the cuticle surface (arrows). Scale bar = 1 μm.

the expense of the nematode (Bird and Riddle, 1984). This interaction has a mildly pathological effect on the infective larva of the nematode, penetrating the cuticle and slowing down movement of the nematode, its vector. Thus, what is food for one group of nematodes (*Acrobeloides–Clavibacter*) becomes a pathogen for another (*Anguina–Clavibacter*).

A more extreme example of a pathogenic interaction occurs between the mycelial endospore-forming bacterium *Pasteuria penetrans* and a wide range of nematodes, including the genus *Meloidogyne*. The parasitic relationship between *Pasteuria* and *Meloidogyne* (Bird, 1986) initially involves adhesion of the bacterial endospore to the L2 whereby it is carried into the plant, where it germinates, penetrates the nematode and reproduces in the posterior of its host, leaving the anterior portion intact. The nematode feeds normally and provides food for the developing endospores, which destroy the nematode's reproductive system as they fill the body.

An equally intimate but non-pathogenic relationship between nematodes and bacteria is that between various nematode parasites of insects belonging to the genus *Steinernema* and the bacterium *Xenorhabdus*. The bacteria become housed monoxenically in a specialized intestinal vesicle in the nematode after they have been ingested. They are released when the nematode invades the haemocoel of the insect and multiply rapidly, causing septicaemia and death. It has been shown (Götz *et al.*, 1981) that the nematodes secrete a substance that inhibits the antibacterial proteins produced by the insect so that *Xenorhabdus* can grow rapidly. It in turn inhibits the growth of other bacteria and provides nutrients that aid the growth and reproduction of *Steinernema*.

The most intimate of the bacterial–nematode associations involves *Wolbachia*, a genus of rickettsia-like, alpha proteobacteria found in obligate intracellular association with a wide variety of arthropods, and a now well-characterized association with *B. malayi* and *Dirofilaria immitis* (Chapter 2). Although not formally identified as *Wolbachia*, the presence of rickettsia-like organisms has been observed, principally in the reproductive tracts of *Globodera* females (Shepherd *et al.*, 1973) and males (Walsh *et al.*, 1983) and in *Heterodera* larvae (Endo, 1979).

As an endosymbiont, *Wolbachia* is not unique in insects. For example, all aphids appear to have symbiotic bacteria of the genus *Buchnera*, believed to have entered an aphid ancestor as a free-living eubacterium 200–250 million years ago (Baumann *et al.*, 1995) and tephritid flies host an endosymbiont, until recently classified in the genus *Erwinia* (Drew and Lloyd, 1991). What makes these insect–microbe interactions particularly interesting is that the insects involved induce galls on the plants from which they feed. A good example is the galls produced on grape roots by the phylloxera aphid *Daktulosphaira vitifoliae*. Whether or not induction of galls by insects involves the same host pathways as feeding site induction by nematodes is unknown, but it is a reasonable hypothesis that phytohormones play a role. *Buchnera* are capable of synthesizing large quantities of tryptophan, an auxin precursor and most aphids are known or thought to contain auxins (Baumann *et al.*, 1995). Similarly, the tephritid endo-symbiont is capable of synthesizing cytokinins (Drew and Lloyd, 1991). It is an intriguing possibility that the insect galling genes also have a bacterial origin, and perhaps this is true also for the cytokinin synthesis genes in *Meloidogyne*. Might the *Mj-CM-1* product (Lambert *et al.*, 1999) also have a role in phytohormone biosynthesis?

Based on extant nematode–microbe relationships, various obviously highly conjectural models for the transfer of bacterial genes to nematodes can be proposed. In the simplest instance, a bacterivorous nematode acquires the enzymes required to invade vascular plants directly by ingesting phytopathogenic bacteria. It is not inconceivable that an endo-symbiotic relationship might evolve, not unlike the *Steinernema–Xenorhabdus* association, with the subventral pharyngeal glands serving the role of the *Steinernema* intestinal vesicle. Alternatively, microbial genes might have entered the nematode via an invasive organism such as *Pasteuria* or *Wolbachia*. In each instance extreme reduction of the bacterium would need to be envisaged to the point where the only remaining remnants were a small suite of genes, the size of which will ultimately be revealed as plant-parasitic nematodes are subjected to extensive DNA sequence analyses.

Concluding Remarks

It is clear that parasitism, whether it be of nematodes or by nematodes, has evolved in close association with bacteria, initially as a food source and then in a finely balanced relationship. The fact that similar associations exist between bacteria and gall-forming insects may be coincidence, or it may reflect some underlying universal mechanism(s) involving host-plant modification. Defining those mechanisms whereby a plant is chosen as being a suitable host, and a feeding site selected and established, has been a goal of plant nematologists for many decades; we sense that enlightenment is near.

References

Atkinson, H.J., Lilley, C.J., Urwin, P.E. and McPherson, M.J. (1998) Engineering resistance to plant-parasitic nematodes. In: Perry, R.N. and Wright, D.J. (eds) *The Physiology and Biochemistry of Free-Living and Plant-Parasitic Nematodes*. CAB International, Wallingford, UK, pp. 381–413.

Baumann, P., Baumann, L., Lai, C.Y., Rouhbakhsh, D., Moran, N.A. and Clark, M.A. (1995) Genetics, physiology, and evolutionary relationships of the genus *Buchnera*: intracellular symbionts of aphids. *Annual Review of Microbiology* 49, 55–94.

Bird, A.F. (1961) The ultrastructure and histochemistry of a nematode-induced giant cell. *Journal of Biophysical and Biochemical Cytology* 11, 701–715.

Bird, A.F. (1962) The inducement of giant cells by *Meloidogyne javanica*. *Nematologica* 8, 1–10.

Bird, A.F. (1967) Changes associated with parasitism in nematodes. I. Morphology and physiology of preparasitic and parasitic larvae of *Meloidogyne javanica*. *Journal of Parasitology* 53, 768–776.

Bird, A.F. (1968) Changes associated with parasitism in nematodes. IV. Cytochemical studies on the ampulla of the dorsal esophageal gland of *Meloidogyne javanica* and on exudations from the buccal stylet. *Journal of Parasitology* 54, 879–890.

Bird, A.F. (1969) Changes associated with parasitism in nematodes. V. Ultrastructure of the stylet exudation and dorsal esophageal gland contents of female *Meloidogyne javanica*. *Journal of Parasitology* 55, 337–345.

Bird, A.F. (1971) Quantitative studies on the growth of syncytia induced in plants by root-knot nematodes. *International Journal for Parasitology* 2, 157–170.

Bird, A.F. (1974) Plant response to root-knot nematode. *Annual Review of Phytopathology* 12, 69–85.

Bird, A.F. (1983) Changes in the dimensions of the oesophageal glands in root-knot nematodes during the onset of parasitism. *International Journal for Parasitology* 13, 343–348.

Bird, A.F. (1986) The influence of the actinomycete, *Pasteuria penetrans*, on the host–parasite relationship of the plant parasitic nematode, *Meloidogyne javanica*. *Parasitology* 93, 571–580.

Bird, A.F. and Bird, J. (1991) *The Structure of Nematodes*, 2nd edn. Academic Press, San Diego, 316 pp.

Bird, A.F. and Loveys, B.R. (1980) The involvement of cytokinins in a host–parasite relationship between the tomato (*Lycopersicon esculentum*) and a nematode (*Meloidogyne javanica*). *Parasitology* 80, 497–505.

Bird, A.F. and Riddle, D.L. (1984) Effect of attachment of *Corynebacterium rathayi* on movement of *Anguina agrostis* larvae. *International Journal for Parasitology* 14, 503–511.

Bird, A.F. and Ryder, M.H. (1993) Feeding of the nematode *Acrobeloides nanus* on bacteria. *Journal of Nematology* 25, 493–499.

Bird, A.F., Downton, J.S. and Hawker, J.S. (1975) Cellulase secretion by second stage larvae of the root-knot nematode (*Meloidogyne javanica*). *Marcellia* 38, 165–169.

Bird, A.F., Stynes, B.A. and Thomson, W.W. (1980) A comparison of nematode and bacteria-colonized galls induced by *Anguina agrostis* in *Lolium rigidum*. *Phytopathology* 70, 1104–1109.

Bird, D.M. (1992) Mechanisms of the *Meloidogyne*–host interaction. In: Gommers, F.J. and Maas, P.W.T. (eds) *Nematology: from Molecule to Ecosystem*. ESN, Dundee, UK, pp. 51–59.

Bird, D.M. (1996) Manipulation of host gene expression by root-knot nematodes. *Journal of Parasitology* 82, 881–888.

Bird, D.M. and Opperman, C.H. (1998) *Caenorhabditis elegans*: a genetic guide to parasitic nematode biology. *Journal of Nematology* 30, 299–308.

Bird, D.M. and Wilson, M.A. (1994) DNA sequence and expression analysis of root-knot nematode elicited giant cell transcripts. *Molecular Plant–Microbe Interactions* 7, 419–424.

Bird, D.M., Opperman, C.H., Jones, S.J.M. and Baillie, D.L. (1999) The *Caenorhabditis elegans* genome: a guide in the post genomics age. *Annual Review of Phytopathology* 37, 247–265.

Blaxter, M. (1998) *Caenorhabditis elegans* is a nematode. *Science* 282, 2041–2046.

Blaxter, M. and Bird, D.M. (1997) Parasitic nematodes. In: Riddle, D.L., Blumenthal, T., Meyer, B.J. and Priess, J.R. (eds) *C. elegans II*. Cold Spring Harbor Laboratory Press, Cold Spring Harbor, New York, pp. 851–878.

Blaxter, M.L., DeLey, P., Garey, J., Liu, L.X., Scheldeman, P., Vierstraete, A., Vanfleteren, J., Mackey, L.Y., Dorris, M., Frisse, L.M., Vida, J.T. and Thomas, W.K. (1998) A molecular evolutionary framework for the phylum Nematoda. *Nature* 392, 71–75.

Blevc-Zacheo, T. and Melillo, M.T. (1997) The biology of giant cells. In: Fenoll, C., Grundler, F.M.W. and Ohl, S.A. (eds) *Cellular and Molecular Aspects of Plant–Nematode Interactions*. Kluwer Academic Publishers, Dordrecht, The Netherlands, pp. 65–79.

Bolla, R.I. (1987) Axenic culture of plant-parasitic nematodes: problems and perspectives. In: Veech, J.A. and Dickson, D.W. (eds) *Vistas on Nematology*. Society of Nematology, Hyattsville, Maryland, pp. 401–407.

Bradley, E.B. and Duffy, M. (1982) *The Value of Plant Resistance to the Soybean Cyst Nematode: a Case Study of Forrest Soybean* (AGES820929). USDA-ARS Staff Report, Washington.

C. elegans Genome Sequencing Consortium (1998) Genome sequence of the nematode *Caenorhabditis elegans*: a platform for investigating biology. *Science* 282, 2012–2018.

Cai, D., Kleine, M., Kifle, S., Harloff, H.J., Sandal, N.N., Marcker, K.A., Klein-Lankhorst, R.M., Salentijn, E.M.J., Lange, W., Stiekema, W.J., Wyss, U., Grundler, F.M.W. and Jung, C. (1997) Positional cloning of a gene for nematode resistance in sugar beet. *Science* 275, 832–834.

Christie, J.R. (1936) The development of root-knot nematode galls. *Phytopathology* 26, 1–22.

Coomans, A. (1983) General principles for the phylogenetic systematics of nematodes. In: Stone, A.R., Platt, H.M. and Khalil, L.F. (eds) *Concepts in Nematode Systematics*. Academic Press, London, pp. 1–10.

de Boer, J.M., Smant, G., Goverse, A., Davis, E.L., Overmars, H.A., Pomp, H., van Gent-Pelzer, M., Zilverentant, J.F., Stokkermans, J.P.W.G., Hussey, R.S., Gommers, F.J., Bakker, J. and Schots, A. (1996) Secretory granule proteins from the subventral esophageal glands of the potato cyst nematode identified by monoclonal antibodies to a protein fraction from second-stage juveniles. *Molecular Plant–Microbe Interactions* 9, 39–46.

De Giorgi, C., De Luca, F. and Lamberti, F. (1996) A silent trans-splicing signal in the cuticlin-encoding gene of the plant-parasitic nematode *Meloidogyne artiellia*. *Gene* 170, 261–265.

Dong, K. and Opperman, C.H. (1997) Genetic analysis of parasitism in the soybean cyst nematode *Heterodera glycines*. *Genetics* 146, 1311–1318.

Drew, R.A.I. and Lloyd, A.C. (1991) Bacteria in the life cycle of tephritid fruit flies. In: Barbosa, P., Krischik, V.A. and Jones, C.G. (eds) *Microbial Mediation of Plant–Herbivore Interactions.* John Wiley & Sons, New York, pp. 441–465.

Dropkin, V.H. (1969) Cellular responses of plants to nematode infections. *Annual Review of Phytopathology* 7, 101–122.

Dropkin, V.H. and Bird, A.F. (1978) Physiological and morphological studies on secretion of a protein–carbohydrate complex by a nematode. *International Journal for Parasitology* 8, 225–232.

Dropkin, V.H. and Boone, W.R. (1966) Analysis of host–parasite relations of root-knot nematodes by single larva inoculations of excised tomato roots. *Nematologica* 12, 225–236.

Endo, B.Y. (1979) The ultrastructure and distribution of an intracellular bacterium-like microorganism in the tissues of larvae of the soybean cyst nematode *Heterodera glycines. Journal of Ultrastructural Research* 67, 1–14.

Endo, B.Y. (1984) Ultrastructure of esophagus of larvae of the soybean cyst nematode, *Heterodera glycines. Proceedings of the Helminthological Society of Washington* 51, 1–24.

Endo, B.Y. (1987) Ultrastructure of the esophagus of larvae of the soybean cyst nematode, *Heterodera glycines. Journal of Nematology* 19, 469–483.

Endo, B.Y. (1993) Ultrastructure of subventral gland secretory granules in parasitic juveniles of the soy bean cyst nematode, *Heterodera glycines. Proceedings of the Helminthological Society of Washington* 60, 22–34.

Endo, B.Y. and Wergin, W.P. (1988) Ultrastructure of the second stage juvenile of the root knot nematode, *Meloidogyne incognita. Proceedings of the Helminthological Society of Washington* 55, 286–316.

Epstein, H.F. and Shakes, D.C. (1995) Caenorhabditis elegans*: Modern Biological Analysis of an Organism.* Academic Press, San Diego, California, 659 pp.

Fenoll, C., Aristizabal, F.A., Sanz-Alferez, S. and del Campo, F.F. (1997a) Regulation of gene expression in feeding sites. In: Fenoll, C., Grundler, F.M.W. and Ohl, S.A. (eds) *Cellular and Molecular Aspects of Plant–Nematode Interactions.* Kluwer Academic Publishers, Dordrecht, The Netherlands, pp. 133–149.

Fenoll, C., Grundler, F.M.W. and Ohl, S.A. (1997b) *Cellular and Molecular Aspects of Plant–Nematode Interactions.* Kluwer Academic Publishers, Dordrecht, The Netherlands, 286 pp.

Frugis, G., Giannino, D., Mele, G., Nicolodi, C., Innocenti, A.M., Chiappetta, A., Bitonti, M.B., Dewitte, W., van Onckelen, H. and Mariotti, D. (1999) Are homeobox knotted-like genes and cytokinins the leaf architects? *Plant Physiology* 119, 371–374.

Gheysen, G., de Almeida Engler, J. and van Montague, M. (1997) Cell cycle regulation in nematode feeding sites. In: Fenoll, C., Grundler, F.M.W. and Ohl, S.A. (eds) *Cellular and Molecular Aspects of Plant–Nematode Interactions.* Kluwer Academic Publishers, Dordrecht, The Netherlands, pp. 120–132.

Golden, J.W. and Riddle, D.L. (1984) The *Caenorhabditis elegans* dauer larva: developmental effects of pheromone, food, and temperature. *Developmental Biology* 102, 368–378.

Golinowski, W., Sobczak, M., Kurek, W. and Grymaszewska, G. (1997) The structure of syncytia. In: Fenoll, C., Grundler, F.M.W. and Ohl, S.A. (eds) *Cellular and Molecular Aspects of Plant–Nematode Interactions.* Kluwer Academic Publishers, Dordrecht, The Netherlands, pp. 80–97.

Götz, P., Boman, A. and Boman, H.G. (1981) Interactions between insect immunity and an insect-pathogenic nematode with symbiotic bacteria. *Proceedings of the Royal Society of London, series B,* 212, 333–350.

Grenache, D.G., Caldicott, I., Albert, P.S., Riddle, D.L. and Politz, S.M. (1996) Environmental induction and genetic control of surface antigen switching in the nematode *Caenorhabditis elegans. Proceedings of the National Academy of Sciences USA* 93, 12388–12393.

Grundler, F.M.W. and Böckenhoff, A. (1997) Physiology of nematode feeding and feeding sites. In: Fenoll, C., Grundler, F.M.W. and Ohl, S.A. (eds) *Cellular and Molecular Aspects of Plant-Nematode Interactions.* Kluwer Academic Publishers, Dordrecht, The Netherlands, pp. 107–119.

Hussey, R.S. (1989) Disease-inducing secretions of plant-parasitic nematodes. *Annual Review of Phytopathology* 27, 123–141.

Hussey, R.S. and Grundler, F.M.W. (1998) Nematode parasitism of plants. In: Perry, R.N. and Wright, D.J. (eds) *The Physiology and Biochemistry of Free-Living and Plant-Parasitic Nematodes.* CAB International, Wallingford, UK, pp. 213–243.

Hussey, R.S. and Mims, C.W. (1990) Ultrastructure of esophageal glands and their secretory granules in the root-knot nematode *Meloidogyne incognita. Protoplasma* 156, 9–18.

Hussey, R.S. and Mims, C.W. (1991) Ultrastructure of feeding tubes formed in giant-cells induced in plants by the root-knot nematode *Meloidogyne incognita. Protoplasma* 162, 99–107.

Hussey, R.S., Mims, C.W. and Westcott, S.W.I. (1992) Ultrastructure of root cortical cells parasitized by the ring nematode *Criconemella xenoplax. Protoplasma* 167, 55–65.

Hussey, R.S., Davis, E.L. and Ray, C. (1994) *Meloidogyne* stylet secretions. In: Lamberti, F., De Giorgi, C. and Bird, D.M. (eds) *Advances in Molecular Plant Nematology.* Plenum Press, New York, pp. 233–249.

Jones, M.G.K. and Northcote, D.H. (1972) Nematode-induced syncytium – a multinucleate transfer cell. *Journal of Cell Science* 10, 789–809.

Jones, M.G.K. and Payne, H.L. (1978) Early stages of nematode-induced syncytium giant-cell formation in roots of *Impatiens balsamina. Journal of Nematology* 10, 70–84.

Keen, N.T. and Roberts, P.A. (1998) Plant parasitic nematodes: digesting a page from the microbe book. *Proceedings of the National Academy of Sciences USA* 95, 4789–4790.

Koltai, H., Chejanovsky, N., Raccah, B. and Spiegel, Y. (1997) The first isolated collagen gene of the root-knot nematode *Meloidogyne javanica* is developmentally regulated. *Gene* 196, 191–199.

Krall, E.L. (1991) Wheat and grass nematodes: *Anguina, Subanguina* and related genera. In: Nickle, W.R. (ed.) *Manual of Agricultural Nematology.* Marcel Dekker, New York, pp. 721–760.

Lambert, K.N., Allen, K.D. and Sussex, I.M. (1999) Cloning and characterization of an esophageal-gland-specific chorismate mutase from the phytoparasitic nematode *Meloidogyne javanica. Molecular Plant–Microbe Interactions* 12, 328–336.

Landels, S. (1989) Fumigants and nematicides. In: *Chemical Economics Handbook.* Stanford Research Institute International, California.

Lewis, W.J., van Lenteren, J.C., Phatak, S.C. and Tumlinson III, J.H. (1997) A total system approach to sustainable pest management. *Proceedings of the National Academy of Sciences USA* 94, 12243–12248.

Linford, M.B. (1937) The feeding of the root-knot nematode in root tissue and nutrient solution. *Phytopathology* 27, 824–835.

Needham, J.T. (1743) Concerning certain chalky concretions, called malm; with some microscopical observations on the farina of Red Lilly, and worms discovered in Smuthy Corn. *Philosophical Transactions of the Royal Society of London* 42, 634.

Nickle, W.R. (1991) *Manual of Agricultural Nematology.* Marcel Dekker, New York, 1035 pp.

Nickle, W.R. and Hooper, D.J. (1991) The Aphelenchina: bud, leaf, and insect nematodes. In: Nickle, W.R. (ed.) *Manual of Agricultural Nematology.* Marcel Dekker, New York, pp. 465–507.

Ophel, K.M., Bird, A.F. and Kerr, A. (1993) Association of bacteriophage particles with toxin production by *Clavibacter toxicus*, the causal agent of annual ryegrass toxicity. *Phytopathology* 83, 676–681.

Opperman, C.H. and Bird, D.M. (1998) The soybean cyst nematode, *Heterodera glycines*: a genetic model system for the study of plant-parasitic nematodes. *Current Opinion in Plant Biology* 1, 342–346.

Perry, R.N. and Wright, D.J. (1998) *The Physiology and Biochemistry of Free-Living and Plant-Parasitic Nematodes.* CAB International, Wallingford, UK, 438 pp.

Poinar, G.O. Jr (1983) *The Natural History of Nematodes.* Prentice Hall, Englewood Cliffs, New Jersey, 323 pp.

Ray, C., Abbott, A.G. and Hussey, R.S. (1994) Trans-splicing of a *Meloidogyne incognita* mRNA encoding a putative esophageal gland protein. *Molecular and Biochemical Parasitology* 68, 93–101.

Reynolds, H.W., Carter, W.W. and O'Bannon, J.H. (1970) Symptomless resistance of alfalfa to *Meloidogyne incognita acrita. Journal of Nematology* 2, 131–134.

Riddle, D.L. and Albert, P.S. (1997) Genetic and environmental regulation of dauer larva development. In: Riddle, D.L., Blumenthal, T., Meyer, B.J. and Priess, J.R. (eds) *C. elegans II.* Cold Spring Harbor Laboratory Press, Cold Spring Harbor, New York, pp. 739–768.

Riddle, D.L. and Bird, A.F. (1985) Responses of *Anguina agrostis* to detergent and anesthetic treatment. *Journal of Nematology* 17, 165–168.

Riddle, D.L. and Georgi, L.L. (1990) Advances in research on *Caenorhabditis elegans* – application to plant parasitic nematodes. *Annual Review of Phytopathology* 28, 247–269.

Riddle, D.L., Blumenthal, T., Meyer, B.J. and Priess, J.R. (1997) *C. elegans II.* Cold Spring Harbor Laboratory Press, Cold Spring Harbor, New York, 1222 pp.

Rosso, M.N., Favery, B., Piotte, C., Arthaud, L., de Boer, J.M., Hussey, R.S., Bakker, J., Baum, T.J. and Abad, P. (1999) Isolation of a cDNA encoding a beta-1,4-endoglucanase in the root-knot nematode *Meloidogyne incognita* and expression analysis during plant parasitism. *Molecular Plant–Microbe Interactions* 12, 585–591.

Sasser, J.N. and Freckman, D.W. (1987) A world perspective on nematology: the role of the Society. In: Veech, J.A. and Dickson, D.W. (eds) *Vistas on Nematology.* Society of Nematology, Hyattsville, Maryland, pp. 7–14.

Schuerger, A.C. and McClure, M.A. (1983) Ultrastructural changes induced by *Scutellonema brachyurum* in potato roots. *Phytopathology* 73, 70–81.

Shepherd, A.M,, Clark, S.A. and Kempton, A. (1973) An intracellular micro-organism associated with tissues of *Heterodera* spp. *Nematologica* 19, 31–34.

Sijmons, P.C., Atkinson, H.J. and Wyss, U. (1994) Parasitic strategies of root nematodes and associated host cell responses. *Annual Review of Phytopathology* 32, 235–259.

Smant, G., Stokkermans, J.P.W.G., Yan, Y., de Boer, J.M., Baum, T.J., Wang, X., Hussey, R.S., Gommers, F.J., Henrissat, B., Davis, E.L., Helder, J., Schots, A. and Bakker, J. (1998) Endogenous cellulases in animals: isolation of β-1,4-endoglucanase genes from two species of plant-parasitic cyst nematodes. *Proceedings of the National Academy of Sciences USA* 95, 4906–4911.

Souza, R.M. and Baldwin, J.G. (1998) Changes in esophageal gland activity during the life cycle of *Nacobbus aberrans* (Nemata: Pratylenchidae). *Journal of Nematology* 30, 275–290.

Sturhan, D. and Brzeski, M.W. (1991) Stem and bulb nematodes, *Ditylenchus* spp. In: Nickle, W.R. (ed.) *Manual of Agricultural Nematology*. Marcel Dekker, New York, pp. 423–464.

Stynes, B.A. and Bird, A.F. (1982) Development of galls induced in *Lolium rigidum* by *Anguina agrostis*. *Phytopathology* 72, 336–346.

Thiery, M.G., Greene, A.E. and Bird, D.M. (1999) The *Lycopersicon esculentum* orthologue of *PHANTASTICA/rough sheath2* (Accession No. AF148934) is expressed in feeding sites induced by root-knot nematodes (PGR99–099). *Plant Physiology* 120, 934.

Thomason, I.J. (1987) Challenges facing nematology: environmental risks with nematicides and the need for new approaches. In: Veech, J.A. and Dickson, D.W. (eds) *Vistas on Nematology*. Society of Nematology, Hyattsville, Maryland, pp. 469–476.

Timmermans, M.C.P., Hudson, A., Becraft, P.W. and Nelson, T. (1999) *ROUGH SHEATH2*: A Myb protein that represses knox homeobox genes in maize lateral organ primordia. *Science* 284, 151–153.

Townsley, F.M. and Ruderman, J.V. (1998) Proteolytic ratchets that control progression through mitosis. *Trends in Cell Biology* 8, 238–244.

Tsiantis, M., Schneeberger, R., Golz, J.F., Freeling, M. and Langdale, J.A. (1999) The maize *rough sheath2* gene and leaf development programs in monocot and dicot plants. *Science* 284, 154–156.

Waites, R.H., Selvadurai, R.N., Oliver, I.R. and Hudson, A. (1998) The *PHANTASTICA* gene encodes a myb transcription factor involved in growth and dorsoventrality of lateral organs in *Antirrhinum*. *Cell* 93, 779–789.

Wallace, H.R. (1963) *The Biology of Plant Parasitic Nematodes*. Arnold, London, 280 pp.

Walsh, J.A., Lee, D.L. and Shepherd, A.M. (1983) The distribution and effect of intracellular rickettsia-like micro-organisms infecting adult males of the potato cyst nematode, *Globodera rostochiensis*. *Nematologica* 29, 227–239.

Williamson, V.M. (1998) Root-knot nematode resistance genes in tomato and their potential for future use. *Annual Review of Phytopathology* 36, 277–293.

Williamson, V.M. (1999) Nematode resistance genes. *Current Opinion in Plant Biology* 2, 327–331.

Williamson, V.M. and Hussey, R.S. (1996) Nematode pathogenesis and resistance in plants. *Plant Cell* 8, 1735–1745.

Wilson, M.A., Bird, D.M. and van der Knaap, E. (1994) A comprehensive subtractive cDNA cloning approach to identify nematode-induced transcripts in tomato. *Phytopathology* 84, 299–303.

Womersley, C.Z., Wharton, D.A. and Higa, L.M. (1998) Survival biology. In: Perry, R.N. and Wright, D.J. (eds) *The Physiology and Biochemistry of Free-Living and Plant-Parasitic Nematodes.* CAB International, Wallingford, UK, pp. 271–302.

Wood, W.B. (1988) *The Nematode* Caenorhabditis elegans. Cold Spring Harbor Laboratory Press, Cold Spring Harbor, New York, 667 pp.

Wyss, U. (1997) Root parasitic nematodes: an overview. In: Fenoll, C., Grundler, F.M.W. and Ohl, S.A. (eds) *Cellular and Molecular Aspects of Plant–Nematode Interactions.* Kluwer Academic Publishers, Dordrecht, The Netherlands, pp. 5–22.

Wyss, U., Grundler, F.M.W. and Münch, A. (1992) The parasitic behaviour of 2nd-stage juveniles of *Meloidogyne incognita* in roots of *Arabidopsis thaliana. Nematologica* 38, 98–111.

Yan, Y., Smant, G., Stokkermans, J., Qin, L., Helder, J., Baum, T., Schots, A. and Davis, E. (1998) Genomic organization of four β-1,4-endoglucanase genes in plant-parasitic cyst nematodes and its evolutionary implications. *Gene* 220, 61–70.

Yeates, G.W. (1998) Feeding in free-living soil nematodes: a functional approach. In: Perry, R.N. and Wright, D.J. (eds) *The Physiology and Biochemistry of Free-Living and Plant-Parasitic Nematodes.* CAB International, Wallingford, UK, pp. 245–269.

The Nematode Cuticle: Synthesis, Modification and Mutants

Antony P. Page

Wellcome Centre of Molecular Parasitology, The Anderson College, The University of Glasgow, 56 Dumbarton Road, Glasgow G11 6NU, UK

Nematode Cuticle: Structure

Nematodes form a diverse phylum composed of both free-living and parasitic species. Parasitic species are of medical, veterinary and agricultural significance, and include species causing diseases that are amongst the most prevalent and debilitating known to mankind (Bird and Bird, 1991). A critical structure of all nematodes is the surface cuticle, which acts as a hydroskeleton, maintains post-embryonic body shape and permits mobility, elasticity and interaction with the external environment; and in parasitic species it represents the site of contact with the host's immune response. Throughout the Nematoda phylum this extracellular matrix has a well-ordered cytoarchitecture, characterized at the electron microscope level by distinct layers and transverse structures (Bird and Bird, 1991) (Fig. 9.1).

The nematode cuticle is a tough, flexible structure, being composed of up to six layers, namely the epicuticle, cortex (inner and outer), medial, fibre and basal layers. The presence of different layers of different thickness is dependent on the stage and species of nematode analysed; for example, a medial layer with struts is present only in adult stage *Caenorhabditis elegans*. The basic structure, synthesis and composition of this exoskeleton is however relatively conserved throughout the Nematoda phylum. The cuticle is composed of highly cross-linked, soluble and insoluble structural proteins, namely the collagens, cuticulins and other minor proteins, and lipids and carbohydrates. The major functions of this resilient structure are to act as an impervious barrier to the environment, allow movement via

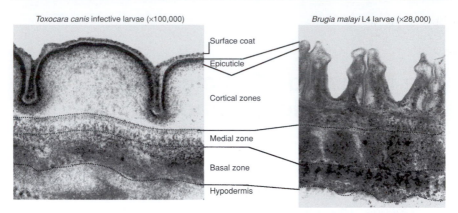

Toxocara canis infective larvae (×100,000)   Brugia malayi L4 larvae (×28,000)

Surface coat
Epicuticle
Cortical zones
Medial zone
Basal zone
Hypodermis

Fig. 9.1. Transmission electron micrographs of parasitic nematode cuticles in transverse section. The structurally distinct layers and the underlying hypodermal syncytia are indicated. Nematodes depicted are the infective larval stage of the canid parasite *Toxocara canis* and the fourth larval stage of the human filarial parasite *Brugia malayi*.

opposed muscles, determine body shape and permit growth through the larval moults.

All nematodes undergo four post-embryonic moults, characterized by the synthesis of a new extracellular proteinaceous cuticle from the predominantly syncytial hypodermis. Five separate cuticles are synthesized by all nematodes by a process referred to as moulting or ecdysis.

Moulting

The moulting process is defined by the cyclical synthesis of structural proteins, including the collagens, and has been well characterized in the free-living nematode *C. elegans* (Singh and Soulston, 1978). Prior to moulting, the lateral hypodermal seam cells become areas of intense biosynthetic activity, which can be gauged by ribosome, endoplasmic reticulum, golgi and mitochondrial changes. There are three major steps in the moulting process:

1. *Apolysis (lethargus)* corresponds to a period of inactivity and results in the separation of connections between the old cuticle basal zone and the underlying hypodermis. The inactivity is a direct result of associated changes in musculature structure and function.
2. *Late lethargus* is characterized by the formation of the new cuticle, which arises externally to the cell membrane of the hypodermis, and thus represents a true extracellular matrix. The epicuticular and cortical layers are the first to be formed and these layers are enriched in the highly

insoluble cross-linked cuticlins. The loosening of the old cuticle is achieved when the nematodes spin and flip around their long axis.

3. *Ecdysis* is the final stage and results in the shedding of the old cuticle. Immediately prior to ecdysis the pharynx begins twitching and gland secretions, predominantly composed of proteases, are released to aid removal of the old cuticle. Finally the pharynx lining is replaced and the animal then pushes with its head to break out of the old cuticle.

The newly synthesized cuticle is highly folded and convoluted and thus allows rapid growth after the moult. Certain stages of some parasitic nematodes retain the old cuticle as a protective sheath.

Proteases play essential roles in cuticle moulting, being involved in the digestion of the cuticle anchoring proteins during apolysis and, in some instances, resorption of the old cuticle proteins during ecdysis. Proteases may also play a role in the moulting cycle by processing associated proenzymes. Moulting enzymes include the leucine aminopeptidases (Rogers, 1982), zinc metalloprotease (Gamble *et al.*, 1989) and cysteine proteases (Richer *et al.*, 1992; Lustigman, 1993). It has been hypothesized that protease inhibitors, such as the cysteine protease cystatins (Lustigman *et al.*, 1992), function by regulating many of these moulting-related enzymes. Moulting enzymes and their inhibitors have thus been proposed as being selective chemotherapy targets in the parasitic nematodes (Lustigman, 1993).

Control of moulting

The control of the moulting process in nematodes is presently not well characterized but is hypothesized as being associated with nuclear hormone-type steroids, as is the case in insects (Mangelsdorf *et al.*, 1995). Recent studies in *C. elegans* lend support to this hypothesis. The nuclear hormone receptors (NHRs) are a large family of transcription factors, with as many as 260 having been identified by the *C. elegans* genome project (*C. elegans* Genome Sequencing Consortium, 1998). Of these potential genes the majority are expected to be functional, with 54 having been directly expressed from mRNA (Sluder *et al.*, 1999). All receptors have typical zinc-finger DNA binding domains, and are presumably induced by ligands such as steroid/thyroid, retinoic acid and ecdysone hormones. Steroid hormones have been detected in various parasitic and free-living nematode species, but it is not yet clear if these are host, vector or culture media derived (Barker and Rees, 1990). The direct synthesis of ecdysteroids does not, however, occur in *C. elegans* cultured in defined media (Barker *et al.*, 1990; Chitwood and Feldlaufer, 1990), and the numerous NHRs therefore fall into the class for which no ligands are known, being referred to as orphan NHRs. *C. elegans* does have a dietary requirement for cholesterol, which is essential for proper development and moulting of the cuticle (Yochem *et al.*, 1999). A role for steroid hormones in the moulting process was reinforced recently when a gene encoding a 'megalin-type' low density lipoprotein receptor (*lrp-1*) was mutated and resulted in moulting defects. These mutants were similar in appearance to cholesterol-starved nematodes and led to the hypothesis that LRP-1 may endocytose sterols from the extracellular fluids (Yochem *et al.*, 1999).

A recent study of one *C. elegans* orphan NHR, CHR3 or *nhr-23*, which is homologous to a moulting-related ecdysteroid receptor of *Drosophila* (DHR3), established an important moulting-related function (Kostrouchova *et al.*, 1998). This NHR is expressed in the cuticle-synthesizing hypodermis of late embryo and early larval stages. A specific gene knockout via double stranded RNA interference (RNAi) resulted in larval hypodermal defects, including the inability to moult and shed the old cuticle and thus represents the first mutation to physically affect the moulting process. Additionally, a second NHR mutant (*daf-12*) has been characterized, and alleles of this mutant fail to develop properly in late larval life by repeating earlier larval stages with associated hypodermal defects (Antebi *et al.*, 1998).

The Cuticle Collagens

The major components of the nematode cuticle are the covalently cross-linked collagens, which constitute > 80% of the soluble proteins (Kingston, 1991). These reducible proteins are also characteristically sensitive to clostridial collagenase (Selkirk *et al.*, 1989). Collagens were initially defined in vertebrates, where they constitute up to 25% of the total body protein (Vuorio and DeCrombrugghe, 1990). These are distinctive triple helix-forming molecules with extensive regions of a glycine-rich repeat [(Gly-X-Y)n, where X and Y are most often proline (Pro) and hydroxyproline (Hyp), respectively], which is also characteristic of the nematode collagens. In *C. elegans* the soluble adult-stage cuticle proteins are composed of 26% Gly, 11% Pro and 12% Hyp (Cox *et al.*, 1981); a similar ratio is found in the adult stages of the parasitic nematode *Onchocerca volvulus* (Sakwe and Titanji, 1997). The hydroxyproline content of cuticle collagens is stage-specific in *C. elegans* (Cox *et al.*, 1981) and sex-dependent in *O. volvulus* adults (Sakwe and Titanji, 1997).

The nematode collagens are synthesized prior to each moult initially as pre-procollagens, which are joined by reducible and non-reducible bonds to yield stable trimers. These structures are then extensively cross-linked by the formation of reducible disulphide bridges and non-reducible di- and isotrityrosine cross-links (Fetterer *et al.*, 1993), resulting in the final cuticular matrix. Although collagens are found in every metazoan phylum analysed, studies of nematode cuticular collagens indicate that they differ markedly from vertebrate interstitial collagens in many features, including structure, assembly and mode of cross-linking (Cox *et al.*, 1981; Fetterer *et al.*, 1993; Johnstone, 1994). Major differences include: (i) the small size of the majority of nematode cuticle collagens (Kramer, 1997); (ii) interrupted Gly-X-Y domains; (iii) processing by subtilisin-like protease (Yang and Kramer, 1994; Thacker *et al.*, 1995); and (iv) the fact that they are extensively cross-linked by disulphide bonds (Cox *et al.*, 1981). Even though there is a great diversity in the morphology between different nematode species, and also between different stages of the same species, the basic biochemical composition of nematode collagens appears to be highly conserved.

The cuticle collagens have been most extensively studied at the biochemical and genetic levels in *C. elegans* both through the cloning of the many collagen genes (Cox, 1992; *C. elegans* Genome Sequencing Consortium, 1998) and by detailed analysis of collagen gene defective mutants (Johnstone, 1994; Kramer, 1994) (Fig. 9.2). *C. elegans* cuticle collagens are encoded by a large multigene family comprising 154 genes (Johnstone, 1999). These genes are generally dispersed throughout the genome (Cox, 1990) and represent approximately 1% of its content. Similar high figures

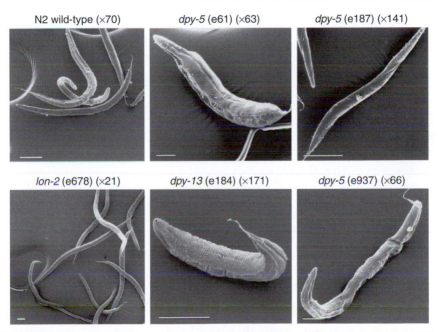

Fig. 9.2. Cuticle collagen and morphological mutants of *C. elegans*. A selection of adult-stage *C. elegans* morphological mutants were analysed by scanning electron microscopy and compared with the wild-type N2 strain. Mutant alleles depicted are: *dpy*, dumpy; *rol*, roller; *lon*, long; *bli*, blister. Scale bar represents 0.1 mm.

(100-plus genes) are also present in the parasitic nematodes *Haemonchus contortus* (Shamansky *et al.*, 1989) and *Ostertagia circumcincta* (Johnstone *et al.*, 1996). The *C. elegans* cuticle collagens are composed of approximately 50% Gly-X-Y residues and characteristically contain three conserved clusters of cysteine residues in the non-Gly-X-Y repeat regions. Major differences between collagens exist in the precise numbering and spacing of the conserved cysteine residue. On this basis, the collagens can be subdivided into four major families, with two additional smaller families. The conserved cysteine blocks are probably involved in collagen chain associations via disulphide bonding and this may relate to their sensitivity to reducing agents. The representative genes *sqt-1* (I), *col-10* (Ia), *col-12* (II) *col-2* (III), *dpy-7* and *dpy-10* belong to these different families (Johnstone, 1999).

The nematode collagen genes in general are small (2 kb) and contain few introns. The encoded proteins are also short (26–35 kDa) and the interruptions in the Gly-X-Y repeats presumably result in increased longitudinal flexibility in these normally rigid proteins, making them most like vertebrate FACIT (fibril-associated collagens with interrupted triple helices), such as type IX cartilage collagen (Prockop and Kivirikko, 1995). All predicted nematode collagen sequences contain non-repetitive amino and carboxyl terminal domains. Collagen gene structure and presumed function are shared between *C. elegans* and the parasitic nematodes, with similar short interrupted collagens now having been sequenced from a wide variety of animal and plant parasitic nematodes (Kingston *et al.*, 1989; Shamansky *et al.*, 1989; Kingston, 1991; Bisoffi and Betschart, 1996; Johnstone *et al.*, 1996; Jones *et al.*, 1996; Koltai *et al.*, 1997; Wang *et al.*, 1998). Indeed, the parasitic nematodes *O. circumcincta* and *Meloidogyne javanica* express orthologues of the *C. elegans* genes *col-12/col-13* (Johnstone *et al.*, 1996) and *dyp-7* (Koltai *et al.*, 1997), respectively.

Collagen Gene Expression

Vertebrate collagens are assembled both as homo- and heterotrimers (Prockop and Kivirikko, 1995), whereas it remains to be conclusively established how the nematode cuticle collagens associate. It is presumed that collagens belonging to the same families, based on the alignment of conserved residues, may have the ability to form heterotrimers and that any collagen could potentially homotrimerize. Evidence exists that two closely related *C. elegans* collagens, ROL-6 and SQT-1, form an association, as both belong to the same structural family (Johnstone, 1999), are coordinately expressed in a ratio of two *sqt-1* to one *rol-6* (Park and Kramer, 1994) and their association is supported by genetic interactions. The *rol-6* phenotypes are suppressed in the *sqt-1* null background whereas *sqt-1* phenotypes are visible in a *rol-6* null background, indicating that ROL-6 requires SQT-1 to function properly but SQT-1 functions independently (Kramer and Johnson, 1993).

Cuticle collagen gene expression in *C. elegans* was originally found to be developmentally regulated (Cox and Hirsh, 1985; Kingston *et al.*, 1989) with collagens being synthesized at high rates during moults and lower rates between moults. A high degree of complexity was noted in the expression of individual larval collagen genes, which follows a precise temporal programme of expression (Johnstone and Barry, 1996). Collagen genes are expressed in waves of early (*dpy-7*), middle (*sqt-1*) and late (*col-12*) expression patterns, as measured on synchronous worm populations by RT-PCR (Johnstone and Barry, 1996). A correlation between structural collagen gene families (see later) and these temporal waves of expression is evident, supporting the hypothesis that collagens within a family may indeed heterotrimerize. It can also be hypothesized that collagens expressed at different times in the moulting cycle may constitute different cuticular structures, such that collagens expressed early in the moulting cycle would form the first cuticle layers.

C. elegans Cuticle Collagen-related Mutants

Collagen gene mutants aid in the understanding of collagen function and the roles that individual collagens play in the structure of the cuticle. Mutations affecting the *C. elegans* cuticle collagens and their associated enzymes normally have phenotypes that affect body shape and have been given names to reflect their morphology (Brenner, 1974; Kusch and Edgar, 1986). Examples of cuticle defect mutants are depicted in Fig. 9.2, and a detailed review of cuticle mutants is to be found in Kramer (1997).

The largest family of cuticle mutants is the short and fat Dumpy (*dpy*) mutants. There are presently 27 unique *dpy* genes (a small proportion being X-chromosome dosage compensation-related). Four of these mutants are collagen mutations and, of these, *dpy-10*, *dpy-7* and *dpy-13* all have mutant alleles resulting from glycine substitutions in the repeat regions (Johnstone, 1994; Kramer, 1997).

Perhaps the most dramatic family of mutants is the Roller (*rol*) mutants, which are helically twisted in a RightROL or LeftROL relative to their longitudinal axis. In these animals the internal organs are also helically twisted and as a result they move in circles instead of the normal sinusoidal motion. There are six unique *rol* genes, with *rol-6* (*su1006*) being the best characterized, encoding a cuticle collagen with an amino acid substitution affecting the procollagen N-terminal cleavage site (Kramer *et al.*, 1990). It has been hypothesized that ROL-6 collagen may be associated with the cuticle fibrous basal layers, which are composed of closely opposed collagen fibres that run at 65° (mirror image) from the longitudinal axis, an angle consistent with the helical twist noted with this mutant (Bergmann *et al.*, 1998).

The three Squat (*sqt*) mutants have a characteristic dominant-dumpy/ recessive-roller phenotype, and of the two genes characterized (*sqt-1* and *sqt-3*), both encode cuticle collagens (Kramer, 1997).

Another dramatic cuticular phenotype is displayed by the Blister (*bli*) mutants, which have fluid-filled cuticle blisters. There are six *bli* genes, two of which encode cuticle collagens, namely *bli-1* and *bli-2*. It is predicted that *bli-1* encodes a strut collagen of the medial layer of the adult cuticle, since *bli-1* mutants have no struts when viewed by electron microscopy and the blister phenotype is significantly restricted to the adult stage (J. Crew and J. Kramer, Chicago, 1999, personal communication). A third *bli* gene, *bli-4*, is indirectly linked to the cuticle collagens. This mutant was characterized and found to encode a kex2 subtilisin-like endoprotease which is potentially involved in processing the N-propeptide domain of the cuticle collagens (Thacker *et al.*, 1995). All the *C. elegans* cuticle collagens have an N-terminal homology block (HB-A) (Kramer, 1994) consisting of a basic kex2-like protease processing site preceding the conserved cysteines and the Gly-X-Y repeats. The BLI-4 enzyme would therefore process the N-procollagen domain (step 4 in Fig. 9.3). Most alleles of *bli-4* are embryonically lethal, whereas the viable allele encodes a partially functional enzyme and results in the cuticle blistering phenotype (Peters *et al.*, 1991; Thacker *et al.*, 1995). The importance of this basic HB-A processing site is further supported by the fact that point mutations in this site are associated with mutant alleles of *sqt-1*, *rol-6* and *dpy-10* (Levy *et al.*, 1993; Yang and Kramer, 1994).

To date, nine characterized genes encode collagen mutants: *sqt-1* (Kramer and Johnson, 1993), *sqt-3* (Vanderkeyl *et al.*, 1994), *rol-6* (Kramer and Johnson, 1993), *dpy-13* (VonMende *et al.*, 1988), *dpy-2* (Levy *et al.*, 1993), *dpy-10* (Levy *et al.*, 1993), *dpy-7* (Johnstone *et al.*, 1994), *bli-1* and *bli-2* (J. Crew and J. Kramer, Chicago, 1999, personal communication). Glycine substitutions are a particularly common feature in these mutants, resulting in more severe phenotypes than nulls, indicating that the abnormal collagens must be interacting and interfering with other cuticle-associated collagens. Amino acid substitutions can cause a variety of effects, including aberrant trimer formation, delayed secretion and over-modification. Many null mutations have no effect, e.g. certain mutant alleles of *sqt-1* and *rol-6* (Yang and Kramer, 1994). However, null mutations in *dpy-10* (Levy *et al.*, 1993) and *dpy-13* (VonMende *et al.*, 1988) have strong phenotypes, indicating that these collagens must be required for normal cuticle assembly.

Collagen Folding

The principles of collagen folding differ markedly from other known proteins since single monomers cannot fold. Triple helix folding is a multi-step process involving chain association, registration, nucleation and propagation (Engel and Prockop, 1991) (Fig. 9.3). The association and registration steps occur via cysteine-linked disulphide bond formation between the C-propeptides of the monomer procollagens, allowing nucleation of the triple helix, and then propagation of the left-handed helix from the C to N terminus in a 'zipper-like' mechanism (Engel and Prockop, 1991). Structural constraints require that every third amino acid must be glycine, which, being the smallest amino acid, fits well into the centre of the helix. This requirement is further demonstrated by mutations replacing glycine with bulkier amino acids (Prockop and Kivirikko, 1995) (see earlier). Such substitutions distort the triple helical structure and in some cases vertebrate triple helices containing these altered peptide chains have reduced thermal stability – for example, type I collagen glycine mutations cause osteogenesis imperfecta (Engel and Prockop, 1991). The second major constraint in collagen folding is the need for correct registration of all three chains. The third constraint is that proline and hydroxyproline are essential to make the triple helix structure rigid (Kivirikko and Pihlajaniemi, 1998). Although any amino acid can occupy the X and Y positions, the imino acids, proline and hydroxyproline, usually predominate. Recent evidence confirmed that Gly-Pro-Hyp is the most common and stabilizing tripeptide found in collagens (Ackerman *et al.*, 1999). In the unfolded state, most proline residues in the Y position are enzymatically hydroxylated, and the resultant 4-hydroxyproline residues are essential for the formation and thermal stabilization of the triple helix. The presence of Gly-Pro-Y and Gly-X-Hyp, however, predisposes the chain to spontaneous formation of *cis*-peptide bonds. Since all peptides must be in the *trans* conformer in the native triple helix, the slow *cis–trans* isomerization becomes a rate-limiting propagation step in the folding of the collagens (Bachinger *et al.*, 1980). Thus it is envisaged that folding *in vivo* must require the action of proline *cis–trans* isomerase and prolyl 4-hydroxylase enzymes. The structural requirement for prolyl residues is further illustrated by the fact that collagens of all organisms examined have melting temperatures in solution consistently 3–4°C above their body temperatures. This is directly related to the content of proline and especially hydroxyproline residues (Engel and Prockop, 1991).

Catalysts of Collagen Folding, Co- and Post-translational Modification

It is becoming apparent that, for cuticle collagen assembly to proceed accurately, specification of which collagen monomers trimerize together might be critical, and this is likely to be directed by the collagen primary structure (Johnstone and Barry, 1996; see earlier) and by co- and post-translational enzyme action (Winter and Page, 2000). Enzymes involved in the co- and post-translational modification and folding of nematode cuticle

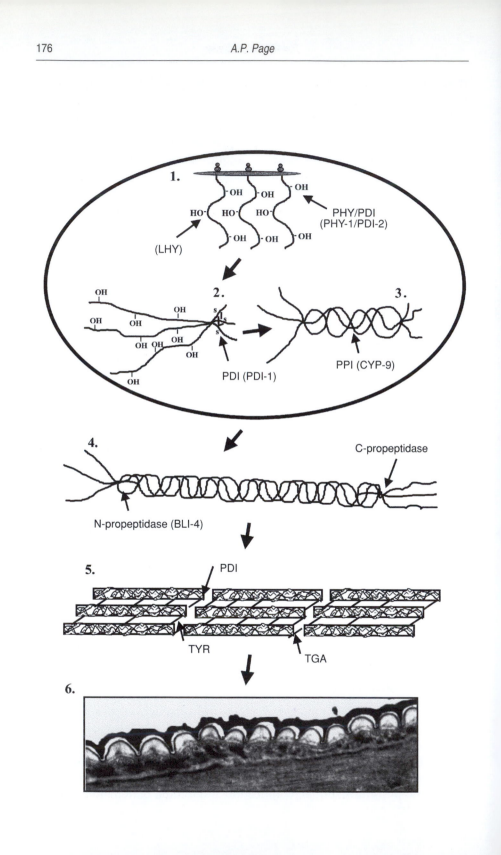

collagens are therefore excellent targets for chemotherapeutic attack against parasitic nematodes, and study of such enzymes may provide new insights into potentially novel methods for parasite control. Target enzymes include peptidyl-prolyl *cis-trans* isomerase (PPI, cyclophilin/ FKBP), prolyl 4-hydroxylase, protein disulphide isomerase (PDI) and lysyl hydroxylase (LHY). The potential roles these enzymes play in the modification of collagens is shown in Fig. 9.3. There may also be important molecular chaperones involved in the folding of collagens, and these may include the collagen-specific heat shock protein, HSP47 or colligin (Nagata, 1996). An HSP47 has recently been described as working in association with a cyclophilin in the folding of vertebrate procollagen I (Smith *et al.*, 1995).

Peptidyl-prolyl cis–trans *isomerase (PPI, cyclophilins/FKBPs) (EC 5.3.4.1)*

PPI or rotamase activity is catalysed by two major classes of structurally divergent enzymes: the cyclophilins (CYPs) and the FK506 binding proteins (FKBPs/FKBs) (Gothel and Marahiel, 1999). These enzymes catalyse the otherwise slow rate-limiting isomerization of *Xaa*-proline peptide bonds and can accelerate the refolding of several proline-containing poly-peptides, including collagens (Bachinger, 1987; Steinmann *et al.*, 1991) (Fig. 9.1). The best example to date of a cyclophilin having a unique sub-strate is the *ninaA* cyclophilin of *Drosophila*, which is involved in rhodopsin folding. Mutations in the *ninaA* gene result in partially folded rhodopsin, causing visual defects (Colley *et al.*, 1991; Stammes *et al.*, 1991). A role for cyclophilin in collagen folding has been demonstrated indirectly, as an active isoform recognized unfolded type III collagens as substrate, catalysed the isomerization of the peptide bonds and resulted in a threefold increase in the rate of *in vitro* folding (Bachinger, 1987). Similar effects of PPI were demonstrated *in vivo* on the folding of chick embryo tendon fibroblast type I procollagens and human fibroblast types I and III collagens (Steinmann *et al.*, 1991).

Fig. 9.3. (Opposite) The proposed roles that enzymes play in the synthesis, modification and cross-linking of the nematode cuticle collagens. Steps 1–3 occur within the endoplasmic reticulum, and include: (1) co-translational lysine and proline hydroxylation; (2) procollagen monomer registration via disulphide bond formation; (3) rate-limiting proline *cis–trans* isomerization of *Xaa*-pro peptide bonds and chain propagation. Steps 4–6 occur outside the endoplasmic reticulum and include: (4) post-translational processing to remove N- and C-propeptide domains; (5) collagen cross-linking, via tyrosine–tyrosine, glutamine–lysine and cysteine–cysteine linked disulphide bonding; (6) the final assembled cuticular matrix. LHY, lysyl hydroxylase; PDI, protein disulphide isomerase; PHY, prolyl 4-hydroxylase; PPI, peptidyl-prolyl *cis–trans* isomerase (cyclophilin/FKBP); TGA, transglutaminase; TYR, tyrosinase.

In addition to possessing PPI enzymatic activity, the cyclophilins were independently shown to be the cellular receptors for the immuno-suppressive agent cyclosporin A (CsA) (Takahashi *et al.*, 1989). This drug has also been demonstrated to inhibit PPI activity (Takahashi *et al.*, 1989) and slows down the *in vitro* folding of collagens (Steinmann *et al.*, 1991). CsA also induced changes in collagen metabolism similar to those observed in several variants of osteogenesis imperfecta caused by impaired helix formation (Bonadio and Byers, 1985) and may therefore explain some of the side-effects observed in CsA-treated animals and patients, such as severe osteopenia (Movsowitz *et al.*, 1988) and gingival hyperplasia (Steinmann *et al.*, 1991). Interestingly, this drug has widespread but, to date, unexplained anti-parasitic effects both *in vitro* and *in vivo*, including stage-specific effects against a number of parasitic nematodes (Chappell and Wastling, 1992; Page *et al.*, 1995a). Likewise, the less characterized class of PPIs, the FKBPs, are receptors for a second class of structurally unrelated immunosuppressive drugs, the FK506 or rapamycin compounds (Gothel and Marahiel, 1999). These drugs also inhibit the FKBP-specific PPI activity but these shared activities are not responsible for the immunosuppression (Schreiber and Crabtree, 1992).

The cyclophilins and FKBPs are large multi-member families in nematodes (Page *et al.*, 1996; A.P. Page, 2000, unpublished results). In *C. elegans*, 11 CYP members (*cyp-1–11*) have been characterized, and a further five members have since been identified by the genome project (*C. elegans* Genome Sequencing Consortium, 1998). Enzymatic (PPI assays) and structural examination of the nematode CYPs supports the existence of four separate classes: [A] cytosolic, [B] secretory, [C] mitochondrial and [D] divergent cyclophilin-like proteins (Page *et al.*, 1996). Direct homo-logues of representative members of the *cyp* genes have been identified in the filarial nematode *Brugia malayi* and other parasitic nematodes (Page *et al.*, 1995b; Ma *et al.*, 1996; Hong *et al.*, 1998a,b; Page and Winter, 1998, 1999; A.P. Page, 2000, unpublished observations).

The potential role in cuticle synthesis of this large group of enzymes was addressed via reporter gene expression pattern analysis in *C. elegans*. This study revealed an unexpected diversity of tissues expressing the differ-ent CYPs, including the gut cells (Page and Winter, 1999), the body wall muscles cells (Page and Winter, 1998), the excretory cell (Dornan *et al.*, 1999) and the hypodermis (Page, 1997). Of the 11 isoforms examined, *cyp-9* was most promising with regards to a role in cuticle collagen folding, being located immediately downstream of a second gene (*pdi-1*), which encodes an enzyme also postulated to play a role in collagen synthesis, namely protein disulphide isomerase (PDI) (Page, 1997). The transcripts *cyp-9* and *pdi-1* are trans-spliced in an operon-specific fashion (Blumenthal and Steward, 1997), with the upstream message (*pdi-1*) being trans-spliced by the ubiquitous trans-spliced leader, SL1, and the downstream message (*cyp-9*) by the operon-specific trans-spliced leader, SL2. Both messages were transcribed coordinately from a single upstream promoter, having an identical spatial and temporal expression pattern: transcripts were expressed in the hypodermis in an oscillating temporal pattern corresponding to the cuticle moulting cycle (Page, 1997). This currently represents one of the best examples of a functional operon in *C. elegans*, being especially significant since it has been postulated that both enzymes play important roles in collagen folding (Fig. 9.3). Significantly, PPI has also been reported to improve the efficiency of PDI as a catalyst in collagen folding (Schonbrunner and Schmid, 1992). An orthologue of this potential collagen-folding operon has now been identified in the closely related nematode species *C. briggsae*. This has identified highly conserved domains in the 5′ promoter region, which may be involved in expression and regulation of this operon (Page, 1999). It can be hypothesized that these enzymes may cooperate at a similar step in the early events of collagen folding in the hypodermis. The PDI-1 enzyme may be involved in the C-terminal nucleation step (step 2, Fig. 9.3), either acting as a catalyst of disulphide bond formation, thereby initiating trimer association, or functioning as a chaperone retaining the collagen monomers in the correct conformation, thus permitting trimerization to proceed in an orderly, non-aggregated fashion. The PPI enzyme encoded by the downstream gene *cyp-9* would then catalyse the rate-limiting proline isomerization step converting the *cis* proline and *cis* 4-hydroxyproline to the trimer accepting *trans* isomers.

The *C. elegans* genome project (*C. elegans* Genome Sequencing Consortium, 1998) has also identified eight putative FK506-binding protein isoforms (FKBs). Initial examination of these enzymes indicated a similar diverse tissue distribution, with at least two multi-domain isoforms being expressed in a cyclical fashion in the hypodermis (A.P. Page, 2000, unpublished results). Interestingly, large multi-domain (65 kDa) isoforms of FKBPs have been demonstrated to participate in the folding of vertebrate collagens (Zeng *et al.*, 1998) and bovine tropoelastin (Davis *et al.*, 1998). The roles played by the nematode FKBs in the PPI folding of the cuticle collagens however remains to be established.

Prolyl 4-hydroxylase (PHY) (EC 1.14.11.2)

Prolyl 4-hydroxylase is an important multi-component ER-resident catalyst involved in the early stages of collagen biosynthesis, being responsible for the oxidation of proline to 4-hydroxyproline in nascent collagen polypeptides (Fig. 9.3). Almost all prolines in the Y position of the Gly-X-Y repeat are hydroxylated (Kivirikko and Pihlajaniemi, 1998). Hydroxylation occurs co-translationally prior to chain registration and the triple helical conformation of collagen actually blocks this reaction. An additional function of this enzyme is to act as a chaperone by retaining unfolded procollagen chains in the ER, releasing them for secretion only when they have folded correctly (Walmsley *et al.*, 1999). In vertebrates this enzyme is an $\alpha_2\beta_2$ tetramer; the α-subunit forms the binding site for the peptide substrate to be hydroxylated (Veijola *et al.*, 1994), whereas the β-subunit is identical to PDI (Pihlajaniemi *et al.*, 1987). Hydroxylase activity is absolutely dependent on this multi-component association. In vertebrates, two isoforms of the α-subunit exist, α(I) and α(II) (Annunen *et al.*, 1997). The α-subunits do not form mixed α(I)α(II)β_2 tetramers and both sets of α-subunits associate with the same β-subunit.

Co-expression of the human α- and β-subunits in the yeast *Pichea pastoralis* produces only trace amounts of active tetramer, with the majority being present in an unassembled form. Co-expression with human type III collagens, however, increases this assembly level tenfold. This indicates that collagen synthesis and the formation of an active prolyl 4-hydroxylase complex are mutually dependent processes (Vuorela *et al.*, 1997). A similar observation has been noted for baculovirus encoded enzymes in insect cells (Lamberg *et al.*, 1996). These findings support the hypothesis that this unusual control mechanism may be a common feature of collagen synthesis in all cell types.

Two conserved *phy* α-subunit encoding genes (*phy-1* and *phy-2*) have been identified in the *C. elegans* genome project (*C. elegans* Genome Sequencing Consortium, 1998). The gene encoding PHY-1 has been cloned from cDNA and expressed as an active recombinant enzyme using the baculovirus system (Veijola *et al.*, 1994). Surprisingly, the PHY-1 α-subunit was found to interact and form an active $\alpha\beta$ complex with the human PDI/β-subunit (Veijola *et al.*, 1994) and, likewise, an $\alpha\beta$ complex with the *C. elegans* PDI isoform PDI-2 (Veijola *et al.*, 1996). This unusual attribute was due to a unique C-terminal extension present on PHY-1 and absent from mammalian α-subunits (Veijola *et al.*, 1994). This was further supported by the observation that PDI-2 formed an active $\alpha_2\beta_2$ complex with the human α-subunit (Veijola *et al.*, 1996). The catalytic properties of the *C. elegans* PHY-1/PDI-2 dimer are similar to that of the vertebrate type II tetramer (Veijola *et al.*, 1994). Neither the human PHY α-subunit nor the *C. elegans* PHY-1 isoforms were able to form either an $\alpha_2\beta_2$ or an $\alpha\beta$ complex with the second *C. elegans* PDI-isoform, PDI-1 (Veijola *et al.*, 1996).

Spatial and temporal expression pattern analysis of *lacZ* reporter constructs in transgenic animals and semi-quantitative RT-PCR, respectively, revealed that both *C. elegans* α-subunit encoding genes were expressed in the hypodermal cells in a cyclical fashion that mirrored the moulting cycle (Winter and Page, 2000). These results indirectly supported a cuticle collagen modifying role for both encoded enzymes. Conclusive evidence supporting this role was obtained genetically, as both genes were shown to be essential for embryonic development and the maintenance of nematode morphology (Winter and Page, 2000). The *phy-1* knock-out nematodes display a cuticle-associated dumpy phenotype, and genetic complementation confirmed *phy-1* to be identical to the molecularly uncharacterized strain *dpy-18*. The *phy-1* gene was cloned from two separate alleles of *dpy-18* (*e364* and *e1096*) revealing a single point mutation and a deletion mutant, respectively. These two strains also synthesized a cuticle with a significantly reduced hydroxyproline content. The *phy-2* single gene disruptions produced no visible phenotype, whereas the combined deletion of *phy-1* and *phy-2* resulted in an embryonic lethal phenotype that was consistent with the inability of the cuticle to maintain normal worm morphology. These mutant embryos elongated normally and synthesized the first cuticle but then reverted to a disorganized twofold shape and eventually died (Winter and Page, 2000). This range of phenotypes is strikingly similar to the noted effects of prolyl 4-hydroxylase 2-oxoglutarate co-factor analogues, such as pyridine 2,5-dicarboxylate. This latter compound causes embryonic arrest, severe dumpy phenotypes and moulting defects in *C. elegans* strains (A.P. Page, 2000, unpublished observations). This confirms that proline hydroxylation is an essential modification of the nematode cuticle collagens.

Protein disulphide isomerase (PDI) (EC 5.2.1.8)

In addition to its role as the β-subunit of PHY, PDI acts independently by catalysing thiol/protein disulphide interchange. The role of PDI as the β-subunit in prolyl 4-hydroxylase is not related to its disulphide isomerase activity and experiments where the vertebrate PDI was mutated in both thioredoxin-like active domains had no effect on tetramer assembly (Vuori *et al.*, 1992). PDI appears to function as a molecular chaperone, retaining the α-subunits in the correct catalytically active, non-aggregated form in the ER-lumen (John *et al.*, 1993). Dissociation of the β-subunits results in insoluble aggregates of the α-subunits, analogous to α-subunits expressed in the absence of PDI. An additional function of PDI in the complex is to maintain the ER luminal location of the α-subunits, since deletion of the ER retention signal from PDI results in the secretion of the complex (Vuori *et al.*, 1992).

The formation and rearrangement of disulphide bridges is a major feature of collagen assembly, both at the early monomer registration stage prior to trimerization, and the latter cross-linking and gelling of the final cuticular matrix (Fig. 9.3). PDI may be involved in the correct registration and nucleation of the triple helix and would thus prevent mismatching of the collagen triple helix (Prockop and Kivirikko, 1995). This enzyme catalyses both intra- and interchain disulphide bond formation. Intrachain disulphide bonds are formed within the N and C propeptide regions and as the propagation proceeds some interchain bonds are also formed. This is especially true for nematode collagens that have conserved groups of cysteine residues which interrupt the Gly-X-Y repeats (see earlier). PDI has also been demonstrated to act as a molecular chaperone in collagen folding, since it interacts with procollagen chains, preventing premature assembly and aggregation (Wilson *et al.*, 1998). PDI therefore plays numerous roles in the folding and trimerization of collagen, acting as a disulphide bond catalyst, as the β-subunit for prolyl 4-hyroxylase and as a molecular chaperone during chain assembly.

The *C. elegans* genome contains two conserved isoforms of PDI (*C. elegans* Genome Sequencing Consortium, 1998). There are, however, many divergent homologues present which contain the associated thioredoxin-like active domains. The PDI homologue, PDI-1, is an excellent candidate in having a collagen-folding role in association with CYP-9 (Page, 1997). The second PDI homologue (*pdi-2*), has also been cloned and expressed from *C. elegans* mRNA, and possesses the associated disulphide isomerase activity (Veijola *et al.*, 1996).

As stated, PDI-2 was found to associate with the human α-subunit as an $\alpha_2\beta_2$ tetramer and the *C. elegans* PHY-1 α-subunit as an αβ dimer (Veijola *et al.*, 1996). In accordance with this biochemical association, genetic disruption of *pdi-2* via RNAi resulted in an identical phenotype to the combined *phy-1/phy-2* knockout, namely embryonic lethality which corresponded to a defective cuticle unable to maintain the shape of the elongated embryo (Winter and Page, 2000). Likewise, PDI-2 has an exclusive hypodermal spatial expression pattern, which temporally mirrors that of PHY-1 and PHY-2 (Winter and Page, 2000). This supports the defined functional role of PDI-2 in the complex formed with the hypodermal α-subunit PHY-1 (Veijola *et al.*, 1996), and establishes this to be an essential enzyme/chaperone in *C. elegans*.

PDI isoforms have also been described in parasitic nematodes, including *O. volvulus* (Wilson *et al.*, 1994), *Nippostrongylus brasiliensis* (Tweedie *et al.*, 1993), *Ancylostoma caninum* (Epe *et al.*, 1998) and *Dirofilaria immitis* (Chandrashekar *et al.*, 1998). However, their involvement in collagen-folding events in the respective organisms remains to be determined.

Lysyl hydroxylase (LHY) (EC 1.14.11.4)

LHY is active as a homodimer, catalysing the hydroxylation of lysine at an early stage in vertebrate collagen biosynthesis (Fig. 9.3) (Kivirikko and Pihlajaniemi, 1998), which is important for carbohydrate attachment and intermolecular collagen cross-link formation (Kivirikko *et al.*, 1992). Hydroxylysine has only been detected in cuticle extracts from the resistant dauer stage larvae of *C. elegans* (Cox *et al.*, 1981) but a single LHY encoding gene has been identified by the genome project (*C. elegans* Genome Sequencing Consortium, 1998), which has been established as having an essential role in the folding of basement membrane type IV collagens (K. Norman and D. Moerman, Vancouver, 1999, personal communication). This indicates that, perhaps, a different mode of cross-linking may exist in the cuticle collagens, e.g. disulphide bonding and tyrosine linkages. Significantly, the cuticle collagens of the parasitic nematode *Ascaris lumbricoides* do not contain cross-links through either lysine or hydroxylysine residues (Adams, 1978).

Non-collagenous Cuticle Structural Components: the Cuticulins

The second major class of cuticle structural components is the insoluble cuticulins. These highly cross-linked proteins constitute the insoluble material remaining after boiling cuticle preparations in strong reducing agents. Like the collagens, they possess a high content of glycine, proline and hydroxyproline (Cox *et al.*, 1981), but they are insensitive to collagenase and instead share many characteristics with the insoluble cuticle proteins of arthropods (Lassandro *et al.*, 1994; Bisoffi *et al.*, 1996). These structural proteins are located in the external cortical and epicuticular layers of the cuticle (Selkirk *et al.*, 1989; Bisoffi *et al.*, 1996), where they form a highly resilient impervious barrier, and in parasitic nematodes may be involved in the observed immunological inertness of these cuticle layers to the host.

The genes encoding these cuticle proteins (*cut* genes) have been characterized in *C. elegans* and two of them have been examined in detail. The first, *cut-1*, was found to encode a novel cysteine- and tyrosine-rich 40 kDa protein, which is expressed exclusively in the dauer stage. The encoded protein becomes insoluble once assembled into the cuticle and is expressed exclusively above the lateral seam cells in the cuticle cortex (Sebastiano *et al.*, 1991). This may contribute to the noted thickening of the dauer cuticle, making it impervious to 1% SDS (Cassada and Russell, 1975). Direct homologues of *cut-1* exist in the plant-parasitic nematode *Meloidogyne artiella* (DeGiorgi *et al.*, 1997) and the human parasites *A. lumbricoides* (Timinouni and Bazzicalupo, 1997; Favre *et al.*, 1998), *B. malayi* and *B. pahangi* (Lewis *et al.*, 1999).

The second characterized cuticulin gene, *cut-2*, encodes a novel 231 amino acid tyrosine-rich protein and is expressed in all stages of *C. elegans* (Lassandro *et al.*, 1994). A partial cuticulin gene sequence with a similar repeat motif to CUT-2 has been identified from cuticle preparations of the parasitic nematode *Ascaris suum* (Bisoffi *et al.*, 1996), but no direct homologues of CUT-2 have so far been detected in other parasitic nematodes. Recombinant CUT-2 can be crosslinked *in vitro* via dityrosine residues in the presence of horseradish peroxidase and H_2O_2, confirming the importance of di- and tri-tyrosine residues to the insoluble nature of these proteins (Fetterer and Rhoads, 1990; Lassandro *et al.*, 1994; Parise and Bazzicalupo, 1997). It may be speculated that these reactions are catalysed by tyrosinases, transglutaminase, PDIs and perhaps some of the abundant cuticular peroxidases and phenol oxidases common to parasitic nematodes, namely glutathione peroxidase (Cookson *et al.*, 1992), the filarial peroxidoxins (Lu *et al.*, 1998) and phenol oxidase (Fetterer and Hill, 1993). In addition to *cut-1* and *cut-2*-like genes, other potential cuticulins have now been detected in the *C. elegans* (*C. elegans* Genome Sequencing Consortium, 1998), *B. malayi* and *O. volvulus* genomes (Filarial Genome Project, 1999).

Cuticle Collagen/Cuticulin Cross-linking Enzymes

Tyrosinase (EC 1.14.18.1) and phenol oxidase (EC 1.10.3.1)

The nematode 2-ME soluble cuticle extracts (collagens) and more significantly the 2-ME insoluble extracts (cuticulins) contain tyrosine-derived cross-links. These have been most extensively studied in the parasitic nematodes *A. suum* (Fujimoto *et al.*, 1981; Sakura and Fujimoto, 1984) and *H. contortus* (Fetterer and Rhoads, 1990), taking the form of di-, tri- and isotrityrosine, and they contribute directly to the insoluble nature of these proteins. Isotrityrosine is a particularly unusual residue; it is unique to the nematode cuticle (Sakura and Fujimoto, 1984), and is hypothesized to be involved in linkage of three peptides (Fetterer and Rhoads, 1990). It is interesting to note that the available cuticulin sequences are particularly tyrosine-rich and that many of the numerous cuticle collagens have highly conserved C-terminal tyrosine residues, which may participate in such linkages. The mode of synthesis of these linkages is currently unknown but is presumed to be catalysed by enzymes such as tyrosinases, phenol oxidases or peroxidases. Interestingly, peroxidases are abundant cuticular enzymes in parasitic nematodes (McGonigle *et al.*, 1998; Selkirk *et al.*, 1998) and may be involved in this cross-linking process. Phenol oxidase activity has also been detected in the parasitic nematode *Trichuris suis* (Fetterer and Hill, 1993), but was shown to be involved in eggshell tanning. At least three tyrosinases and one phenol oxidase homologue are present in the *C. elegans* genome (M. Blaxter, Edinburgh, 1999, personal communication; *C. elegans* Genome Sequencing Consortium, 1998) but their role in collagen/ cuticulin cross-linking remains to be firmly established.

Transglutaminases (TGA) (EC 2.3.2.13)

The transglutaminases are calcium-dependent enzymes that catalyse the cross-linking of proteins by promoting the formation of isopeptide bonds between the γ-carboxyl group of a glutamine in one polypeptide chain and the ε-amino group of a lysine in the second (Greenberg *et al.*, 1991). These enzymes have previously been established as being important in the cross-linking of structural proteins such as fibrinogen (Murthy *et al.*, 1991) and collagen (Juprelle-Soret *et al.*, 1988). In nematodes this class of enzymes has been proposed to be involved in the cross-linking of the insoluble cuticulins (Lustigman, 1993), though the cuticular collagens may also be potential substrates. The importance of these enzymes to nematode cuticle development has been determined by examining the effects of specific inhibitors (Lustigman, 1993; Lustigman *et al.*, 1995). TGase inhibitors have indeed been found to affect nematode development, by inhibiting the moulting process in *O. volvulus* larvae (Lustigman *et al.*, 1995), microfilaria release in *B. malayi* (Mehta *et al.*, 1992) and growth and development of *B. malayi* and *Acanthocheilonema viteae* (Rao *et al.*, 1991). TGase activity has subsequently been detected in extracts of *B. malayi* (Singh and Mehta, 1994), *D. immitis* (Singh *et al.*, 1995) and *C. elegans* (Madi *et al.*, 1998), although to date no TGase homologues have been identified in either *B. malayi* or *C. elegans* by their respective genome projects.

A recent study in *D. immitis* isolated a TGase active enzyme that surprisingly had no homology to other TGases in the databases, being instead a PDI homologue. This enzyme exhibited both disulphide bond isomerase activity and inhibitable Ca-dependent TGase activity (Chandrashekar *et al.*, 1998). This represents an additional cuticle-specific role for the multifunctional PDI enzymes and establishes them as central enzymes in the folding and cross-linking of this exoskeleton.

Concluding Remarks

The nematode cuticle is a highly complex extracellular matrix, which predominantly comprises a complex mixture of collagens. This resilient structure has undoubtedly led to the widespread success of the Nematoda as a phylum. A further degree of complexity exists in the expression of the collagen-encoding genes and in their ultimate control. The modification of the collagens, most specifically the proline isomerization, proline hydroxylation and ultimate cross-linking, together with the enzymes involved in these processes, may represent the weak link in the nematode's otherwise impregnable coat of armour.

Acknowledgements

The author thanks Alan Winter (Glasgow) for critical comments and discussions regarding this chapter. This work is funded by the Medical Research Council through the award of an MRC Senior Fellowship.

References

Ackerman, M.S., Bhate, M., Shenoy, N., Beck, K., Ramshaw, J.A.M. and Brodsky, B. (1999) Sequence dependence of the folding of collagen-like peptides – single amino acids affect the rate of triple-helix nucleation. *Journal of Biological Chemistry* 274, 7668–7673.

Adams, E. (1978) Invertebrate collagens. *Science* 202, 591–598.

Annunen, P., Helaakoski, T., Myllyharju, J., Veijola, J., Pihlajaniemi, T. and Kivirikko, K.I. (1997) Cloning of the human prolyl 4-hydroxylase alpha subunit isoform alpha(II) and characterization of the type II enzyme tetramer – the alpha(I) and alpha(II) subunits do not form a mixed alpha(I)alpha(II)beta(2) tetramer. *Journal of Biological Chemistry* 272, 17342–17348.

Antebi, A., Culotti, J.G. and Hedgecock, E.M. (1998) *daf-12* regulates developmental age and the dauer alternative in *Caenorhabditis elegans*. *Development* 125, 1191–1205.

Bachinger, H.P. (1987) The influence of peptidyl-prolyl *cis-trans* isomerase on the *in vitro* folding of type III collagens. *Journal of Biological Chemistry* 262, 17144–17148.

Bachinger, H.P., Bruckner, P., Timpl, R., Prockop, D.J. and Engel, J. (1980) Folding mechanism of the triple helix in type-III collagen and type-III pN-collagen: role of disulfide bridges and peptide bond isomerisation. *European Journal of Biochemistry* 106, 619–632.

Barker, G.C. and Rees, H.H. (1990) Ecdysteroids in nematodes. *Parasitology Today* 16, 384–387.

Barker, G.C., Chitwood, D.J. and Rees, H.H. (1990) Ecdysteroids in helminths and annelids. *Invertebrate Reproduction and Development* 18, 1–11.

Bergmann, D.C., Crew, J.R., Kramer, J.M. and Wood, W.B. (1998) Cuticle chirality and body handedness in *Caenorhabditis elegans*. *Developmental Genetics* 23, 164–174.

Bird, A.F. and Bird, J. (1991) *The Structure of Nematodes*, 2nd edn. Academic Press, San Diego.

Bisoffi, M. and Betschart, B. (1996) Identification and sequence comparison of a cuticular collagen of *Brugia pahangi*. *Parasitology* 113, 145–155.

Bisoffi, M., Marti, S. and Betschart, B. (1996) Repetitive peptide motifs in the cuticlin of *Ascaris suum*. *Molecular and Biochemical Parasitology* 80, 55–64.

Blumenthal, T. and Steward, K. (1997) RNA processing and gene structure. In: Riddle, D.L., Blumenthal, T., Meyer, B.J. and Priess, J.R. (eds) *C. elegans II*. Cold Spring Harbor Laboratory Press, Cold Spring Harbor, New York, pp. 117–145.

Bonadio, J. and Byers, P.H. (1985) Subtle structural alterations in the chains of type-I procollagen produce osteogenesis imperfecta type-II. *Nature* 316, 363–366.

Brenner, S. (1974) The genetics of *Caenorhabditis elegans*. *Genetics* 77, 71–94.

C. elegans Genome Sequencing Consortium (1998) Genome sequence of the nematode *C-elegans*: a platform for investigating biology. *Science* 282, 2012–2018.

Cassada, R. and Russell, R. (1975) The dauer larva, a post-embryonic developmental variant of the nematode *Caenorhabditis elegans*. *Developmental Biology* 46, 326–342.

Chandrashekar, R., Tsuji, N., Morales, T., Ozols, V. and Mehta, K. (1998) An ERp60-like protein from the filarial parasite *Dirofilaria immitis* has both transglutaminase and protein disulfide isomerase activity. *Proceedings of the National Academy of Sciences USA* 95, 531–536.

Chappell, L.H. and Wastling, J.L. (1992) Cyclosporin A: antiparasitic drug, modulator of the host–parasite relationship and immunosuppressant. *Parasitology* 105, S25–S40.

Chitwood, D.J. and Feldlaufer, M.F. (1990) Ecdysteroids in axenically propagated *Caenorhabditis elegans* and culture-medium. *Journal of Nematology* 22, 598–607.

Colley, N., Baker, E., Stamnes, M. and Zuker, C. (1991) The cyclophilin homolog *ninaA* is required in the secretory pathway. *Cell* 67, 255–263.

Cookson, E., Blaxter, M. and Selkirk, M. (1992) Identification of the major soluble cuticular glycoprotein of lymphatic filarial nematode parasites (gp29) as a secretory homolog of glutathione-peroxidase. *Proceedings of the National Academy of Sciences USA* 89, 5837–5841.

Cox, G.N. (1990) Molecular biology of the cuticle collagen gene families of *Caenorhabditis elegans* and *Haemonchus contortus*. *Acta Tropica* 47, 269–281.

Cox, G.N. (1992) Molecular and biochemical aspects of nematode cuticle collagens. *Journal of Parasitology* 78, 1–15.

Cox, G.N. and Hirsh, D. (1985) Stage-specific patterns of collagen gene-expression during development of *Caenorhabditis elegans*. *Molecular and Cellular Biology* 5, 363–372.

Cox, G.N., Kusch, M. and Edgar, R.S. (1981) Cuticle of *Caenorhabditis elegans*: its isolation and partial characterisation. *Journal of Cell Biology* 90, 7–17.

Davis, E.C., Broekelmann, T.J., Ozawa, Y. and Mecham, R.P. (1998) Identification of tropoelastin as a ligand for the 65-kD FK506-binding protein, FKBP65, in the secretory pathway. *Journal of Cell Biology* 140, 295–303.

DeGiorgi, C., DeLuca, F., DiVito, M. and Lamberti, F. (1997) Modulation of expression at the level of splicing of *cut-1* RNA in the infective second-stage juvenile of the plant parasitic nematode *Meloidogyne artiellia*. *Molecular and General Genetics* 253, 589–598.

Dornan, J., Page, A.P., Taylor, P., Wu, S., Winter, A.D., Husi, H. and Walkinshaw, M.D. (1999) Biological, biochemical and structural characterisation of a 'divergent loop' cyclophilin from *Caenorhabditis elegans*. *Journal of Biological Chemistry* 274, 34877–34883.

Engel, J. and Prockop, D.J. (1991) The zipper-like folding of collagen triple helices and the effects of mutations that disrupt the zipper. *Annual Review of Biophysics and Biophysical Chemistry* 20, 137–152.

Epe, C., Kohlmetz, C. and Schnieder, T. (1998) A recombinant protein disulfide isomerase homologue from *Ancylostoma caninum*. *Parasitology Research* 84, 763–766.

Favre, R., Cermola, M., Nunes, C.P., Hermann, R., Muller, M. and Bazzicalupo, P. (1998) Immuno-cross-reactivity of CUT-1 and cuticlin epitopes between *Ascaris lumbricoides, Caenorhabditis elegans,* and *Heterorhabditis*. *Journal of Structural Biology* 123, 1–7.

Fetterer, R.H. and Hill, D.E. (1993) The occurrence of phenol oxidase activity in female *Trichuris suis*. *Journal of Parasitology* 79, 155–159.

Fetterer, R.H. and Rhoads, M.L. (1990) Tyrosine-derived cross-linking amino-acids in the sheath of *Haemonchus contortus* infective larvae. *Journal of Parasitology* 76, 619–624.

Fetterer, R.H., Rhoads, M.L. and Urban, J.F. (1993) Synthesis of tyrosine-derived cross-links in *Ascaris suum* cuticular proteins. *Journal of Parasitology* 79, 160–166.

Filarial Genome Project (1999) Deep within the filarial genome: progress of the filarial genome project. *Parasitology Today* 15, 219–224.

Fujimoto, D., Horiuchi, K. and Hirama, M. (1981) Isotrityrosine, a new crosslinking amino-acid isolated from *Ascaris* cuticle collagen. *Biochemical and Biophysical Research Communications* 99, 637–643.

Gamble, H.R., Purcell, J.P. and Fetterer, R.H. (1989) Purification of a 44 kilodalton protease which mediates the ecdysis of infective *Haemonchus contortus* larvae. *Molecular and Biochemical Parasitology* 33, 49–58.

Gothel, S.F. and Marahiel, M.A. (1999) Peptidyl-prolyl *cis-trans* isomerases, a superfamily of ubiquitous folding catalysts. *Cellular and Molecular Life Sciences* 55, 423–436.

Greenberg, C.S., Birckbichler, P.J. and Rice, R.H. (1991) Transglutaminases – multifunctional cross-linking enzymes that stabilize tissues. *FASEB Journal* 5, 3071–3077.

Hong, X., Ma, D. and Carlow, K.S.C. (1998a) Cloning expression and characterisation of a new filarial cyclophilin. *Molecular and Biochemical Parasitology* 91, 353–358.

Hong, X.Q., Ma, D., Page, A.P., Kumar, S. and Carlow, C.K.S. (1998b) Highly conserved large molecular weight cyclophilin of filarial parasites. *Experimental Parasitology* 88, 246–251.

John, D.C.A., Grant, M.E. and Bulleid, N.J. (1993) Cell-free synthesis and assembly of prolyl 4-hydroxylase – the role of the beta-subunit (pdi) in preventing misfolding and aggregation of the alpha-subunit. *EMBO Journal* 12, 1587–1595.

Johnstone, I.L. (1994) The cuticle of the nematode *Caenorhabditis elegans* – a complex collagen structure. *BioEssays* 16, 171–178.

Johnstone, I.L. (1999) http://www.worms.gla.ac.uk/collagen/cecolgenes.htm

Johnstone, I.L. and Barry, J.D. (1996) Temporal reiteration of a precise gene expression pattern during nematode development. *EMBO Journal* 15, 3633–3639.

Johnstone, I.L., Shafi, Y. and Barry, J.D. (1992) Molecular analysis of mutations in the *Caenorhabditis elegans* collagen genes *dpy-7*. *EMBO Journal* 11, 3857–3863.

Johnstone, I.L., Shafi, Y., Majeed, A. and Barry, J.D. (1996) Cuticular collagen genes from the parasitic nematode *Ostertagia circumcincta*. *Molecular and Biochemical Parasitology* 80, 103–112.

Jones, J.T., Curtis, R.H., Wightman, P.J. and Burrows, P.R. (1996) Isolation and characterization of a putative collagen gene from the potato cyst nematode *Globodera pallida*. *Parasitology* 113, 581–588.

Juprelle-Soret, M., Wattiaux-Deconinck, S. and Wattiaux, R. (1988) Subcellular-localization of transglutaminase – effect of collagen. *Biochemical Journal* 250, 421–427.

Kingston, B.I. (1991) Nematode collagen genes. *Parasitology Today* 7, 11–15.

Kingston, B.I., Wainwright, S.M. and Cooper, D. (1989) Comparison of collagen gene-sequences in *Ascaris suum* and *Caenorhabditis elegans*. *Molecular and Biochemical Parasitology* 37, 137–146.

Kivirikko, K.I. and Pihlajaniemi, T. (1998) Collagen hydroxylases and the protein disulfide isomerase subunit of prolyl 4-hydroxylases. *Advances in Enzymology and Related Areas of Molecular Biology* 72, 325–400.

Kivirikko, K.I., Myllyla, R. and Pihlajaniemi, T. (1992) *Post Translational Modification of Proteins*. CRC Press, Boca Raton, Florida, pp. 1–51.

Koltai, H., Chejanovsky, N., Raccah, B. and Spiegel, Y. (1997) The first isolated collagen gene of the root-knot nematode *Meloidogyne javanica* is developmentally regulated. *Gene* 196, 191–199.

Kostrouchova, M., Krause, M., Kostrouch, Z. and Rall, J.E. (1998) CHR3: a *Caenorhabditis elegans* orphan nuclear hormone receptor required for proper epidermal development and molting. *Development* 125, 1617–1626.

Kramer, J.M. (1994) Structures and functions of collagens in *Caenorhabditis elegans*. *FASEB Journal* 8, 329–336.

Kramer, J.M. (1997) Extracellular matrix. In: Riddle, D.L., Blumenthal, T., Meyer, B.J. and Priess, J.R. (eds) *C. elegans II*. Cold Spring Harbor Laboratory Press, Cold Spring Harbor, New York, pp. 471–500.

Kramer, J.M. and Johnson, J.J. (1993) Analysis of mutations in the *sqt-1* and *rol-6* collagen genes of *Caenorhabditis elegans*. *Genetics* 135, 1035–1045.

Kramer, J.M., French, R.P., Park, E.C. and Johnson, J.J. (1990) The *Caenorhabditis elegans rol-6* gene, which interacts with the *sqt-1* collagen gene to determine organismal morphology, encodes a collagen. *Molecular and Cellular Biology* 10, 2081–2089.

Kusch, M. and Edgar, R.S. (1986) Genetic studies of unusual loci that affect body shape of the nematode *Caenorhabditis elegans* and may code for cuticle structural proteins. *Genetics* 113, 621–639.

Lamberg, A., Helaakoski, T., Myllyharju, J., Peltonen, S., Notbohm, H., Pihlajaniemi, T. and Kivirikko, K.I. (1996) Characterization of human type III collagen expressed in a baculovirus system – production of a protein with a stable triple helix requires coexpression with the two types of recombinant prolyl 4-hydroxylase subunit. *Journal of Biological Chemistry* 271, 11988–11995.

Lassandro, F., Sebastiano, M., Zei, F. and Bazzicalupo, P. (1994) The role of dityrosine formation in the cross-linking of *cut-2*, the product of a 2nd cuticlin gene of *Caenorhabditis elegans*. *Molecular and Biochemical Parasitology* 65, 147–159.

Levy, A.D., Yang, J. and Kramer, J.M. (1993) Molecular and genetic analyses of the *Caenorhabditis elegans dpy-2* and *dpy-10* collagen genes – a variety of molecular alterations affect organismal morphology. *Molecular Biology of the Cell* 4, 803–817.

Lewis, E., Hunter, S., Tetley, L., Nunes, C., Bazzicalupo, P. and Devaney, E. (1999) *cut-1*-like genes are present in the filarial nematodes, *Brugia pahangi* and *Brugia malayi*, and, as in other nematodes, code for components of the cuticle. *Molecular and Biochemical Parasitology* 101, 173–183.

Lu, W.H., Egerton, G.L., Bianco, A.E. and Williams, S.A. (1998) Thioredoxin peroxidase from *Onchocerca volvulus*: a major hydrogen peroxide detoxifying enzyme in filarial parasites. *Molecular and Biochemical Parasitology* 91, 221–235.

Lustigman, S. (1993) Molting, enzymes and new targets for chemotherapy of *Onchocerca volvulus*. *Parasitology Today* 9, 294–297.

Lustigman, S., Brotman, B., Huima, T., Prince, A.M. and McKerrow, J.H. (1992) Molecular-cloning and characterization of onchocystatin, a cysteine proteinase-inhibitor of *Onchocerca volvulus*. *Journal of Biological Chemistry* 267, 17339–17346.

Lustigman, S., Brotman, B., Huima, T., Castelhano, A.L., Singh, R.N., Mehta, K. and Prince, A.M. (1995) Transglutaminase-catalyzed reaction is important for molting of *Onchocera volvulus* 3rd-stage larvae. *Antimicrobial Agents and Chemotherapy* 39, 1913–1919.

Ma, D., Hong, X., Raghavan, N., Scott, A.L., McCarthy, J.S., Nutman, T.B., Williams, S.A. and Carlow, C.K.S. (1996) A cyclosporin A-sensitive small molecular weight cyclophilin of filarial parasites. *Molecular and Biochemical Parasitology* 79, 235–241.

Madi, A., Punyiczki, M., DiRao, M., Piacentini, M. and Fesus, L. (1998) Biochemical characterization and localization of transglutaminase in wild-type and cell-death mutants of the nematode *Caenorhabditis elegans*. *European Journal of Biochemistry* 253, 583–590.

Mangelsdorf, D.J., Thummel, C., Beato, M., Herrlich, P., Schutz, G., Umesono, K., Blumberg, B., Kastner, P., Mark, M., Chambon, P. and Evans, R.M. (1995) The nuclear receptor superfamily – the 2nd decade. *Cell* 83, 835–839.

McGonigle, S., Dalton, J.P. and James, E.R. (1998) Peroxidoxins: a new antioxidant family. *Parasitology Today* 14, 139–145.

Mehta, K., Rao, U.R., Vickery, A.C. and Fesus, L. (1992) Identification of a novel transglutaminase from the filarial parasite *Brugia malayi* and its role in growth and development. *Molecular and Biochemical Parasitology* 53, 1–15.

Movsowitz, C., Epstein, S., Fallon, M., Ismail, F. and Thomas, S. (1988) Cyclosporin-A *in vivo* produces severe osteopenia in the rat – effect of dose and duration of administration. *Endocrinology* 123, 2571–2577.

Murthy, S.N.P., Wilson, J., Guy, S.L. and Lorand, L. (1991) Intramolecular cross-linking of monomeric fibrinogen by tissue transglutaminase. *Proceedings of the National Academy of Sciences USA* 88, 10601–10604.

Nagata, K. (1996) Hsp47: a collagen-specific molecular chaperone. *Trends in Biochemical Sciences* 21, 23–26.

Page, A.P. (1997) Cyclophilin and protein disulphide isomerase genes are co-transcribed in a functionally related manner in *Caenorhabditis elegans*. *DNA and Cell Biology* 16, 1335–1343.

Page, A.P. (1999) A highly conserved nematode protein folding operon in *Caenorhabditis elegans* and *Caenorhabditis briggsae*. *Gene* 230, 267–275.

Page, A.P. and Winter, A.D. (1998) A divergent multi-domain cyclophilin is highly conserved between parasitic and free-living nematode species and is important in larval muscle development. *Molecular and Biochemical Parasitology* 95, 215–227.

Page, A.P. and Winter, A.D. (1999) Expression pattern and functional significance of a divergent nematode cyclophilin in *Caenorhabditis elegans*. *Molecular and Biochemical Parasitology* 99, 301–306.

Page, A.P., Kumar, S. and Carlow, C.K.S. (1995a) Parasite cyclophilins and antiparasite activity of cyclosporin A. *Parasitology Today* 11, 385–388.

Page, A.P., Landry, D., Wilson, G.G. and Carlow, C.K.S. (1995b) Molecular characterization of a cyclosporin A-insensitive cyclophilin from the parasitic nematode *Brugia malayi*. *Biochemistry* 34, 11545–11550.

Page, A.P., MacNiven, K. and Hengartner, M.O. (1996) Cloning and biochemical characterisation of the cyclophilin homologues from the free-living nematode *Caenorhabditis elegans*. *Biochemical Journal* 317, 179–185.

Parise, G. and Bazzicalupo, P. (1997) Assembly of nematode cuticle: role of hydrophobic interactions in CUT-2 cross-linking. *Biochimica et Biophysica Acta – Protein Structure and Molecular Enzymology* 1337, 295–301.

Park, Y.S. and Kramer, J.M. (1994) The *C. elegans sqt-1* and *rol-6* collagen genes are coordinately expressed during development, but not at all stages that display mutant phenotypes. *Developmental Biology* 163, 112–124.

Peters, K., McDowall, J. and Rose, A.M. (1991) Mutations in the *bli-4* (I) locus of *Caenorhabditis elegans* disrupt both adult cuticle and early larval development. *Genetics* 129, 95–102.

Pihlajaniemi, T., Helaakoski, T., Tasanen, K., Myllyla, R., Huhtal, M.-L., Koivu, J.K. and Kivirikko, K.I. (1987) Molecular cloning of the β-subunit of human prolyl 4-hydroxylase. This subunit and protein disulphide isomerase are products of the same gene. *EMBO Journal* 6, 643–649.

Prockop, D.J. and Kivirikko, K.I. (1995) Collagens: molecular biology, diseases, and potential for therapy. *Annual Review of Biochemistry* 64, 403–434.

Rao, U.R., Mehta, K., Subrahmanyam, D. and Vickery, A.C. (1991) *Brugia malayi* and *Acanthocheilonema viteae* – antifilarial activity of transglutaminase inhibitors *in vitro*. *Antimicrobial Agents and Chemotherapy* 35, 2219–2224.

Richer, J.K., Sakanari, J.A., Frank, G.R. and Grieve, R.B. (1992) *Dirofilaria immitis* – proteases produced by 3rd-stage and 4th-stage larvae. *Experimental Parasitology* 75, 213–222.

Rogers, W. (1982) Enzymes in the exsheathing fluid of nematodes and their biological significance. *International Journal for Parasitology* 12, 495–502.

Sakura, S. and Fujimoto, D. (1984) Absorbtion and fluorescence studies of tyrosine derived crosslinking amino acids from collagen. *Photochemistry and Photobiology* 40, 731–734.

Sakwe, A.M. and Titanji, V.P.K. (1997) Evidence for increased hydroxylation of pyrrolidone amino acid residues in the cuticle of mature *Onchocerca volvulus*. *Biochimica et Biophysica Acta – Molecular Basis of Disease* 1360, 196–202.

Schonbrunner, E.R. and Schmid, F.X. (1992) Peptidyl-prolyl *cis-trans* isomerase improves the efficiency of protein disulfide isomerase as a catalyst of protein folding. *Proceedings of the National Academy of Sciences USA* 89, 4510–4513.

Schreiber, S.L. and Crabtree, G.R. (1992) The mechanism of action of cyclosporin A and FK506. *Immunology Today* 13, 136–142.

Sebastiano, M., Lassandro, F. and Bazzicalupo, P. (1991) *Cut-1* a *Caenorhabditis elegans* gene coding for a dauer-specific noncollagenous component of the cuticle. *Developmental Biology* 146, 519–530.

Selkirk, M.E., Nielsen, L., Kelly, C., Partono, F., Sayers, G. and Maizels, R.M. (1989) Identification, synthesis and immunogenicity of cuticular collagens from the filarial nematodes *Brugia malayi* and *Brugia pahangi*. *Molecular and Biochemical Parasitology* 32, 229–246.

Selkirk, M.E., Smith, V.P., Thomas, G.R. and Gounaris, K. (1998) Resistance of filarial nematode parasites to oxidative stress. *International Journal for Parasitology* 28, 1315–1332.

Shamansky, M.L., Pratt, P., Boisvenue, R.J. and Cox, G.N. (1989) Cuticle collagen genes of *Haemonchus contortus* and *Caenorhabditis elegans* are highly conserved. *Molecular and Biochemical Parasitology* 37, 73–86.

Singh, N. and Soulston, J. (1978) Some observations on molting in *C. elegans*. *Nematologica* 24, 63–71.

Singh, R.N. and Mehta, K. (1994) Purification and characterization of a novel transglutaminase from filarial nematode *Brugia malayi*. *European Journal of Biochemistry* 225, 625–634.

Singh, R.N., Chandrashekar, R. and Mehta, K. (1995) Purification and partial characterization of a transglutaminase from dog filarial parasite, *Dirofilaria immitis*. *International Journal of Biochemistry and Cell Biology* 27, 1285–1291.

Sluder, A.E., Mathews, S.W., Hough, D., Yin, V.P. and Maina, C.V. (1999) The nuclear receptor superfamily has undergone extensive proliferation and diversification in nematodes. *Genome Research* 9, 103–120.

Smith, T., Ferreira, L.R., Herbert, C., Norris, K. and Sauk, J.J. (1995) Hsp47 and cyclophilin B traverse the endoplasmic reticulum with procollagen into pre-golgi intermediate vesicles; a role for Hsp47 and cyclophilin B in the export of procollagen from the endoplasmic reticulum. *Journal of Biological Chemistry* 270, 18323–18328.

Stammes, M.A., Shieh, B.H., Chuman, L., Harris, G.L. and Zuker, C.S. (1991) The cyclophilin homolog *ninaA* is a tissue-specific integral membrane protein required for the proper synthesis of a subset of *Drosophila* rhodopsins. *Cell* 65, 219–227.

Steinmann, B., Bruckner, P. and Superti-Furga, A. (1991) Cyclosporin A slows collagen triple-helix formation *in vivo*: indirect evidence for a physiological role of peptidyl prolyl *cis-trans* isomerase. *Journal of Biological Chemistry* 266, 1299–1303.

Takahashi, N., Hayano, T. and Suzuki, M. (1989) Peptidyl-prolyl *cis-trans* isomerase is the cyclosporin A-binding protein cyclophilin. *Nature* 337, 473–475.

Thacker, C., Peters, K., Srayko, M. and Rose, A.M. (1995) The *bli-4* locus of *Caenorhabditis elegans* encodes structurally distinct kex2/subtilisin-like endoproteases essential for early development and adult morphology. *Genes and Development* 9, 956–971.

Timinouni, M. and Bazzicalupo, P. (1997) *cut-1*-like genes of *Ascaris lumbricoides*. *Gene* 193, 81–87.

Tweedie, S., Grigg, M.E., Ingram, L. and Selkirk, M.E. (1993) The expression of a small heat-shock protein homolog is developmentally-regulated in *Nippostrongylus brasiliensis*. *Molecular and Biochemical Parasitology* 61, 149–154.

Vanderkeyl, H., Kim, H., Espey, R., Oke, C.V. and Edwards, M.K. (1994) *Caenorhabditis elegans sqt-3* mutants have mutations in the *col-1* collagen gene. *Developmental Dynamics* 201, 86–94.

Veijola, J., Koivunen, P., Annunen, P., Pihlajaniemi, T. and Kivirikko, K. (1994) Cloning, baculovirus expression, and characterization of the alpha-subunit of prolyl 4-hydroxylase from the nematode *Caenorhabditis elegans* – this alpha-subunit forms an active alpha-beta dimer with the human protein disulfide-isomerase beta-subunit. *Journal of Biological Chemistry* 269, 26746–26753.

Veijola, J., Annunen, P., Koivunen, P., Page, A.P., Philajaniemi, T. and Kivirikko, K.I. (1996) Baculovirus expression of two protein disulphide isomerase isoforms from *Caenorhabditis elegans* and characterisation of prolyl 4-hydroxylases containing one of these polypeptides as their B subunit. *Biochemical Journal* 317, 721–729.

VonMende, N., Bird, D.M., Albert, P.S. and Riddle, D.L. (1988) *dpy-13* – a nematode collagen gene that affects body shape. *Cell* 55, 567–576.

Vuorela, A., Myllyharju, J., Nissi, R., Pihlajaniemi, T. and Kivirikko, K.I. (1997) Assembly of human prolyl 4-hydroxylase and type III collagen in the yeast *Pichia pastoris*: formation of a stable enzyme tetramer requires coexpression with collagen and assembly of a stable collagen requires coexpression with prolyl 4-hydroxylase. *EMBO Journal* 16, 6702–6712.

Vuori, K., Pihlajaniemi, T., Myllyla, R. and Kivirikko, K.I. (1992) Site-directed mutagenesis of human protein disulfide isomerase – effect on the assembly, activity and endoplasmic-reticulum retention of human prolyl 4-hydroxylase in *Spodoptera frugiperda* insect cells. *EMBO Journal* 11, 4213–4217.

Vuorio, E. and DeCrombrugghe, B. (1990) The family of collagen genes. *Annual Review of Biochemistry* 59, 837–872.

Walmsley, A.R., Batten, M.R., Lad, U. and Bulleid, N.J. (1999) Intracellular retention of procollagen within the endoplasmic reticulum is mediated by prolyl 4-hydroxylase. *Journal of Biological Chemistry* 274, 14884–14892.

Wang, T.Y., Deom, C.M. and Hussey, R.S. (1998) Identification of a *Meloidogyne incognita* cuticle collagen gene and characterization of the developmental expression of three collagen genes in parasitic stages. *Molecular and Biochemical Parasitology* 93, 131–134.

Wilson, R., Lees, J.F. and Bulleid, N.J. (1998) Protein disulfide isomerase acts as a molecular chaperone during the assembly of procollagen. *Journal of Biological Chemistry* 273, 9637–9643.

Wilson, W.R., Tuan, R.S., Shepley, K.J., Freedman, D.O., Greene, B.M., Awadzi, K. and Unnasch, T.R. (1994) The *Onchocerca volvulus* homologue of the multifunctional polypeptide protein disulfide isomerase. *Molecular and Biochemical Parasitology* 68, 103–117.

Winter, A.D. and Page, A.P. (2000) Prolyl 4-hydroxylase is an essential procollagen modifying enzyme required for exoskeleton formation and the maintenance of body shape in the nematode *Caenorhabditis elegans*. *Molecular and Cellular Biology* 20, 4084–4093.

Yang, J. and Kramer, J.M. (1994) *In vitro* mutagenesis of *Caenorhabditis elegans* cuticle collagens identifies a potential subtilisin-like protease cleavage site and demonstrates that carboxyl domain disulfide bonding is required for normal function but not assembly. *Molecular and Cellular Biology* 14, 2722–2730.

Yochem, J., Tuck, S., Greenwald, I. and Han, M. (1999) A gp330/megalin-related protein is required in the major epidermis of *Caenorhabditis elegans* for completion of molting. *Development* 126, 597–606.

Zeng, B.F., MacDonald, J.R., Bann, J.G., Reck, K., Gambee, J.E., Boswell, B.A. and Bachinger, H.P. (1998) Chicken FK506-binding protein, FKBP65, a member of the FKBP family of peptidylprolyl *cis-trans* isomerases, is only partially inhibited by FK506. *Biochemical Journal* 330, 109–114.

Chitinases of Filarial Nematodes

Ralf Adam, Birgit Drabner and Richard Lucius

Department of Molecular Parasitology, Institute of Biology, Humboldt University, Berlin, Germany

Introduction

Chitinases are a group of enzymes that hydrolyse chitin, a homopolymer of poly-β(1–4)-linked *N*-acetylglucosamine monomers. Chitin is part of the exoskeleton of arthropods and is also found in various fungi and bacteria. It has been described as a constituent of the nematode eggshell and eggshell-derived structures (Wharton 1983; Brydon *et al.*, 1987). Thus, it is mandatory that nematodes possess enzymes to cleave chitin during the escape of the first larval stage from the egg and it was indeed shown that a chitinase is active in this process (Rogers, 1958). Chitinase activity is in fact associated with nematode eggs (Ward and Fairbairne, 1972), uterine stages (Justus and Ivey, 1969) and female worms (Gooday *et al.*, 1988). The description of filarial chitinases as prominent antigens with a possible role in protective immunity provided a stimulus to study these proteins further at the molecular level. Filarial chitinases of *Brugia malayi, Wuchereria bancrofti, Acanthocheilonema viteae* and *Onchocerca volvulus* were cloned, expressed and tested in immunization experiments. Some of these experiments established a protective potential for recombinant filarial chitinases, but much has to be done to evaluate this potential further.

The Role of Filarial Chitinases in Microfilariae

The hatching of the first larval stage of filarial nematodes, the microfilaria (mf), from the eggshell is quite different from the process in other nematodes, since these parasites are viviparous; in addition, the timing of the eclosion from the eggshell differs between species. Some filarial species produce mf which remain fully surrounded by a modified eggshell (the sheath) until they are taken up by a suitable arthropod host. In other species, the structure corresponding to the eggshell is a thin envelope that is cast by nearly mature mf in the proximal part of the uterus before birth. These mf are termed unsheathed. Eggshells of filarial worms have been proved to contain chitin (Brydon *et al.*, 1987). If chitinases were involved in the process of hatching, they should appear at time points when chitin degradation is likely to occur. In fact, chitinases from species with sheathed mf were described from blood mf (Fuhrman, 1995a) while chitinases of species with unsheathed mf have been described from uterine stages in *Onchocerca* spp. and *A. viteae* (Brydon *et al.*, 1987; Adam *et al.*, 1996).

A protein that was later identified as mf chitinase was first characterized by Canlas *et al.* (1984) from *B. malayi*. The authors described the 70 and 75 kDa target proteins of a monoclonal antibody (mAb), MF1, on the surface of the sheath of *B. malayi* mf. The expression of the target antigens was dependent on the age of the mf, as intrauterine or newly born mf contained low amounts of the protein at best, while the protein increased in abundance in mf after several days of residence in the vertebrate host (Fuhrman *et al.*, 1987). The onset of chitinase production coincided with the ability of the mf to penetrate the midgut of the mosquito vector after transmission and to develop subsequently to infective third-stage larvae (L3) (Fuhrman *et al.*, 1987). The cDNA sequence of the MF1 target antigen was described by Fuhrman *et al.* (1992) and the homologies to bacterial and yeast chitinases indicated that the protein was an mf endochitinase. This function was confirmed by the demonstration of chitinolytic activity.

The coincidence between the appearance of chitinase on the mf sheath and the ability to infect the arthropod vector is indicative of a role for the protein in eclosion of the mf from its sheath. In addition, chitinases could theoretically be required to infect the mosquito vector, but it seems that mf penetrate the mosquito midgut before the chitinous peritrophic membrane is established (Perrone and Spielman, 1986) and so do not have to traverse a chitin barrier. An alternative function would be the diversion of host defence mechanisms within the arthropod host by production of *N*-acetylglucosamine (GlcNAc). An interesting experiment demonstrated a higher penetration rate of *Brugia pahangi* mf from the midgut into the haemocoel of *Aedes aegypti* after supplementation of the blood meal with GlcNAc (Ham *et al.*, 1991). The authors postulated that a potential antimicrofilarial GlcNAc-specific lectin in the midgut could have been saturated and thus rendered inefficient by the GlcNAc supplement. This would represent an analogy to the function of chitinases of rickettsia-like endosymbionts of tsetse flies which enhance the infectivity of the trypanosomes within the blood meal (Welburn *et al.*, 1993).

The pattern of mf chitinase expression of *A. viteae*, a filarial species with unsheathed mf, was characterized with two mAbs (24-4 and 2H2) raised against *A. viteae* L3 chitinase by Adam *et al.* (1996). Immunofluorescent antibody tests (IFATs) with cryostat sections and uterine contents of adult female worms revealed that the target epitopes of 24-4 were accessible in the distal parts of the uterus, on the surface of nearly mature uterine mf which were about to cast the eggshell (Fig. 10.1). The surface of younger uterine stages, intact newborn mf or blood mf did not bear the target epitope, while degenerated newborn mf stained positively. The expression of chitinase exclusively on the surface of nearly mature uterine mf is

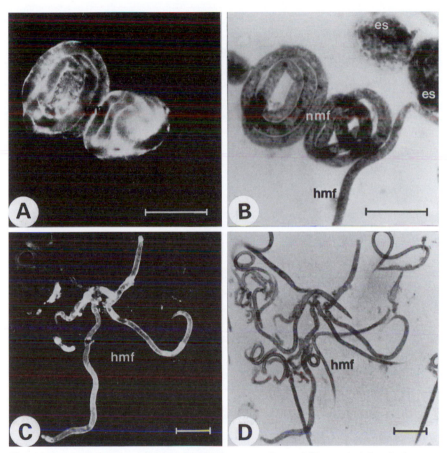

Fig. 10.1. Localization of *Acanthocheilonema viteae* chitinase in intrauterine and newborn mf by IFAT with mAb 24-4 (left panels) and corresponding light microscope photographs (right panels). (A), (B) Uterine contents of female *A. vitoae* with nearly mature mf inside the eggshell, younger embryonic stages and a mature, hatched mf. (C), (D) Newborn mf. Note that fluorescent mf are swollen and stumpy. Bars represent 50 μm; es, embryonic stages; nmf, nearly mature mf; hmf, hatched mf.

compatible with a role in cleaving the eggshell. Immunogold localization on ultra-thin sections showed that chitinase epitopes were present within the cuticle, but not on the surface of younger uterine stages (Fig. 10.2). The chitinase within the cuticle of immature uterine mf could represent a storage form which is transported to the mf surface prior to the hatching process. MAb 2H2 recognized target proteins of 220 kDa and 26 kDa in uterine contents of female *A. viteae*, but chitinase activity was detectable only in the 220 kDa molecule, suggesting that the smaller protein is a degradation product. An absence of chitinase in circulating mf of sheathless filarial species was also shown for *Onchocerca* spp. and *Dirofilaria immitis* (Fuhrmann, 1995a,b).

The Role of Chitinases in the Infective Larva Stage

L3 chitinases have been characterized from *A. viteae* (Adam *et al.*, 1996; Wu *et al.*, 1996) and *O. volvulus* (Wu *et al.*, 1996). The *A. viteae* L3 chitinase was studied using the same anti-*A. viteae* chitinase mAbs, 24-4 and 2H2,

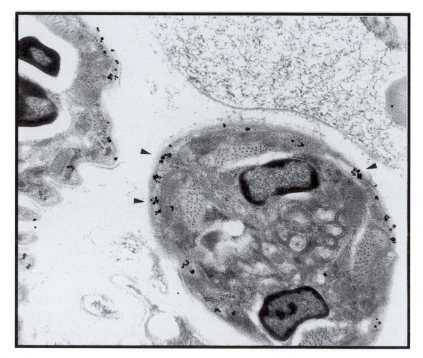

Fig. 10.2. Immunogold staining of an ultra-thin section of an immature *Acanthocheilonema viteae* uterine mf with mAb 24-4. Note that *A. viteae* chitinase is present in the cuticle (arrowheads), but not on the surface. (Photograph: W. Rudin.)

as described above (Adam *et al.*, 1996). MAb 2H2 recognized bands of 68 kDa and 205 kDa in immunoblots with L3 antigen and in L3 culture supernatant, while the other mab, 24-4, only recognized the 205 kDa band. The N-terminal protein sequence of both proteins revealed 20 identical amino acids with homology to the endochitinase of *B. malayi* (Fuhrman *et al.*, 1992). The quantity of 205 kDa material decreased with a concomitant increase of 68 kDa material after extensive treatment of L3 proteins with reducing agents, suggesting that the 205 kDa form is a trimer of disulphide-linked 68 kDa monomers. In substrate activity gels (Trudel and Asselin, 1989) only the 68 kDa protein was found to be enzymatically active. Thus, mAb 2H2 recognizes an epitope accessible on both, the active and the inactive form of chitinase, while the target epitope of mAb 24-4 is accessible only on the inactive presumptive L3 chitinase trimer. Immunogold staining of ultra-thin sections of L3 with mAbs 24-4 and 2H2 revealed *A. viteae* chitinase in the cellular cytoplasm and in the lumen of the glandular oesophagus of vector-derived L3 (Adam *et al.*, 1996). Kaltmann (1990) determined that vector-derived *A. viteae* L3 released the target epitope of 24-4 under vertebrate culture conditions at 37°C mainly during the first 24 h after isolation from ticks. Such release was reduced at 28°C and no release occurred when the L3s were cultured in insect medium at 37°C or 28°C. Supernatants of moulting L3 also contained chitinase, whereas fourth-stage larvae (L4) did not contain the protein. These results are consistent with the data of Wu *et al.* (1996) who found that *A. viteae* L3s secrete an enzymatically active 75 kDa form and an inactive 220 kDa form of chitinase during 3–6 days in culture under vertebrate conditions up to the L3/L4 moult, when the release ceases abruptly. The *de novo* synthesis of the 75 kDa form of *A. viteae* L3 chitinase, as quantified by biosynthetic labelling of L3s within the vector ticks, was shown to occur only at 27°C and not at 37°C. Together, these data suggest that L3 chitinase is produced and stored in the glandular oesophagus of the L3 within the intermediate host, while secretion is triggered by the environmental conditions of the vertebrate host, and occurs during the early phase of infection and during moulting.

The role of L3 chitinase is not fully understood and different scenarios are conceivable, either alternatively or in combination with each other. Firstly, the protein could contribute to egress of the L3 from the chitinous mouthparts of the vector during the blood meal, as soon as the host blood provides the necessary temperature shift and other required stimuli. A chitinolytic activity could facilitate this process but no data are available on this aspect. Secondly, the release of chitinase during the first days of culture under vertebrate conditions is compatible with the view that chitinase acts on host molecules during an early stage of infection. It is possible that the enzyme interacts with, for example, elements of the extracellular matrix, facilitating the migration through host tissues. A similar role in degradation of vertebrate tissue structures was discussed for the chitinase-like vertebrate protein HC gp-39 (Hakala *et al.*, 1993). Further studies on the substrate specificity of filarial chitinases are necessary to obtain precise information relating to this. Thirdly, the release of chitinase during the L3/L4 moult indicates a role in moulting. The *A. viteae* L3 chitinase is located in the glandular oesophagus, which leads into the buccal cavity. These pharyngeal glands of nematodes deliver products that contribute to the moult (Bird and Bird, 1991) and are secreted around the time of moulting in *O. volvulus* (Strote and Bonow, 1991). It is possible that chitinases are released into the buccal cavity and act in degrading the L3 cuticle at its anterior pole, allowing the hatching of L4 through the ruptured exuvia. Recently, McKerrow *et al.* (1999) indicated that material of the glandular oesophagus could be transported to the surface of L3 via a network of intercellular channels. Such a mechanism could allow chitinase to reach the cuticle of the developing L4, where it could have a role in restructuring the new cuticle and in degrading the L3 cuticle.

To date, the relationship between the described mf chitinases and L3 chitinases remains unclear. Arnold *et al.* (1996) described the presence of multiple bands hybridizing with a chitinase probe in the genome of *B. malayi*. It is not clear whether these sequences are functional genes and which proteins such presumptive genes would encode. It is conceivable that the chitinase of each stage is encoded by a distinct gene. However, it cannot be ruled out that one gene is differentially expressed in mf and L3, while the other members of the gene family could fulfil other functions.

Biochemistry and Molecular Biology of Filarial Chitinases

The filarial chitinases are members of the family of 18 glycosyl hydrolases which contain a characteristic motif at the active centre (Henrissat, 1990; Watanabe *et al.*, 1993). Based on structural similarities, it is expected that the published filarial chitinases are endochitinases which produce two to six chito-oligomers. These might be further degraded to monomers via β-N-acetylglucosaminidases in analogy with findings in insects. Native *B. malayi* chitinase purified by lectin chromatography showed K_m values of 77 μM for 4-methylumbelliferyl (4-MU) GlcNAc$_2$, and 27 μM for 4-MU GlcNAc$_3$ (Fuhrman *et al.*, 1995), which is similar to the data for native *Heligmosomoides polygyrus* chitinase (Arnold *et al.*, 1993). The recombinant chitinases of *B. malayi*, *A. viteae* and *O. volvulus* have been shown to be enzymatically active (Venegas *et al.*, 1996; Drabner *et al.*, unpublished results). The K_m (μM) values for recombinant full-length *B. malayi* chitinase were 45 for 4-MU GlcNAc$_2$, and 22 for 4-MU GlcNAc$_3$ (Venegas *et al.*, 1996).

The published filarial chitinases show a distinct modular domain structure (Fig. 10.3). The first 17–22 amino acids serve as a cleavable signal sequence, indicating that the enzymes are secreted or are components of the outer membrane. The catalytic domain, essential for the degradation of

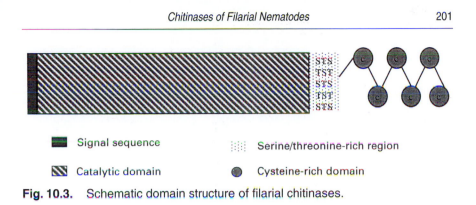

Signal sequence

Serine/threonine-rich region

Catalytic domain

Cysteine-rich domain

Fig. 10.3. Schematic domain structure of filarial chitinases.

the substrate, includes the active centre of this type of chitinase (LIVMFY-DN-G-LIVMF-DN-LIVMF-DN-X-E; Henrissat, 1990; Watanabe *et al.*, 1994). It is followed by a Ser/Thr-rich domain which comprises three to four repeats of each of the 14 amino acids. These sites are supposed to be the target for extensive O-glycosylation (Kuranda and Robbins, 1991). The presence of several prolines suggests that the Ser/Thr-rich domain acts as a hinge-like linker region. The C terminus of filarial chitinases comprises a domain with six cysteine residues spaced in a conserved pattern. This pattern also occurs in the chitin binding domains of insect chitinases, and Tellam (1996) therefore suggested that the domain facilitates the hydrolysation process via juxtaposing the β 1–4 glycosidic linkages of chitin and the active site of the catalytic domain. However, truncated *B. malayi* chitinase which lacks the chitin-binding domain nevertheless has the full enzymatic activity (Venegas *et al.*, 1996).

The described enzymatically active filarial chitinases have a high degree of identity and similarity at the level of nucleic acids and amino acids, respectively (Fig. 10.4). While the nucleic acid sequences are up to 75% identical, the amino acid sequences show an identity of 70–80% with a further 10–15% similar amino acids. The chitinase of *C. elegans* (Swiss Prot Q11174) is less related to the filarial chitinases, exhibiting 42% identity of nucleotides, and 42% identity and 48% similarity at the amino acid level compared with the *O. volvulus* L3 chitinase. The catalytic domain shows high homology to filarial chitinases and the active centre is relatively conserved, with only two out of nine amino acids exchanged. The chitinase of *Manduca sexta* (Kramer *et al.*, 1993) is relatively homologous to the nematode chitinases (64% nucleotide identity, and 35% identity and 45% similarity to the L3 chitinase of *O. volvulus* at the amino acid level), while the enzymes of *Streptomyces* spp., bacteria and fungi are less similar (Blaak *et al.*, 1993).

Chitinase-like Proteins in Vertebrates

Because vertebrates do not produce chitin, it was postulated that the enzymes involved in the formation and turnover of chitin could be a target

```
                    10        20        30        40        50        60
                                                    signal sequence
OvL3 Chit   ------------------EVDNNFCKVMRIGAMLIPFIILGNAI----IAYGYVRGCYYT
AvL3 Chit   --------------------VLQGDMN-WIMITLFIILANAIT---VVNGYVRGCYYT
B.m. Chit   ------------VLPKFEKWTTLSQSDMNRTTLILFFIILSNTIT---VIHGYVRGCYYT
C.eleg. Chit MLLGKFLLVASFILPIAYTWTGATIRNHPADVVAARNKITSRSVARSEPTNSYIRPCYFT
                                              *     ..       *. * **.*

OvL3 Chit   NWAQYRQGEGKFLPEDIPKGLCTHILYAFAKVDQSGTSLPFEWNDEDTNWS-KGMYSRVT
AvL3 Chit   NWAQYRQGEGKFLPEDIPIGLCTHILYAFAKVDEKGTSMAFEWNDEDTEWS-KGMYSRVI
B.m. Chit   NWAQYRDGEGKFLPGNIPNGLCTHILYAFAKVDELGDSKPFEWNDEDTEWS-KGMYSAVT
C.eleg. Chit NWAQYRQGRAKFVPEDYTPGLCTHILFAFGWMNADYTVRAYDPADLPNDWAGEGMYRRVN
            ****** *   **.*      *******.**   .      ..  * . *.  *** *

OvL3 Chit   KLKENDPEMKILLSYGGYNFGSSTFTAIRNRAEKRKHFIKSAIAFLRKNKFDGFDFDWEY
AvL3 Chit   KLRENDPTLKILLSYGGYNFGSSTFAAIAKSAEKRKHFIQSATTFLRKHKFDGFDLDWEY
B.m. Chit   KLRETNPGLKVLLSYGGYNFGSAIFTGIAKSAQKTERFIKSAIAFLRKNNFDGFDLDWEY
C.eleg. Chit KLKVTDTQLKTLLSFGGWSFGTALFQGMAASSASRKVFIDSAITFVRTWGFDGIDIDWEY
            **.    .*  ***.**  **.   .        ** ** ** *   **   *  .****

OvL3 Chit   PIGMA--QEYAKLVNEMKVAFVEEAKKSDSEQLLLTAAVSAGKHTIDQSYNVQSLGENFD
AvL3 Chit   PTGVA--EEHAKLVEEMKAAFVEEAKESGKQQLLLTAAVSAGKMTIDESYNVQSLGKSLD
B.m. Chit   PVGVA--EEHAKLVEAMKTAFVEEAKTSGKQRLLLTAAVSAGKGTIDGSYNVESLGKNFD
C.eleg. Chit PSGATDMANYVALVKELKAACESEAGSTGKDRLLVTAAVAAGPATIDAGYDIPNLAPNFD
            * *  .     **  .*  *    **     .****.***  *** .   * *
                                    catalytic domain
OvL3 Chit   LLSLMSYDFHGSWEMNVDLHAKLHPTKGETSGTGIFNTEFAANYWLSKGMPKQKIIIGIP
AvL3 Chit   LLFLMSYDLHGSWEKNVDLHGELRPTERETSGTGVCNTEFAANYWAEKGMPKQKIIIGIP
B.m. Chit   LLFLMSYDLHGSWDLHGKLHPTKGEVSGIGIFNTEFAADYWASKGMPKEKIIIGIP
C.eleg. Chit FILLMSYDFFGAWASLVGFNSPLYATTELPAEWNGWNVDSSARYWNQKGMPKEKIIVGMP
             . ***** * .  *     *     .*  . * ** ***** ***.*.*

OvL3 Chit   TYGRGWTLRDSSKTTIGAEGISPSSPSTTNPAGGTAAYWEICKYLKEGGKETIDEQGVGA
AvL3 Chit   AYSRGWTLSNPSETAIGAEGDRPSSPSTTNPAGGTAAYWEICKYVKEGGKETVDKKSVGA
B.m. Chit   MYAQGWTLDNPSETAIGAAASRPSSASKTNPAGGTASYWEICKYLKEGGKETVHQEGVGA
C.eleg. Chit TYGRGWTLNNASAINPGTSGS-PAKITQYVQEAGVGAYFEFCEMLANGATRYWDSQSQVP
             *  .****   *    *.   *     *.      . * *   . *

OvL3 Chit   CMVQGSQWYGYDNEETIRMKMRWLKEKGYGGAFIWTLDFDDF-KGTSCGEG-PYPLLSAI
AvL3 Chit   YMVKGDQWHGYDNEETIKIKMKWLKEEDYGGAFMWTLDFDDF-KGTSCGKG-PYPLLNAI
B.m. Chit   YMVKGDQWYGYDNEETIRIKMKWLKEKGYGGAFIWALDFDDF-TGKSCGKG-PYPLLNAI
C.eleg. Chit YLVQGNQWWSYDDEESFANKMAYVKREGYGGAFVWTLDFDDFNAGCSNSNGQLYPLISVI
             .*.* **  ** **.    **    .*  *****.*.****** .          *
                                    S/T rich domain
OvL3 Chit   NHELKGEATATTRSLRT--------TITQSSTIGSTKFETTTTASEITKNNKIKTTTIAV
AvL3 Chit   NNGLESEQTPSTRAVPETTEDTEVEAMTEAPGTANDTEMETAEMPETTEDIEIETSTETS
B.m. Chit   SSELEGE-----SENPE--ITTEEPSITETEAYETDETEETSET-EAYDTDETEETSETE
C.eleg. Chit AKELGGVIIPKKGGVTT--------APTTVATTVTTGRPPMTSAVTTTTAATTTTTRAAT
             *                                       .   .
                                    chitin-binding domain
OvL3 Chit   EPTGESS---DIK---------------CPESFGLFRHPNDCHLFIHCAHDHPYVKLCP
AvL3 Chit   STTAAPE---TTEGTEVETTTSVVKPGEECPEPDGLFPHHSDCHLFINCANNYPHIMECP
B.m. Chit   ATTYDTD---ETE-------------GQECPERDGLFPHPTDCHLFIQCANNIAYVMQCP
C.eleg. Chit TTTASNTNV------------------CSGKSDGFYPNSNNCGLFVLCLSSKSYSMSCP
             *                           *  .   . * **. *         **

OvL3 Chit   PNTFFNDKIKVCDHFGECDE----------------------------------------
AvL3 Chit   VGTFFDDTIKVCNYMRNAPDTCKFGNFMYIYLLEKSR-----------------------
B.m. Chit   ATTFFNDAIKVCDHMTNAPDTCIFVVLCNVLVVEKFRVKKKKKKKKKKKKKK--------
C.eleg. Chit SGLQYSASLKYCTTSTASGCSVTTTRAPTTTTKSAPTVTTTTRAPTTTTPAFKCTKDGFF
                 .   .* * .

OvL3 Chit   ---------------------------------------------
AvL3 Chit   ---------------------------------------------
B.m. Chit   ---------------------------------------------
C.eleg. Chit GVPSDCLKFIRCVNGISYNFECPNGLSFHADTMMCDRPDPSKCAK
```

for drug development in organisms that contain chitin. However, it has to be considered that chitinase-like proteins from vertebrates have recently been described. Two prominent members of the human chitinase family, termed YKL-40 (or HC gp-39) and chitotriosidase, respectively, are secreted by chondrocytes, synovial fibroblasts and macrophages (Boot *et al.*, 1995; Renkema *et al.*, 1998). HC gp-39 was reported to have a role in the remodelling or degrading of extracellular matrix and thus might contribute to the pathogenesis of rheumatoid arthritis (Volck *et al.*, 1998) or colorectal cancer (Cintin *et al.*, 1999). Furthermore, HC gp-39 was described as acting as a chitin-specific lectin (Renkema *et al.*, 1998), suggesting a possible role in defence against microorganisms. While no chitinolytic activity has been described for HC gp-39/YKL-40, Boot *et al.* (1995) localized enzymatically active chitotriosidase to human macrophages. This enzyme was initially found due to a several hundredfold increase in enzyme activity in patients with Gaucher disease (Hollak *et al.*, 1994; Renkema *et al.*, 1995), which is characterized by an accumulation of glycosyl ceramide within the lysosomes of macrophages, which interferes with the function of the cells. Human HC gp-39 (Gene Bank accession no. P36222) and the human chitotriosidase (Gene Bank accession no. AAC50246) show overall amino acid identities of 41% and 52% and similarities of 45% and 55%, respectively, with the L3 chitinase of *O. volvulus*. Furthermore, a group of oviduct-associated chitinase-like proteins, the 'oviductins', have been described from mammals (mice, baboons, humans) and other vertebrates (e.g. *Xenopus*). Oviductins are secreted into the lumen of the oviduct under hormonal control during oestrus (Sendai *et al.*, 1995). These proteins have a characteristic domain structure like other chitinases, including a domain with significant similarities to the catalytic domain, though it is not enzymatically active, as well as Ser/Thr-rich repeats of 15 amino acids each (Malette *et al.*, 1995). The carboxy-terminal Ser/Thr-rich domains are heavily O-glycosylated and show features of mucins. It was thus suggested that the chitinase-like domain targets the oviductins to the oocyte via interaction with specific oligosaccharide moieties of the zona pellucida and that the mucin-type glycoprotein domains act as a protective shield surrounding the oocyte. Such a hormonal control of expression as described for the oviductins is reminiscent of induction of insect chitinases which are active during moulting through ecdysone-like hormones (Fukamizo and Kramer, 1987). By analogy, it is possible that the expression of filarial chitinases secreted during moulting is also regulated by hormones, as has been shown

---

Fig. 10.4. (Opposite) Alignments of nematode chitinases with the active centre boxed. Identical amino acids are marked with an asterisk (*); similar amino acids are marked with a dot (.). Conserved cysteine residues in the chitin-binding domain are printed in bold. Gene Bank accession numbers: C.eleg. Chit (*C. elegans*), Q11174; OvL3 Chit (*O. volvulus*), L42021; AvL3 Chit (*A. viteae*), U14638; B.m. Chit (*B. malayi*), M73689.

to be the case for other helminths (Barker *et al.*, 1990; Warbrick *et al.*, 1993). Further studies must show whether the respective regulatory elements are present in filarial chitinase genes.

Immunological Aspects of Filarial Chitinases

It is conspicuous that published work on filarial chitinases has been in the context of protective immunity. The first filarial chitinase was described by Canlas *et al.* (1984), who reported that a mAb, MF1, binding to the sheath of *B. malayi* mf, mediated the *in vitro* adherence of spleen cells of jirds (*Meriones unguiculatus*) and human buffy coat cells. Passive immunization of infected jirds with MF1 decreased the number of circulating *B. malayi* mf. The MF1 target epitope was also recognized by sera of individuals infected with Brugian filariasis who had remained amicrofilaraemic despite exposure to transmission. Fuhrman *et al.* (1992, 1995) identified the target antigens of MF1, two proteins of 70 and 75 kDa, as isoforms of a chitinase. Immunization of jirds with a recombinant fusion protein with maltose binding protein, expressed in *E. coli*, significantly delayed the onset of microfilaraemia and reduced the number of mf in the systemic circulation, but did not reduce the worm burden (Wang *et al.*, 1997). This partial protection against mf was obtained after intraperitoneal or subcutaneous immunization employing alum or Freund's complete adjuvant, irrespective of whether the challenge infection was given intraperitoneally or subcutaneously. Immunization studies with a truncated polypeptide revealed that the protective epitope was located at the carboxyl terminus of the protein, which also bears the target epitope of MF1. These data convincingly show that mf chitinase of *B. malayi* induces partial anti-mf immunity in jirds via an antibody-mediated mechanism. Interestingly, antibodies raised against full-length *B. malayi* mf chitinase and antibodies of lymphatic filariasis patients bind predominantly to the C-terminal portion of the protein (Wang *et al.*, 1997). The same holds true for *O. volvulus* L3 chitinase, where antibodies in sera of onchocerciasis patients reacted distinctly more strongly with the C terminus as compared with the recombinant full-length chitinase (Drabner *et al.*, unpublished results).

A filarial chitinase-like protein was identified when Freedman *et al.* (1989) compared the pattern of IgG antibodies of individuals from the Cook Islands infected with *W. bancrofti* versus individuals exposed to transmission without being infected ('putatively immunes'). All sera of a carefully defined group of seven putatively immune people reacted with a 43 kDa protein of the L3 stage of the closely related filarial parasite *B. malayi* in immunoblots. In contrast, only one out of 12 (8%) sera of microfilaraemic patients recognized this protein. This association between differential recognition and resistance against infection suggested a role for the antigen in protective immunity. A clone was isolated using a rabbit antiserum against the 43 kDa protein from a *W. bancrofti* genomic expression library. An alignment of the coding sequence revealed 46% identity to the chitinase of *Serratia marcescens* but only 23% identity to *B. malayi* mf chitinase. The protein was present in L3 and uterine mf of *B. malayi* (Raghavan *et al.*, 1994). The recombinant protein was differentially recognized by the antibodies of putatively immune individuals, but not by infected patients from the same endemic area. So far, no immunization studies with the recombinant protein have been published.

A similar analysis of antibody responses in an animal model of filariasis revealed an L3 chitinase recognized by sera of animals resistant to infection. Lucius *et al.* (1991) found that sera of jirds vaccinated with irradiation-attenuated L3s of *A. viteae* differentially recognized L3 antigens of 205 kDa and 67 kDa in immunoblots (Fig. 10.5). Adam *et al.* (1996) identified these proteins as two forms of an L3 chitinase after protein sequencing and cloning from an L3 cDNA expression library. Wu *et al.* (1996) identified the same chitinase (molecular weight of the bands: 220 and 75 kDa) as an L3 specific protein by metabolic labelling of *A. viteae* L3 in ticks and cloned the protein and the homologous *O. volvulus* L3

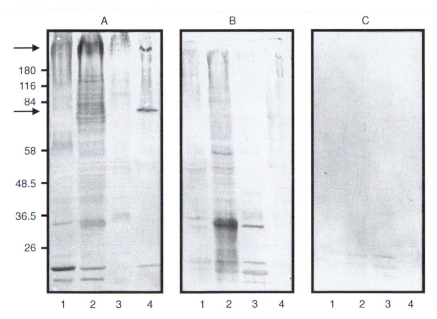

Fig. 10.5. Immunoblotting analysis of the antibody response of jirds against various stages of *A. viteae*. (A) Reaction with sera of jirds vaccinated with irradiated L3; (B) reaction of sera of *A. viteae*-infected jirds; (C) reaction with sera of naive jirds. 1, Male antigens; 2, female antigens; 3, mf antigens; 4, L3 antigens. Note that the reaction of vaccinated jird sera is predominantly directed against chitinase bands of 205 kDa and 67 kDa (arrows) as well as against a 17 kDa protein.

chitinase from cDNA libraries. In our laboratory, immunization of jirds with enzymatically active *E. coli*-expressed full-length *A. viteae* L3 chitinase in combination with the co-blockpolymer adjuvant STP (Byars and Allison, 1987) did not induce statistically significant protection in two experiments. Similarly, native chitinase purified from *A. viteae* L3s by lectin affinity chromatography used in combination with STP did not induce protection (two experiments). However, immunization of jirds with live attenuated *Salmonella typhimurium* expressing active *A. viteae* chitinase reduced the number of adult worms compared with the control group (48.6% protection, $P < 0.05$) in one experiment, but not in a repeat experiment (Adam *et al.*, unpublished results).

Recombinant *O. volvulus* L3 chitinase has been used for immunization studies in rodent models. Due to the strict primate specificity of *O. volvulus* the protective potential had to be tested either by vaccination of jirds and subsequent challenge infection with *A. viteae* L3s or by vaccination of mice and subsequent challenge by *O. volvulus* L3s implanted in micropore chambers. It has to be kept in mind that these models do not cover all parameters of a natural vaccination model. Several experiments using *E. coli*-expressed *O. volvulus* L3 chitinase together with Freund's complete adjuvant (Y. Wu *et al.*, personal communication), with STP adjuvant (Drabner *et al.*, unpublished results) or with a DNA vaccine (R. Harrison *et al.*, personal communication) did not protect against challenge with *A. viteae* L3s. In many of these experiments there was a trend towards protection, but the levels did not reach statistical significance. *E. coli*-expressed *O. volvulus* chitinase did not induce protection against chamber-implanted *O. volvulus* L3s in mice, while a DNA vaccine yielded 53% reduction ($P < 0.05$) in one experiment (Harrison *et al.*, 1999). A second DNA vaccination experiment showed a trend towards protection but no statistical significance.

Together, these data suggest that L3 chitinase has potential as a vaccine against filarial infective third-stage larvae. Considering the fact that L3 chitinase is an antigen exposed at critical time points of the parasite life cycle (host invasion and first moult within the definitive host), this protein has appealing properties as a vaccine candidate. However, studies on the effect of adjuvants, on antigen delivery systems and on the immune effector mechanisms generated will be necessary, as only a better understanding of the immunobiology of the early filarial infection will allow optimization of the immunization protocols. Among other tasks, it will be important to study the T-cell response against filarial chitinases in more detail. Our own analyses show that the deduced amino acid sequence of *O. volvulus* L3 chitinase comprises 24 putative T-cell epitopes, as predicted by the algorithm of Rothbard and Taylor (1988). We tested synthetic peptides comprising these T-cell epitopes in proliferation assays with spleen cells of chitinase-immunized BALB/c mice and found that 11 epitopes were able to restimulate spleen cells. Most of the predicted or demonstrated T-cell epitopes were clustered in the central part of the protein (Drabner *et al.*, unpublished results). Furthermore, tests in our laboratory revealed that the pattern of T-cell and of IgG subclass responses against *O. volvulus* L3 chitinase in mice can be influenced by the choice of adjuvants, which drive the immune response into the Th_1 or Th_2 direction, respectively (Drabner *et al.*, unpublished results). Because the choice of adjuvants and the mode of antigen delivery seems to influence the protective capacity of L3 chitinase, such information will be crucial to the design of optimized vaccination strategies.

Acknowledgements

We are grateful for unpublished information from Professor A.E. Bianco, Dr Y. Wu and Dr R. Harrison, Liverpool School of Tropical Medicine. We thank Dr W. Rudin, Swiss Tropical Institute in Basel, for providing Fig. 10.2. We thank Dr B. Kalinna (Humboldt-University) for reading the manuscript. Our studies on L3 chitinases were supported by the Edna McConnell Clark Foundation and the Commission of the European Communities to RL (grant IC18-CT95-0017).

References

Adam, R., Kaltmann, B., Rudin, W., Friedrich, T., Marti, T. and Lucius, R. (1996) Identification of chitinase as the immunodominant filarial antigen recognized by sera of vaccinated rodents. *Journal of Biological Chemistry* 271, 1441–1447.

Arnold, K., Brydon, L.J., Chappell, L.H. and Gooday, G.W. (1993) Chitinolytic activities in *Heligmosomoides polygyrus* and their role in egg hatching. *Molecular and Biochemical Parasitology* 58, 317–323.

Arnold, K., Venegas, A., Houseweart, C. and Fuhrman, J.A. (1996) Discrete transcripts encode multiple chitinase isoforms in Brugian microfilariae. *Molecular and Biochemical Parasitology* 80, 149–158.

Barker, G., Chitwood, D. and Rees, H. (1990) Ecdysteroids in helminths and annelids. *Invertebrate Reproduction and Development* 18, 1–11.

Bird, A. and Bird, J. (1991) *The Structure of Nematodes*. Academic Press, San Diego, California.

Blaak, H., Schnellmann, J., Walter, S., Henrissat, B. and Schrempf, H. (1993) Characteristics of an exochitinase from *Streptomyces olivaceoviridis*, its corresponding gene, putative protein domains and relationship to other chitinases. *European Journal of Biochemistry* 214, 659–669.

Boot, R.G., Renkema, G.H., Strijland, A., van Zonneveld, A.J. and Aerts, J.M. (1995) Cloning of a cDNA encoding chitotriosidase, a human chitinase produced by macrophages. *Journal of Biological Chemistry* 270, 26252–26256.

Brydon, L.J., Gooday, G.W., Chappell, L.H. and King, T.P. (1987) Chitin in egg shells of *Onchocerca gibsoni* and *Onchocerca volvulus*. *Molecular and Biochemical Parasitology* 25, 267–272.

Byars, N.E. and Allison, A.C. (1987) Adjuvant formulation for use in vaccines to elicit both cell mediated and humoral immunity. *Vaccine* 5, 223–228.

Canlas, M., Wadee, A., Lamontagne, L. and Piessens, W.F. (1984) A monoclonal antibody to surface antigens on microfilariae of *Brugia malayi* reduces microfilaremia in infected jirds. *American Journal of Tropical Medicine and Hygiene* 33, 420–424.

Cintin, C., Johansen, J.S., Christensen, I.J., Price, P.A., Sorensen, S. and Nielsen, H.J. (1999) Serum YKL-40 and colorectal cancer. *British Journal of Cancer* 79, 1494–1499.

Freedman D.O., Nutman, T.B. and Ottensen, E.A. (1989) Protective immunity in bancroftian filariasis. Selective recognition of a 43-kDa larval stage antigen by infection-free individuals in an endemic area. *Journal of Clinical Investigation* 83, 14–22.

Fuhrman, J.A. (1995a) Filariasis: the role of chitinase in larval development and transmission. In: Komuniecki, R. (ed.) *Molecular Approaches to Parasitology.* Wiley-Liss, New York, pp. 77–87.

Fuhrman, J.A. (1995b) Filarial chitinases. *Parasitology Today* 11, 259–261.

Fuhrman, J., Urioste, S., Hamill, B., Spielman, A. and Piessens, W. (1987) Functional and antigenic maturation of *Brugia malayi* microfilariae. *American Journal of Tropical Medicine and Hygiene* 36, 70–74.

Fuhrman, J.A., Lane, W.S., Smith, R.F., Piessens, W.F. and Perler, F.B. (1992) Transmission-blocking antibodies recognize microfilarial chitinase in Brugian lymphatic filariasis. *Proceedings of the National Academy of Sciences USA* 89, 1548–1552.

Fuhrman, J.A., Lee, J. and Dalamagas, D. (1995) Structure and function of a family of chitinase isozymes from Brugian microfilariae. *Experimental Parasitology* 80, 672–680.

Fukamizo, T. and Kramer, K.J. (1987) Effect of 20-hydroxyecdysone on chitinase and β-*N*-acetylglucosaminidase during the larval–pupal transformation on *Manduca sexta* (L.). *Insect Biochemistry* 17, 547–550.

Gooday, G.W., Brydon, L.J. and Chappel, L.H. (1988) Chitinase in female *Onchocerca gibsoni* and its inhibition by allosamidin. *Molecular and Biochemical Parasitology* 29, 223–225.

Hakala, B.E., White, C. and Recklies, A.D. (1993) Human cartilage gp-39, a major secretory product of articular chondrocytes and synovial cells, is a mammalian member of a chitinase protein family. *Journal of Biological Chemistry* 268, 25803–25810.

Ham, P., Phiri, J. and Nolan, G. (1991) Effect of *N*-acetyl-D-glucosamine on the migration of *Brugia pahangi* microfilariae into the haemocoel of *Aedes aegypti*. *Medical and Veterinary Entomology* 5, 485–493.

Harrison, R.A., Wu, Y., Egerfou, G. and Bianco, A.E. (1999) DNA immunisation with *Onchocerca volvulus* chitinase induces partial protection against infection with L3 in mice. *Vaccine* 18, 647–655.

Henrissat, B. (1990) Weak sequence homologies among chitinases detected by clustering analysis. *Protein Sequence Data Analysis* 3, 523–526.

Hollak, C.E.M., van Weely, S., van Oers, M.H.J. and Aerts, J.M.F.G. (1994) Marked elevation of plasma chitotriosidase activity. A novel hallmark of Gaucher disease. *Journal of Clinical Investigation* 93, 1288–1292.

Justus, D. and Ivey, M. (1969) Chitinase activity in developmental stages of *Ascaris suum* and its inhibition by antibody. *Journal of Parasitology* 55, 472–476.

Kaltmann, B. (1990) PhD thesis, Faculty of Biology, University of Heidelberg.

Kuranda, M.J. and Robbins, P.W. (1991) Chitinase is required for cell separation during growth of *Saccharomyces cerevisiae*. *Journal of Biological Chemistry* 266, 19758–19767.

Kramer, K.J., Corpuz, L., Choi, H.K. and Muthukrishnan, S. (1993) Sequence of a cDNA and expression of the gene encoding epidermal and gut chitinases of *Manduca sexta*. *Insect Biochemistry and Molecular Biology* 23, 691–701.

Lucius, R., Textor, G., Kern, A. and Kirsten, C. (1991) *Acanthocheilonema viteae*: Vaccination of jirds with irradiation-attenuated stage-3-larvae and with exported larval antigens. *Experimental Parasitology* 73, 184–196.

Malette, B., Paquette, Y., Merlen, Y. and Bleau, G. (1995) Oviductins possess chitinase and mucin-like domains: a lead in the search for biological function of these oviduct-specific ZP-associating glycoproteins. *Molecular Reproduction and Development* 41, 384–397.

McKerrow, J.H., Huima, T. and Lustigman, S. (1999) Do filarial nematodes have a vascular system? *Parasitology Today* 15, 123.

Perrone, J. and Spielman, A. (1986) Microfilarial penetration of the midgut of a mosquito. *Journal of Parasitology* 72, 723–727.

Raghavan, N., Freedman, D.O., Fitzgerald, P.C., Unnasch, T.R., Ottesen, E.A. and Nutman, T.B. (1994) Cloning and characterization of a potentially protective chitinase-like recombinant antigen from *Wuchereria bancrofti*. *Infection and Immunity* 62, 1901–1908.

Renkema, G.H., Boot, R.G., Muijsers, A.O., Donker-Koopman, W.E. and Aerts, J.M.F.G. (1995) Purification and characterization of human chitotriosidase, a novel member of the chitinase family of proteins. *Journal of Biological Chemistry* 270, 2198–2202.

Renkema, G.H., Boot, R.G., Au, F.L., Donker-Koopman, W.E., Strijland, A., Muijsers, A.O., Hrebicek, M. and Aerts, J.M. (1998) Chitotriosidase, a chitinase, and the 39-kDa human cartilage glycoprotein, a chitin-binding lectin, are homologues of family 18 glycosyl hydrolases secreted by human macrophages. *European Journal of Biochemistry* 251, 504–509.

Rogers, W.P. (1958) Physiology of the hatching of eggs of *Ascaris lumbricoides*. *Nature* 181, 1410–1411.

Rothbard, J.B. and Taylor, W.R. (1988) A sequence pattern common to T cell epitopes. *EMBO Journal* 7, 93–100.

Sendai, Y., Abe, H., Kikuchi, M., Satoh, T. and Hoshi, H. (1995) Purification and molecular cloning of bovine oviduct specific glycoprotein. *Biology of Reproduction* 50, 927–934.

Strote, G. and Bonow, I. (1991) Morphological demonstration of essential functional changes after *in vitro* and *in vivo* transition of infective *Onchocerca volvulus* to the post-infective stage. *Parasitology Research* 77, 526–535.

Tellam, R.L. (1996) Protein motifs in filarial chitinases: an alternative view. *Parasitology Today* 12, 291–292.

Trudel, J. and Asselin, A. (1989) Detection of chitinase activity after polacrylamide gel electrophoresis. *Analytical Biochemistry* 178, 362–366.

Venegas, A., Goldstein, J.C., Beauregard, K., Oles, A., Abdulhayoglu, N. and Fuhrman, J.A. (1996) Expression of recombinant microfilarial chitinase and analysis of domain function. *Molecular and Biochemical Parasitology* 78, 149–159.

Volck, B., Price, P.A., Johansen, J.S., Sorensen, O., Benfield, T.L., Nielsen, H.J., Calafat, J. and Borregaard, N. (1998) YKL-40, a mammalian member of the chitinase family, is a matrix protein of specific granules in human neutrophils. *Proceedings of the Association of American Physicians* 110, 351–360.

Wang, S.H., Zheng, H.J., Dissanayake, S., Cheng, W.F., Tao, Z.H., Lin, S.Z. and Piessens, W.F. (1997) Evaluation of recombinant chitinase and SXP1 antigens as antimicrofilarial vaccines. *American Journal of Tropical Medicine and Hygiene* 56, 474–481.

Warbrick, E., Barker, G., Rees, H. and Howells, R. (1993) The effect of invertebrate hormones and potential hormone inhibitors on the third larval moult of the filarial nematode, *Dirofilaria immitis, in vitro. Parasitology* 107, 459–463.

Ward, K. and Fairbairn, D. (1972) Chitinase in developing eggs of *Ascaris suum* (Nematoda). *Journal of Parasitology* 58, 546–549.

Watanabe, T., Kobori, K., Miyashita, K., Fujii, T., Sakai, H., Uchida, M. and Tanaka, H. (1993) Identification of glutamic acid 204 and aspartic acid 200 in chitinase A1 of *Bacillus circulans* WL-12 as essential residues for chitinase activity. *Journal of Biological Chemistry* 268, 18567–18572.

Watanabe, T., Uchida, M., Kobori, K. and Tanaka, H. (1994) Site-directed mutagenesis of the Asp-197 and Asp-202 residues in chitinase A1 of *Bacillus circulans* WL-12. *Bioscience Biotechnology and Biochemistry* 58, 2283–2285.

Welburn, S.C., Arnold, K., Maudlin, I. and Gooday, G.W. (1993) Rickettsia-like organisms and chitinase production in relation to transmission of trypanosomes by tsetse flies. *Parasitology* 107, 141–145.

Wharton, D. (1983) The production and functional morphology of helminth egg-shells. *Parasitology* 86, 85–97.

Wu, Y., Adam, R., Williams, S.A. and Bianco, A.E. (1996) Chitinase genes expressed by infective larvae of the filarial nematodes, *Acanthocheilonema viteae* and *Onchocerca volvulus. Molecular and Biochemical Parasitology* 75, 207–219.

Acetylcholinesterase Secretion by Nematodes

Murray E. Selkirk, Siân M. Henson, Wayne S. Russell and Ayman S. Hussein

Department of Biochemistry, Imperial College of Science, Technology and Medicine, London SW7 2AY, UK

Introduction and Historical Perspective

Acetylcholine (ACh) is the major excitatory neurotransmitter which regulates motor functions in both free-living and parasitic nematodes (Segerberg and Stretton, 1993; Rand and Nonet, 1997). Consistent with the use of cholinergic motor neurons, acetylcholinesterase (AChE) has been localized to nerve fibres and neuromuscular junctions in a range of species. Paradoxically, however, many parasitic nematodes also secrete AChE from specialized amphidial and secretory glands. This was initially discovered via cytochemical staining of sections of *Nippostrongylus brasiliensis* by Lee (1970), and led to the rapid documentation of AChE secretion in a number of species by Ogilvie *et al.* (1973). Despite the fact that these original studies were performed some 30 years ago, we still do not understand the physiological function of these enzymes. This chapter reviews the current information on the properties of nematode secreted AChEs and attempts to distil past and current thoughts on their potential function.

Cholinesterase Structure and Function

The classical role of AChEs is to terminate transmission of neuronal impulses by rapid hydrolysis of ACh. The closely related butyrylcholinesterases (BuChEs) or pseudocholinesterases have a less stringent substrate specificity but their function remains ill-defined. In mammals, BuChE is found at high concentration in the plasma and the gut, where it has been postulated to play a role in detoxification of plant esters ingested in the diet (Taylor, 1991). Both AChEs and BuChEs are expressed at other sites in the body such as haematopoeitic cells and the developing nervous system, in which a variety of roles such as regulation of differentiation and morphogenesis have been proposed (reviewed by Massoulié *et al.*, 1993).

Cholinesterases exist in multiple molecular forms distinguished by their subunit interactions and hydrodynamic properties. In vertebrates, alternative splicing of a single AChE gene generates distinct catalytic subunits which may be assembled into asymmetric (A) or globular (G) forms, the latter consisting of monomers (G_1), dimers (G_2) and tetramers (G_4) of a catalytic subunit. Asymmetric forms are composed of one to three tetramers linked to Q subunits which associate with the basal lamina of neuromuscular junctions. Globular forms may be hydrophilic or amphiphilic, the latter associating with cell membranes via glycolipid anchors or a non-catalytic subunit bearing covalently attached fatty acids (Massoulié *et al.*, 1993). In contrast, invertebrates appear exclusively to express globular forms of AChE, which invariably show considerable activity against BuCh (Toutant, 1989).

Resolution of the structure of AChE from the electric ray *Torpedo californica* highlighted the surprising observation that the catalytic triad (Ser-200, His-440, Glu-327) was located at the base of a narrow gorge extending approximately halfway (20 nm) into the enzyme (Sussman *et al.*, 1991). This was unexpected, given the extremely high turnover rate of the enzyme. ACh is oriented in the active site by interaction of the quaternary nitrogen of choline with Trp-84, and is thought to be attracted to this site by a strong electrostatic dipole aligned with the gorge, which is lined with 14 aromatic residues (Ripoll *et al.*, 1993).

Nematode Acetylcholinesterases

In contrast to vertebrates and insects, nematode AChEs are encoded by separate genes. Unsurprisingly, this is best defined in *Caenorhabditis elegans*, which is now known to possess four AChE genes (Grauso *et al.*, 1998). The two major classes of enzyme, A and B, encoded by *ace-1* and *ace-2*, are required for normal motility but appear to have overlapping functions. Class C AChE accounts for less than 5% of the total activity in the worm (Kolson and Russell, 1985), whereas class D represents less than 0.1% of activity. Homozygous mutants in *ace-1*, *ace-2* or *ace-3* have no visible phenotype, whereas *ace-1/ace-2* mutants are severely uncoordinated, and the triple mutation is lethal. Recent studies indicate that *ace-1* is expressed in the musculature of the body wall, anal sphincter, vulva and pharynx, in addition to cephalic sensory neurons (Culetto *et al.*, 1999). In the closely related nematode *Steinernema carpocapsae*, *ace-1* encodes a hydrophilic catalytic subunit which is assembled into an amphiphilic tetramer via disulphide bonding to a hydrophobic (non-catalytic) subunit, and *ace-2* encodes an amphiphilic catalytic subunit which assembles into a glycosyl phosphatidylinositol-linked amphiphilic dimer (Arpagaus *et al.*, 1992).

Neuronal AChEs from parasitic nematodes are less well defined. A single dimeric amphiphilic (G_2^a) form has been described in *Parascaris equorum* (Talesa *et al.*, 1997), and two forms of AChE have been detected in *Trichinella spiralis* which sedimented at 5.3 S and 13 S in sucrose gradients (deVos and Dick, 1992). We have identified a detergent soluble enzyme in somatic extracts of *N. brasiliensis* which resolved at 10.2 S and was shifted to 9.4 S in the presence of Triton X-100, suggestive of a tetrameric amphiphilic (G_4^a) form (Hussein *et al.*, 1999b). Although it has been suggested that parasitic nematodes might express a more restricted repertoire of AChEs than free-living forms (Talesa *et al.*, 1997), we see no compelling reason for this assumption. The existence of four AChEs in *C. elegans* was determined with the aid of genetic analysis and the genome project, resources largely unavailable to those working on parasitic species. It is worth pointing out that 95% of AChE activity in extracts of *C. elegans* is accounted for by two gene products, and thus any analogous minor variants in parasitic species are most likely to have been overlooked at present by the biochemical assays employed.

Acetylcholinesterase Secretion by Parasitic Nematodes

The unusual phenomenon of AChE secretion by parasitic nematodes has been largely documented by analysis of products secreted during *in vitro* culture. This is the case for most nematode excretion/secretion (ES) products, and assumes that this is indicative of secretion *in vivo*. The potent antigenicity of these enzymes in experimental infections suggested that they were indeed secreted (Bremner *et al.*, 1973; McKeand *et al.*, 1994) and we have recently confirmed this by recovery of AChE from the jejunal lumen of rats infected with *N. brasiliensis* (W.S. Russell *et al.*, 2000, unpublished results). Secretion of AChE is largely restricted to parasites of the alimentary tract, though there are a number of exceptions. The cattle lungworm *Dictyocaulus viviparus*, and *Stephanurus dentatus*, which colonizes the ureters of porcine kidneys, both secrete AChE (Rhoads, 1981; McKeand *et al.*, 1994). Conversely, the phenomenon of AChE secretion is not shown by all nematodes that colonize the gastrointestinal tract (Ogilvie *et al.*, 1973), raising questions about the biological function and necessity of these enzymes for parasite establishment. There appears to be no obvious discrimination between secretors and non-secretors based upon whether the parasites reside in the lumen or are firmly anchored to or invade mucosal tissue. Thus, *N. brasiliensis* and *Ascaris* spp. are both luminal dwellers, although only the former parasite secretes AChE. *Necator americanus* secretes large quantities of the enzyme, whereas *T. spiralis* does not.

Table 11.1 illustrates the stage- and species-specificity of this phenomenon in selected examples.

An interesting twist to the story is provided by studies on *N. brasiliensis*, which secretes three distinct isoforms of AChE, designated A, B and C (Ogilvie *et al.*, 1973). These enzymes can be easily separated by non-denaturing electrophoresis due to their distinct pIs, and this is illustrated in Fig. 11.1, which also shows the distinct electrophoretic properties of the amphiphilic enzyme (arrowed) found only in somatic extracts and therefore presumably associated with neuromuscular function. The overall amount of AChE produced by this parasite increases dramatically following establishment in the jejunum, and a switch in isoform expression occurs,

Table 11.1. Selected examples of AChE secretion. Data taken from Ogilvie *et al.* (1973) and our observations.

Species	Host	Stage	AChE secreted hour^{-1} (units g^{-1} wet weight)
Trichostrongylus colubriformis	Sheep	L3	0.0025
		L4	12.6
		Adult	6.5
Oesophagostomum radiatum	Calf	L4	10.9
		Adult	0.6
Nippostrongylus brasiliensis	Rat	Adult	0.4
Ostertagia circumcincta	Sheep	Adult	0.2
Haemonchus contortus	Sheep	L3	0.0005
		Adult	0.017
Trichinella spiralis	Rat	L3	nd

One unit of AChE activity is defined as 1 μmol of substrate hydrolysed min^{-1} at 37°C.
nd = not detectable.

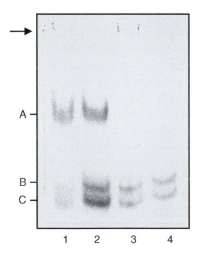

Fig. 11.1. Somatic and secreted AChEs in *N. brasiliensis*. Extracts of parasites collected 4 days (track 1) and 8 days (track 3) post-infection of rats were resolved by non-denaturing PAGE alongside secreted products from parasites also collected 4 days (track 2) and 8 days (track 4) post-infection, and AChE activity visualized by cytochemical staining. Secretory enzymes designated as forms A, B and C are indicated, and the non-secreted isoform is arrowed. Reproduced from Hussein *et al.* (1999b), with permission.

from initial predominant expression of form A to subsequent exclusive expression of forms B and C (Edwards *et al.*, 1971; Blackburn and Selkirk, 1992a). Intriguingly, it has also been shown that the amount of AChE produced by these parasites appears to be regulated in part by the immune status of the mammalian host. Transfer of parasites into immunologically naive hosts results in reduced levels of nematode AChE (Sanderson *et al.*, 1972). Administration of the broad anti-inflammatory agent cortisone to infected rats leads to a similar reduction of parasite AChE, whereas passive transfer of immune serum results in up-regulation of enzyme expression (Jones and Ogilvie, 1972). These data suggest that expression of parasite AChEs may be regulated by an unidentified component of the host immune response and may contribute to maintenance of their position in the gastrointestinal tract.

Forms and Properties of the Secreted Enzymes

In contrast to the neuromuscular enzymes, results to date suggest that all nematode secretory AChEs are non-amphiphilic, although both monomeric and dimeric forms have been documented. Analysis of the AChE activity secreted by *N. americanus* indicated that it existed as a single G_2^{na} form (Pritchard *et al.*, 1994), whereas both monomeric and dimeric hydrophilic-secreted AChEs have been described for *Trichostrongylus colubriformis* (Griffiths and Pritchard, 1994). The three isoforms of AChEs secreted by *N. brasiliensis* are all monomeric, between 69 and 74 kDa in apparent mass, and analysis of substrate specificity and sensitivity to a panel of cholinesterase inhibitors indicates that they can all be classified as true AChEs rather than BuChEs (Grigg *et al.*, 1997).

Like the neuronal AChEs of *C. elegans*, the three secreted enzymes of *N. brasiliensis* are encoded by separate genes. We have complete cDNA sequences of forms B and C (Hussein *et al.*, 1999a, 2000) and are currently finalizing the sequence of form A (A.S. Hussein *et al.*, 2000, unpublished results). The key features of the primary structure of AChE B and AChE C are illustrated in Fig. 11.2 in comparison with those of the *Torpedo* AChE. Both nematode enzymes lack the carboxyl-terminal cysteine implicated in dimer formation in vertebrate AChEs, which explains their monomeric nature. The carboxyl terminus is severely truncated in comparison with vertebrate enzymes, aligning at a position approximating to the end of the catalytic domain of *Torpedo* AChE. In this respect the *Nippostrongylus*-secreted AChEs are similar to another hydrophilic monomeric enzyme with a truncated carboxyl terminus found in venom from the krait *Bungarus fasciatus* (Cousin *et al.*, 1996). The sequences of AChE B and AChE C otherwise show features consistent with AChEs from diverse species. The residues that constitute the catalytic triad are conserved, as are the six cysteine residues implicated in disulphide bond formation. An additional two

Torpedo californica

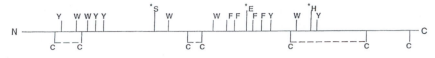

Nippostrongylus brasiliensis

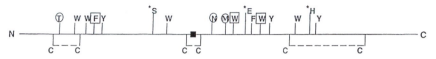

Fig. 11.2. Schematic representation of the primary structure of secreted AChE B of *N. brasiliensis* in comparison with that of *Torpedo californica*, for which the three-dimensional structure has been resolved. The residues in the catalytic triad (Ser-His-Glu) are depicted with an asterisk, and the position of cysteine residues and the predicted intramolecular disulphide bonding pattern common to cholinesterases is indicated. An insertion of 17 amino acids relative to the *Torpedo* sequence, which would predict a novel loop at the molecular surface, is marked with a black box. The 14 aromatic residues lining the active-site gorge of the *Torpedo* enzyme are illustrated. Identical residues in the nematode enzyme are indicated in plain text, conservative substitutions are boxed, and non-conservative substitutions are circled. The amino acid sequence of AChE C is 90% identical to AChE B, and differs only in the features illustrated in that Thr-70 is substituted by Ser.

cysteine residues are present, but it is not known at present whether they contribute to an additional intramolecular disulphide bridge. Eleven of the 14 aromatic residues lining the gorge are either conserved or show conservative substitutions in both of the *N. brasiliensis* enzymes. Two of the aromatic residues (Tyr-70 and Phe-288 in *Torpedo*), which are substituted in the nematode sequences by Thr/Ser-65 and Met-301, are also substituted by non-aromatic residues in mammalian BuChE (Lockridge *et al.*, 1987) and *C. elegans* ACE-1 (Arpagaus *et al.*, 1994).

Mutagenesis studies on *Torpedo* and human AChE have shown that Phe-288 and Phe-290 dictate substrate specificity, most probably via steric occlusion, but also possibly by stabilizing the substrate in an optimal position for catalysis (Harel *et al.*, 1992; Ordentlich *et al.*, 1993). The intermediate substrate specificity of certain invertebrate enzymes such as *C. elegans* ACE-1 and *Drosophila melanogaster* AChE (both enzymes hydrolyse BuSCh at approximately 50% the rate of ASCh) has been suggested to be due to the substitution of Phe-288 by glycine and leucine, respectively (Gnagey *et al.*, 1987; Arpagaus *et al.*, 1994). Replacement of Phe-288 in *Torpedo* and human AChE by non-aromatic residues greatly enhanced the ability of these enzymes to hydrolyse butyrylthiocholine (BuSCh), in addition to conferring sensitivity to inhibition by the BuChE-specific inhibitor iso-OMPA (Harel *et al.*, 1992; Ordentlich *et al.*, 1993). Both AChEs secreted by *N. brasiliensis* have a methionine residue in the position corresponding to Phe-288 in the *Torpedo* enzyme. The triple mutant (M300G/W302F/W345F) showed good activity with propionylthiocholine (PSCh) and BuSCh, indicating that the collective effect of larger residues in all these three positions acted to restrict access of larger substrates (Hussein *et al.*, 2000). These experiments suggest that the differences in substrate specificity between the nematode-secreted AChEs and other invertebrate neuronal enzymes can be explained by relatively simple substitutions in residues lining the active-site gorge. Nevertheless, anomalies exist in that both the wild type and mutant AChEs secreted by *N. brasiliensis* are insensitive to iso-OMPA (Hussein *et al.*, 1999a, 2000) in contrast with analogous mutants of vertebrate AChEs and invertebrate enzymes, including the *N. brasiliensis* somatic AChE (Hussein *et al.*, 1999b). It will therefore be informative to determine the primary structure of the somatic (neuronal) *N. brasiliensis* AChE(s), as it is not only inhibited by iso-OMPA but also displays significantly greater activity than the secreted enzymes against butyrylcholine (BuCh) (Fig. 11.3).

Putative Functions for Secreted Acetylcholinesterases

The unusual phenomenon of AChE secretion by parasitic nematodes has naturally provoked considerable hypotheses concerning their physiological function (Rhoads, 1984; Lee, 1996), though most of these have not yet been systematically investigated. There has been a recent minor revival of interest in this subject, and we therefore hope that the coming years might provide answers to what has been a long-standing conundrum. Putative functions of the nematode-secreted AChEs are discussed in the following sections.

Regulation of intestinal peristalsis or local spasm

This was the original hypothesis put forward by Lee (1970) and expanded by Ogilvie *et al.* (1973). Secretory products of *N. brasiliensis* do indeed decrease the amplitude of contractions of segments of uninfected rat intestine maintained in an organ bath, but a role for AChE in this phenomenon was discounted due to the heat stability of the parasite factor, and the inability to duplicate the effect with AChE from the electric eel (Foster *et al.*, 1994). Subsequent investigations demonstrated that the suppression of contraction could be duplicated by a 30–50 kDa fraction of secreted products, which contained a protein of 30 kDa that was immunologically cross-reactive with mammalian vasoactive intestinal peptide (VIP). Moreover, an antibody to porcine VIP significantly reduced the inhibitory effect of parasite-secreted products on contraction *in vitro* (Foster and Lee, 1996).

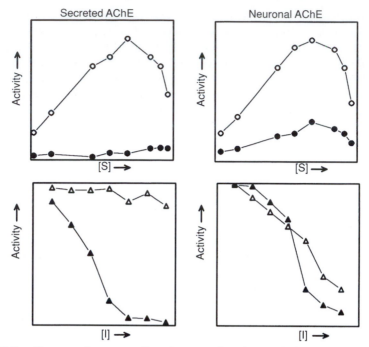

Fig. 11.3. Comparative properties of secreted and somatic (i.e. neuronal) AChEs of *N. brasiliensis*. The properties of the three secreted AChEs are broadly similar, and this figure represents the net activity of a mixture of these enzymes. The top panels depict substrate specificities, and show activity against acetyl-thiocholine (open circles) and butyrylthiocholine (solid circles) over a range of substrate concentrations [S] between 0.05 and 20 mM (log scale). The bottom panels show sensitivity to the AChE-specific inhibitor BW284c51 (solid triangles) and the BuChE-specific inhibitor iso-OMPA (open triangles) over a range of inhibitor concentrations [I] between 10^{-9} and 10^{-3} M (log scale).

The ability of such a protein to access the muscularis externa and thereby influence peristalsis *in vivo* has to be called into question, but a role for regulation of local spasm via inhibition of the muscularis mucosa is feasible. Similarly, a role for parasite AChEs in the latter phenomenon cannot yet be discounted.

Regulation of intestinal transport processes

It has been known for some time that the enteric nervous system does not simply regulate smooth muscle contraction, but is intimately involved in the control of transport processes in enterocytes. Nerve fibres in the mucosa terminate subjacent to the basement membrane of epithelial and entero-endocrine cells, on which muscarinic acetylcholine receptors (mAChRs) have been localized. Muscarinic AChR agonists evoke chloride secretion from intestinal epithelial cells and the electrogenic flux creates an osmotic gradient, resulting in movement of water into the lumen (Cooke, 1984). In addition, these agonists evoke mucus secretion from goblet cells. This can be elicited by electrical field stimulation of rat ileum, and is blocked by preincubation in tetrodotoxin or atropine, indicating regulation by cholinergic elements of the enteric nervous system (Phillips *et al.*, 1984). Direct application of ACh results in rapid mucus secretion from intestinal crypt goblet cells maintained in an organ culture system (Specian and Neutra, 1980). Additionally, ACh stimulates active chloride secretion from enterocytes, and this effect can be blocked by atropine, indicating the involvement of muscarinic cholinergic receptors (Cooke, 1984). It is likely that these secretory events contribute to expulsion of pathogens and noxious agents from the gastrointestinal tract. Fluid and mucus secretion are stimulated during primary infection with nematode parasites, and are rapidly and intensely stimulated upon secondary infection. Anaphylaxis-induced chloride secretion from enterocytes of rats infected with *T. spiralis* can be blocked by histamine and serotonin receptor antagonists, and is also substantially inhibited by tetrodotoxin and atropine, implicating both mast cells and the enteric nervous system in the regulation of this response (Castro and Arntzen, 1993). Serotonin, histamine and prostaglandin E2 are known to evoke substantial release of ACh in intestinal preparations, and this has also been documented following intestinal anaphylaxis (Javed *et al.*, 1992). It is therefore an attractive proposition that AChEs secreted by nematode parasites of the gastrointestinal tract act to inhibit secretory responses by hydrolysing ACh released from the enteric nervous system.

Regulation of lymphoid/myeloid cell functions

Rhoades (1984) first suggested that nematode secretory AChE might modulate the host inflammatory response, in a thought-provoking review in which she noted that ACh had been recorded to have numerous effects on leucocytes, including stimulation of chemotaxis and lysosomal enzyme secretion by neutrophils, histamine and leukotriene release by mast cells, and augmentation of lymphocyte-mediated cytotoxicity. More recently, muscarinic agonists have been shown to stimulate exocytosis in Paneth cells, epithelial granulocytes located at the base of the crypts of Lieberkühn (Satoh *et al.*, 1992). The granules of these cells contain a variety of anti-microbial products, including an array of pore-forming proteins termed cryptdins or crypt defensins (Ouellette and Selsted, 1996), although to our knowledge these have not been tested for toxicity against macroparasites such as nematodes. The presence of muscarinic ACh receptors (mAChR) on lymphocytes has been inferred from radioligand binding studies using [^{3}H]quinuclidinyl benzilate or [^{3}H]*N*-methyl scopolamine (Eva *et al.*,

1989). These reports should be treated with caution, as although muscarinic agonists induce interleukin-2 secretion in a murine T cell hybrid transfected with a gene encoding a muscarinic subtype 1 (m1) AChR (Desai *et al.*, 1990), the demonstration of naturally occurring AChRs on lymphocytes has relied predominantly on radioligand binding, and is complicated by the finding that, in lymphocytes, antagonists show a lower affinity than in other tissues (Costa *et al.*, 1995). Conclusive determination of the expression of AChRs on lymphoid or myeloid cells should now be easily demonstrable with the array of antibody and DNA probes available. Messenger RNA for m3, m4 and m5 subtypes has been detected by RT-PCR in human blood mononuclear cells, and m3 and m4 receptors have similarly been detected in rat blood mononuclear cells, although at levels approximately 10^2 and 10^5 times lower than in cells of the cerebral cortex (Costa *et al.*, 1994, 1995). The significance of such low-level expression of AChRs is unclear but it should be emphasized that these data were obtained with quiescent cells. An immediate imperative, therefore, is to correlate alterations in levels of expression of these receptors as a result of nematode infection with functional assays.

Hydrolysis of alternative substrates

It is possible that nematode-secreted AChEs act on alternative substrates to ACh. We had previously suggested, on the basis of structural similarity, that platelet-activating factor (PAF), a potent phospholipid mediator of inflammation, might represent such an alternative substrate (Blackburn and Selkirk, 1992b) but subsequent studies demonstrated that purified AChEs did not cleave PAF, and the enzyme responsible for this activity in secreted products of *N. brasiliensis*, PAF acetylhydrolase, was purified and defined as a distinct heterodimeric protein (Grigg *et al.*, 1996). Although an open mind on the subject sould be kept, the strict substrate specificity of the nematode-secreted AChEs suggests that they most likely act on ACh alone.

Binding to ingested AChE inhibitors

A number of natural inhibitors of cholinesterases exist in the form of plant alkaloids such as physostigmine from the calabar bean, or chaconine and solanine from green potatoes (Jbilo *et al.*, 1994). BuChEs may therefore act as a first line of defence against such toxins that are eaten or inhaled, providing protection against sensitive neuronal AChEs by acting as a systemic 'sink' for absorption of these compounds, and this suggestion is supported by a study in which high levels of BuChE mRNA were found in the liver and lungs (Jbilo *et al.*, 1994). In a similar manner, plant-derived inhibitors ingested via the alimentary tract might have presented sufficient threat of interference with parasite neuromuscular function as to drive the expression of secreted AChEs as a front-line defence. There are logistical problems with this argument, however – not least the natural avoidance of such foodstuffs by animals, and the presence of high concentrations of host-derived BuChE, which would seem to render the action of parasite AChEs redundant. Expression of BuChEs is certainly affected by exposure of individuals to cholinesterase inhibitors such as organophosphates (Soreq and Zakut, 1990), although it is unlikely that nematode-secreted AChEs have evolved to provide protection against such artificial compounds. Nevertheless, it would be interesting to evaluate the resistance of nematodes to ChE inhibitors with respect to their capacity to secrete AChEs.

Acetylcholine Receptor Expression in the Intestinal Tract

We are currently engaged in examining the expression of muscarinic AChRs in rat intestinal mucosa throughout infection with *N. brasiliensis*. Immunocytochemistry with the monoclonal antibody M35, which is pan-specific for all muscarinic receptor subtypes (Matsuyama *et al.*, 1988), has yielded staining of the basolateral domain of mucosal epithelial cells in the jejunum as expected, but also revealed expression of mAChRs on unidentified cells in the lamina propria, observed only in infected animals. The intensity of staining increased progressively, following entry of parasites into the jejunum, reaching a plateau around day 10 post-infection (Russell and Henson, 2000, unpublished results). These alterations in AChR expression may provide a lead to understanding the reasons for AChE secretion by parasitic nematodes. It is therefore a priority to identify the AChR-positive cells in the lamina propria, and we are currently addressing this issue. If these experiments implicate AChR expression on lymphoid or myeloid cells, then this would open up a new area for investigation of neural regulation of the immune system. Possible points of interaction with mucosal cellular functions stimulated by ACh are summarized in Fig. 11.4.

Several of the postulated roles for nematode-secreted AChEs assume that they gain access to the intestinal mucosa. Several possibilities exist for transport of parasite AChE across the epithelial cell barrier, such as: (i) utilization of existing pathways for receptor-mediated transcytosis; (ii) a paracellular route facilitated by parasite-secreted proteases as observed for a bacterial elastase (Azghani *et al.*, 1993); and (iii) increased paracellular permeability resulting from inflammatory events in the mucosa. We consider the latter suggestion most likely, as this has been duplicated by *ex vivo* perfusion with rat mast cell protease II (Scudamore *et al.*, 1995). Moreover, cholinergic stimulation attenuates epithelial barrier properties to macromolecules in rat ileal crypts (Phillips *et al.*, 1987).

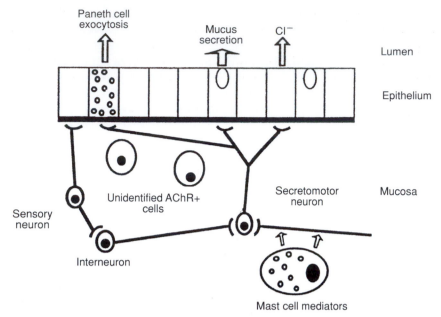

Fig. 11.4. Model for cholinergic signalling in the intestinal mucosa, providing a possible rationale for AChE secretion by parasitic nematodes. ACh released from enteric cholinergic motor neurons stimulates chloride secretion, mucus secretion and Paneth cell exocytosis through muscarinic receptors. Secretory responses may be modulated by mast cell mediators, either directly or via the induction of neural reflex programmes. The role of muscarinic receptor-positive cells in the lamina propria of rats infected with *N. brasiliensis* is undetermined, as are potential mechanisms of trans-epithelial transport of the enzymes. Adapted from Cooke (1984).

Nematode Acetylcholinesterases as Drug Targets

Effective broad-spectrum anthelmintics have been available for many years and, with the notable exception of the benzimidazoles, tend to be targeted towards neuromuscular functions. The avermectins promote opening of glutamate-gated chloride channels, exerting a major effect via paralysis of pharyngeal pumping. Drugs such as levamisole and pyrantel are nicotinic acetylcholine receptor agonists, and act at neuromuscular junctions causing spastic paralysis, whereas piperazine is a GABA (gamma-amino-butyric acid) receptor agonist and causes flaccid paralysis. AChE inhibitors such as haloxon, metrifonate and aldicarb result in accumulation of ACh at neuromuscular junctions and also lead to paralysis via sustained contraction (Martin, 1997; see also Chapter 21).

There is a need for development of new anthelmintics, as resistance in livestock has been recorded against benzimidazoles, imidothiazoles and the avermectin/macrocyclic lactones, in addition to older drugs such as organophosphates. Although the overall incidence is difficult to determine, there is a clear trend towards development of resistance on a global scale (Sangster, 1999). There is a natural tendency to scepticism when considering AChEs as targets for anthelmintics, as the AChE inhibitors employed to date have been restricted to organophosphates and carbamates, irreversible or quasi-irreversible inhibitors that act by phos-phorylating or carbamylating the active-site serine residue of the enzyme. In general, these compounds are non-selective towards nematode AChEs and thus have a low therapeutic index. Variable levels of toxicosis may result from treatment of animals with organophosphates, and they also rep-resent a significant health hazard to handlers of preparations for control of ectoparasites. Environmental contamination is a major concern, as both classes of compounds have been widely used as insecticides in addition to control of plant-parasitic nematodes. Thus, although these compounds are still in use, they are being progressively withdrawn from the market.

In contrast, reversible inhibitors of AChE act in a competitive manner, effecting blockade of substrate at the active site, or non-competitive bind-ing to a region of the enzyme that has been termed the 'peripheral' site, near the lip of the active-site gorge. There are extensive panels of different classes of these inhibitors, developed not only for use as insecticides but also for the clinical treatment of Alzheimer's disease (Giacobini, 1998). Peripheral site ligands can be highly selective (Eichler *et al.*, 1994). It is therefore quite feasible that compounds with selective activity against nematode enzymes might be identified or designed, and this possibility is particularly attractive given that the enzymes are well-validated targets for anthelmintics. The primary potential inhibitor selectivity determinants are the eight residues clustered near the opening to the active-site gorge. Five of these residues (Y70, Q74, S81, W279 and G335) in vertebrate AChEs are substituted in the nematode enzymes that we have sequenced to date by S/T, T, A, N and Y, respectively (Hussein *et al.*, 1999a, 2000). In addition, there are substantial differences in 'second shell' residues, which do not flank the gorge but interact with gorge residues, and these residues are known to play a role in selectivity to inhibitors (Mutero *et al.*, 1994). We therefore intend to screen both secreted and neuronal nematode enzymes for selective inhibitors, correlating this information with differences in primary and ultimately tertiary structure of the enzymes.

Concluding Remarks

Cholinesterases secreted by parasitic nematodes of (predominantly) the alimentary tract or other mucosal tissues are authentic AChEs when ana-lysed by substrate specificity, inhibitor sensitivities and primary structure. In the first two respects, they resemble vertebrate AChEs, whereas somatic (and therefore presumably neuronal) enzymes of nematodes analysed to date display enzymatic properties similar to those of other invertebrate AChEs. The function of the secreted enzymes remains undefined, but may be related to regulation of physiological responses that promote expulsion of parasites by cholinergic elements of the enteric nervous system. These potentially include fluid and mucus secretion from enterocytes and degranulation of Paneth cells, though the possibility of activation of alternative lymphoid cells cannot be discounted until further information is available. Although primary structures of neuronal AChEs have not yet been fully reported for animal parasitic nematodes, the secreted AChEs show sufficient significant differences in residues at the entrance to the active-site gorge to suggest that development of novel, selective inhibitors should be feasible.

Acknowledgements

This work was supported by the Wellcome Trust. We acknowledge the valuable contributions of Matilde Chacón and Angela Smith to these studies. Figures 11.1 and 11.4 have been reproduced or modified from previous publications with permission.

References

Arpagaus, M., Richier, P., Bergé, J.-B. and Toutant, J.-P. (1992) Acetylcholinesterases of the nematode *Steinernema carpocapsae*. Characterization of two types of amphiphilic forms differing in their mode of membrane association. *European Journal of Biochemistry* 207, 1101–1108.

Arpagaus, M., Fedon, Y., Cousin, X., Chatonnet, A., Bergé, J.-B., Fournier, D. and Toutant, J.-P. (1994) cDNA sequence, gene structure, and *in vitro* expression of *ace-1*, the gene encoding acetylcholinesterase of class A in the nematode *Caenorhabditis elegans*. *Journal of Biological Chemistry* 269, 9957–9965.

Azghani, A.O., Gray, L.D. and Johnson, A.R. (1993) A bacterial protease perturbs the paracellular barrier function of transporting epithelial cell monolayers in culture. *Infection and Immunity* 61, 2681–2686.

Blackburn, C.C. and Selkirk, M.E. (1992a) Characterisation of the secretory acetylcholinesterases from adult *Nippostrongylus brasiliensis*. *Molecular and Biochemical Parasitology* 53, 79–88.

Blackburn, C.C. and Selkirk, M.E. (1992b) Inactivation of platelet activating factor by a putative acetylhydrolase from the gastrointestinal nematode parasite *Nippostrongylus brasiliensis*. *Immunology* 75, 41–46.

Bremner, K.C., Ogilvie, B.M. and Berie, D.A. (1973) Acetylcholinesterase secretion by parasitic nematodes III. *Oesophagostomum* spp. *International Journal for Parasitology* 3, 609–618.

Castro, G.A. and Arntzen, C.J. (1993) Immunophysiology of the gut: a research frontier for integrative studies of the common mucosal immune system. *American Journal of Physiology* G599–G610.

Cooke, H.J. (1984) Influence of enteric cholinergic neurons on mucosal transport in guinea pig ileum. *American Journal of Physiology* 246, G263–G267.

Costa, P., Traver, D.J., Auger, C.B. and Costa, L.G. (1994) Expression of cholinergic muscarinic receptor subtypes mRNA in rat blood mononuclear cells. *Immunopharmacology* 28, 113–123.

Costa, P., Auger, C.B., Traver, D.J. and Costa, L.G. (1995) Identification of m3, m4 and m5 subtypes of muscarinic receptor mRNA in human blood mononuclear cells. *Journal of Neuroimmunology* 60, 45–51.

Cousin, X., Bon, S., Duval, N., Massoulié, J. and Bon, C. (1996) Cloning and expression of acetylcholinesterase from *Bungarus fasciatus* venom. *Journal of Biological Chemistry* 271, 15099–15108.

Culetto, E., Combes, D., Fedon, Y., Roig, A., Toutant, J.P. and Arpagaus, M. (1999) Structure and promoter activity of the 5′ flanking region of *ace-1*, the gene encoding acetylcholinesterase of class A in *Caenorhabditis elegans*. *Journal of Molecular Biology* 290, 951–966.

Desai, D.M., Newton, M.E., Kadlecek, T. and Weiss, A. (1990) Stimulation of the phosphatidyl-inositol pathway can induce T-cell activation. *Nature* 348, 66–69.

deVos, T. and Dick, T.A. (1992) Characterization of cholinesterases from the parasitic nematode *Trichinella spiralis*. *Comparative Biochemistry and Physiology C* 103, 129–134.

Edwards, A.J., Burt, J.S. and Ogilvie, B.M. (1971) The effect of host immunity upon some enzymes of the parasitic nematode *Nippostrongylus brasiliensis*. *Parasitology* 62, 339–347.

Eichler, J., Anselment, A., Sussman, J.L., Massoulié, J. and Silman, I. (1994) Differential effects of 'peripheral' site ligands on *Torpedo* and chicken acetylcholinesterase. *Molecular Pharmacology* 45, 335–340.

Eva, C., Ferrero, P., Rocca, P., Funaro, A., Bergamasco, B., Ravizza, L. and Genazzani, E. (1989) [^{3}H] N-methylscopolamine binding to muscarinic receptors in human peripheral blood lymphocytes: characterization, localization on T-lymphocyte subsets and age-dependent changes. *Neuropharmacology* 28, 719–726.

Foster, N. and Lee, D.L. (1996) A vasoactive intestinal polypeptide-like protein excreted/secreted by *Nippostrongylus brasiliensis* and its effect on contraction of uninfected rat intestine. *Parasitology* 112, 97–104.

Foster, N., Dean, E.J. and Lee, D.L. (1994) The effect of homogenates and excretory/secretory products of *Nippostrongylus brasiliensis* and of acetylcholinesterases on the amplitude and frequency of contraction of uninfected rat intestine *in vitro*. *Parasitology* 108, 453–459.

Giacobini, E. (1998) Invited review: cholinesterase inhibitors for Alzheimer's disease therapy: from tacrine to future applications. *Neurochemistry International* 32, 413–419.

Gnagey, A.L., Forte, M. and Rosenberry, T.L. (1987) Isolation and characterization of acetylcholinesterase from *Drosophila*. *Journal of Biological Chemistry* 262, 13290–13298.

Grauso, M., Culetto, E., Combes, D., Fedon, Y., Toutant, J.-P. and Arpagaus, M. (1998) Existence of four acetylcholinesterase genes in the nematodes *Caenorhabditis elegans* and *Caenorhabditis briggsae*. *FEBS Letters* 424, 279–284.

Griffiths, G. and Pritchard, D.I. (1994) Purification and biochemical characterisation of acetylcholinesterase (AChE) from the excretory/secretory products of *Trichostrongylus colubriformis*. *Parasitology* 108, 579–586.

Grigg, M.E., Gounaris, K. and Selkirk, M.E. (1996) Characterisation of a platelet-activating factor acetylhydrolase secreted by the nematode parasite *Nippostrongylus brasiliensis*. *Biochemical Journal* 317, 541–547.

Grigg, M.E., Tang, L., Hussein, A.S. and Selkirk, M.E. (1997) Purification and properties of monomeric (G1) forms of acetylcholinesterase secreted by *Nippostrongylus brasiliensis*. *Molecular and Biochemical Parasitology* 90, 513–524.

Harel, M., Sussman, J.L., Krejci, E., Bon, S., Chanal, P., Massoulié, J. and Silman, I. (1992) Conversion of acetylcholinesterase to butyrylcholinesterase: modelling and mutagenesis. *Proceedings of the National Academy of Sciences USA* 89, 10827–10831.

Hussein, A.S., Chacón, M.R., Smith, A.M., Tosado-Acevedo, R. and Selkirk, M.E. (1999a) Cloning, expression and properties of a non-neuronal secreted acetylcholinesterase from the parasitic nematode *Nippostrongylus brasiliensis*. *Journal of Biological Chemistry* 274, 9312–9319.

Hussein, A.S., Grigg, M.E. and Selkirk, M.E. (1999b) *Nippostrongylus brasiliensis*: characterisation of a somatic amphiphilic acetylcholinesterase with properties distinct from the secreted enzymes. *Experimental Parasitology* 91, 144–150.

Hussein, A.S., Smith, A.M., Chacón, M.R. and Selkirk, M.E. (2000) A second non-neuronal secreted acetylcholinesterase from the parasitic nematode *Nippostrongylus brasiliensis*: determinants of substrate specificity. *European Journal of Biochemistry* 267, 2276–2282.

Javed, N.H., Wang, Y.Z. and Cooke, H.J. (1992) Neuroimmune interactions: role for cholinergic neurons in intestinal anaphylaxis. *American Journal of Physiology* 263, G847–G852.

Jbilo, O., Bartels, C.F., Chatonnet, A., Toutant, J.P. and Lockridge, O. (1994) Tissue distribution of human acetylcholinesterase and butyrylcholinesterase messenger RNA. *Toxicon* 32, 1445–1457.

Jones, V.E. and Ogilvie, B.M. (1972) Protective immunity to *Nippostrongylus brasiliensis* in the rat. III. Modulation of worm acetylcholinesterase by antibodies. *Immunology* 22, 119–129.

Kolson, D.L. and Russell, R.L. (1985) A novel class of acetylcholinesterase, revealed by mutations, in the nematode *Caenorhabditis elegans*. *Journal of Neurogenetics* 2, 93–110.

Lee, D.L. (1970) The fine structure of the excretory system in adult *Nippostrongylus brasiliensis* (Nematoda) and a suggested function for the 'excretory glands'. *Tissue and Cell* 2, 225–231.

Lee, D.L. (1996) Why do some nematode parasites of the alimentary tract secrete acetylcholinesterase? *International Journal for Parasitology* 26, 499–508.

Lockridge, O., Bartels, C.F., Vaughan, T.A., Wong, C.K., Norton, S.E. and Johnson, L.L. (1987) Complete amino acid sequence of human serum cholinesterase. *Journal of Biological Chemistry* 262, 549–557.

Martin, R.J. (1997) Modes of action of anthelmintic drugs. *Veterinary Journal* 154, 11–34.

Massoulié, J., Pezzementi, L., Bon, S., Krejci, E. and Vallette, F.-M. (1993) Molecular and cellular biology of cholinesterases. *Progress in Neurobiology* 41, 31–91.

Matsuyama, T., Luiten, P.G.M., Spencer, J. and Strosberg, A.D. (1988) Ultra-structural localisation of immunoreactive sites for muscarinic acetylcholine receptor proteins in the rat cerebral cortex. *Neuroscience Research Communications* 2, 69–76.

McKeand, J.B., Knox, D.P., Duncan, J.L. and Kennedy, M.W. (1994) The immunogenicity of the acetylcholinesterases of the cattle lungworm, *Dictyocaulus viviparus. International Journal for Parasitology* 24, 501–510.

Mutero, A., Pralavorio, M., Bride, J.M. and Fournier, D. (1994) Resistance-associated point mutations in insecticide-insensitive acetylcholinesterase. *Proceedings of the National Academy of Sciences USA* 91, 5922–5926.

Ogilvie, B.M., Rothwell, T.L.W., Bremner, K.C., Schitzerling, H.J., Nolan, J. and Keith, R.K. (1973) Acetylcholinesterase secretion by parasitic nematodes. 1. Evidence for secretion of the enzyme by a number of species. *International Journal for Parasitology* 3, 589–597.

Ordentlich, A., Barak, D., Kronman, C., Flashner, Y., Leitner, M., Segall, Y., Ariel, N., Cohen, S., Velan, B. and Shafferman, A. (1993) Dissection of the human acetylcholinesterase active center determinants of substrate specificity. *Journal of Biological Chemistry* 268, 17083–17095.

Ouellette, A.J. and Selsted, M.E. (1996) Paneth cell defensins: endogenous peptide components of intestinal host defense. *FASEB Journal* 10, 1280–1289.

Phillips, T.E., Phillips, T.H. and Neutra, M.H. (1984) Regulation of intestinal goblet cell secretion. IV. Electrical field stimulation *in vitro. American Journal of Physiology* 247, G682–G687.

Phillips, T.E., Phillips, T.L. and Neutra, M.R. (1987) Macromolecules can pass through occluding junctions of rat ileal epithelium during cholinergic stimulation. *Cell and Tissue Research* 247, 547–554.

Pritchard, D.I., Brown, A. and Toutant, J.-P. (1994) The molecular forms of acetylcholinesterase from *Necator americanus* (Nematoda), a hookworm parasite of the human intestine. *European Journal of Biochemistry* 219, 317–323.

Rand, J.B. and Nonet, M.L. (1997) Synaptic transmission. In: Riddle, D.L., Blumenthal, T., Meyer, B.J. and Priess, J.R. (eds) *C. elegans II.* Cold Spring Harbor Press, Cold Spring Harbor, New York, pp. 611–643.

Rhoads, M.L. (1981) Cholinesterase in the parasitic nematode *Stephanurus dentatus.* Characterisation and sex dependence of a secretory cholinesterase. *Journal of Biological Chemistry* 256, 9316–9323.

Rhoads, M.L. (1984) Secretory cholinesterases of nematodes: possible functions in the host–parasite relationship. *Tropical Veterinarian* 2, 3–10.

Ripoll, D.R., Faerman, C.H., Axelson, P.H., Silman, I. and Sussman, J.L. (1993) An electrostatic mechanism for substrate guidance down the aromatic gorge of acetylcholinesterase. *Proceedings of the National Academy of Sciences USA* 90, 5128–5132.

Sanderson, B.E., Jenkins, D.C. and Ogilvie, B.M. (1972) *Nippostrongylus brasiliensis:* relation between immune damage and acetylcholinesterase levels. *International Journal for Parasitology* 2, 227–232.

Sangster, N.C. (1999) Anthelmintic resistance: past, present and future. *International Journal for Parasitology* 29, 115–124.

Satoh, Y., Ishikawa, K., Oomori, Y., Takeda, S. and Ono, K. (1992) Bethanechol and a G-protein activator, NaF/AlCl3, induce secretory response in Paneth cells of mouse intestine. *Cell and Tissue Research* 269, 213–220.

Scudamore, C.L., Thornton, E.M., McMillan, L., Newlands, G.F.J. and Miller, H.R.P. (1995) Release of the mucosal mast cell granule chymase, rat mast cell protease-II, during anaphylaxis is associated with the rapid development of paracellular permeability to macromolecules in rat jejunum. *Journal of Experimental Medicine* 182, 1871–1881.

Segerberg, M.A. and Stretton, A.O. (1993) Actions of cholinergic drugs in the nematode *Ascaris suum*. Complex pharmacology of muscle and motorneurons. *Journal of General Physiology* 101, 271–296.

Soreq, H. and Zakut, H. (1990) *Cholinesterase Genes: Multileveled Regulation.* Karger, Basel, 108 pp.

Specian, R.D. and Neutra, M.R. (1980) Mechanism of rapid mucus secretion in goblet cells stimulated by acetylcholine. *Journal of Cell Biology* 85, 626–640.

Sussman, J.L., Harel, M., Frolow, F., Oefner, C., Goldman, A., Toker, L. and Silman, I. (1991) Atomic structure of acetylcholinesterase from *Torpedo californica*: a prototypic acetylcholine-binding protein. *Science* 253, 872–879.

Talesa, V., Romani, R., Grauso, M., Rosi, G. and Giovanni, E. (1997) Expression of a single dimeric membrane-bound acetylcholinesterase in *Parascaris equorum*. *Parasitology* 115, 653–660.

Taylor, P. (1991) The cholinesterases. *Journal of Biological Chemistry* 266, 4025–4028.

Toutant, J.-P. (1989) Insect acetylcholinesterase: catalytic properties, tissue distribution and molecular forms. *Progress in Neurobiology* 32, 423–446.

The Surface and Secreted Antigens of *Toxocara canis*: Genes, Protein Structure and Function

Rick M. Maizels and Alex Loukas

Institute of Cell, Animal and Population Biology, University of Edinburgh, Edinburgh EH9 3JT, UK

Introduction

Toxocara canis is a remarkable nematode parasite, commonly found in dogs but able to invade a wide range of other hosts, including humans (Glickman and Schantz, 1981; Lewis and Maizels, 1993). It displays an array of striking features that excite interest: a tissue-dwelling phase that can endure for many years; an ability to cross the placenta and infect unborn pups; a tropism for neurological tissue in paratenic hosts; survival *in vitro* for many months in serum-free medium; the secretion of a set of biologically active glycoproteins *in vivo* and *in vitro*; and the possession of a surface glycocalyx that is jettisoned under immune attack. For all these reasons, *T. canis* presents an important model system for parasitic nematodes (Maizels and Robertson, 1991; Maizels *et al.*, 1993). In addition, toxocariasis causes a significant pathology in humans (Gillespie, 1987, 1993) as well as in dogs (Lloyd, 1993), and elimination of this infection would be a highly desirable goal.

This chapter summarizes the biological features of this parasite, and focuses on the molecular structure and biological function of the major glycoproteins found on the surface coat and secreted by larvae maintained *in vitro*. Since this subject was last reviewed (Maizels *et al.*, 1993), substantial progress has been made in defining the primary sequence, glycosylation and antigenicity of secreted proteins. In particular, sequence information has led to demonstration of functional properties that may explain how this extraordinary organism defuses host immunity at the molecular level.

Developmental Biology of *Toxocara*

T. canis infection is initiated when embryonated eggs are ingested into the stomach. The resistant eggshell is broken down and infective larvae emerge, ready to penetrate the intesinal mucosa. Invasion occurs in all mammalian species, but only in canid hosts do larvae progress along a typical ascarid nematode developmental pathway by migrating through the lungs, trachea and oesophagus back to the gastrointestinal tract. In other species, the larvae remain in the tissue-migratory phase without ever developing (Sprent, 1952); they are locked in the larval stage from which they are only released if the paratenic host is carnivorously consumed by canid species (Warren, 1969). In dogs, there is a further fascinating property: many invading larvae are restrained from development, arrested in the tissues until reactivated during the third trimester of pregnancy; then, transplacental invasion of the gestating pups can occur, and subsequently additional larvae pass into the colostrum to cause postnatal infection (Sprent, 1958; Griesemer *et al.*, 1963; Scothorn *et al.*, 1965; Burke and Roberson, 1985). Thus female dogs contain a reservoir of infection, producing waves of larvae for each new litter of offspring, and the majority of pups are infected at birth with *T. canis*.

Dormant tissue larvae have been recorded as long as 9 years after infection in a mammalian host (Beaver, 1962, 1966), while incubation *in vitro* permits larvae to survive for up to 18 months (de Savigny, 1975). In neither context is any morphological development observed. However, the parasites maintain a high metabolic rate, being dependent (*in vitro*, at least) on a regular supply of glucose and amino acids. Cultured larvae secrete copious quantities of glycoprotein antigens, which are described more fully below as *T. canis* excreted/secreted (TES) products.

Immune Evasion

The ability of arrested-stage larval parasites to survive in the tissues for many years must depend on potent immune-evasive and anti-inflammatory mechanisms operated by the parasite (Ghafoor *et al.*, 1984; Smith, 1991). Secreted macromolecules are the primary candidates for immune evasion mediators, and indeed larvae are found to release large quantities of glycoproteins *in vitro*. Antibodies to secreted TES glycoproteins also detect antigens *in vivo* which appear to have been released from parasites. For this reason, we have analysed the secreted proteins in detail, as set out below. Several routes have been adopted: the direct, biochemical approach of protein analysis; an immunological path of identifying antigenic determinants; and most recently a molecular biological strategy. In the latter respect, we have undertaken a broader survey of the major products encoded by this parasite, expecting that it must devote a large part of its energy to evading the immune response (Tetteh *et al.*, 1999). A range of interesting molecules has emerged, some of which can be predicted from their primary sequence structure to fulfil certain functions: mucins, proteases, enzyme inhibitors, etc. Another group of molecules is even more intriguing: these are unique to the parasite, with no similar proteins found amongst the free-living nematodes, and are produced in very large quantities by the invasive parasite. It is likely that these play a crucial, if untold, role in parasite success.

Excreted/Secreted Glycoproteins (TES)

It has long been considered that the products secreted by nematodes hold the key to their success as parasites (Schwartz, 1921; Stirewalt, 1963). De Savigny (1975) first demonstrated that *T. canis* larvae could be cultured for long periods of time, and that these culture supernatants contained antigens that were specific in diagnosing human toxocariasis. Since then, these products have been analysed with increasingly sophisticated techniques (Sugane and Oshima, 1983; Maizels *et al.*, 1984, 1993; Badley *et al.*, 1987). Metabolic labelling with different precursor amino acids and pulse-chase studies have defined steps in protein synthesis (Page and Maizels, 1992), while the glycan moieties have been studied with lectins (Meghji and Maizels, 1986; Page *et al.*, 1992c), glycanases (Page and Maizels, 1992) and mass spectrometry (Khoo *et al.*, 1991, 1993). Most recently, proteomic tools have been applied to distinguish closely migrating mucin species and to correlate them to a sequence database for *T. canis* (Loukas *et al.*, 1999).

   T. canis larvae secrete at least 50 distinct macromolecules, as enumerated by two-dimensional gel analysis (Page *et al.*, 1991, 1992a), although multiple components may be derived from individual gene products. The major TES proteins are designated TES-32, TES-55, TES-70, etc. according to their apparent molecular weights (in kilodaltons) on SDS-PAGE (Maizels *et al.*, 1984; Meghji and Maizels, 1986; Maizels and Robertson, 1991), and many now have individual gene names according to the function of the protein (Fig. 12.1). The major TES products are all glycosylated, some extensively, and the total secreted material contains 400 µg carbohydrate mg^{-1} protein (Meghji and Maizels, 1986). As described below, the predominant mode of glycosylation is *O*-linked mainly with *N*-acetylgalactosamine and galactose.

Recombinant Excreted/Secreted Proteins

Over the past 5 years, the genes encoding the major surface and secreted antigens of *T. canis* have been cloned and expressed. These are summarized below, giving both the TES designation (indicating molecular weight of mature product) and the new gene name based on deduced or

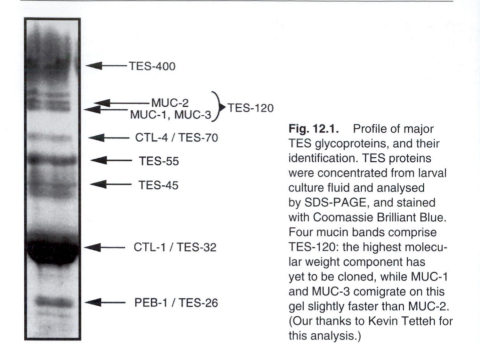

Fig. 12.1. Profile of major TES glycoproteins, and their identification. TES proteins were concentrated from larval culture fluid and analysed by SDS-PAGE, and stained with Coomassie Brilliant Blue. Four mucin bands comprise TES-120: the highest molecular weight component has yet to be cloned, while MUC-1 and MUC-3 comigrate on this gel slightly faster than MUC-2. (Our thanks to Kevin Tetteh for this analysis.)

confirmed biological function. As a result of these investigations, *T. canis* stands as the best-characterized helminth parasite in terms of the panel of proteins that are secreted, and understanding of its surface is also relatively advanced (Page *et al.*, 1992b).

TES-26, Tc-PEB-1, phosphatidylethanolamine-binding protein

The gene for TES-26 is one of the most highly expressed in the infective larval stage, and it recently accounted for 1.9% of the clones randomly selected from a cDNA library (Tetteh *et al.*, 1999). However, no mRNA for TES-26 could be detected in adult-stage parasites (Gems *et al.*, 1995). The sequence shows it to be a genetic fusion beween two modules of disparate ancestry, separated by a transmembrane segment. The C-terminal 138 amino acids show significant similarity (36% amino acid identity) to mammalian phosphatidylethanolamine (PE)-binding proteins, such as those highly expressed in sperm (Gems *et al.*, 1995). Binding studies with recombinant TES-26 showed it to be a functional PE-binding protein (Gems *et al.*, 1995). Homologues have also been identified in the filarial nematodes *Onchocerca volvulus* (Lobos *et al.*, 1991) and *Brugia malayi* (Blaxter *et al.*, 1996). TES-26, or *Tc*-PEB-1 as it is now designated, is unique among this gene family, however, in containing a 72 amino acid (aa) N-terminal extension (aa 22–93) consisting of two 36-aa six-cysteine domains, termed NC6 or SXC motifs (discussed in more detail below). Despite the abundance of its mRNA, TES-26 is a relatively minor TES antigen, and is not readily apparent on the surface. It is not known whether the N-terminal SXC region is extracellular, but this seems likely because the activity of PE-binding proteins in mammalian cells is often on the cytoplasmic face of the plasma membrane.

TES-32, Tc-CTL-1, C-type lectin-1

TES-32 is the most abundant single protein product secreted by the parasite. It is also heavily labelled by surface iodination of live larvae (Maizels *et al.*, 1984, 1987), and is known by monoclonal antibody reactivity to be expressed in the cuticular matrix of the larval parasite (Page *et al.*, 1992a). TES-32 was cloned by matching peptide sequence derived from gel-purified protein to an expressed sequence tag (EST) dataset of randomly selected clones from a larval cDNA library (Loukas *et al.*, 1999). Because of the high level of expression of TES-32 mRNA, clones encoding this protein were repeatedly sequenced and deposited in the dataset (Tetteh *et al.*, 1999). Full sequence determination showed a major domain with similarity to mammalian C-type (calcium-dependent) lectins (C-TLs), together with shorter N-terminal tracts rich in cysteine and threonine residues. Native TES-32 was then shown to bind to immobilized monosaccharides in a calcium-dependent manner (Loukas *et al.*, 1999).

The identity of TES-32 and CTL-1 was confirmed by polyclonal antibodies to recombinant CTL-1, which bound to native TES-32, and by monoclonal antibody Tcn-3, raised to native TES-32 (Maizels *et al.*, 1987), which specifically recognized recombinant CTL-1. The CTL-1 sequence also contained three sites for *N*-glycosylation, which had previously been shown to be present on TES-32 (Page and Maizels, 1992). Both Tcn-3 and polyclonal antibody to the recombinant CTL-1 protein localize to the cuticle of the infective larvae by immunoelectron microscopy (Fig. 12.2).

Binding studies confirmed that TES-32/*Tc*-CTL-1 binds to carbohydrates, and moreover that it has an unusual specificity in adhering to both mannose and galactose-type sugars. Molecular modelling predicts a structure very similar to mammalian mannose-binding lectin, but with crucial differences around the binding site (Fig. 12.3). Thus, while the mannose-specific lectin is impeded by a histidine packing residue from binding galactose, the *Toxocara* lectin has a cysteine loop resulting in a much more open binding site (Loukas *et al.*, 1999). Most members of the C-TL superfamily are mammalian immune system receptors such as selectin, CD23 and DEC-205. Other C-type lectins include mannose-binding protein-A (MBP-A), found in mammalian serum, and invertebrate innate defence molecules (Weis *et al.*, 1998). Thus, this was the first example of a protein related to host defence being used by a parasite,

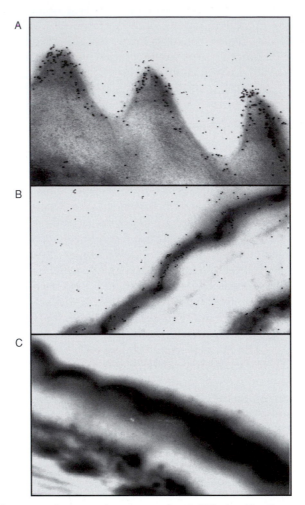

Fig. 12.2. Immuno-electron microscopy of anti-CTL-1 antibodies on sections of infective larvae of *Toxocara canis*: (A) monoclonal antibody Tcn-3; (B) polyclonal antibody to recombinant CTL-1; (C) normal mouse serum. *T. canis* larvae were fixed in 2% glutaraldehyde/0.5% paraformaldehyde for 30 min. Fixed larvae were then incubated sequentially for 6 h in each of 0.6 M, 1.2 M, 1.8 M and 2.3 M sucrose in PBS at 4°C, sectioned at −100°C and trimmed at −80°C using an FCS cryo-attachment for an Ultracut UCT ultramicrotome (Leica). Sections were placed on grids and blocked with 10% BSA followed by incubation in primary antibody (mouse anti-rCTL-1, monoclonal antibody Tcn-3 or NMS) diluted 1 : 50 in PBS-T/1% BSA overnight at 4°C. Grids were then washed with PBS-T and incubated with gold (10 nm particle size) conjugated goat anti-mouse IgG (British BioCell) diluted 1 : 50 for 1 h at room temperature. Grids were washed with distilled water and stained with uranyl acetate for 5 min before viewing with a Philips CM120 Biotwin electron microscope.

presumably to subvert the immune system (Loukas *et al.*, 1999). TES-32 has close equivalents in a 33 kDa *T. cati* protein (Kennedy *et al.*, 1987) and a *T. vitulorum* product of 30 kDa (Page *et al.*, 1991), so the expression of such proteins may be a common feature of these organisms.

Several other C-type lectins have since been found in *Toxocara* (Fig. 12.4). Two are simple variants of TES-32, differing by 13–17% in amino acid sequence but with identical ligand binding sites; these have been termed CTL-2 and CTL-3. It has yet to be determined whether these additional lectins are alleles or represent different coding loci, and it has not been established whether they are also secreted. A fourth lectin, CTL-4, corresponds to TES-70 (Loukas *et al.*, 2000), as described below. In addition, there is some evidence that TES-45 and TES-55 are lectins, as detailed below.

TES-45 and TES-55

TES-45 and TES-55 are two glycoproteins that have yet to be identified at a genetic level, but evidence has been obtained that they may also be lectins. Carbohydrate affinity chromatography with mannose-agarose shows that TES-32 selectively binds as expected, but that TES-45 is also present in small amounts (Loukas *et al.*, 1999); unlike TES-32, TES-45 does not bind to *N*-acetylgalactosamine. No sequence information has yet been obtained on TES-45, but it is recognized by polyclonal antibodies generated to TES-32,

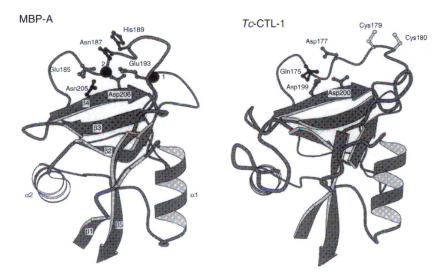

Fig. 12.3. Model of the structure of Tc-CTL-1, based on the known crystal structure of rat MBP-A. Model created by Dr Nick Mullin and published in Loukas *et al.* (1999).

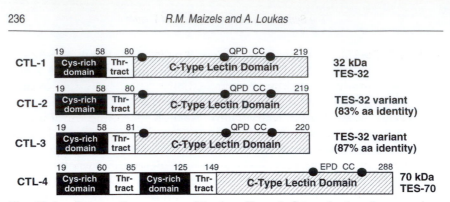

Fig. 12.4. Schematic summary of the four *T. canis* C-type lectins discovered. All have an N-terminal predicted signal sequence of 18 aa, with mature proteins starting at amino acid 19. Other numbers above the bars are those of the last amino acid in each domain. The key carbohydrate-binding residues (QPD or EPD) are shown, as is the position of the unusual double cysteine in the adjacent loop. Solid circles represent *N*-glycosylation sites.

indicating a close sequence similarity. TES-55 has been purified and tryptic peptides have been sequenced (M. Hintz, University of Giessen, Germany). These sequences are similar but not identical to TES-32, indicating that this is an as yet uncloned new lectin.

TES-70, Tc-CTL-4, C-type lectin-4

TES-70 is a relatively abundant secreted glycoprotein, recently found to be a new C-type lectin. This novel lectin gene was isolated by screening a larval cDNA expression library with antibodies to total parasite secretions (TES). Parallel purification and peptide analysis identified a sequence corresponding to this new gene, which proved to encode the TES-70 product. Full sequence of this gene revealed that it is indeed a C-type lectin, now designated CTL-4, and antibodies to recombinant CTL-4 specifically react to TES-70. The predicted protein is longer than CTL-1/TES-32, containing a duplication of the N-terminal cysteine- and threonine-rich tracts (Fig. 12.4), but the sequence is still substantially shorter than would be expected for a protein of 70 kDa. It seems likely that the threonine-rich segments are heavily *O*-glycosylated as well as *N*-glycosylated, as previously reported. Although native TES-70 does not bind to immobilized monosaccharide, it does exhibit a lectin-like activity in specifically binding to the surface of epithelial cells (MDCK, canine kidney) in a calcium-dependent manner (Loukas *et al.*, 2000). This is an important demonstration that the ligands for *T. canis* lectins include determinants on the host cell surface. Lectin-like activity of TES-70 may also explain why it selectively reacts with normal immunoglobulins in immunoprecipitation reactions (Loukas *et al.*, 2000).

TES-120, Tc-MUC-1, MUC-2 and MUC-3 mucins

The major surface coat component of *Toxocara* larvae runs as a set of four closely migrating bands with apparent mobility of 120 kDa on SDS-PAGE. One of these was cloned and sequenced, identified as a serine-rich mucin and designated MUC-1 (Gems and Maizels, 1996). We have now established that there are at least five distinct mucin genes in this parasite, which bear general similarity but important distinctions. Thus, MUC-2, MUC-3, MUC-4 and MUC-5 are all threonine-rich rather than serine-rich, and all five differ in the repeat motifs within the mucin domains. All have similar non-mucin, cysteine-rich domains originally termed NC6 (nematode six-cysteine) domains, and since renamed SXC (six-cysteine), as described below. All mucins have a pair of SXC domains at their C-terminus, while MUC-3 and MUC-5 also have paired N-terminal SXC domains.

A combination of mass spectrometry on purified bands, with limited protein sequence, and generation of antibodies to mucin-specific sequences has established that MUC-1, MUC-2 and MUC-3 are all present in the complex of secreted TES-120 proteins (Loukas *et al.*, 2000). In addition, a further band migrating slightly more slowly than TES-120 represents an as yet unidentified threonine-rich mucin-like gene product. Two further mucins, MUC-4 and MUC-5, have also been cloned (A. Doedens *et al.*, 2000, unpublished results; Loukas *et al.*, 2000), but as yet there are no data to suggest that these are secreted by the larvae. In the case of MUC-5, release seems unlikely because it has a relatively high lysine content and it is known that TES-120 does not incorporate lysine (Gems and Maizels, 1996).

TES-400

TES-400 is a high molecular weight, diffuse component which has so far eluded analysis, and is distinct in many respects from the other TES molecules. Unlike the smaller proteins, TES-400 is not detected on the larval surface by any labelling technique; it appears to be predominantly carbohydrate as judged by gel staining, lectin binding (Meghji and Maizels, 1986) and proteolytic sensitivity. It does not incorporate [^{35}S]-methionine (Meghji and Maizels, 1986; Badley *et al.*, 1987), and incorporates [^{14}C]-serine very poorly (Page and Maizels, 1992). TES-400 is resistant to most proteases, other than pronase, treatment with which increases its apparent molecular weight on SDS-PAGE gels, presumably by removing the SDS-binding peptide moiety (Page and Maizels, 1992). Interestingly, TES-400 reacts with polyclonal antibodies to Tc-CTL-1, but not with the CTL-1-specific monoclonal antibody Tcn-3 (Loukas *et al.*, 1999). This suggests that TES-400 may be analogous to mammalian proteoglycans such as aggrecan and brevican, which contain lectin domains within a highly glycosylated structure.

Novel ES proteins?

The TES proteins described above were initially defined in biochemical terms, and subsequently reconciled to molecular databases. The reverse route has also been pursued: we have characterized the most highly expressed mRNAs in the infective larva, and examined whether they encode secreted proteins. Of the eight most abundant transcripts, three are for known TES proteins (TES-32, TES-26 and TES-120/MUC-1) and one is for a mitochondrial protein. The remaining four highly expressed mRNAs are all large, novel sequences, which we have termed *abundant novel transcript* genes. Together these four transcripts account for 18% of the parasite mRNA represented in a cDNA library. Although these genes bear no coding sequence homology to each other or to any known sequence (including the *C. elegans* genome), they share 3′ untranslated region sequences, indicating that common control elements may underlie their very high rate of transcription. All ANT sequences bear potential *N*-glycosylation sites, and preliminary evidence points to one of them (ANT-005) encoding a protein of 40 kDa in TES (K. Tetteh, 1999, unpublished results).

Secreted Enzymes

Proteases

It is known that *Toxocara* larvae secrete a serine protease (Robertson *et al.*, 1989), which migrated around 120 kDa on substrate gels, but we have been unable to assign this activity to a particular molecule with any certainty. Other protease types have also been found. One, a cathepsin L, is interesting because it has a substrate specificity more reminiscent of a cathepsin B, but its sequence outside the active site is much more related to the cathepsin L subgroup (Loukas *et al.*, 1998). We have also characterized an asparaginyl endopeptidase (legumain), a member of a newly emerging family of cysteine proteases with no structural homology to the papain-like enzymes (Maizels *et al.*, 2000), and a cathepsin Z, a member of a recently defined subgroup with characteristic insertions close to the active site residues (Falcone *et al.*, 2000). So far, none of these cysteine proteases has been found to be secreted by larvae, and no serine protease gene has been isolated. It is likely that many more proteases from this parasite will yet be discovered to play important roles in larval biology.

Superoxide dismutases

The secretion of superoxide dismutase (SOD) by cultured larvae has been detected by conventional biochemical tests, and a recombinant protein

(Tc-SOD-1) has been cloned and expressed (Matzilevich *et al.*, 1999, unpublished results). While antibodies to this protein recognize a dimer of 20/22 kDa in larval extracts, evidence has yet to be obtained that this gene product is represented in larval TES. In addition, several more SOD genes have been identified by an EST project from a cDNA library from this stage. It seems likely, but has yet to be demonstrated, that SODs represent an important part of the parasite's defence mechanism against immune attack, as has been shown for other nematodes (James, 1994; Ou *et al.*, 1995).

The Surface Coat

The external surface of the *T. canis* larva is covered by a carbohydrate-rich glycocalyx termed the surface coat (Maizels and Page, 1990; Page *et al.*, 1992b), a fuzzy envelope 10 nm thick and detached by a similar distance from the epicuticle. The surface coat appears to play a primary role in immune evasion, as it is shed when the parasite is bound by granulocytes (Fattah *et al.*, 1986) or antibodies (Smith *et al.*, 1981; Page *et al.*, 1992b). Surface coats are a common feature of nematode organisms, both parasitic and free-living (Blaxter *et al.*, 1992); hence this very direct means of immune evasion may be a simple adaptation to parasitism.

Biochemical analysis of the surface coat, coupled with electron microscopy, showed that its principal constituent is TES-120, the set of secreted mucins. Other TES glycoproteins, such as TES-32 and TES-70, are not found in the surface coat, but remain associated with the nematode body cuticle. The surface coat carries a strong negative charge, binding to cationic stains such as ruthenium red and cationized ferritin (Page *et al.*, 1992b), but the identity of this charge group has yet to be determined.

The coat mucins are thought to be secreted from two major glands in the larval body: the oesophageal and secretory glands. The latter, previously termed the excretory cell, is directly connected to the cuticle by a duct opening at the secretory pore (Nichols, 1956). It is not certain whether the panel of secreted mucins are all represented in the coat, or whether there are important differences between the two compartments.

The NC6/SXC Domain in *T. canis*

The nematode six-cysteine domain (NC6/SXC) was first recognized in PEB-1/TES-26 (Gems *et al.*, 1995), and then in Tc-MUC-1 (Gems and Maizels, 1996). No fewer than eight different *T. canis* proteins are now known to include NC6/SXC motifs in their structure (Fig. 12.5), and at least five of these are associated with the parasite surface or its secretions.

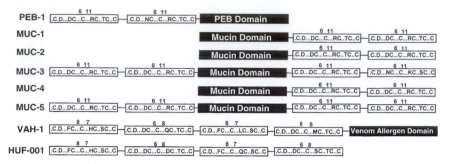

Fig. 12.5. Schematic summary of the eight *T. canis* proteins containing pre-dicted SXC (NC6) domains. The consensus is shown in the N-terminal domain of PEB-1 (phosphatidylethanolamine-binding protein-1) as xCxDxxxDC(6x)C(11x) RCxxTCxxC. This consensus is faithfully repeated in MUC-1 (mucin-1), MUC-2, MUC-4 and MUC-5, and in all but the C-terminal domain of MUC-3. This domain (and the C-terminal SXC domain of PEB-1) show consensus spacing but some variation in consensus residues. Two additional proteins with quadrupled SXC domains differ in spacing between cysteines-2, -3 and -4, and show more variation in consensus residues. These are VAH-1 (venom allergen homologue) and HUF-001 (homologue of unknown function-001).

These include TES-26/PEB-1, MUC-1, MUC-2 and MUC-3, and a relative of the *Ancylostoma*-secreted protein termed venom allergen homologue (VAH-1). It is quite possible that MUC-4 is also secreted, as may be a novel protein consisting solely of four SXC domains which we have named homologue of unknown function-001, or HUF-001 (Tetteh *et al.*, 1999). More recently, the same domains have been found in *C. elegans* (Blaxter, 1998), *Necator americanus* (Daub *et al.*, 2000) and also in some cnidarian animals (the phylum that includes sea anemones). The motif has therefore been renamed SXC (six-cysteine) (Blaxter, 1998). The structure of a single SXC domain from a potassium channel-blocking toxin BgK from the sea anemone *Bunodosoma granulifera* has been solved (Dauplais *et al.*, 1997) and provides a framework for modelling *T. canis* SXC domains (Loukas *et al.*, 2000).

Venom Allergen Homologues: Tc-CRISP (Tc-VAH-1) and Tc-VAH-2

A prominently expressed gene family in many parasitic nematodes bears similarity to allergens present in vespid (e.g. wasp) venom. For example, the canine hookworm *Ancylostoma caninum* expresses high levels of this protein when larvae are activated to invade (Hawdon *et al.*, 1996). Similar genes are found in *B. malayi* (R.M. Maizels and J. Murray, 1999, unpublished results) and in *T. canis*. Remarkably, there are two forms in *T. canis*, one containing no fewer than four SXC domains (R.M. Maizels *et al.*, 1999, unpublished results), and the other being more similar to the structures in other nematodes. No function is yet ascribed to this product, but there is a distant similarity to cysteine-rich salivary protein (CRISP) from mammals (L.X. Liu and R.M. Maizels, 1999, unpublished results).

Carbohydrate Moieties

Ascarid nematodes were shown more than 60 years ago to contain high levels of carbohydrate antigens (Campbell, 1936). Analyses of whole TES products revealed a surprisingly high level of glycosylation, mostly accounted for by galactose and *N*-acetylgalactosamine (Meghji and Maizels, 1986). As the core sugars for asparagine (*N*)-linked oligosaccharides are mannose and *N*-acetylglucosamine, this composition indicated a predominance of *O*-linked glycosylation. Analysis by mass spectrometry (Khoo *et al.*, 1991, 1993) defined two major *O*-linked glycans, both variants of a trisaccharide containing *N*-acetylgalactosamine, galactose and fucose. The fucose is invariably *O*-methylated, and the galactose is methylated in 50% of the molecules. In *T. cati* the galactose is fully methylated, so that the trisaccharide containing a single methylation site on fucose alone represents a *T. canis*-specific structure. The *N*-linked oligosaccharide structure which predominates has also been determined to be a biantennary trimannosylated structure linked to a chitobiose-asparagine core, resembling products found in insects (Khoo *et al.*, 1993).

Carbohydrate specificities were also pre-eminent in a panel of monoclonal antibodies generated to TES (Maizels *et al.*, 1987). MAbs Tcn-2 and Tcn-8 recognize a spectrum of TES glycoproteins, including TES-400, TES-120, TES-70, TES-55 and TES-32, through a periodate-sensitive determinant presumed to be conjugated to various peptide backbones. There is a striking contrast between the two antibodies, however, in that Tcn-2 reacts only to *T. canis* glycoproteins, while Tcn-8 cross-reacts fully with products from *T. cati*. It remains to be determined whether Tcn-2 recognizes the species-specific carbohydrate structure described above, but this seems a likely explanation.

The *T. canis* trisaccharide can be considered to be a mammalian blood group antigen H (or type O) to which has been added additional methyl groups that represent a substantial modification. There is, however, some evidence for minor unmodified bloodgroup-like oligosaccharides (Khoo *et al.*, 1993). Moreover, the presence of tetrasaccharides containing *N*-acetylgalactosamine linked to the galactose is indicated by the blood group A-like reactivity of TES in antibody and lectin binding analyses (Smith *et al.*, 1983; Meghji and Maizels, 1986).

Concluding Remarks

Toxocara has an exceptional ability to withstand attack by the immune system, most probably due to the specific glycoproteins that constitute the surface and secreted compartments of the parasite. Tissue-dwelling larvae appear to devote a large part of their metabolic energy into production of the TES proteins. We have discovered that one of the major products released by the infective larvae of the nematode *T. canis* is similar at both structural and functional levels to host C-type lectins, a family of proteins that are pivotal in the immune system. These sugar-binding proteins mediate key events in the immune response to infection, particularly in inflammation when cells and harmful molecules are focused on the site of invasion. Host lectins are required to bind to carbohydrate 'danger' signals in order to initiate inflammatory influx around the parasite, and the release of a parasite lectin may block this process.

The molecular characterization of the TES proteins now provides an opportunity for immunological investigations at a much finer level of definition. Significant contrasts have been observed between infected patients with different syndromes, with respect to surface binding, proportions of anti-peptide and anti-carbohydrate specificities, and isotype (Smith, 1993), and it may now be possible to relate these patterns to recognition of individual antigens. The T cell response to TES is also known to be remarkably skewed towards the Th2 phenotype (Del Prete *et al.*, 1991) and large quantities of TES given to mice induce eosinophilia (Sugane and Oshima, 1984). These biological activities may soon be attributed to identified molecular products of the parasite.

There are practical potentials for much of this work. As the parasite is essentially a successful tissue transplant, residing in muscle or other soft tissue without rejection, we believe that parasite products could prove useful in enhancing tissue graft success in clinical medicine. On a more immediate basis, we have isolated a range of parasite proteins that can be synthesized in the laboratory, promising a much better diagnostic reagent for determining parasite infection whether in dogs, the natural host, or on the occasions when human infection occurs. Finally, and perhaps most importantly, the isolation of genes encoding the major larval antigens paves the way for development of a vaccine against canine toxocariasis, which may in future lead to the eradication of this threat to both human and veterinary health.

Acknowledgements

The studies described in this chapter have been supported by the Medical Research Council, the Leverhulme Trust and the Wellcome Trust.

References

Badley, J.E., Grieve, R.B., Bowman, D.D., Glickman, L.T. and Rockey, J.H. (1987) Analysis of *Toxocara canis* larval excretory–secretory antigens: physicochemical characterization and antibody recognition. *Journal of Parasitology* 73, 593–600.

Beaver, P.C. (1962) Toxocarosis (visceral larva migrans) in relation to tropical eosinophilia. *Bulletin de la Société de Pathologie Exotique* 55, 555–576.

Beaver, P.C. (1966) Zoonoses, with particular reference to parasites of veterinary importance. In: Soulsby, E.J.L. (ed.) *Biology of Parasites*. Academic Press, New York, pp. 215–227.

Blaxter, M.L. (1998) *Caenorhabditis elegans* is a nematode. *Science* 282, 2041–2046.

Blaxter, M.L., Page, A.P., Rudin, W. and Maizels, R.M. (1992) Nematode surface coats: actively evading immunity. *Parasitology Today* 8, 243–247.

Blaxter, M.L., Raghavan, N., Ghosh, I., Guiliano, D., Lu, W., Williams, S.A., Slatko, B. and Scott, A.L. (1996) Genes expressed in *Brugia malayi* infective third stage larvae. *Molecular and Biochemical Parasitology* 77, 77–93.

Burke, T.M. and Roberson, E.L. (1985) Prenatal and lactational transmission of *Toxocara canis* and *Ancylostoma caninum*: experimental infection of the bitch before pregnancy. *International Journal for Parasitology* 15, 71–75.

Campbell, D.H. (1936) An antigenic polysaccharide fraction of *Ascaris lumbricoides* (from hog). *Journal of Infectious Diseases* 59, 266–280.

Daub, J., Loukas, A., Pritchard, D. and Blaxter, M.L. (2000) A survey of genes expressed in adults of the human hookworm *Necator americanus*. *Parasitology* 120, 171–184.

Dauplais, M., Lecoq, A., Song, J., Cotton, J., Jamin, N., Gilquin, B., Roumestand, C., Vita, C., de Medeiros, C.L.C., Rowan, E.G., Harvey, A.L. and Menez, A. (1997) On the convergent evolution of animal toxins. Conservation of a diad of functional residues in potassium channel-blocking toxins with unrelated structures. *Journal of Biological Chemistry* 272, 4302–4309.

de Savigny, D.H. (1975) *In vitro* maintenance of *Toxocara canis* larvae and a simple method for the production of *Toxocara* ES antigen for use in serodiagnosis test for visceral larva migrans. *Journal of Parasitology* 61, 781–782.

Del Prete, G., De, C.M., Mastromauro, C., Biagiotti, R., Macchia, D., Falagiani, P., Ricci, M. and Romagnani, S. (1991) Purified protein derivative of *Mycobacterium tuberculosis* and excretory–secretory antigen(s) of *Toxocara canis* expand human T cells with stable and opposite (type 1 T helper or type 2 T helper) profile of cytokine production. *Journal of Clinical Investigation* 88, 346–350.

Falcone, F., Tetteh, K.K.A., Hunt, P., Blaxter, M.L., Loukas, A.C. and Maizels, R.M. (2000) The new subfamily of cathepsin Z-like protease genes includes *Tc-cpz-1*, a cysteine protease gene expressed in *Toxocara canis* adults and infective larvae. *Experimental Parasitology* 94, 201–203.

Fattah, D.I., Maizels, R.M., McLaren, D.J. and Spry, C.J.F. (1986) *Toxocara canis*: interaction of human eosinophils with infective larvae *in vitro*. *Experimental Parasitology* 61, 421–433.

Gems, D.H. and Maizels, R.M. (1996) An abundantly expressed mucin-like protein from *Toxocara canis* infective larvae: the precursor of the larval surface coat glycoproteins. *Proceedings of the National Academy of Sciences USA* 93, 1665–1670.

Gems, D.H., Ferguson, C.J., Robertson, B.D., Page, A.P., Blaxter, M.L. and Maizels, R.M. (1995) An abundant, *trans*-spliced mRNA from *Toxocara canis* infective larvae encodes a 26 kDa protein with homology to phosphatidylethanolamine binding proteins. *Journal of Biological Chemistry* 270, 18517–18522.

Ghafoor, S.Y.A., Smith, H.V., Lee, W.R., Quinn, R. and Girdwood, R.W.A. (1984) Experimental ocular toxocariasis: a mouse model. *British Journal of Ophthalmology* 68, 89–96.

Gillespie, S.H. (1987) Human toxocariasis. *Journal of Applied Bacteriology* 63, 473–479.

Gillespie, S.H. (1993) The clinical spectrum of human toxocariasis. In: Lewis, J.W. and Maizels, R.M. (eds) *Toxocara and Toxocariasis: Clinical, Epidemiological and Molecular Perspectives*. Institute of Biology, London, pp. 55–61.

Glickman, L.T. and Schantz, P.M. (1981) Epidemiology and pathogenesis of zoonotic toxocariasis. *Epidemiologic Reviews* 3, 230–250.

Griesemer, R.A., Gibson, J.P. and Elsasser, D.S. (1963) Congenital ascariasis in gnotobiotic dogs. *Journal of the American Veterinary Medical Association* 143, 962–964.

Hawdon, J.M., Jones, B.F., Hoffman, D.R. and Hotez, P.J. (1996) Cloning and characterization of *Ancylostoma*-secreted protein. A novel protein associated with the transition to parasitism by infective hookworm larvae. *Journal of Biological Chemistry* 271, 6672–6678.

James, E.R. (1994) Superoxide dismutase. *Parasitology Today* 10, 481–484.

Kennedy, M.K., Maizels, R.M., Meghji, M., Young, L., Qureshi, F. and Smith, H.V. (1987) Species-specific and common epitopes on the secreted and surface antigens of *Toxocara cati* and *Toxocara canis* infective larvae. *Parasite Immunology* 9, 407–420.

Khoo, K.-H., Maizels, R.M., Page, A.P., Taylor, G.W., Rendell, N. and Dell, A. (1991) Characterisation of nematode glycoproteins: the major *O*-glycans of *Toxocara* excretory secretory antigens are methylated trisaccharides. *Glycobiology* 1, 163–171.

Khoo, K.-H., Morris, H.R. and Dell, A. (1993) Structural characterisation of the major glycans of *Toxocara canis* ES antigens. In: Lewis, J. and Maizels, R.M. (eds) *Toxocara and Toxocariasis: Clinical, Epidemiological and Molecular Perspectives*. Institute of Biology, London, pp. 133–140.

Lewis, J.W. and Maizels, R.M. (eds) (1993) *Toxocara and toxocariasis. Clinical, Epidemiological and Molecular Perspectives*. Institute of Biology, London.

Lloyd, S. (1993) *Toxocara canis*: the dog. In: Lewis, J.W. and Maizels, R.M. (eds) *Toxocara and Toxocariasis: Clinical, Epidemiological and Molecular Perspectives*. Institute of Biology, London, pp. 11–24.

Lobos, E., Weiss, N., Karam, M., Taylor, H.R., Ottesen, E.A. and Nutman, T.B. (1991) An immunogenic *Onchocerca volvulus* antigen: a specific and early marker of infection. *Science* 251, 1603–1605.

Loukas, A. and Maizels, R.M. (1999) Helminth C-type lectins and host–parasite interactions. *Parasitology Today* 16, 333–339.

Loukas, A., Selzer, P.M. and Maizels, R.M. (1998) Characterisation of *Tc-cpl-1*, a cathepsin L-like cysteine protease from *Toxocara canis* infective larvae. *Molecular and Biochemical Parasitology* 92, 275–289.

Loukas, A.C., Doedens, A., Hintz, M. and Maizels, R.M. (2000) Identification of a new C-type lectin, TES-70, secreted by infective larvae of *Toxocara canis*, which binds to host endothelial cells. *Parasitology* (in press).

Loukas, A.C., Mullin, N.P., Tetteh, K.K.A., Moens, L. and Maizels, R.M. (1999) A novel C-type lectin secreted by a tissue-dwelling parasitic nematode. *Current Biology* 9, 825–828.

Loukas, A.C., Hintz, M., Linder, D., Mullin, N.P., Parkinson, J., Tetteh, K.K.A. and Maizels, R.M. (2000) A family of secreted mucins from the parasitic nematode *Toxocara canis* bear diverse mucin domains but share similar flanking six-cysteine (SXC) repeat motifs. *Journal of Biological Chemistry* (in press).

Maizels, R.M. and Page, A.P. (1990) Surface associated glycoproteins from *Toxocara canis* L2 parasites. *Acta Tropica* 47, 355–364.

Maizels, R.M. and Robertson, B.D. (1991) *Toxocara canis*: secreted glycoconjugate antigens in immunobiology and immunodiagnosis. In: Kennedy, M. (ed.) *Parasitic Nematodes – Antigens, Membranes and Genes.* Taylor and Francis Ltd, London, pp. 95–115.

Maizels, R.M., de Savigny, D. and Ogilvie, B.M. (1984) Characterization of surface and excretory–secretory antigens of *Toxocara canis* infective larvae. *Parasite Immunology* 6, 23–37.

Maizels, R.M., Kennedy, M.W., Meghji, M., Robertson, B.D. and Smith, H.V. (1987) Shared carbohydrate epitopes on distinct surface and secreted antigens of the parasitic nematode *Toxocara canis*. *Journal of Immunology* 139, 207–214.

Maizels, R.M., Gems, D.H. and Page, A.P. (1993) Synthesis and secretion of TES antigens from *Toxocara canis* infective larvae. In: Lewis, J.W. and Maizels, R.M. (eds) *Toxocara and Toxocariasis: Clinical, Epidemiological and Molecular Perspectives.* Institute of Biology, London, pp. 141–150.

Maizels, R.M., Tetteh, K.K.A. and Loukas, A.C. (2000) *Toxocara canis*: genes expressed by the arrested infective larval stage of a parasitic nematode. *International Journal of Parasitology* 30, 495–508.

Meghji, M. and Maizels, R.M. (1986) Biochemical properties of larval excretory–secretory (ES) glycoproteins of the parasitic nematode *Toxocara canis*. *Molecular and Biochemical Parasitology* 18, 155–170.

Nichols, R.L. (1956) The etiology of visceral larva migrans. I. Diagnostic morphology of infective second-stage *Toxocara* larvae. *Journal of Parasitology* 42, 349–362.

Ou, X., Tang, L., McCrossan, M., Henkle-Dührsen, K. and Selkirk, M.E. (1995) *Brugia malayi*: localisation and differential expression of extracellular and cytoplasmic CuZn superoxide dismutases in adults and microfilariae. *Experimental Parasitology* 80, 515–529.

Page, A.P. and Maizels, R.M. (1992) Biosynthesis and glycosylation of serine/threonine-rich secreted proteins from *Toxocara canis* larvae. *Parasitology* 105, 297–308.

Page, A.P., Richards, D.T., Lewis, J.W., Omar, H.M. and Maizels, R.M. (1991) Comparison of isolates and species of *Toxocara* and *Toxascaris* by biosynthetic labelling of somatic and ES proteins from infective larvae. *Parasitology* 103, 451–464.

Page, A.P., Hamilton, A.J. and Maizels, R.M. (1992a) *Toxocara canis*: monoclonal antibodies to carbohydrate epitopes of secreted (TES) antigens localize to different secretion-related structures in infective larvae. *Experimental Parasitology* 75, 56–71.

Page, A.P., Rudin, W., Fluri, E., Blaxter, M.L. and Maizels, R.M. (1992b) *Toxocara canis*: a labile antigenic coat overlying the epicuticle of infective larvae. *Experimental Parasitology* 75, 72–86.

Page, A.P., Rudin, W. and Maizels, R.M. (1992c) Lectin binding to secretory structures and the surface coat of *Toxocara canis* infective larvae. *Parasitology* 105, 285–296.

Robertson, B.D., Bianco, A.E., McKerrow, J.H. and Maizels, R.M. (1989) *Toxocara canis*: proteolytic enzymes secreted by the infective larvae *in vitro*. *Experimental Parasitology* 69, 30–36.

Schwartz, B. (1921) Effects of secretions of certain parasitic nematodes on coagulation of the blood. *Journal of Parasitology* 7, 144–150.

Scothorn, M.W., Koutz, F.R. and Groves, H.F. (1965) Prenatal *Toxocara canis* infection in pups. *Journal of the American Veterinary Medical Association* 146, 45–48.

Smith, H.V. (1991) Immune evasion and immunopathology in *Toxocara canis* infection. In: Kennedy, M.W. (ed.) *Parasitic Nematodes – Antigens, Membranes and Genes*. Taylor and Francis, London, pp. 116–139.

Smith, H.V. (1993) Antibody reactivity in human toxocariasis. In: Lewis, J.M. and Maizels, R.M. (eds) *Toxocariasis: Epidemiological, Clinical and Molecular Perspectives*. Institute of Biology, London, pp. 91–109.

Smith, H.V., Quinn, R., Kusel, J.R. and Girdwood, R.W.A. (1981) The effect of temperature and antimetabolites on antibody binding to the outer surface of second stage *Toxocara canis* larvae. *Molecular and Biochemical Parasitology* 4, 183–193.

Smith, H.V., Kusel, J.R. and Girdwood, R.W.A. (1983) The production of human A and B blood group like substances by *in vitro* maintained second stage *Toxocara canis* larvae: their presence on the outer larval surfaces and in their excretions/secretions. *Clinical and Experimental Immunology* 54, 625–633.

Sprent, J.F.A. (1952) On the migratory behavior of the larvae of various *Ascaris* species in white mice. I. Distribution of larvae in tissues. *Journal of Infectious Diseases* 90, 165–176.

Sprent, J.F.A. (1958) Observations on the development of *Toxocara canis* (Werner, 1782) in the dog. *Parasitology* 48, 184–210.

Stirewalt, M.A. (1963) Chemical biology of secretions of larval helminths. *Annals of the New York Academy of Sciences* 113, 36–53.

Sugane, K. and Oshima, T. (1983) Purification and characterization of excretory and secretory antigen of *Toxocara canis* larvae. *Immunology* 50, 113–120.

Sugane, K. and Oshima, T. (1984) Induction of peripheral blood eosinophilia in mice by excretory and secretory antigen of *Toxocara canis* larvae. *Journal of Helminthology* 58, 143–147.

Tetteh, K.K.A., Loukas, A., Tripp, C. and Maizels, R.M. (1999) Identification of abundantly-expressed novel and conserved genes from infective stage larvae of *Toxocara canis* by an expressed sequence tag strategy. *Infection and Immunity* 67, 4771–4779.

Warren, E.G. (1969) Infections of *Toxocara canis* in dogs fed infected mouse tissues. *Parasitology* 59, 837–841.

Weis, W.I., Taylor, M.E. and Drickamer, K. (1998) The C-type lectin superfamily in the immune system. *Immunological Reviews* 163, 19–34.

Nematode Gut Peptidases, Proteins and Vaccination

D.P. Knox, P.J. Skuce, G.F. Newlands and D.L. Redmond

Moredun Research Institute, Pentlands Science Park, Bush Loan, Penicuik, Midlothian EH26 0PZ, UK

Introduction

Anthelmintic resistance in nematode populations and concerns about the effects of drug residues on consumer health and the environment have focused attention on the prospect of developing effective anti-nematode vaccines. These studies focus on characterizing the host anti-parasite immune response, the nature of the protective components of the response, the parasite life-cycle stage targeted and, ultimately, from a vaccine development point of view, identifying the antigens stimulating it. Technological developments in molecular biology have accelerated the process with the prospect of being able to produce even the least abundant antigen in limitless quantity. In addition, functional genomics approaches are used for antigen identification while rapid developments in immunology allow the more precise definition of protective as opposed to irrelevant or harmful immune responses. These developments have, in turn, led to enhanced and very selective means of antigen delivery while recent advances in 'naked' DNA vaccination point to a future where antigen is delivered to the host immune system along with the cell signals required to initiate the desired immune response. Therefore, vaccination is feasible – or is it?

Host and parasite have coevolved over millions of years and exist in mutual tolerance, disease only being apparent when the balance is upset by the host being exposed to unusually high parasite burdens at times when immunocompetence is diminished by age, physiological state, disease or nutritional status. The host immune response is multifaceted and it is difficult to dissect it to the extent of claiming that a particular component solely mediates protective immunity. Imbalances in this immune response are often the physical manifestations of parasite-induced disease. Susceptibility and resistance to parasite infection certainly have a host genetic basis and are likely to be influenced by parasite genetics in that some parasite genotypes may be more virulent than others. Therefore, vaccine development based on antigens that stimulate protective host immune responses during the course of infection may not be as straightforward as might be anticipated. An alternative approach is initially to ignore the host immune response and focus on parasite proteins that may be essential to parasite survival, such as excretory/secretory (ES), cuticular or gut-expressed proteins. This chapter will focus on vaccination against nematode parasites based on proteins isolated from the surface of the gut.

The Nematode Gut

The nematode gut comprises, from anterior to posterior, a buccal capsule, a muscular pharynx, a fairly straight intestine which is a single cell thick and, finally, a rectum or cloaca. The pharynx and cloaca are, in part, lined with cuticle while the surface of the intestinal cells facing the lumen have microvilli (Bird, 1971). Oesophageal glands, one dorsal and two subventral, open into the base of the buccal capsule, while the oesophagus is separated from the intestine by the oesophago-intestinal valve (Colam, 1971). The length of the microvilli varies with nematode species from, for example, 1 μm in adult *Nippostrongylus brasiliensis* (Jamaur, 1966) to 5.5 μm to 7.5 μm in *Haemonchus contortus* (Fig. 13.1) (Munn, 1977). In strongylid nematodes, there is a fibrous submicrovillar layer to which the filamentous cores of the microvilli attach (Munn, 1977) and this layer, together with the microvilli, can be dissected from the remainder of the intestine. The product is tubular and has been termed the endotube because of its intracellular origin (Munn, 1981). Endotubes are present in 12 different species of strongylid (Munn and Greenwood, 1983) and are generally absent in non-strongylid orders. The endotube is presumed to be cytoskeletal but can undergo DTT/ATP-induced contraction (Munn and Greenwood, 1983).

Proteins at the Microvillar Surface

Contortin

The microvillar surface is coated with a layer of electron-dense amorphous material (glycocalyx). In *H. contortus,* helical filaments composed of contortin are associated with this layer and fill the spaces between the microvilli. There can be up to ten strands of contortin in each microvillus;

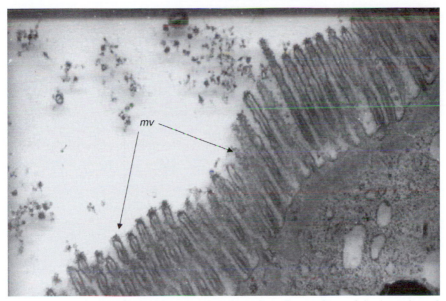

Fig. 13.1. Electronmicrograph showing the microvillar surface of the intestine of *Haemonchus contortus*. mv, microvilli.

each is continuous from the base to the tip of the microvillus and many extend into the gut lumen (Munn, 1977). Contortin is not attached to the plasma membrane and some lies free in the lumen of the intestine; it is also present in large amounts in the pharynx, though this may not be the case *in vivo* (Munn, 1977). Contortin can be purified from phosphate-buffered saline extracts of adult parasites by ultra-centrifugation and is not, apparently, a glycoprotein. Over 90% of the material purified in this way has an apparent molecular weight of 60 kDa as judged by SDS-PAGE, the remainder being composed of contaminating membrane fragments. The spatial arrangement of contortin in the intestine is illustrated in a series of elegant electronmicrographs in Munn (1977).

Vaccination of lambs with a contortin-enriched preparation gave a mean reduction in worm burdens of 78% (Table 13.1) (Munn *et al.*, 1987). This result was particularly significant because it showed that proteins expressed on the surface of the gut, albeit from a blood-feeding nematode, could induce high levels of protective immunity when used as an immunogen. These proteins are not normally accessible to the host immune system during the course of infection; they are termed hidden or concealed antigens and the immunity conferred by them is described as artificial immunity.

Contortin-like material was also observed in *Teladorsagia circumcincta*, another stomach worm of ruminants, while material separating the microvilli of the rodent nematodes *N. brasiliensis* and *Syphacia obvelata* was

Table 13.1. Summary of the protection induced against *Haemonchus contortus* using proteins purified from the intestine.

Antigen	Molecular weight (kDa)	Identity	Sheep breed	Mean protection (%)		References
				Faecal egg output	Worms	
Contortin-enriched	60	Undefined	Clun Forest cross-breed	No data	78	Munn et al., 1987
H11	110	Aminopeptidase	Various	93	77	Newton and Munn, 1999
H-gal-GP	35 to 230 (unreduced)	Digestive protease complex?	Suffolk Cross	93	72	Smith et al., 1994; Smith and Smith, 1996
TSBP	37 to 97	Cysteine proteases	Suffolk Cross	77	47	Knox et al., 1999
P1	45, 49, 53	Undefined	Dorset	69	NS	Munn et al., 1997
GA1	46, 52	Transport protein?	Goats	50	60	Jasmer et al., 1993
AC-1	35	Cysteine protease[a]	Dorset Cross	93	87	Boisvenue et al., 1992

[a]Presumed (but not proven) to be derived from the gut.

quite distinct in appearance compared with contortin. Contortin is quite an unusual biopolymer, being a flexible, helical and extracellular protein (Munn, 1977).

H11

Contortin-enriched preparations also contained a 110 kDa major antigenic contaminant as judged by Western blotting, despite only faint staining being evident in protein gels (Smith and Munn, 1990). This protein has been purified using concanavalin A lectin-affinity chromatography combined with, on some occasions, Mono Q anion-exchange chromatography and it migrates as a doublet at 110 kDa as judged by SDS-PAGE, hence the short-hand designation, H11. It is the most effective immunogen isolated from a parasitic nematode to date, inducing high levels of protection against challenge in recipient lambs (Table 13.1). It is highly effective in inducing protection in a range of sheep breeds, in very young lambs and in pregnant ewes, and is effective against anthelmintic-resistant *H. contortus* (reviewed by Newton and Munn, 1999). The protection persists for at least 23 weeks (Andrews *et al.*, 1997). Moderate levels of protection are transferred to the newborn lamb via colostrum (Andrews *et al.*, 1995) indicating that protective immunity is antibody mediated. It is highly effective when administered in the ethically acceptable adjuvant, QuilA. Despite H11 being defined as a hidden antigen, it is interesting to note that the immune response in sheep, primed by immunization, is boosted by H11 released from dead or dying parasites (Andrews *et al.*, 1997). For these reasons, intensive work is now in progress to develop a recombinant vaccine against haemonchosis.

H11 is an integral membrane glycoprotein which is only expressed on the intestinal microvilli of the parasitic stages. It shows microsomal aminopeptidase A and M activities and is expressed as three isoforms (Graham *et al.*, 1993; Smith *et al.*, 1997). H11 has the predicted structure of a type II integral membrane protein with a short N-terminal cytoplasmic tail, a transmembrane region and an extracellular region organized into four domains (Newton and Munn, 1999) with three N-linked glycosylation sites (Smith *et al.*, 1997). There is an HEXXHXW sequence motif followed by a glutamic acid which is a characteristic of the zinc-binding sequence of microsomal aminopeptidases. Enzyme activity is localized exclusively in the microvilli and is inhibited by the aminopeptidase-specific inhibitors bestatin and amastatin and by the chelating agent phenanthroline (Smith *et al.*, 1997). H11 antisera inhibit the aminopeptidase activity *in vitro* (Munn *et al.*, 1997) and the level of inhibition observed is closely correlated to protection. Protection levels are reduced by dissociation and denaturation (Munn *et al.*, 1997), observations that indicate the involvement of conformational epitopes in the induction of immunity. The three isoforms have been expressed as enzymically active recombinant proteins using the baculovirus-S*f*9 insect cell system and the outcome of protection trials is awaited. In addition, full-length cDNAs and defined fragments have been expressed in *Escherichia coli* (Newton and Munn, 1999).

Homologues of H11 are present in *T. circumcincta* and *Ostertagia ostertagi* and are currently being evaluated in protection trials. Moreover, aminopeptidase activity is present in the intestinal microvilli of the human hookworm *Necator americanus* (McLaren *et al.*, 1974) and this activity can be solubilized from the intestinal brush border using detergents. This material also contains a protein doublet at 110 kDa which may be a homologue of H11 and may have utility in developing molecular vaccines against hookworm infection in humans (Smith *et al.*, 1997). In addition, pigs immunized with aminopeptidase purified from the gut of adult *Ascaris suum* were protected against challenge infection, as judged by a 50% reduction in larval counts compared with challenge controls (Ferguson *et al.*, 1969). A recent study reported that sheep vaccinated with leucine aminopeptidase purified from adult *Fasciola hepatica* were highly (89%) protected against metacercarial challenge (Piacenza *et al.*, 1999). All the experiments cited emphasize the potential generic utility of gut-expressed aminopeptidases as anti-nematode vaccine components.

GA1 antigens

Monoclonal antibodies were used to probe gut surface antigens from *H. contortus* which may be associated with protective immunity and which may be phylogenetically conserved. One of the monoclonal antibodies recognized a carbohydrate epitope on several proteins and it was also identified in related adult and larval nematodes, including *O. ostertagi*, *Trichostrongylus colubriformis* and *Caenorhabditis elegans* but not *Trichinella spiralis* (Jasmer *et al.*, 1993). This monoclonal antibody was subsequently used, in a generic manner, to identify other membrane and secreted proteins from *Haemonchus*, several of which were novel and others that had been described previously (Rehman and Jasmer, 1998). A different monoclonal antibody, with specificity for *H. contortus* only, was used to identify and purify a group of proteins, M_r 46, 52 and 100 kDa, which, collectively, induced reductions of 60% and 50% in worm and faecal egg outputs, respectively, in immunized goats.

N-terminal protein sequence analyses of the three proteins, termed p46[GA1], p52[GA1] and p100[GA1], and cDNA library immunoscreening showed that all three were encoded by the same GA1 gene and are initially expressed as a polyprotein (p100[GA1]). The individual proteins could be released by serine protease-mediated cleavage (Jasmer *et al.*, 1996); p100[GA1] and p46[GA1] shared the same N-terminal sequence while that of p52[GA1] was located midway through the predicted protein sequence derived from a full-length cDNA encoding p100^{GA1}. Western blot analysis indicated that the GA1 proteins were expressed in adult worms but not in infective larvae, a result essentially confirmed by Northern blot analysis. The GA1 gene product showed closest homology with bacterial Tolb proteins, which are associated with the bacterial membrane and periplasm and may be involved in transport. All GA1 proteins and the Tolb protein shared a consensus Ser-Pro-Asp-Gly sequence, which was repeated in both proteins with identical spacing between the repeats (Jasmer *et al.*, 1996). p52^{GA1} contained a glycosylinositolphospholipid anchor indicative of an integral membrane protein while p46^{GA1} lacked obvious sequences normally associated with membrane insertion, indicating apical gut membrane association via a distinct mechanism (Jasmer *et al.*, 1996).

GA1-derived proteins were detected in ES products from adult worms and host abomasal mucus, indicative of release from the microvillar surface. Following from this, protective immunity stimulated by immunization with these proteins may involve anamnestic and mucosal immune responses. This suggestion was supported in a later study (Karanu *et al.*, 1997a), which provided evidence for a contribution from CD4$^+$ lymphocytes to gut antigen-induced immunity.

P150

During the course of developing procedures for the purification of H11 to homogeneity, Smith *et al.* (1993) identified a group of three peptides that were separated from ConA lectin-binding H11 by ion-exchange chromatography. The constituent peptides (45, 49 and 53 kDa) showed some similarities to the GA1 proteins and induced mean reductions of 69% in faecal egg output and 38% and 20% reductions in female and male worm numbers, respectively (Smith *et al.*, 1993). The proteins, designated P45, P49 and P53, form a complex (P150) which is a ubiquitous constituent of the microvillar membrane of the intestinal cells. P53 has a membrane anchor and associates non-covalently with a disulphide bridged dimer of P45 and P49. All three components share peptide epitopes (Rocha and Munn, 1997).

H-gal-GP

Smith *et al.* (1994) used several different techniques, including lectin screening of worm sections and radio-labelling, to target integral membrane glycoproteins on the luminal membrane of *H. contortus* gut. These proteins were then purified from detergent extracts of whole worms and evaluated in protection trials. One fraction, which selectively bound to lectins with specificity for *N*-acetylgalactosamine, reduced mean challenge worm burdens by up to 72% and mean faecal egg counts by up to 93%.

This fraction was termed *Haemonchus* galactose-containing glycoprotein complex (H-gal-GP). The microvillar surface of the intestinal cells of worms retrieved from vaccinated lambs was coated with sheep immunoglobulin and protection observed, over a series of trials, was correlated with systemic antibody titre (Smith *et al.*, 1999).

H-gal-GP can be visualized as a single diffusely staining band after Blue Native PAGE with an estimated molecular weight of about 1000 kDa and resolves as four major protein zones, designated A, B, C and D, at approximately 230, 170, 45 and < 35 kDa, respectively, under non-reducing conditions (Smith *et al.*, 1999). Each zone resolved as more than one band after reduction. Biochemical analyses indicated that H-gal-GP exhibited aspartyl, metallo- and, on occasion, cysteine protease activities. N-terminal sequence analysis of the individual peptide components from the four major protein zones showed clear homology with pepsin and metalloprotease cDNAs recently isolated from *Haemonchus* (Fig. 13.2) (Longbottom *et al.*, 1997; Redmond *et al.*, 1997; Smith *et al.*, 1999).

The pepsin, activated by cleavage of a proenzyme, has two putative active site domains comprising hydrophobic-hydrophobic-Asp-Thr-Gly amino acids, is potentially glycosylated and has a free cysteine residue which may allow it to form dimers, as in the case of human and *Plasmodium falciparum*-derived aspartyl proteases (Longbottom *et al.*, 1997). However,

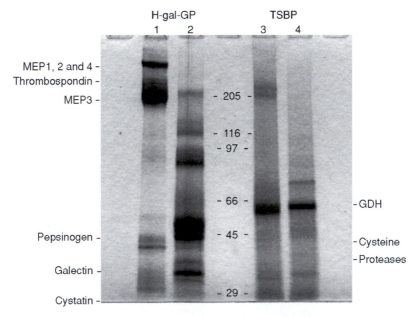

Fig. 13.2. The peptide components of H-gal-GP and TSBP visualized by Coomassie Blue staining of non-reducing (lanes 1 and 3) and reducing (lanes 2 and 4) SDS-PAGE gels.

this suggestion is not supported by non-reducing SDS-PAGE analysis. It also contains additional amino acid insert sequences, in particular one of 23 amino acids, as well as a five amino acid deletion, compared with human and porcine pepsin. The functional significance of these sequence changes is undefined at present.

The metalloprotease MEP1 showed closest homology to mammalian neutral endopeptidases (Redmond *et al.*, 1997), which are type II integral membrane proteins and have diverse tissue localization patterns, including the microvillar surfaces of the kidney and intestine (Rocques *et al.*, 1993). Neutral endopeptidases are zinc-dependent proteases and are responsible for the degradation of pharmacologically active peptides in nervous and peripheral tissues. MEP1 contains all the critical amino acids and motifs associated with the neutral endopeptidases, including a consensus active site sequence VxxHExxH which, in MEP1, is VVGHELVH, where the two histidines are putative zinc-coordinating ligands and the glutamate plays a role in catalysis by polarizing a water molecule. Southern blot analysis indicated that *Haemonchus* contained more than one MEP and this has been confirmed with a combination of cDNA library screening and N-terminal amino acid sequencing, which has shown that H-gal-GP contains at least four distinct MEPs with 38–54% identity at the amino acid level.

The N-terminal sequence of one peptide from the 35 kDa zone of H-gal-GP showed some homology to cathepsin B-like cysteine proteases. Molecular cloning has also identified a thrombospondin homologue associated with the diffusely staining region between zones A and B, a galectin associated with zone D (Newlands *et al.*, 1999) and a low molecular weight (approximately 13 kDa) cysteine protease inhibitor, cystatin.

H-gal-GP has proved resistant to fractionation using a variety of chromatographic techniques under native conditions. Various analyses consolidate the view that H-gal-GP is a genuine complex of proteins with zones A to D being held together electrostatically and subunits of three of the four zones associating by disulphide bonding (Smith *et al.*, 1999). With one exception, the subunits contain N-linked glycans and, despite behaving like an integral membrane protein, it remains unclear how H-gal-GP is anchored to the membrane.

So, what does the complex do? It could be a multi-protease complex crucial for digesting the blood meal and protease components are clearly localized to the microvillar surface of the intestinal cells (Fig. 13.3). Certainly, the pepsin shows strong haemoglobinase activity with optimal activity at pH 4, which would be consistent with a worm living in the highly acidic environment of the ruminant true stomach. The MEPs are also likely to be involved in digestion but could also degrade peptides ingested with the blood meal, which may have adverse effects on the worm. It is worthy of note that MEPs and aminopeptidases co-localize in the mammalian kidney brush border and are both required to deactivate biologically active peptides. H11 is, of course, an amino-peptidase. Perhaps the proteases of

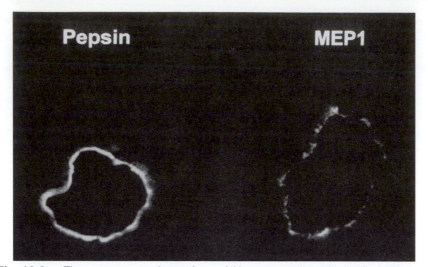

Fig. 13.3. Fluorescence on the surface of *H. contortus* intestinal cells following incubation of transverse sections of the worm with fluorescein-labelled antibody probes.

H-gal-GP and H11 do not simply have a digestive function but protect the parasite from ingested toxic proteins.

Thrombospondins are multi-domain glycoproteins which, amongst other roles, have been implicated in attachment of *Plasmodium*-infected erythrocytes to endothelial cells (Roberts *et al.*, 1985) whilst a *Plasmodium* sporozoite-derived thrombospondin homologue (TRAP) participates in invasion of host liver cells (Robson *et al.*, 1988). Could the thrombospondin component of H-gal-GP immobilize or trap red blood cells in the worm intestine to facilitate proteolytic lysis of the red blood cell, releasing haemoglobin for digestion by pepsin and gut-expressed cysteine proteases (see below)? The predicted thrombospondin amino acid sequence contains a number of repeat sequences with homology to Kunitz-type protease inhibitors and anticoagulants such as snake venom. It is possible (but we have no evidence for this yet) that the inhibitors are cleaved and may serve an anticoagulant function in the gut lumen. However, a Kunitz-type molecule localized to the gut, parenchymal tissue and tegument of the trematode *F. hepatica* has recently been characterized (Bozas *et al.*, 1995) and did not inhibit key serine proteases of the blood-coagulation pathway. Another possibility is that the inhibitors play some role in regulating parasite serine protease activities and it is worthy of note that the Thiol-Sepharose binding proteins, described below, include a gut-derived serine protease. The galectins are soluble lectins which specifically bind β-galactoside sugars and have diverse roles in cell adhesion, immune function and apoptosis. The H-gal-GP-associated galectins are tandem repeat-type galectins, having two carbohydrate recognition domains (Newlands *et al.*, 1999). These could act as a molecular glue hold-ing elements of the complex together, though this idea is not supported by the observation that removal of the galectin with lactose did not affect the mobility of H-gal-GP in native gels (Newlands *et al.*, 1999). Galectin thus purified was ineffective in a protection trial. In contrast, galectins were dominant proteins in a fraction from *T. circumcincta* which stimulated protective immunity in lambs (Meeusen, 1995). The cystatin has been cloned and expressed as a functionally active recombinant protein in *E. coli* using the plasmid pET22b⁺ (G.F.J. Newlands, P.J. Skuce and D.P. Knox, 2000, unpublished results). It is a very potent inhibitor of the cysteine proteases expressed in the gut of adult *Haemonchus* and possibly regulates their activity. Again, however, it proved ineffective in a protection trial.

None of the individual components of H-gal-GP which have been purified or recombinant proteins encoding it have, to date, individually conferred protection levels against challenge infection as high as those observed following vaccination with the whole complex. The protection conferred by vaccination with H-gal-GP was relatively unaffected by dissoci-ation or reduction of the immunogen prior to administration, indicating that protection was not dependent on conformational epitopes (Smith and Smith, 1996). The same authors tested the main components of H-gal-GP, which were simply excised from non-reducing SDS-PAGE gels prior to use as potential immunogens. Variable degrees of protection were observed between individual lambs when different excised bands were compared but in no case was the protection significant.

Homologues of H-gal-GP have been identified in *O. ostertagi, T. circumcincta* and *Trichostrongylus vitrinus* (Smith *et al.*, 1993) and are being evaluated as protective antigens in the former two species. Preliminary results encourage cautious optimism, with moderate levels of protection being observed in some trials (W.D. Smith, 2000, unpublished results).

Cysteine proteases

Cysteine proteases have long been implicated in the digestion of the blood meal in schistosomes, Fasciolidae and hookworms (reviewed by Tort *et al.*, 1999) and, in the case of *F. hepatica*, are practically useful protective immunogens for bovine fasciolosis (Dalton *et al.*, 1996) and, in combina-tion with leucine aminopeptidase (Piacenza *et al.*, 1999), for ovine fasciolosis. Earlier work (Knox *et al.*, 1993) showed that extracts from adult *H. contortus* contained several strong cysteine protease activities and that these proteases had the capacity to degrade haemoglobin and albumin, major constituents of the blood meal. They also degraded fibrinogen and plasminogen, indicating a possible anticoagulant role. These studies have been elegantly extended by the demonstration that ES cysteine proteases from the adult parasite also degraded fibrinogen, haemoglobin, collagen,

IgG and albumin (Rhoads and Fetterer, 1995), degradation being attributable to cathepsin L-like proteases. Moreover, L4 and adult *H. contortus*
maintained in the presence of a [^{3}H]-labelled extracellular matrix, readily
degraded glycoprotein, collagen and elastin components of the matrix,
degradation being blocked by a specific cysteine protease inhibitor
(Rhoads and Fetterer, 1997). However, *in vitro* uptake of radiolabelled
haemoglobin by adult parasites was not inhibited by the cysteine protease
inhibitor, although haemoglobin breakdown in the culture medium was
reduced by 50% (Fetterer and Rhoads, 1997). These observations suggest
that cysteine proteases are functional in the extracorporeal digestion of
the blood meal but are not required for uptake of the products. The latter
function may be mediated by metallo- and serine proteases, given that
uptake is markedly reduced in the presence of 1, 10 phenanthroline, an
inhibitor of metalloproteases, and AEBSF, an inhibitor of serine proteases.
Are these metalloproteases the MEPs found in H-gal-GP and is the serine
protease a component of TSBP, described below?

Water-soluble extracts of adult *H. contortus* were shown to contain a
35 kDa cysteine protease which degraded fibrinogen and increased clotting
time in sheep plasma *in vitro* (Cox *et al.*, 1990). Lambs immunized with
extracts from the adult parasite enriched for this activity were substantially
protected against challenge infection (Boisvenue *et al.*, 1992). The cDNA
(AC1) encoding this protease was isolated and showed 42% amino acid
sequence identity to human cathepsin B. It encoded a protease predicted
to comprise 342 amino acids, including a 15 amino acid hydrophobic
N-terminal signal sequence. Genomic DNA analysis showed that the gene
was a member of a small gene family (*AC*-1 to 5) expressed in the adult
parasite, two of which are tandemly linked (Pratt *et al.*, 1990, 1992a).
Primary structure comparisons suggest that members of the family may
have differing substrate specificities and associated physiological functions.
All contain potential signal sequences and may be secreted. However, no
localization data have been reported, though Rehman and Jasmer (1999)
were able to amplify AC-2 and AC-4 from a total gut RNA pool as well as a
novel sequence encoding a cysteine protease designated hc.gcp7. A similar
small gene family has been demonstrated in *O. ostertagi* (CP1–3) and,
again, two of these are tandemly linked and show closer homology to the
Haemonchus AC sequences than to cathepsin B (Pratt *et al.*, 1992b).

Membrane extracts from adult *H. contortus* were enriched 24-fold for
cysteine protease activity by passage over a Thiol-Sepharose affinity column
and the proteins obtained (abbreviated as TSBP) were clearly localized to
the microvillar surface of the intestinal cells (Knox *et al.*, 1995, 1999). TSBP
comprised a prominent 60 kDa protein and several minor bands between
35 and 45 kDa and 97 to 120 kDa (Fig. 13.2). Protease activity at 38, 52
and 70 kDa was attributable to cysteine proteases and at 70 and 88 kDa to
serine/metalloproteases, as judged by inhibition analyses. Lectin-binding
studies showed that most of the TSBPs were glycosylated. Expression library

(lambda gt11 and lambda ZAP) immunoscreening has revealed that the prominent 60 kDa component of TSBP is a glutamate dehydrogenase (GDH) homologue (Skuce *et al.*, 1999a) and that the cysteine protease activity can be attributed to the protein products of three distinct genes, cDNAs derived from which have been designated hmcp1, 4 and 6 (Skuce *et al.*, 1999b).

Lambs immunized with TSBP were substantially protected against a single challenge infection with *H. contortus*, with reductions in daily faecal egg outputs of 77% and final worm burdens of 47% over three trials. Again, the microvillar surface of worms surviving in vaccinated lambs was coated with sheep immunoglobulin and recent experiments in our laboratory show that antibody harvested from vaccinated lambs functionally inhibits the cysteine protease components of TSBP (D.P. Knox, 2000, unpublished results). Of interest, cysteine protease-enriched fractions prepared in the same way from water-soluble parasite extracts were ineffective immunogens. These extracts are likely to contain ES proteases and those present free in the gut lumen. To conclude that these proteases are not effective immunogens on the basis of this result is unsound because Thiol-Sepharose only interacts with free –SH groups and it is likely that ES cysteine proteases would only be present in whole worm extracts as inactive precursors.

Following identification of the GDH component by molecular cloning, enzyme activity was readily demonstrated in TSBP using a standard commercial kit. GDH expression is developmentally regulated, being restricted to the blood-feeding parasite stages. Immunolocalization studies showed that the enzyme was present in the cytoplasm of the intestinal cells, not the microvillar surface. This would argue against the GDH being the protective component of TSBP, as it would be inaccessible to antibody, and this, indeed, appears to be the case. TSBP can be further fractionated by anion exchange chromatography into a fraction that contains the bulk of the GDH protein and activity, fractions that contain the cysteine proteases and very little GDH, and a protein fraction that does not bind to the column (Knox *et al.*, 1995). Lambs immunized with the GDH fraction, or the unbound protein fraction, were completely unprotected against challenge, whilst lambs immunized with the pooled protease-containing fractions were protected against challenge to the same degree as vaccination controls that received unfractionated TSBP (Knox *et al.*, 1995), supporting the view that protection is attributable to the cysteine proteases.

The expression of the cysteine protease-encoding genes *hmcp* 1, 4 and 6 also coincides with the onset of blood feeding, and immunolocalization studies suggest that each was expressed at the luminal surface of the intestinal cells (Skuce *et al.*, 1999b). Like the other cysteine protease-encoding genes characterized from *H. contortus* (Cox *et al.*, 1990; Pratt *et al.*, 1990, 1992a; Rehman and Jasmer, 1998), *hmcp* 1, 4 and 6 showed closest homology to cathepsin B, with consensus signal and propeptide cleavage sites, a conserved Cys, His, Asn catalytic triad, 14 cysteine residues and a number of proline residues, all of which contribute to tertiary structure and suggesting that the tertiary structure should be similar to cathepsin B. This was confirmed by molecular modelling, an example of which is shown in Fig. 13.4. Like cathepsin B, each contained a structural element, the occluding loop, which occludes the opening into the active-site cleft containing the catalytic triad. Of particular interest, key residues that determine substrate specificity differ between the hmcp sequences and cathepsin B (Skuce *et al.*, 1999), possibly conferring cathepsin L-like characteristics, a finding in accord with existing biochemical definitions of *H. contortus* cysteine proteases in ES (Rhoads and Fetterer, 1995) and TSBP (D.P. Knox, 2000, unpublished results). Substitutions in active-site cleft residues which contribute to substrate binding have been suggested to contribute to the cathepsin L-like specificities of *Ancylostoma caninum* ES proteases (Brinkworth *et al.*, 1996) and, conversely, cathepsin B-like specificity of a cathepsin L homologue from *Toxocara canis* (Loukas *et al.*, 1998). It remains to be seen whether this paradoxical substrate specificity is a common feature amongst nematode cysteine proteases.

For the development of anti-nematode vaccines with worldwide utility, geographical strain differences leading to antigenic diversity would be a serious concern. We have been unable to isolate any of the AC sequences (Cox *et al.*, 1990; Pratt *et al.*, 1992) or gcp7 (Rehman and Jasmer, 1998) from UK isolates of *H. contortus*, either by gene-specific PCR or by immunoscreening. However, recent EST data (D. Jasmer, EMBL database) identified sequences encoding hmcp 1 to 4, AC3 and gcp7 from a USA strain. Inter- and intra-geographical isolate heterogeneity amongst proteases found in adult worm ES has also been reported (Karanu *et al.*, 1997b). Clearly, this diversity of protease expression requires further investigation before its possible implications for vaccine development can be established.

Gut Antigen-based Vaccination and Non-blood-feeders

Experiments to date, where homologues of H11, H-gal-GP and TSBP extracted from *O. ostertagi* or *T. circumcincta* have been evaluated in protection trials conducted in cattle and sheep, as appropriate, have generated inconclusive data. For example, in one trial, lambs vaccinated with TSBP purified from *T. circumcincta* were highly protected (71% reduction in egg output, 65% reduction in worms retrieved compared with challenge controls) against a single challenge infection but this level of protection could not be reproduced in two subsequent trials. An obvious explanation is that these parasites are ingesting insufficient antibody for the vaccine to be effective. Another possibility is that protection is absolutely dependent on the cysteine protease components of the vaccine being enzymically

HMCP1

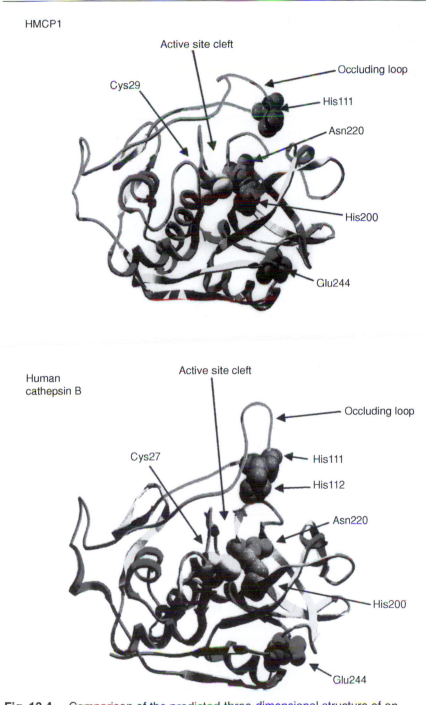

Fig. 13.4. Comparison of the predicted three-dimensional structure of an *H. contortus* gut-derived cysteine protease (HMCP1) with that of human cathepsin B.

active and it is known that TSBPs are unstable. Another explanation lies in individual variation in the level of innate immunity. Lambs within a control experimental group, where age, sex and breed are the same and food intake is controlled, often show highly variable levels of susceptibility to challenge infection.

Of course, a protein target identified by homology to another nematode species may be less than appropriate. *T. circumcincta* and *O. ostertagi* do express equivalents of H11, H-gal-GP and TSBP but neither parasite is an obligate blood-feeder. Comparisons of ES proteases from these species and those from *H. contortus* (D.P. Knox and H.D.F. Schallig, 2000, unpublished results) indicate that the former are much less dependent on cysteine proteases, with metallo- and serine proteases being more prominent (Young *et al.*, 1995). Intestinal nematodes such as *Trichostrongylus vitrinus* tend to contain and secrete predominantly serine and metalloproteases (MacLennan *et al.*, 1997) so perhaps successful vaccination with gut proteins may require species-specific selection of the protein target(s).

Glycosylation of Parasite Protein

Virtually all of the proteins described above are known to be, or predicted from gene sequence data to be, glycosylated. Because several (H11, H-gal-GP and GA1) can be separated from each other by lectin-affinity chromatography, glycosylation must vary depending on the protein. The nature of the *N*-linked oligosaccharides associated with proteins in detergent extracts of the adult parasite and with H11 specifically has been investigated in detail (Haslam *et al.*, 1996; see also Chapter 15). These experiments identified a core fucosylation of a type not previously observed in eukaryotic glycoproteins. The major *N*-linked glycans had up to three fucose residues attached to chitobiose cores. These were found at the 3- and/or 6-positions of the proximal *N*-acetylglucosamine (GlcNAc) and at the 3-position of the distal GlcNAc. The latter substitution was stated to be unique in *N*-glycans (Haslam *et al.*, 1996). It was proposed that these multifucosylated core structures could be highly immunogenic.

Integral Membrane Proteins and Membrane Anchors

Of the integral membrane proteins referred to above, H11 has a distinct transmembrane region (Smith *et al.*, 1997), the GA1 proteins are anchored via a glycosylinositolphospholipid (GPI) membrane anchor on p52, with p46 being held in place by as yet undefined interactions (Jasmer *et al.*, 1996). The P150 complex is anchored by a membrane (GPI) anchor on the 53 kDa component which forms non-covalent associations with the 45 and 53 kDa components (Rocha and Munn, 1997). The means by which

H-gal-GP and TSBP are anchored in the membrane is less clear. In the case of H-gal-GP, most components contain a predicted signal peptide, which is subsequently cleaved, and some contain potential transmembrane regions (though the latter require experimental confirmation). GPI anchoring is not involved (Smith *et al.*, 1999). Hydrophobicity analyses suggest that MEP2, MEP3 and MEP4 all contain at least one short hydrophobic region which could serve as transmembrane helices (Fig. 13.5). No regions of particular hydrophobicity were identified in MEP1, suggesting that this protein is held at the membrane by association with another membrane-anchored protein (Redmond *et al.*, 1997). A homologue of MEP1, designated MEP1b (Rehman and Jasmer, 1999), has a signal peptide indicative of secretion. Perhaps MEP1 is, in fact, intracellular? The same type of analysis has identified two short regions in the pepsin homologue that may serve to anchor this molecule to the surface of the membrane. In the case of the cysteine protease components of TSBP, each contains consensus signal and pro-enzyme cleavage sites with no evidence of a membrane anchor or any significant hydrophobic regions (Skuce *et al.*, 1999b). These features would indicate that they should be processed into the water-soluble phase, yet this is apparently not the case – a hypothesis

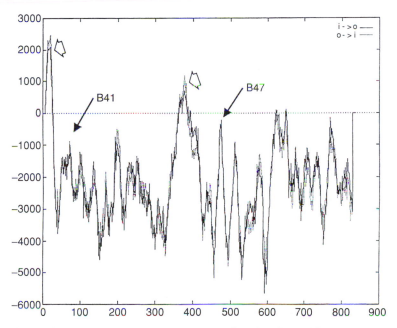

Fig. 13.5.   Hydrophobicity analysis of the predicted amino acid sequence from a metalloprotease component of H-gal-GP (MEP3) showing two potential transmembrane domains (indicated by open arrows). B41 and B47 indicate the relative positions of two N-terminal sequences determined from bands present when H-gal-GP is reduced.

supported by the observation that protease-enriched preparations prepared from the water-soluble and membrane-associated phases of adult parasite extracts did not stimulate any degree of protective immunity in vaccine trials (Knox *et al.*, 1999). One possibility is that these proteases are held in the membrane by association with the cystatin component of H-gal-GP.

Concluding Remarks

Clearly the commercial production of a vaccine against haemonchosis is quite feasible provided that the antigen(s) can be produced as recombinant proteins where the protective epitopes are expressed in an appropriate manner. To date, in the case of H11, H-gal-GP and TSBP, recombinant proteins expressed in *E. coli* have not induced significant protective immunity (E.A. Munn, D.P. Knox and W.D. Smith, 2000, unpublished results), probably due to incorrect folding or a lack of glycosylation, or because more than one component was required. In the case of H11, protection obtained by vaccination with the native antigen is markedly reduced by dissociation and reduction. All three isoforms of H11 have been expressed as enzymically active recombinant proteins in *Baculovirus* and testing is ongoing. If successful, these trials will establish the principle of vaccinating against a gut nematode with a recombinant protein. However, *Baculovirus* is not an appropriate system for commercial vaccine production, hence there will still be a requirement to seek alternative expression systems.

The indications are that gut antigen-based vaccines, at least based on the antigens described above and mediated by high systemic antibody responses, are unlikely to be wholly effective against non-blood-feeding nematodes. Different antigens may be required and may need to be delivered in such a way as to stimulate local mucosal immune responses.

Future work will, inevitably, explore the possibility of DNA vaccination, a topic that has recently been the subject of a comprehensive review (Alarcon *et al.*, 1999). DNA vaccines are stable indefinitely and, potentially, obviate the need to perform much of the recombinant protein expression work referred to in the discussions above. DNA vaccines can induce protective cytotoxic T cell responses, helper T cell and humoral immunity and the immune response required can be modulated by co-administration with cytokine cDNAs and by using different modes and sites of delivery, including mucosal delivery. DNA vaccines can be based on single genes or entire expression libraries. DNA vaccination is certainly an attractive potential solution to the difficulties of developing recombinant vaccines based on multi-protein complexes such as H-gal-GP. We have already established that intramuscular delivery into sheep of *hmcp*6, as a cDNA cloned into pcDNA3 (Invitrogen), primed the immune system to produce a moderate antibody response to a single intramuscular injection of the corresponding recombinant protein. We now need to enhance this effect and to evaluate the responses to DNA vaccines containing all the components of H-gal-GP or TSBP, or combinations thereof.

References

Alarcon, J.B., Waine, G.W. and McManus, D.P. (1999) DNA vaccines: technology and application as anti-parasite and anti-microbial agents. *Advances in Parasitology* 42, 344–410.

Andrews, S.J., Hole, N.J.K., Munn, E.A. and Rolph, T.P. (1995) Vaccination of sheep against haemonchosis with H11 – prevention of the periparturient rise and colostral transfer of protective immunity. *International Journal for Parasitology* 25, 839–846.

Andrews, S.J., Rolph, T.P. and Munn, E.A. (1997) Duration of protective immunity against ovine haemonchosis following vaccination with the nematode gut membrane antigen H11. *Research in Veterinary Science* 62, 223–227.

Bird, A.F. (1971) *The Structure of Nematodes*. Academic Press, London.

Boisvenue, R.J., Stiff, M.I., Tonkinson, L.V., Cox, G.N. and Hageman, R. (1992) Fibrinogen-degrading proteins from *Haemonchus contortus* used to vaccinate sheep. *American Journal of Veterinary Research* 53, 1263–1265.

Bozas, S.E., Panaccio, M., Creaney, J., Dosen, M., Parsons, J.C., Vlasuk, G.V., Walker, I.D. and Spithill, T.W. (1995) Characterisation of a novel Kunitz-type molecule from the trematode *Fasciola hepatica*. *Molecular and Biochemical Parasitology* 74, 19–29.

Brinkworth, R.I., Brindley, P.J. and Harrop, S.A. (1996) Structural analysis of the catalytic site of ACCP-1, a cysteine proteinase secreted by the hookworm *Ancylostoma caninum*. *Biochimica et Biophysica Acta* 1298, 4–8.

Colam, J.B. (1971) Studies on gut ultra-structure and digestive physiology in *Cosmocerca ornata* (Nematoda: Ascaridida). *Parasitology* 62, 259–272.

Cox, G.N., Pratt, D., Hageman, R. and Boisvenue, R.J. (1990) Molecular cloning and sequencing of a cysteine protease expressed by *Haemonchus contortus* adult worms. *Molecular and Biochemical Parasitology* 41, 25–34.

Dalton, J.P., McGonigle, S., Rolph, T.P. and Andrews, S.J. (1996) Induction of protective immunity in cattle against infection with *Fasciola hepatica* by vaccination with cathepsin L proteases and haemoglobin. *Infection and Immunity* 64, 5066–5074.

Ferguson, D.L., Rhodes, M.B., Marsh, C.L. and Payne, L.C. (1969) Resistance of immunised animals to infection by the larvae of the large roundworm of swine (*Ascaris suum*). *Federation Proceedings* 28, 497.

Fetterer, R.H. and Rhoads, M.L. (1997) The *in vitro* uptake and incorporation of haemoglobin by adult *Haemonchus contortus*. *Veterinary Parasitology* 69, 77–87.

Graham, M., Smith, T.S., Munn, E.A. and Newton, S.E. (1993) Recombinant DNA molecules encoding aminopeptidase enzymes and their use in the preparation of vaccines against helminth infections. Patent No. WO 93/23542.

Haslam, S.M., Coles, G.C., Munn, E.A., Smith, T.S., Smith, H.F., Morris, H.R. and Dell, A. (1996) *Haemonchus contortus* glycoproteins contain oligosaccharides with novel highly fucosylated core structures. *Journal of Biological Chemistry* 271, 30561–30570.

Jamaur, M.P. (1966) Cytochemical and electron microscopic studies on the pharynx and intesinal epithelium of *Nippostrongylus brasiliensis*. *Journal of Parasitology* 52, 1116–1128.

Jasmer, D.P., Perryman, L.P., Conder, G.A., Crow, S. and McGuire, T.C. (1993) Protective immunity to *Haemonchus contortus* induced by immunoaffinity isolated antigens that share a phylogenetically conserved carbohydrate gut surface epitope. *Journal of Immunology* 151, 5450–5460.

Jasmer, D.P., Perryman, L.P. and McGuire, T.C. (1996) *Haemonchus contortus* GA1 antigens: related phospholipase C-sensitive, apical gut membrane proteins encoded as a polyprotein and released from the nematode during infection. *Proceedings of the National Academy of Sciences USA* 93, 8642–8647.

Karanu, F.N., McGuire, T.C., Davis, W.C., Besser, T.E. and Jasmer, D.P. (1997a) CD4+T lymphocytes contribute to protective immunity induced in sheep and goats by *Haemonchus contortus* gut antigens. *Parasite Immunology* 19, 435–445.

Karanu, F.N., Rurangirwa, F.R., McGuire, T.C. and Jasmer, D.P. (1997b) *Haemonchus contortus*: inter- and intrageographic isolate heterogeneity of proteases in adult worm excretory-secretory products. *Experimental Parasitology* 86, 88–91.

Knox, D.P., Redmond, D.L. and Jones, D.G. (1993) Characterisation of proteases in extracts of adult *Haemonchus contortus*, the ovine abomasal nematode. *Parasitology* 106, 395–404.

Knox, D.P., Smith, S.K., Smith, W.D., Redmond, D.L. and Murray, J.M. (1995) Thiol Binding Proteins. Patent Application No. PCT/GB95/00665.

Knox, D.P., Smith, S.K. and Smith, W.D. (1999) Immunization with an affinity purified protein extract from the adult parasite protects lambs against *Haemonchus contortus*. *Parasite Immunology* 21, 201–210.

Longbottom, D., Redmond, D.L., Russell, M., Liddell, S., Smith, W.D. and Knox, D.P. (1997) Molecular cloning and characterisation of an aspartate protease associated with a highly protective gut membrane protein complex from adult *Haemonchus contortus*. *Molecular and Biochemical Parasitology* 88, 63–72.

Loukas, A., Selzer, P.M. and Maizels, R.M. (1998) Characterisation of Tc-cp-1, a cathepsin L-like cysteine protease from *Toxocara canis* infective larvae. *Molecular and Biochemical Parasitology* 92, 275–289.

MacLennan, K., Gallagher, M.P. and Knox, D.P. (1997) Stage-specific serine and metallo-proteinase release by adult and larval *Trichostrongylus vitrinus*. *International Journal for Parasitology* 27, 1031–1036.

McLaren, D.J., Burt, J.S. and Ogilvie, B.M. (1974) The anterior glands of *Necator americanus*. II. Cytochemical and functional studies. *International Journal for Parasitology* 4, 39–46.

Meeusen, E.N.T. (1995) Production of antigens. Patent No. WO 95/09182.

Munn, E.A. (1977) A helical, polymeric extracellular protein associated with the luminal surface of *Haemonchus contortus* intestinal cells. *Tissue and Cell* 9, 23–34.

Munn, E.A. (1981) The endotube, a macroscopic intracellular structure from the syncytial intestine of the parasitic nematode *Haemonchus contortus*. *Journal of Physiology* 319, 7–8P.

Munn, E.A. and Greenwood, C.A. (1983) Endotube-brush border complexes dissected from the intestines of *Haemonchus contortus* and *Ancylostoma caninum*. *Parasitology* 87, 129–137.

Munn, E.A., Graham, M. and Coadwell, W.J. (1987) Vaccination of young lambs by means of a protein fraction extracted from adult *Haemonchus contortus. Parasitology* 94, 385–397.

Munn, E.A., Smith, T.S., Smith, H., Smith, F. and Andrews, S.J. (1997) Vaccination against *Haemonchus contortus* with denatured forms of the protective antigen H11. *Parasite Immunology* 19, 243–248.

Newlands, G.F.J., Skuce, P.J., Knox, D.P., Smith, S.K. and Smith, W.D. (1999) Cloning and characterisation of a β-galactoside-binding protein (galectin) from the gastrointestinal nematode *Haemonchus contortus. Parasitology* 119, 483–490.

Newton, S.E. and Munn, E.A. (1999) The development of vaccines against gastrointestinal nematodes, particularly *Haemonchus contortus. Parasitology Today* 15, 116–122.

Piacenza L., Acosta, D., Basmadjan, I., Dalton, J.P. and Carmona, C. (1999) Vaccination with cathepsin L proteases and with leucine aminopeptidase induces high levels of protection against fascioliasis in sheep. *Infection and Immunity* 67, 1954–1961.

Pratt, D., Cox, G.N., Milhausen, M.J. and Boisvenue, R.J. (1990) A developmentally regulated cysteine protease gene family in *Haemonchus contortus. Molecular and Biochemical Parasitology* 43, 181–192.

Pratt, D., Armes, L.G., Hagemen, R., Reynolds, V., Boisvenue, R. and Cox, G.N. (1992a) Cloning and sequence comparison of four distinct cysteine proteases expressed by *Haemonchus contortus* adult worms. *Molecular and Biochemical Parasitology* 51, 209–218.

Pratt, D., Boisvenue, R.J. and Cox, G.N. (1992b) Isolation of putative cysteine protease genes of *Ostertagia ostertagi. Molecular and Biochemical Parasitology* 56, 39–48.

Redmond, D.L., Knox, D.P., Newlands, G.F.J. and Smith, W.D. (1997) Molecular cloning of a developmentally regulated putative metallo-peptidase present in a host protective extract of *Haemonchus contortus. Molecular and Biochemical Parasitology* 85, 77–87.

Rehman, A. and Jasmer, D.P. (1999) A tissue specific approach for analysis of membrane and secreted proteins from *Haemonchus contortus* gut and its application to diverse nematode species. *Molecular and Biochemical Parasitology* 97, 55–68.

Rhoads, M.L. and Fetterer, R.H. (1995) Developmentally regulated secretion of cathepsin L-like proteases by *Haemonchus contortus. Journal for Parasitology* 81, 505–512.

Rhoads, M.L. and Fetterer, R.H. (1997) Extracellular matrix: a toll for defining extracorporeal digestion of parasite proteases. *Parasitology Today* 13, 119–122.

Roberts, D.D., Sherwood, J.A., Spitalnik, S.L., Panton, L.J., Howard, R.J., Dixit, V.M., Frazier, W.A., Miller, L.H. and Ginsburg, V. (1985) Thrombospondin binds *falciparum* malaria parasitized erythrocytes and may mediate cytoadherence. *Nature* 318, 64–66.

Robson, K.J., Hall, J.R., Jennings, M.W., Harris, T.J., Marsh, K., Newbold, C.I., Tate, V.E. and Weatherall, D.J. (1988) A highly conserved amino acid sequence in thrombospondin, properdin and in proteins from sporozoites and blood-stages of the human malaria parasite. *Nature* 335, 79–82.

Rocha, J. and Munn, E.A. (1997) P150, a protective glycoprotein complex of the microvillar membrane of *Haemonchus contortus*. Conference Abstract from *Parasitic Helminths – from Genomes to Vaccines*, Edinburgh, UK, 6–9 September 1997.

Rocques, B.P., Noble, F., Dauge, V., Fournie-Zaluski, M. and Beaumont, A. (1993) Neutral endopeptidase 24.11: structure, inhibition and experimental and clinical pharmacology. *Pharmacological Reviews* 45, 87–146.

Skuce, P.J., Stewart, E.M., Smith, W.D. and Knox, D.P. (1999a) Cloning and characterisation of glutamate dehydrogenase (GDH) from the gut of *Haemonchus contortus. Parasitology* 118, 297–304.

Skuce, P.J., Redmond, D.L., Liddell, S., Stewart, E.M., Newlands, G.F.J., Smith, W.D. and Knox, D.P. (1999b) Molecular cloning and characterisation of gut-derived cysteine proteases associated with a host-protective extract from *Haemonchus contortus. Parasitology* 119, 405–412.

Smith, S.K. and Smith, W.D. (1996) Immunisation of sheep with an integral membrane glycoprotein complex of *Haemonchus contortus* and its major polypeptide components. *Research in Veterinary Science* 60, 1–6.

Smith, S.K., Pettit, D., Newlands, G.F.J., Redmond, D.L., Skuce, P.J., Knox, D.P. and Smith, W.D. (1999) Further immunisation and biochemical studies with a protective antigen complex from the microvillar membrane of the intestine of *Haemonchus contortus. Parasite Immunology* 21, 187–199.

Smith, T.S. and Munn, E.A. (1990) Strategies for vaccination against gastrointestinal nematodes. *Revues of the Scientific and Technical Office for International Epizooitology* 9, 577–595.

Smith, T.S., Graham, M., Munn, E.A., Newton, S.E., Knox, D.P., Coadwell, W.J., McMichael-Phillips, D., Smith, H., Smith, W.D. and Oliver, J.J. (1997) Cloning and characterisation of a microsomal aminopeptidase from the intestine of the nematode *Haemonchus contortus. Biochimica et Biophysica Acta* 1338, 295–306.

Smith, W.D. and Smith, S.K. (1993) Evaluation of aspects of the protection afforded to sheep immunised with a gut membrane protein from *Haemonchus contortus. Research in Veterinary Science* 55, 1–9.

Smith, W.D., Smith, S.K., Murray, J.M., Liddell, S. and Knox, D.P. (1993) Vaccines against metazoan parasites. Patent Application No. PCT/GB/93/01:521.

Smith, W.D., Smith, S.K. and Murray, J.M. (1994) Protection studies with integral membrane fractions of *Haemonchus contortus. Parasite Immunology* 16, 231–241.

Tort, J., Brindley, P.J., Knox, D.P., Wolfe, K.H. and Dalton, J.P. (1999) Proteases and associated genes of parasitic helminths. *Advances in Parasitology* 43, 162–166.

Young, C.J., McKeand, J.B. and Knox, D.P. (1995) Proteinases released *in vitro* by the parasitic stages of *Teladorsagia circumcincta*, an ovine abomasal parasite. *Parasitology* 110, 465–471.

Metabolic Transitions and the Role of the Pyruvate Dehydrogenase Complex During Development of *Ascaris suum*

14

Richard Komuniecki and Patricia R. Komuniecki

Department of Biology, University of Toledo, Toledo, OH 43606-3390, USA

Introduction

The parasitic nematode, *Ascaris suum*, undergoes a number of well-characterized metabolic transitions during its development (Table 14.1), but little is known about the regulation of these events (Barrett, 1976; Komuniecki and Komuniecki, 1995). Adults reside in the porcine small intestine and fertilization takes place under low oxygen tensions. The unembryonated 'egg' that leaves the host is metabolically quiescent, has no detectable cytochrome oxidase activity or ubiquinone and appears to be transcriptionally inactive (Cleavinger *et al.*, 1989; Takamiya *et al.*, 1993). Embryonation requires oxygen and after about 48–72 h is accompanied by

Table 14.1. Metabolic transitions during the development of *Ascaris suum*.

Stage	Environment	Metabolic rate	Temperature
UE	'Egg', microaerobic → aerobic	Low	Ambient
L1	'Egg', aerobic	High	Ambient
L2/L3	'Egg', aerobic	Quiescent, low	Ambient
L3 (hatched)	Gut, microaerobic	High	38–40°C
L3	Liver → lung, aerobic	? (In culture, low)	38–40°C
L4	Gut, microaerobic	High	38–40°C
Adult	Gut, microaerobic → anaerobic	High	38–40°C

a dramatic increase in cytochrome oxidase activity (Oya *et al.*, 1963; Sylk, 1969; Barrett, 1976). Development to a second-stage larva (L2) occurs within the 'eggshell', takes about 21 days (depending on temperature) and is accompanied by a number of well-defined metabolic changes. For example, glycogen is initially utilized and then resynthesized from stored triglyceride and these larvae are one of the few animals to possess a functional glyoxylate cycle (Barrett *et al.*, 1970; Barrett, 1976). Both malate synthase and isocitrate lyase, key enzymes of the cycle, increase dramatically at about day 10 (Barrett *et al.*, 1970). After the moult to the L2, metabolic rate diminishes significantly and this dormant larva can survive for long periods until ingested by a suitable host. Little is known about the regulation of energy generation in this developmentally arrested state. In fact, these larvae closely resemble the well-studied dauer larva of many free-living nematodes (Larsen *et al.*, 1995; Blaxter and Bird, 1997; Riddle and Albert, 1997; Antebi *et al.*, 1998; Rajan, 1998). Since dauer larva formation is linked exclusively with the transition to the third stage (L3), some authors have questioned whether this quiescent *A. suum* larva is actually an L3 and not an L2. In fact, recent evidence strongly suggests that continued larval development within the egg is accompanied by a second moult and that the infective stage is actually an L3, as has been reported for other closely related nematodes (Geenen *et al.*, 1999).

Hatching is triggered in the microaerobic vertebrate gut and the L3 migrates to the liver and the lungs (Saz *et al.*, 1968; Murrell *et al.*, 1997). Hatched L3s are cyanide-sensitive and have a functional tricarboxylic acid cycle, even though hatching occurs in the microaerobic environment of the gut (Barrett, 1976). L3s isolated directly from rabbit lungs also are cyanide-sensitive but, interestingly, they appear to contain many of the enzymes necessary for the anaerobic energy-generating pathways operative in fourth-stage larva (L4) and the adult (Saz *et al.*, 1968; Douvres and Tromba, 1971; Urban and Douvres, 1981; Komuniecki and Vanover, 1987; Vanover and Komuniecki, 1989). Culture of L3s isolated from rabbit lung results in a rapid moult to the L4, the loss of cyanide sensitivity and the appearance of branched-chain fatty acids, characteristic end-products of adult muscle metabolism (Komuniecki and Vanover, 1987; Vanover and Komuniecki, 1989). The relationship of these cultured L3 and L4 to their *in vivo* counterparts is unclear. Certainly, many processes are quite similar, but L3s do not develop to adults in rabbits, and the loss of cytochrome oxidase activity that accompanies the moult to the L4 *in vivo* does not appear to be paralleled *in vitro* (Vanover and Komuniecki, 1989).

Anaerobic Mitochondrial Metabolism

The anaerobic metabolism of adult body wall muscle has been well characterized (Komuniecki and Harris, 1995). Large glycogen stores are converted cytoplasmically to malate, which then enters the mitochondrion and is ultimately converted to a variety of reduced organic acids, including succinate, acetate, propionate and the branched-chain fatty acids (BFA), 2-methylbutyrate and 2-methylvalerate (Fig. 14.1) (Saz and Weil, 1960; Rioux and Komuniecki, 1984; Komuniecki and Harris, 1995). BFAs are abundant and have a significant impact on a number of aspects of *A. suum* physiology. For example, they rival chloride as the major anions of the perienteric fluid, are excreted across the cuticle and comprise a substantial portion of the fatty acids found in stored 'egg' triglycerides (Saz and Lescure, 1966; Sims *et al.*, 1992).

Mitochondria from body wall muscle and probably the pharynx lack a functional TCA cycle and their novel anaerobic pathways rely on reduced organic acids as terminal electron acceptors, instead of oxygen (Saz, 1971; Ma *et al.*, 1993; Duran *et al.*, 1998). Malate and pyruvate are oxidized intramitochondrially by malic enzyme and the pyruvate dehydrogenase complex, respectively, and excess reducing power in the form of NADH drives Complex II and β-oxidation in the direction opposite to that observed in aerobic organelles (Kita, 1992; Duran *et al.*, 1993; Ma *et al.*,

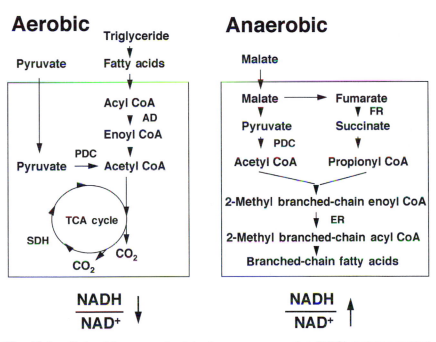

Fig. 14.1. Role of the pyruvate dehydrogenase complex (PDC) during aerobic/anaerobic transitions in the development of *Ascaris suum*. PDC, pyruvate dehydrogenase complex; AD, acyl CoA dehydrogenase; ER, enoyl CoA reductase; FR, fumarate reductase; SDH, succinate dehydrogenase.

1993; Komuniecki and Harris, 1995; Komuniecki and Komuniecki, 1995). Most importantly, the NADH-dependent reductions of fumarate and 2-methyl branched-chain acyl CoAs are coupled to site 1, electron-transport associated, energy generation (Kita, 1992; Ma *et al.*, 1993). Although most of the key enzymes involved in these pathways have been well characterized (Komuniecki and Komuniecki, 1995), little is known about the nature of the proton gradient in these anaerobic mitochondria or the factors regulating carbon flux and energy generation during muscle contraction. More importantly, these processes may differ substantially from their mammalian counterparts. For example, during vertebrate muscle contraction, Ca^{2+} released from the sarcoplasmic reticulum is readily accumulated by muscle mitochondria, where it activates the Ca^{2+}-sensitive pyruvate dehydrogenase phosphatase (PDP) and α-ketoglutarate dehydrogenase complex and effectively links contraction and energy generation (Chen *et al.*, 1996; Roche and Cox, 1996). In contrast, mitochondria from *A. suum* muscle are not uncoupled when incubated in Ca^{2+}, suggesting a limited capacity for high-affinity Ca^{2+} uptake, and helminth PDPs and α-ketoglutarate dehydrogenase complexes appear to be Ca^{2+}-insensitive (Song and Komuniecki, 1994; Diaz and Komuniecki, 1996).

It is clear from the above discussion that metabolism in *A. suum* includes both low activity and high activity states in aerobic, microaerobic and virtually anaerobic environments. Although we are beginning to understand how mitochondria from adult muscle are modified to generate energy in the absence of oxygen, much less is known about mitochondrial biogenesis during these various transitions. A variety of different strategies can be identified from the deletion of non-essential activities to the overexpression of key anaerobic enzymes. More recently, the expression of stage-specific isoforms have been tentatively described. For example, in early aerobic larval stages, Complex II of the mitochondrial electron-transport chain functions as a succinate dehydrogenase, but, in contrast, in adult muscle it functions in the opposite direction as a fumarate reductase (Tielens and Roos, 1994; Saruta *et al.*, 1995). Kinetic differences between the complexes isolated from the two stages have been well characterized and different stage-specific forms of the Fp subunit have been identified (Furushima *et al.*, 1990; Kuramochi *et al.*, 1994). Similar observations have been made about the roles of the acyl CoA dehydrogenase and enoyl CoA reductase in the switch from fatty acid oxidation to branched-chain fatty acid synthesis that accompanies the transition to anaerobic metabolism (Duran *et al.*, 1993, 1998).

This chapter focuses on the developmental regulation of the pyruvate dehydrogenase complex (PDC). The PDC plays diverse and pivotal roles in the entry of glycolytically generated carbon into the TCA cycle in aerobic stages and the metabolism of mitochondrially generated pyruvate in anaerobic stages (Fig. 14.1).

The Role of the PDC During Development

The PDC catalyses the irreversible oxidative decarboxylation of pyruvate (Fig. 14.2). The PDC is a large multi-enzyme complex consisting of three catalytic components: pyruvate dehydrogenase (E1), dihydrolipoyl transacetylase (E2) and dihydrolipoyl dehydrogenase (E3) and, in eukaryotes, an E3-binding protein (E3BP) (Randle, 1986; Roche and Cox, 1996). PDC activity in higher eukaryotes is regulated by the reversible phosphorylation/dephosphorylation of E1 catalysed by a distinct E1 kinase (PDK) and PDP (Patel and Roche, 1990; Song and Komuniecki, 1994; Roche and Cox, 1996; Bowker-Kinley *et al.*, 1998; Huang *et al.*, 1998a). Not surprisingly, the PDC plays an important role during a number of the metabolic transitions that characterize *A. suum* development.

The regulation of PDC activity in the anaerobic body wall muscle mito-chondria of the L4 and adult muscle has been extensively characterized (Thissen *et al.*, 1986; Thissen and Komuniecki, 1988; Klingbeil *et al.*, 1997; Chen *et al.*, 1998; Huang *et al.*, 1998b). Initially, the identification of PDC activity with significant endogenous PDK activity in adult *A. suum* muscle was surprising, given the sensitivity of this complex and its associated PDK to elevated NADH/NAD$^+$ and acetyl CoA/CoA ratios (Thissen *et al.*, 1986; Chen *et al.*, 1998). NADH and acetyl CoA can inhibit PDC activity directly by end-product inhibition, or indirectly by stimulation of the PDK which catalyses the phosphorylation and inactivation of E1. Therefore, elevation of these ratios during the transition to anaerobic metabolism in *A. suum* has the potential to phosphorylate and inactivate the PDC when maximal flux through the complex is necessary to fuel a relatively inefficient fermentative metabolism. In fact, in many facultative anaerobes, such as *Escherichia coli*, the PDC is down-regulated during anaerobiosis and other

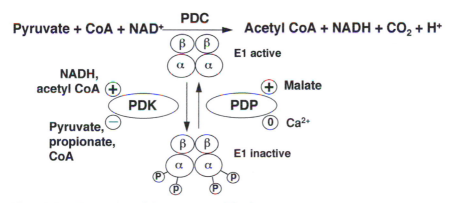

Fig. 14.2. Regulation of the pyruvate dehydrogenase complex (PDC) from adult *A. suum* muscle. PDC, pyruvate dehydrogenase complex; E1, pyruvate dehydrogenase subunit of the PDC; PDK, pyruvate dehydrogenase kinase; PDP, pyruvate dehydrogenase phosphatase.

enzymes better suited to functioning in reduced environments, such as pyruvate : ferridoxin oxidoreductase and pyruvate : formate lyase, are involved in pyruvate decarboxylation (Knappe and Sawers, 1990). In contrast, in obligate anaerobes, such as *Enterococcus faecalis*, the PDC is dramatically overexpressed and more resistant to end-product inhibition (Snoep *et al.*, 1993). Since prokaryotic PDCs are not regulated by covalent modification, activation of a PDK under these conditions is not a problem.

Predictably, the regulation of the PDC in adult *A. suum* body wall muscle appears to be modified to maintain PDC activity under the reducing conditions present in the host gut (Klingbeil *et al.*, 1996; Komuniecki, 1996). The PDC in adult body wall muscle is abundant; in fact the PDC is more abundant in *A. suum* body wall muscle than in any other eukaryotic tissue studied to date and approaches 2% of the total soluble protein (Thissen *et al.*, 1986). The PDC also is less sensitive to end-product inhibition by elevated $NADH/NAD^+$ and acetyl CoA/CoA ratios than PDCs from aerobic organisms (Roche and Cate, 1977; Thissen *et al.*, 1986; Chen *et al.*, 1998). Interestingly, E3, the enzyme responsible for the NAD^+-dependent reoxidation of the reduced lipoyl domains of E2, appears to be identical in both aerobic and anaerobic stages. In contrast, the PDC from adult muscle lacks the terminal lipoyl domain found in E3BPs from all other sources (Klingbeil *et al.*, 1996; Komuniecki, 1996). It appears that the binding of E3 to this 'anaerobic' E3BP significantly reduce the sensitivity of the E3 to NADH inhibition and help to maintain PDC activity in the face of the elevated $NADH/NAD^+$ ratios associated with anaerobiosis (Klingbeil *et al.*, 1996).

Similarly, the regulation of PDK activity is modified in adult muscle PDC. For example, PDK activity is inhibited by pyruvate and propionate (metabolites elevated during anaerobic metabolism) and is less sensitive to stimulation by elevated $NADH/NAD^+$ and acetyl CoA/CoA ratios (Fig. 14.2) (Thissen *et al.*, 1986; Chen *et al.*, 1998). The effects of NADH and acetyl CoA on PDK activity are mediated by the degree of E3-catalysed oxidation and E2-catalysed acetylation of the inner lipoyl domain of E2 (Roche and Cate, 1977; Rahmatullah and Roche, 1985, 1987; Ravindran *et al.*, 1996; Yang *et al.*, 1998), so that the regulation of this phenomenon is complex and involves multiple interacting components.

Not surprisingly, it appears that the stoichiometry of phosphorylation and inactivation of the adult muscle PDC is also altered to prevent its complete inactivation during anaerobiosis (Figs 14.2, 14.3). In mammalian E1s, each E1α of the $\alpha_2\beta_2$ tetramer contains three distinct phosphorylation sites and, given the specificity of the mammalian PDK, inactivation *in vivo* is associated primarily with the phosphorylation of site 1 (Fig. 14.3, Table 14.2) (Yeaman *et al.*, 1978; Korotchkina and Patel, 1995). However, phosphorylation at any of the three phosphorylation sites is sufficient for inactivation. More importantly, inactivation is characterized by half-of-the-site reactivity, and phosphorylation in only one of the two E1α subunits

results in the complete inactivation of the tetramer and prevents phosphorylation in the other E1α subunit (Korotchkina and Patel, 1995). In
contrast, the E1α in the PDC purified from adult body wall muscle contains
only two phosphorylation sites; inactivation is accompanied by substantially

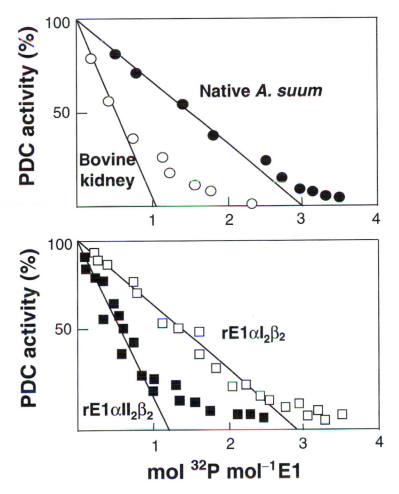

Fig. 14.3. Stoichiometry of phosphorylation/inactivation of the bovine kidney
and *A. suum* E1 isoforms. Adult *A. suum* PDC was depleted of its E1 component
and reconstituted with either bovine kidney E1 or recombinant *A. suum* E1s
containing either the αI or αII isoform (Klingbeil *et al.*, 1997; Huang *et al.*, 1998b).
The hybrid complexes were then assayed for PDC activity and the incorporation
of ^{32}P, as described fully in Thissen *et al.* (1986). Upper panel: ○, bovine kidney
E1; ●, E1 isolated directly from the adult *A. suum* PDC. Lower panel: □, *A. suum*
E1αI isoform; ■, *A. suum* E1αII isoform.

Table 14.2. Phosphorylation of *A. suum* and mammalian E1s.

E1	Activity with PDK	Phosphorylation sites/tetramer	^{32}P incorporation (mol ^{32}P/mol E1)	Half-of-the-site reactivity
A. suum				
E1αI	High[a]	4	~4	No
E1αII	Low[a]	6	~3	No
Bovine kidney	High	6	~3	Yes

[a]Assayed after reconstitution with E1-depleted adult *A. suum* PDC.

more phosphorylation than observed in the mammalian complex and both E1α subunits of the tetramer are phosphorylated (Fig. 14.3) (Thissen and Komuniecki, 1988; Huang *et al.*, 1998b). These differences effectively prevent the complete inactivation of the complex, especially in the presence of PDP activity. Interestingly, in contrast to mammalian PDPs, the *A. suum* PDP is dramatically stimulated by malate, the major mitochondrial substrate in adult *A. suum* muscle. This represents yet another regulatory modification designed to maintain the PDC in an active dephosphorylated state (Fig. 14.2) (Song and Komuniecki, 1994).

Recently, two E1α isoforms (E1αI and E1αII) have been identified in cDNA libraries prepared from adult *A. suum* muscle which are 90% identical at the predicted amino acid level (Johnson *et al.*, 1992; Huang *et al.*, 1998b). However, the predicted amino acid sequence of E1αI contains only two phosphorylation sites (Ser-203 is changed to Ala) and it appears to be identical to the E1α isolated directly from adult muscle (Table 14.2). In contrast, E1αII contains the three phosphorylation sites identified in mammalian E1s and appears to be most abundant in the L3. Both isoforms have been functionally expressed in *E. coli* with a muscle-specific β-subunit and the affinity-purified proteins used to reconstitute PDC activity in an adult *A. suum* muscle E1-deficient PDC (Huang *et al.*, 1998b). Both of the E1 isoforms appear to be modified to decrease the effectiveness of phosphorylation in the inactivation of the complex and to maintain PDC activity in the presence of PDK and the stimulatory reducing conditions encountered in the host gut. For example, substantially more phosphate is incorporated into E1αI than E1αII as inactivation proceeds, as observed in the native PDC isolated from body wall muscle (Fig. 14.3). In contrast, E1αII exhibits a stoichiometry of phosphorylation/inactivation identical to that observed for E1s isolated from aerobic organisms (Fig. 14.3). However, E1αII is phosphorylated more slowly than E1-αI, which minimizes the effects of PDK stimulation on PDC activity. Whether a different PDK isoform is present in aerobic larval stages with greater activity toward E1αII remains to be determined.

PDC and Metabolic Transitions

In contrast to the regulation of PDC activity in anaerobic muscle, much less is known about the PDC during other metabolic transitions. Clearly, during the first few days of embryonic development, prior to the synthesis of ubiquinone and cytochrome oxidase and the switch to lipid metabolism, the PDC must be responsible for the metabolism of pyruvate generated by the rapid glycogen depletion that accompanies these early cell divisions. Substantial lactate formation in the enclosed environment of the 'eggshell' would lead to rapid acidification and could be catastrophic. Although little is known about the regulation of PDC activity in this environment, the absolute amounts of PDC appear to decrease as development proceeds. In addition, immunoblotting with affinity-purified antisera raised against the subunits of the adult muscle PDC indicated that the muscle E3BP is not present in the early aerobic embryonic and larval stages (Fig. 14.4). Further, the apparent mobilities of the E1α and E1β subunits are also different, suggesting the presence of additional, as yet unidentified, stage-specific isoforms (Klingbeil *et al.*, 1996) (Fig. 14.4). Recently, a sperm-specific E1β subunit has been identified which, based on immunoblotting, appears to persist during early embryonic development (Huang and Komuniecki, 1997). Mitochondria from both sperm and oocytes, in contrast to mitochondria from adult muscle, contain dramatically elevated levels of many of the TCA cycle enzymes, although a functional cycle does not appear to be

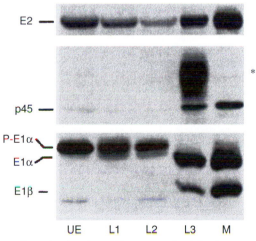

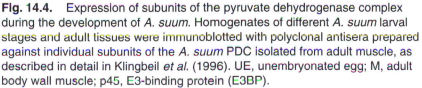

Fig. 14.4. Expression of subunits of the pyruvate dehydrogenase complex during the development of *A. suum*. Homogenates of different *A. suum* larval stages and adult tissues were immunoblotted with polyclonal antisera prepared against individual subunits of the *A. suum* PDC isolated from adult muscle, as described in detail in Klingbeil *et al.* (1996). UE, unembryonated egg; M, adult body wall muscle; p45, E3-binding protein (E3BP).

operative (Komuniecki *et al.*, 1993). Whether these elevated enzyme levels are a 'preadaptation' to aerobic energy generation during early embryogenesis is unclear, but the numerous paternally derived mitochondria that are present in the unembryonated 'egg' do not appear to persist once embryogenesis is initiated. Whether these 'paternal mitochondria' play a metabolic role during the first few days of development remains to be determined (Anderson *et al.*, 1995).

After the L3 hatch in the microaerobic environment of the vertebrate gut, the PDC again must be involved in the anaerobic metabolism of the

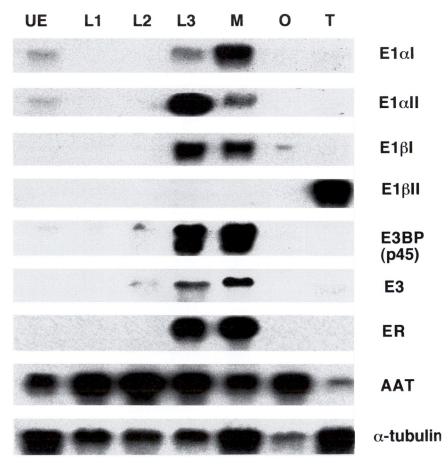

Fig. 14.5. Composite of Northern blots prepared with RNA isolated from different *A. suum* larval stages and adult tissues and probed with *A. suum* cDNAs specific for E1αl (Johnson *et al.*, 1992); E1αll (Johnson *et al.*, 1992); E1βl (Wheelock *et al.*, 1991); E1βll (Huang and Komuniecki, 1997); E3BP, E3-binding protein (p45); E3; ER, enoyl CoA reductase (Duran *et al.*, 1993, 1998); AAT, adenine nucleotide translocase; α-tubulin. UE, unembryonated 'egg'; L1, first-stage larva; L2, second-stage larva; L3, third-stage larva; M, adult muscle; O, ovaries plus oviducts; T, testis.

glycogen resynthesized during the formation of the L1 and prior to the entry of the L3 into the more aerobic environment of the luminal epithelium on its journey to the liver. Again, little is known about the regulation of the PDC during this migration, but both Northern and Western blotting suggest that genes coding for the subunits of the PDC present in adult muscle all appear to be significantly expressed in rabbit lung-derived L3s (Wheelock *et al.*, 1991; Johnson *et al.*, 1992; Huang and Komuniecki, 1997) (Figs 14.4, 14.5). These larvae are aerobic and cyanide-sensitive and their muscle mitochondria contain substantial cytochrome oxidase activity; thus the function of these 'anaerobic' isoforms is unclear. However, recent histochemical studies suggest that both cytochrome oxidase (an enzyme of aerobic pathways) and the 2-methyl branched-chain enoyl CoA reductase (a key enzyme of anaerobic pathways) are present in the same mitochondria (Mei *et al.*, 1997). Whether this pattern of expression represents another preadaptation to the 'anticipated' microenvironment on the lumen or is simply an artifact of altered development in rabbits remains to be determined.

Concluding Remarks

In summary, it is clear that *A. suum* undergoes a number of metabolic transitions during development and, in the case of the pyruvate dehydrogenase complex at least, is constantly fine-tuning the subunit-specific expression and function of the PDC during the course of development.

References

Anderson, T.J., Komuniecki, R., Komuniecki, P.R. and Jaenike, J. (1995) Are mitochondria inherited paternally in *Ascaris*? *International Journal for Parasitology* 25, 1001–1004.

Antebi, A., Culotti, J.G. and Hedgecock, E.M. (1998) daf-12 regulates developmental age and the dauer alternative in *Caenorhabditis elegans*. *Development* 125, 1191–1205.

Barrett, J. (1976) Intermediary metabolism in *Ascaris* eggs. In: Van den Bossche, H. (ed.) *Biochemistry of Parasites and Host–Parasite Relationships*. Elsevier, Amsterdam, pp. 117–123.

Barrett, J., Ward, C.W. and Fairbairn, D. (1970) The glyoxylate cycle and the conversion of triglycerides to carbohydrates in developing eggs of *Ascaris lumbricoides*. *Comparative Biochemistry and Physiology* 35, 577–586.

Blaxter, M. and Bird, D. (1997) Parasitic nematodes. In: Riddle, D.L., Blumenthal, T., Meyer, B. and Priess, J. (eds) *C. elegans II*. Cold Spring Harbor Laboratory Press, Cold Spring Harbor, New York, pp. 851–878.

Bowker-Kinley, M.M., Davis, W.I., Wu, P., Harris, R.A. and Popov, K.M. (1998) Evidence for existence of tissue-specific regulation of the mammalian pyruvate dehydrogenase complex. *Biochemical Journal* 329, 191–196.

Chen, G., Wang, L., Liu, S., Chuang, C. and Roche, T.E. (1996) Activated function of the pyruvate dehydrogenase phosphatase through Ca^{2+}-facilitated binding to the inner lipoyl domain of the dihydrolipoyl acetyltransferase. *Journal of Biological Chemistry* 271, 28064–28070.

Chen, W., Huang, X., Komuniecki, P.R. and Komuniecki, R. (1998) Molecular cloning, functional expression, and characterization of pyruvate dehydrogenase kinase from anaerobic muscle of the parasitic nematode *Ascaris suum*. *Archives of Biochemistry and Biophysics* 353, 181–189.

Cleavinger, P.J., McDowell, J.W. and Bennett, K.L. (1989) Transcription in nematodes: early *Ascaris* embryos are transcriptionally active. *Developmental Biology* 133, 600–604.

Diaz, F. and Komuniecki, R.W. (1996) Characterization of the alpha-ketoglutarate dehydrogenase complex from *Fasciola hepatica*: potential implications for the role of calcium in the regulation of helminth mitochondrial metabolism. *Molecular and Biochemical Parasitology* 81, 243–246.

Douvres, F.W. and Tromba, F.G. (1971) Comparative development of *Ascaris suum* in rabbits, guinea pigs, mice, and swine in 11 days. *Proceedings of The Helminthological Society of Washington* 38, 246–252.

Duran, E., Komuniecki, R.W., Komuniecki, P.R., Wheelock, M.J., Klingbeil, M.M., Ma, Y.C. and Johnson, K.R. (1993) Characterization of cDNA clones for the 2-methyl branched-chain enoyl-CoA reductase. An enzyme involved in branched-chain fatty acid synthesis in anaerobic mitochondria of the parasitic nematode *Ascaris suum*. *Journal of Biological Chemistry* 268, 22391–22396.

Duran, E., Walker, D.J., Johnson, K.R., Komuniecki, P.R. and Komuniecki, R.W. (1998) Developmental and tissue-specific expression of 2-methyl branched-chain enoyl CoA reductase isoforms in the parasitic nematode, *Ascaris suum*. *Molecular and Biochemical Parasitology* 91, 307–318.

Furushima, R., Kita, K., Takamiya, S., Konishi, K., Aoki, T. and Oya, H. (1990) Structural studies on three flavin-interacting regions of the flavoprotein subunit of complex II in *Ascaris suum* mitochondria. *FEBS Letters* 263, 325–328.

Geenen, P.L., Bresciani, J., Boes, J., Pederson, A., Erickson, L., Fagerholm, H.-P. and Nansen, P. (1999) The morphogenesis of *Ascaris suum* to the infective third-stage larvae within the egg. *Journal of Parasitology* 85, 616–622.

Huang, Y.J. and Komuniecki, R. (1997) Cloning and characterization of a putative testis-specific pyruvate dehydrogenase beta subunit from the parasitic nematode, *Ascaris suum*. *Molecular and Biochemical Parasitology* 90, 391–394.

Huang, B., Gudi, R., Wu, P., Harris, R.A., Hamilton, J. and Popov, K.M. (1998a) Isoenzymes of pyruvate dehydrogenase phosphatase. DNA-derived amino acid sequences, expression, and regulation. *Journal of Biological Chemistry* 273, 17680–17688.

Huang, Y.J., Walker, D., Chen, W., Klingbeil, M. and Komuniecki, R. (1998b) Expression of pyruvate dehydrogenase isoforms during the aerobic/anaerobic transition in the development of the parasitic nematode, *Ascaris suum*: altered stoichiometry of phosphorylation/inactivation. *Archives of Biochemistry and Biophysics* 352, 263–270.

Johnson, K.R., Komuniecki, R., Sun, Y. and Wheelock, M.J. (1992) Characterization of cDNA clones for the alpha subunit of pyruvate dehydrogenase from *Ascaris suum*. *Molecular and Biochemical Parasitology* 51, 37–48.

Kita, K. (1992) Electron-transfer complexes of mitochondria in *Ascaris suum*. *Parasitology Today* 8, 155–159.

Klingbeil, M.M., Walker, D.J., Arnette, R., Sidawy, E., Hayton, K., Komuniecki, P.R. and Komuniecki, R. (1996) Identification of a novel dihydrolipoyl dehydrogenase-binding protein in the pyruvate dehydrogenase complex of the anaerobic parasitic nematode, *Ascaris suum*. *Journal of Biological Chemistry* 271, 5451–5457.

Klingbeil, M.M., Walker, D.J., Huang, Y.J. and Komuniecki, R. (1997) Altered phosphorylation/inactivation of a novel pyruvate dehydrogenase in adult *Ascaris suum* muscle. *Molecular and Biochemical Parasitology* 90, 323–326.

Knappe, J. and Sawers, G. (1990) A radical-chemical route to acetyl-CoA: the anaerobically induced pyruvate formate-lyase system of *Escherichia coli*. *FEMS Microbiology Reviews* 75, 383–398.

Komuniecki, P.R. and Vanover, L. (1987) Biochemical changes during the aerobic–anaerobic transition in *Ascaris suum* larvae. *Molecular and Biochemical Parasitology* 22, 241–248.

Komuniecki, P.R., Johnson, J., Kamhawi, M. and Komuniecki, R. (1993) Mitochondrial heterogeneity in the parasitic nematode, *Ascaris suum*. *Experimental Parasitology* 76, 424–437.

Komuniecki, R. (1996) In: Roche, T., Patel, M. and Harris, R.A. (eds) α-*Keto Acid Dehydrogenase Complexes*. Birkhauser Verlag AG, Basel, pp. 96–99.

Komuniecki, R. and Harris, B. (1995) Carbohydrate and energy metabolism in helminths. In: Marr, J. and Mueller, M. (eds) *Biochemistry and Molecular Biology of Parasites*. Academic Press, New York, pp. 49–66.

Komuniecki, R. and Komuniecki, P.R. (1995) Aerobic–anaerobic transitions in energy metabolism during the development of the parasitic nematode *Ascaris suum*. In: Boothroyd, J.C. and Komuniecki, R. (eds) *Molecular Approaches to Parasitology*. Wiley-Liss, New York, pp. 109–121.

Korotchkina, L.G. and Patel, M.S. (1995) Mutagenesis studies of the phosphorylation sites of recombinant human pyruvate dehydrogenase. Site-specific regulation. *Journal of Biological Chemistry* 270, 14297–14304.

Kuramochi, T., Hirawake, H., Kojima, S., Takamiya, S., Furushima, R., Aoki, T., Komuniecki, R.W. and Kita, K. (1994) Sequence comparison between the flavoprotein subunit of the fumarate reductase (complex II) of the anaerobic parasitic nematode, *Ascaris suum*, and the succinate dehydrogenase of the aerobic, free-living nematode, *Caenorhabditis elegans*. *Molecular and Biochemical Parasitology* 68, 177–187.

Larsen, P.L., Albert, P.S. and Riddle, D.L. (1995) Genes that regulate both development and longevity in *Caenorhabditis elegans*. *Genetics* 139, 1567–1583.

Ma, Y.C., Funk, M., Dunham, W.R. and Komuniecki, R. (1993) Purification and characterization of electron-transfer flavoprotein : rhodoquinone oxidoreductase from anaerobic mitochondria of anaerobic mitochondria of the adult parasitic nematode, *Ascaris suum*. *Journal of Biological Chemistry* 268, 20360–20365.

Mei, B., Komuniecki, R. and Komuniecki, P.R. (1997) Localization of cytochrome oxidase and the 2-methyl branched-chain enoyl CoA reductase in muscle and hypodermis of *Ascaris suum* larvae and adults. *Journal of Parasitology* 83, 760–763.

Murrell, K.D., Eriksen, L., Nansen, P., Slotved, H.C. and Rasmussen, T. (1997) *Ascaris suum*: a revision of its early migratory path and implications for human ascariasis. *Journal of Parasitology* 83, 255–260.

Oya, H., Costello, L.C. and Smith, W.N. (1963) The comparative biochemistry of developing *Ascaris* eggs. II. Changes in cytochrome *c* oxidase activity during embryonation. *Journal for Cellular and Comparative Physiology* 62, 287–294.

Patel, M.S. and Roche, T.E. (1990) Molecular biology and biochemistry of pyruvate dehydrogenase complexes. *FASEB Journal* 4, 3224–3233.

Rahmatullah, M. and Roche, T.E. (1985) Modification of bovine kidney pyruvate dehydrogenase kinase activity by CoA esters and their mechanism of action. *Journal of Biological Chemistry* 260, 10146–10152.

Rahmatullah, M. and Roche, T.E. (1987) The catalytic requirements for reduction and acetylation of protein X and the related regulation of various forms of resolved pyruvate dehydrogenase kinase. *Journal of Biological Chemistry* 262, 10265–10271.

Rajan, T.V. (1998) A hypothesis for the tissue specificity of nematode parasites. *Experimental Parasitology* 89, 140–142.

Randle, P.J. (1986) Fuel selection in animals. *Biochemical Society Transactions* 14, 799–806.

Ravindran, S., Radke, G.A., Guest, J.R. and Roche, T.E. (1996) Lipoyl domain-based mechanism for the integrated feedback control of the pyruvate dehydrogenase complex by enhancement of pyruvate dehydrogenase kinase activity. *Journal of Biological Chemistry* 271, 653–562.

Riddle, D.L. and Albert, P.S. (1997) Genetic and environmental regulation of dauer larva development. In: Riddle, D.L., Blumenthal, T., Meyer, B. and Priess, J. (eds) *C. Elegans II*. Cold Spring Harbor Laboratory Press, Cold Spring Harbor, New York, pp. 739–768.

Rioux, A. and Komuniecki, R. (1984) 2-Methylvalerate formation in mitochondria of *Ascaris suum* and its relationship to anaerobic energy generation. *Journal of Comparative Physiology* 154, 349–354.

Roche, T.E. and Cate, R.L. (1977) Purification of porcine liver pyruvate dehydrogenase complex and characterization of its catalytic and regulatory properties. *Archives of Biochemistry and Biophysics* 183, 664–677.

Roche, T.E. and Cox, D.J. (1996) In: Agius, L. and Sherratt, H.S.A. (eds) *Channeling in Intermediary Metabolism*. Portland Press, London, pp. 115–132.

Saruta, F., Kuramochi, T., Nakamura, K., Takamiya, S., Yu, Y., Aoki, T., Sekimizu, K., Kojima, S. and Kita, K. (1995) Stage-specific isoforms of complex II (succinate-ubiquinone oxidoreductase) in mitochondria from the parasitic nematode, *Ascaris suum*. *Journal of Biological Chemistry* 270, 928–932.

Saz, H.J. (1971) Anaerobic phosphorylation in *Ascaris* mitochondria and the effects of anthelmintics. *Comparative Biochemistry and Physiology* 39B, 627–637.

Saz, H.J. and Lescure, O.L. (1966) Interrelationships between the carbohydrate and lipid metabolism of *Ascaris lumbricoides* egg and adult stages. *Comparative Biochemistry and Physiology* 18, 845–857.

Saz, H.J. and Weil, A. (1960) The mechanism of formation of α-methylbutyrate from carbohydrate by *Ascaris lumbricoides* muscle. *Journal of Biological Chemistry* 235, 914–918.

Saz, H.J., Lescure, O.L. and Bueding, E. (1968) Biochemical observations of *Ascaris suum* lung-stage larvae. *Journal of Parasitology* 54, 457–461.

Sims, S.M., Magas, L.T., Barsuhn, C.L., Ho, N.F., Geary, T.G. and Thompson, D.P. (1992) Mechanisms of microenvironmental pH regulation in the cuticle of *Ascaris suum*. *Molecular and Biochemical Parasitology* 53, 135–148.

Snoep, J.L., de Graef, M.R., Westphal, A.H., de Kok, A., Teixeira, de Mattos, M.J. and Neijssel, O.M. (1993) Differences in sensitivity to NADH of purified pyruvate dehydrogenase complexes of *Enterococcus faecalis, Lactococcus lactis, Azotobacter vinelandii* and *Escherichia coli*: implications for their activity *in vivo. FEMS Microbiology Letters* 114, 279–283.

Song, H. and Komuniecki, R. (1994) Novel regulation of pyruvate dehydrogenase phosphatase purified from anaerobic muscle mitochondria of the adult parasitic nematode, *Ascaris suum. Journal of Biological Chemistry* 269, 31573–31578.

Sylk, S.R. (1969) Cytochrome *c* oxidase in migrating larvae of *Ascaris lumbricoides* var. *suum. Experimental Parasitology* 24, 32–36.

Takamiya, S., Kita, K., Wang, H., Weinstein, P.P., Hiraishi, A., Oya, H. and Aoki, T. (1993) Developmental changes in the respiratory chain of *Ascaris* mitochondria. *Biochimica et Biophysica Acta* 1141, 65–74.

Thissen, J. and Komuniecki, R. (1988) Phosphorylation and inactivation of the pyruvate dehydrogenase from the anaerobic parasitic nematode, *Ascaris suum.* Stoichiometry and amino acid sequence around the phosphorylation sites. *Journal of Biological Chemistry* 263, 19092–19097.

Thissen, J., Desai, S., McCartney, P. and Komuniecki, R. (1986) Improved purification of the pyruvate dehydrogenase complex from *Ascaris suum* body wall muscle and characterization of PDHa kinase activity. *Molecular and Biochemical Parasitology* 21, 129–138.

Tielens, A.G. and Roos, M.H. (1994) Differential expression of two succinate dehydrogenase subunit-β genes and a transition in energy metabolism during the development of the parasitic nematode *Haemonchus contortus. Molecular and Biochemical Parasitology* 66, 273–281.

Urban, J.F. Jr and Douvres, F.W. (1981) *In vitro* development of *Ascaris suum* from third- to fourth-stage larvae and detection of metabolic antigens in multi-well culture systems. *Journal of Parasitology* 67, 800–806.

Vanover, L. and Komuniecki, P.R. (1989) Effect of gas phase on carbohydrate metabolism of *Ascaris suum* larvae. *Molecular Biochemical Parasitology* 36, 29–40.

Wheelock, M.J., Komuniecki, R., Duran, E. and Johnson, K.R. (1991) Characterization of cDNA clones for the beta subunit of pyruvate dehydrogenase from *Ascaris suum. Molecular and Biochemical Parasitology* 45, 9–17.

Yang, D., Gong, X., Yakhnin, A. and Roche, T.E. (1998) Requirements for the adaptor protein role of dihydrolipoyl acetyltransferase in the up-regulated function of the pyruvate dehydrogenase kinase and pyruvate dehydrogenase phosphatase. *Journal of Biological Chemistry* 273, 14130–14137.

Yeaman, S.J., Hutcheson, E.T., Roche, T.E., Pettit, F.H., Brown, J.R., Reed, L.J., Watson, D.C. and Dixon, G.H. (1978) Sites of phosphorylation on pyruvate dehydrogenase from bovine kidney and heart. *Biochemistry* 17, 2364–2370.

Novel Carbohydrate Structures

Anne Dell, Stuart M. Haslam and Howard R. Morris

Department of Biochemistry, Imperial College of Science, Technology and Medicine, London SW7 2AY, UK

Introduction

The interface between nematodes and their environment, whether they be parasitic or free-living, is the cuticle surface, the outer layer of which in many species is covered by a carbohydrate-rich glycocalyx or surface coat (Blaxter *et al.*, 1992). In addition, many nematodes excrete or secrete antigenic glycoconjugates (ES antigens) which can either help to form the glycocalyx or dissipate more extensively into the nematode's environment. As well as carrying out or contributing to a purely physical role in the form of a lubricant or protective barrier, the glycocalyx and ES antigens represent the main immunogenic challenge to the host and could there-fore be crucial in determining whether successful parasitism is established (Maizels *et al.*, 1987a,b,c; Maizels and Selkirk, 1988). Considerable experi-mental evidence exists, via measurement of sensitivity to periodate cleavage and/or peptide-*N*-glycosidase F (PNGase F) digestion, to indicate that the carbohydrate component of nematode antigens is important for antibody recognition. In addition, numerous lectin-binding studies have demonstrated the presence of saccharide determinants on the surfaces of many different species of parasitic nematode (reviewed by Maizels and Selkirk, 1988).

Recent advances in glycobiology have significantly increased interest in nematode glycoconjugates (Dell *et al.*, 1999a,b). As detailed structural analysis deciphers the intricate labyrinth of their molecular architecture, functional issues are beginning to be rationally addressed. It is anticipated that a better understanding of the glycobiology of nematode parasitism could ultimately lead to rational drug and vaccine design. Furthermore, the antigenic and frequent stage- or species-specific nature of parasite glycoconjugates can, in principle, be exploited in the search for definitive serodiagnostics. Their often exotic and diverse structures, paradoxically embracing principles of both evolutionary divergence and conservation, represent interesting challenges as we seek to understand the genetic and developmental regulation of the biosynthetic enzymes involved in their production.

This chapter first discusses basic structural principles relating to the main classes of glycoconjugates in order to assist understanding of nematode structures. It then focuses on a selection of parasitic nematodes where detailed structural data on glycoconjugates have been obtained in recent years and where this structural information is starting to provide insights into possible molecular functions.

There are four main classes of glycoconjugates: glycoproteins, glycosphingolipids, glycosylphosphatidyl inositol (GPI)-anchored glycoconjugates (which have been extensively studied in the flagellated kinetoplastid parasites such as those from the genera *Trypanosoma* and *Leishmania*) and proteoglycans. Of these four classes, the first two have received the most attention in nematodes and it is upon these that the remainder of this chapter is focused.

Glycoproteins

There are two main types of protein glycosylation: *N*-glycosylation, in which the glycan is attached to an asparagine residue present in a tripeptide consensus sequon Asn-X-Ser/Thr (where X can be any amino acid except proline); and *O*-glycosylation, in which the glycan is attached to a serine or threonine residue. Glycoproteins can contain just *N*- or *O*-glycans or a combination of both. A tremendous heterogeneity of glycosylation can be produced, depending on whether an individual glycosylation site on the protein is occupied or not (macro-heterogeneity) and on the different types of glycan structures that are present at a particular glycosylation site (micro-heterogeneity). The combination of micro- and macro-heterogeneity can lead to a glycoprotein having numerous different glycoforms. This causes the broad smearing appearance of glycoproteins when they are analysed by SDS-PAGE.

Protein glycosylation occurs in the endoplasmic reticulum (ER) and Golgi compartments of the cell and involves a complex series of enzymatic reactions catalysed by membrane-bound glycosyltransferases and glycosidases. The two main types of protein glycosylation have very different biosynthetic pathways.

The biosynthesis of N-glycans

With the exception of a few truncated structures (see later) all asparagine-linked oligosaccharides share a common trimannosyl-chitobiose core (Fig. 15.1). This is derived from a biosynthetic precursor, composed of three glucose, nine mannose and two N-acetylglucosamine residues ($Glc_3Man_9GlcNAc_2$) (Fig. 15.1), which is added cotranslationally to polypeptides on the luminal side of the ER. Prior to addition to the polypeptide, the precursor glycan is anchored to the ER membrane through a phosphodiester linkage to the terpenoid lipid, dolichol. Soon after glycan attachment to Asn, the three glucosyl residues are removed by ER resident glucosidases and the resulting $Man_9GlcNAc_2$ glycan is then further processed in the ER and Golgi via pathways that, at least in part, are shared by all plant and animal cell types (reviewed by Kornfeld and Kornfeld, 1985; Schachter, 1995). Processing involves stepwise trimming by specific exoglycosidases and is commonly followed by stepwise addition of new sugar residues, the latter reactions being catalysed by specific glycosyl transferases.

Trimming by α-mannosidases, without any subsequent glycosyl addition to the periphery, results in glycans having the composition $Man_xGlcNAc_2$. In mammals, because of the substrate specificity of the mannosidases involved, x is not usually less than five. Glycans of composition $Man_{5-9}GlcNAc_2$ are designated 'high mannose' or 'oligomannose' (see Fig. 15.1 for examples). In invertebrates it is not uncommon for additional α-mannoses to be removed, giving glycans of composition $Man_{1-4}GlcNAc_2$. These small glycans are frequently referred to as 'truncated'.

Trimming by α-mannosidases plus glycosyl addition to the distal side of the core results in the formation of the most abundant class of mammalian glycans, the 'complex type' structures. These are characterized by the presence of variable numbers of antennae (most commonly two to four) whose biosynthesis is initiated in the medial Golgi by the addition of GlcNAc 'stubs' to the two α-mannoses of the core (Fig. 15.1). The β-mannose is also a possible site for GlcNAc attachment and a GlcNAc residue attached to the 4-position of this mannose is referred to as a 'bisecting' residue. In plants and some invertebrates (but not mammals) the β-mannose can be substituted at the 2-position with xylose. The chitobiose portion of the core is often modified by fucosylation on the proximal GlcNAc. In mammals this core fucose is linked to the 6-position of the GlcNAc whilst in plants it occurs at the 3-position. Invertebrates are capable of fucosylating the core at either position.

Processing in the trans-Golgi converts the small pool of 'core plus stubs' into an extensive array of mature oligosaccharides. In mammals the antennae stubs (excluding the bisecting GlcNAc) are usually elongated by the addition of β-Gal (galactose) to give $Gal\beta1$-4GlcNAc (lacNAc). Antennae can be lengthened by the sequential addition of GlcNAc and Gal

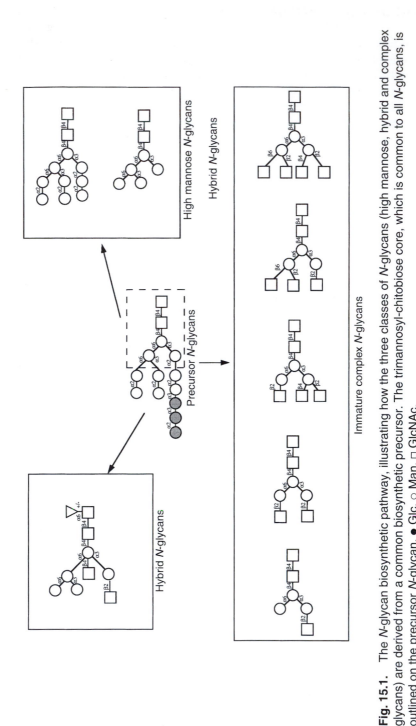

Fig. 15.1. The *N*-glycan biosynthetic pathway, illustrating how the three classes of *N*-glycans (high mannose, hybrid and complex glycans) are derived from a common biosynthetic precursor. The trimannosyl-chitobiose core, which is common to all *N*-glycans, is outlined on the precursor *N*-glycan. ● Glc, ○ Man, □ GlcNAc.

residues to the first lacNAc, resulting in tandem repeats of lacNAc, i.e. 'polylactosamine' structures. In a restricted number of mammalian glycoproteins, β-GalNAc (*N*-acetylgalactosamine) is added to the GlcNAc stubs in place of β-Gal and thus they have GalNAcβ1-4GlcNAc or 'lacdiNAc' antennae. Interestingly, both types of antennae are common in invertebrates.

Biosynthesis of complex-type structures is completed by a variety of 'capping' reactions, the most important in mammals being sialylation and fucosylation. Capping sugars are usually alpha-linked, unlike the backbone residues, which are normally beta-linked. Other common capping moieties include Gal, GalNAc and sulphate. Examples of capped antennae are given in Fig. 15.2. With the exception of sialic acid, all of the above capping moieties are also found in invertebrates. In addition, it is not uncommon for lower animals to modify their antennae in ways that have not yet been observed in mammalian glycoproteins and examples of unusual modifications in nematode glycoproteins are given in later sections.

A fourth family of *N*-glycans are referred to as 'hybrid' glycans. These glycans share structural features of the high mannose and complex-type families. They usually retain two mannoses on the 6-arm of the trimannosyl core whilst complex-type antennae are elaborated on the 3-arm. Hybrid structures are frequently bisected and may also be core fucosylated (Fig. 15.1). It should be noted that whilst core fucosylation is common in complex-type, hybrid and truncated glycans, it is rarely found in high mannose glycans.

The biosynthesis of O-glycans

In contrast to the highly conserved initial stages of *N*-glycan biosynthesis, *O*-glycosylation shows a great deal of species diversity in both core and peripheral structures. This section focuses on the established pathways of *O*-glycan biosynthesis in mammals because the little work that has been published on nematode *O*-glycosylation to date indicates that mammalian-like glycans are present in these animals. However, the reader should be aware that very different pathways for *O*-glycan biosynthesis occur in other systems (e.g. yeast, plants, molluscs, etc.) and it is highly probable that new structures that differ significantly from mammalian *O*-glycans will be found in nematodes.

Unlike *N*-glycosylation, *O*-linked glycans are not preassembled on a dolichol derivative but are added sequentially from sugar nucleotide donors. *O*-glycan biosynthesis is initiated in the cis-Golgi of mammals by the addition of a single GalNAc residue to a serine or threonine residue. A specific amino acid sequon is not required for glycosylation. Nevertheless, as demonstrated by computer analysis of glycoproteins whose sites of attachment are known, it is possible to make predictions as to whether a

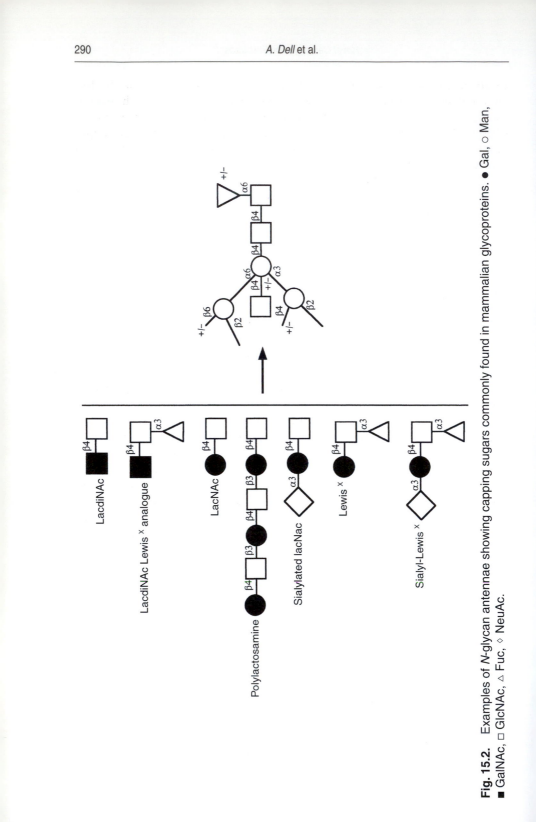

Fig. 15.2. Examples of *N*-glycan antennae showing capping sugars commonly found in mammalian glycoproteins. ● Gal, ○ Man, ■ GalNAc, □ GlcNAc, △ Fuc, ◇ NeuAc.

serine or threonine residue is likely to be glycosylated (http://www.cbs.dtu.
dk/databases/OGLYCBASE/). Usually domains rich in serine, threonine
and proline are heavily *O*-glycosylated. *O*-glycans can vary in size from a
single GalNAc residue (referred to as the Tn antigen) to much larger oligosaccharides. As with complex-type *N*-glycans, it is easiest to view *O*-glycans as
containing a core region which can be elongated with antennae. The
antennae are biosynthesized in a similar manner to *N*-glycans and consequently *N*- and *O*-glycans often carry the same terminal structures. The core
regions are, however, very different. At least seven *O*-glycan core structures
exist, with four (core types 1, 2, 3 and 4) being particularly widespread in
mammalian glycoproteins (Fig. 15.3) (reviewed in Schachter, 1995).

Glycolipids

Glycolipids are composed of one or more carbohydrate residues linked
to a hydrophobic ceramide (*N*-acylsphingosine) moiety. Like glycoproteins,
they are produced in the endomembrane system of the cell, where the
biosynthesis of their oligosaccharide component closely resembles that of
O-glycans, in that individual monosaccharides are added sequentially
from sugar nucleotide donors by the action of various glycosyltransferases.
Although glycolipids can show heterogeneity in both their oligosaccharide
and ceramide constituents, they are most commonly characterized on the
basis of their carbohydrate structure. Again like *O*-glycans, the oligosaccharide portions of glycolipids are composed of a series of cores, which can be
elongated and/or capped by additional residues. There are four basic
cores that form the lacto-, globo-, muco- and ganglio-series of glycolipids
(Fig. 15.4). Variations from these cores have been described in the human
parasitic trematode *Schistosoma mansoni*. Levery *et al.* (1992) and Makaaru

Core 1   Gal(β1-3)GalNAcα-Ser/Thr-R

Core 2   GlcNAc(β1-6)GalNAcα-Ser/Thr-R
                          /
                Gal(β1-3)

Core 3   GlcNAc(β1-3)GalNAcα-Ser/Thr-R

Core 4   GlcNAc(β1-6)GalNAcα-Ser/Thr-R
                          /
                GlcNAc(β1-3)

Fig. 15.3. Structures of the
four most commonly observed
O-glycan cores.

Lacto-series   GlcNAcβ1-3Galβ1-4Glcβ1–1Cer

Globo-series   Galα1-3/4Galβ1-4Glcβ1–1Cer

Muco-series   Galβ1-3/4Galβ1-4Glcβ1–1Cer

Fig. 15.4. Oligosaccharide structures of the core regions of the lacto-, globo-, muco- and ganglio-series of glycolipids.

Ganglio-series   GlcNAcβ1-4Galβ1-4Glcβ1–1Cer

et al. (1992) demonstrated that *S. mansoni* produced a novel 'schisto'-series of glycolipids, which contained the core structure GlcNAc1-3GalNAc1-3GalNAc1-4Glc1-1Cer. More recently Khoo *et al.* (1997a) described another novel core structure from an *S. mansoni* glycolipid comprising GlcNAc1-4GlcNAc1-3GalNAc1-4Glc1-1Cer.

Toxocara Methylated *O*-glycans

The first glycosylated nematode antigens of which the glycan moieties were studied and structurally characterized in any detail, and which still probably best illustrate how structural data can complement and help to explain data from more classical immunological techniques such as monoclonal antibody specificities, are those from the ascarid intestinal parasite of dogs, *Toxocara canis* (Maizels and Robertson, 1990; Khoo *et al.*, 1991, 1993). As well as being a significant veterinary problem, the infective L2 larvae emerging from the soil-transmitted eggs can invade and survive in a broad range of paratenic hosts, including humans. In human infections the larvae become developmentally arrested but remain metabolically active and migrate through many tissues, causing muscular weakness, eosinophilia, hepatosplenomegaly and bronchospasm, as well as optical and neuro-logical lesions (Gillespie, 1988; Taylor *et al.*, 1988).

Surface radioiodination, probing with monoclonal antibodies and lectin binding studies have revealed a set of well-defined, glycosylated, larval stage-specific, surface-exposed antigens (named TES-32, 55, 70, 120), all of which are also released as excretive/secretive (ES) antigens by the cultivated larval parasite (Maizels and Page, 1990). A large proportion of the humoral immune response is directed to carbohydrate epitopes. Cross-reactive carbohydrate determinants as defined by the binding of monoclonal antibodies (Tcn-monoclonals) were shown to be commonly present on most of the surface and ES antigens, as well as on those from the related ascarid species, *T. cati*. Only one monoclonal antibody, Tcn-2, was specific enough to recognize, exclusively, carbohydrate determinants from *T. canis* and not any of the *T. cati* ES products (Kennedy *et al.*, 1987; Maizels *et al.*, 1987a). Detailed mass spectrometric structural analysis of the major glycans present on the heavily *O*-glycosylated ES glycoproteins of *T. canis* and *T. cati* provided a potential explanation for the different monoclonal antibody specificities (Khoo *et al.*, 1991). The major *O*-glycans from *T. canis* are two, approximately equally abundant, core type 1 trisaccharides of sequence 2-*O*-Me-Fuc(α1-2)4-*O*-Me-Gal(β1-3)GalNAc, and 2-*O*-Me-Fuc (α1-2)Gal(β1-3)GalNAc; whereas those from *T. cati* are predominantly the former. Similar studies on the ES antigens of a related ascarid nematode, *Ascaris suum*, showed that their major *O*-glycans, which have compositions of deoxyHex$_1$Hex$_1$HexNAc$_1$ and deoxyHex$_1$Hex$_1$HexNAc$_2$, are not *O*-methylated (Khoo *et al.*, 1993).

The species-specific *O*-methylation found on an otherwise quite common oligosaccharide sequence most likely contributes to the epitope recognized by the Tcn monoclonals. Thus, despite substantial reactivity of anti-*T. canis* ES serum on *A. suum* L3/4 ES products, none of the anti-saccharide Tcn monoclonals recognized the *A. suum* ES antigens (Kennedy *et al.*, 1989), consistent with the absence of *O*-methylation in the latter. An interesting feature of the *Toxocara* trisaccharide is its similarity to the human blood group antigens, in particular the H-determinant of the ABO system (Fuc(α1-2)Gal(β1-R)). Indeed anti-ABO antibodies have been demonstrated to bind to the ES products and surface of *T. canis* (Smith *et al.*, 1983). It is interesting to hypothesize that the parasite could be using such carbohydrate epitopes in order to confer 'self' status and thus evade immune detection and damage. A cDNA from the gene encoding the apoprotein precursor of the most abundant surface coat constituent, TES-120, has been isolated and demonstrated to contain a typical mucin domain of 86 amino acids, 72% of which are serine or threonine residues which are arranged in heptameric repeats interspersed with proline residues, consistent with it being a heavily *O*-glycosylated mucin (Gems and Maizels, 1996). The structure and function of the *T. canis* antigens are discussed in much greater detail in Chapter 12.

Novel Fucosylated *N*-glycan Core Structures in *Haemonchus contortus*

H. contortus is an economically important gastrointestinal nematode that parasitizes domestic ruminants. It is one of the best-characterized parasitic nematodes in terms of its *N*-glycans. Large numbers of eggs are passed in the faeces of infected animals and hatch to produce the non-parasitic L1 and L2 forms. After a further moult the infective third-stage larva is ingested by grazing animals and passes to the abomasum, or true stomach, where it undergoes the third and fourth moults and reaches maturity. The life cycle is similar to the closely related human hookworms (*Ancylostoma duodenale* and *Necator americanus*), except that the infective third-stage larvae of these species can directly penetrate the skin to enter the blood stream, where they migrate to the intestine via the heart and lungs. *H. contortus* is a blood-feeding parasite, which therefore causes anaemia and a loss of animal condition; heavy infections can be fatal.

The first detailed structural studies concentrated on characterizing the *N*-glycans of the adult stage of *H. contortus* glycoproteins. Three families of *N*-glycans were revealed: high mannose structures, complex-type structures with short antennae comprised mostly of lacdiNAc (GalNAcβ1-4GlcNAc), and fucosylated lacdiNAc, and truncated structures with highly unusual core structures. The cores were substituted with up to three fucose residues, including a novel form of fucosylation on the distal *N*-acetylglucosamine of the chitobiose unit (Haslam *et al.*, 1996) (Fig. 15.5). This remarkable degree of core fucosylation has not previously been observed in any eukaryotic glycoprotein. The presence of *N*-glycans with di- and tri-fucosylated cores is immunologically very interesting. The Fucα(1-3) GlcNAc moiety has been shown to be a highly antigenic epitope in both insect (Prenner *et al.*, 1992) and plant glycoproteins (Ramirez-Soto and Poretz, 1991; Wilson *et al.*, 1998), and has also been shown to be the major allergenic determinant in honeybee phospholipase A_2 (Weber *et al.*, 1987; Kubelka *et al.*, 1993). Therefore it could be postulated that *H. contortus* is using these highly antigenic carbohydrate epitopes as components of ES products or dynamic surface coat glycoproteins to divert the host immune response away from the nematode. The novel core fucosylation was also observed on the vaccine glycoprotein H11, which is the major integral membrane protein in the microvillar plasma membrane of *H. contortus* (Munn *et al.*, 1993; Smith *et al.*, 1993; see also Chapter 13). It is a type II membrane protein with four consensus *N*-glycosylation sites on the extracellular region, which accounts for the bulk of its 972 amino acids (Smith *et al.*, 1997). About one-quarter of the antibodies produced in response to injection of H11 are anti-carbohydrate.

The characterization of *H. contortus N*-glycans was extended to the infective third-stage larvae of the parasite. The majority of glycan structures observed in the adult were also found in the L3, indicating that most of the glycosyl transferases expressed in the adult are also expressed at the L3 stage. The exception to this was the lack of the novel fucosylation of the distal *N*-acetylglucosamine residue of the chitobiose core (Haslam *et al.*, 1998), suggesting that the transferase responsible for this glycosylation is not expressed at the L3 stage. It remains to be defined if this stage specificity in core fucosylation is responsible for the larval specific surface glycoprotein antigens that have been immunologically detected (Ashman *et al.*, 1995; Raleigh and Meeusen, 1996; Raleigh *et al.*, 1996). The possibility that expression of stage-specific carbohydrate antigens plays an important

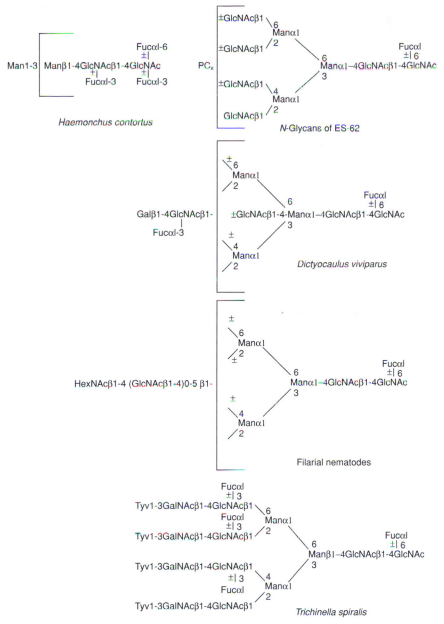

Fig. 15.5. Structures of *N*-linked glycans from several different species of parasitic nematodes, illustrating both similarities with mammalian glycans (compare with Figs 15.1 and 15.2) and features unique to nematodes (e.g. tyvelose and PC capping and novel core fucosylation). The filarial nematode glycans are believed to be substituted with charged residues, which are not yet characterized.

role in the parasite's interaction with the host immune system is an attractive one. Another possibility is that they are involved in the physiological transformation from a free-living mode of life to a parasitic one.

Prior to the above studies, knowledge of *H. contortus* protein glycosylation was limited to results from work employing immunological probes and lectin binding studies. Of particular interest was a 143 kDa immunogenic protein, found in adult-stage worms, which bound concanavalin A and *Helix pomatia* agglutinin but was resistant to *N*- and *O*-glycosidase digestion (Rhoads and Fetterer, 1990, 1994). Other workers have shown that a 70–90 kDa immunogenic protein found on the surface of *H. contortus* larvae binds wheatgerm agglutinin but is resistant to peptide *N*-glycosidase F (Ashman *et al.*, 1995). Jasmer *et al.* (1993) raised a series of monoclonal antibodies to gut surface antigens and two selected monoclonals were shown (by the sensitivity of the epitope to periodate oxidation) to react with carbohydrate antigens. Attempts to remove the carbohydrate epitope with peptide *N*-glycosidase F and *O*-glycanase failed. It is probable that many of the putative carbohydrate epitopes described in the above studies correspond to *N*-glycans with highly fucosylated cores. In particular, the frequently observed insensitivity to peptide *N*-glycosidase F is explained by fucosylation at the 3-position of the proximal GlcNAc of the core, which inhibits the enzyme's action (Tretter *et al.*, 1991).

As detailed structural information is obtained from a greater diversity of parasitic nematodes it is expected that the occurrence of stage-specific carbohydrate antigens, and therefore stage-specific glycosyl transferase activities, will be identified as a common characteristic.

Multi-antennary Lewisˣ *N*-glycans in *Dictyocaulus viviparus*

D. viviparus, commonly referred to as cattle lungworm, is a major pathogen of cattle, with heavy infections being fatal. Until recently, very little was known about glycosylation in *D. viviparus*. An early exploratory study using lectins, PNGase F digestion and metabolic labelling had suggested that few of the ES products have PNGase F-sensitive *N*-glycans (Britton *et al.*, 1993). In another study monoclonal antibodies have been used to probe immunodominant antigens on the surface of *D. viviparus* larvae (Gilleard *et al.*, 1995). Significantly, this work has revealed an immunodominant antigen that migrates as a diffuse band between 29 and 40 kDa on SDS-PAGE. The poor resolution on SDS-PAGE is indicative of glycosylated molecules.

Recent structural studies based on fast atom bombardment mass spectrometry have provided interesting insights into *N*-glycosylation in *D. viviparus* glycoproteins (Haslam *et al.*, 2000). This work has characterized a family of complex-type *N*-glycans in *D. viviparus* which have Lewisˣ (Galβ1-4(Fucα1-3)GlcNAc) antennae on bi-, tri- and tetra-antennary structures (Fig. 15.5). This is the first example of the Lewis[x] structure being a component of a nematode glycoconjugate.

The Lewis[x] epitope is a major component of glycoconjugates of the human parasitic trematodes *S. mansoni, S. japonicum* and *S. haematobium* (Cummings and Nyame, 1996; Khoo *et al.*, 1997b; Nyame *et al.*, 1998). It has been proposed that glycoconjugates containing the Lewis[x] epitope play important immunological roles in schistosomiasis. In particular, glycans containing the Lewis[x] epitope have been implicated in promoting a Th2 immune response (humoral immunity) over a Th1 immune response (cellular immunity), thus potentially limiting the host's cellular immune response to the parasite (Velupillai and Harn, 1994). Infection with schistosomes induces the production of antibodies to the Lewis[x] epitope and as the epitope is normally expressed in many tissues of both rodents and humans, including leucocytes, this antibody production causes the development of autoimmunity and complement-dependent cytolysis of leucocytes (Nyame *et al.*, 1996, 1997). The Lewis[x] epitope has been demonstrated on bovine glycoconjugates (Savage *et al.*, 1990; Siciliano *et al.*, 1993); thus it is possible that a similar phenomenon is occurring during *D. viviparus* infection of cattle.

Phosphorylcholine-substituted *N*-glycans of Filarial Nematodes

Of all the diseases caused by human parasitic nematodes, none are more physically striking than those caused by filarial nematodes. During the parasitism of their vertebrate hosts, adult filarial nematodes secrete glyco-proteins containing phosphorylcholine (PC) (Harnett and Parkhouse, 1995). A characteristic of filarial nematode antigens is their immuno cross-reactivity, which is due to the PC (Maizels *et al.*, 1987c). A number of studies have demonstrated that the PC moiety has an immunomodulatory function and this is dealt with in greater detail in Chapter 19. Attempts to characterize the molecular structure of filarial PC antigens is hampered by difficulties associated with obtaining enough experimental material from human parasites. Therefore a model system was developed using a 62 kDa ES glycoprotein (ES-62) from the rodent filarial nematode *Acanthocheilon-ema viteae*. Observations that peptide *N*-glycosidase F removes all radio-activity from [³H] choline-labelled ES-62 and also removes binding sites for a PC-specific monoclonal antibody, in conjunction with the fact that nematodes cultured with inhibitors of *N*-glycan processing secrete a protein that lacks PC, led to the conclusion that PC is attached to ES-62 via an *N*-linked glycan (Harnett *et al.*, 1993; Houston *et al.*, 1997). Mass spectro-metric structural analysis revealed that the PC-substituted *N*-glycans of ES-62 comprise trimannosyl-chitobiose cores, with and without fucosyla-tion, to which are added from one to four additional *N*-acetylglucosamine residues ($Fuc_{0-1}Man_3GlcNAc_{3-6}$) that are separately attached as antenna 'stubs'. It is probable that PC is attached to one or more of these *N*-acetylglucosamine residues (Haslam *et al.*, 1997) (Fig. 15.5).

Recently, the structural characterization of PC-substituted *N*-glycans has been extended to whole worm extracts of *A. viteae* and the human filarial nematode *Onchocerca volvulus* plus the ES material from the closely related bovine parasite *O. gibsoni*. The two *Onchocerca* species were found to contain PC-substituted *N*-glycans of the type previously characterized on ES-62, indicating that the PC biosynthetic pathway is conserved amongst the filarial nematodes. In addition to the PC-glycans described above, a second family of polar (possibly PC, though the nature of the polar group is not yet known) *N*-glycans was discovered whose members are remarkably rich in GlcNAc. The second group of *N*-glycans have highly unusual chito-oligomeric ($GlcNAc\beta1$-$4GlcNAc\beta1$-)$_X$ antennae containing up to five (possibly six) GlcNAc residues (Fig. 15.5). This is the first report of chito-oligomers in a eukaryotic glycoprotein (Haslam *et al.*, 1999).

The closest homology to these latter structures in nature are the Nod factors, which are signal molecules produced by *Azorhizobium*, *Bradyrhizobium* and *Rhizobium* species that trigger nodule formation in leguminous plants (Geremia *et al.*, 1994). Recently it has been shown that Nod-like molecules are likely to play important roles in vertebrate development. Thus the *Xenopus* protein DG42, which is expressed for a short time during embryo development, has been shown to synthesize chito-oligomers in *in vitro* experiments (Semino and Robbins, 1995). Evidence for a developmental role of short chitin oligosaccharides in vertebrate development has recently been obtained by Bakkers *et al.* (1997), who showed that fertilized zebrafish eggs injected with antiserum against the DG42 protein demonstrated embryos with severe defects in trunk and tail development. It is interesting to speculate that the filarial nematode chitin oligomers might play a role in worm development or parasite–host interactions.

Another interesting observation is that several species of filarial nematodes have been shown to express chitinase (Fuhrman, 1995). Indeed the chitinase of *A. viteae* infective stage larvae (L3) is the main target of the protective humoral immune response when jirds are vaccinated with irradiated attenuated L3s (Adam *et al.*, 1996; see also Chapter 10). It remains to be established whether there is an interaction between the parasite's oligo-chitin *N*-glycans and chitinase and whether such an interaction has a role to play in parasite–host interaction.

PC has also been characterized as a substituent on carbohydrates of glycosphingolipids isolated from the porcine gastrointestinal nematode *A. suum* (Lochnit *et al.*, 1998a). Therefore it can be hypothesized that the substitution of PC on nematode glycoconjugates could be a shared immunosuppressive strategy. Additional interesting characteristics of *A. suum* glycolipids include substitution with phosphoethanolamine and the fact that the oligosaccharide core belongs to the 'arthro' series with a β-linked mannose (Galα1-3GalNAcβ1-4GlcNAcβ1-3Manβ1-4Glcβ1-1Ceramide). The same core sequence was demonstrated on neutral glycosphingolipids of *A. suum* (Lochnit *et al.*, 1997) and on the neutral glycosphingolipids of the free-living nematode *Caenorhabditis elegans* (Gerdt *et al.*, 1997). Acidic *A. suum* glycosphingolipids contained unusual phosphoinositol and 3-sulphogalactosylcerebroside structures (Lochnit *et al.*, 1998b).

The Tyvelose Immunodominant Epitope of *Trichinella spiralis*

T. spiralis has one of the widest ranges of potential hosts of any known parasite and is capable of infecting almost any mammal. Its life cycle is completed within a single host species, without the requirement for an intermediate vector or a free-living larval stage. The commonest mode of transmission to humans is via ingestion of animal muscle tissue (usually pork) containing viable mature L1 larvae. Once inside the human host, the larvae invade the mucosal epithelium of the small intestine, moult to adulthood, mate and reproduce. Despite their relatively large size (males are about 1.5 mm in length and females about 3 mm) the worms inhabit an intracellular niche. The diameter of the adult worms is comparable to that of the cells that they invade, but being many times longer they occupy a number of cells simultaneously, migrating through the epithelial layer in a sinusoidal manner and leaving a trail of dead cells in their wake. Females can produce up to 2000 L1 larvae during their life span. The newborn larvae enter the general circulation, establishing transient infections in various organs before arriving at their final destination in striated skeletal muscle. Once the larva has invaded a muscle cell, de-differentiation begins and conversion to a specialized cell referred to as a nurse cell occurs over a 20 day period post-infection. The epithelial invasion and muscle cell de-differentiation processes are covered in greater detail in Chapters 6 and 7.

The exact mechanisms by which infective larvae of *T. spiralis* recognize, invade and migrate within the intestinal epithelium are unknown but it is likely that carbohydrate structures on the surface or in ES products are important in these events. Evidence for this comes from experiments that have shown that monoclonal antibodies developed against ES antigens effect rapid expulsion of infective larvae in passively immunized neonatal rats (Otubu *et al.*, 1993). The monoclonal antibodies recognize highly unusual carbohydrate epitopes that are present in the ES glycoproteins and also on the cuticular surface of the larvae. The immunodominant glycans of the L1 stage are tri- and tetra-antennary *N*-linked structures composed of GalNAcβ1-4GlcNAc (lacdiNAc) antennae, which are capped with D-tyvelose (3,6-dideoxy-D-*arabino*hexose). The majority of antennae are also fucosylated (Fig. 15.5). Tyvelose is a sugar more typically associated with the cell wall lipopolysaccharides of some pathogenic bacteria, where it is a dominant antigenic determinant. Its presence in a eukaryotic glycoprotein is highly unusual. The structures of the novel ES glycans were established from complementary data acquired from three groups of researchers. Firstly, the dideoxy sugar was isolated and its structure, including absolute stereochemistry, was established by Wisnewski *et al.* (1993), largely using GC-MS technology. Secondly, the complete structures of the tri- and tetra-antennary glycans, with the exception of the tyvelose anomeric stereochemistry, were defined by Reason *et al.* (1994), using a fast atom bombardment–mass spectrometry (FAB-MS) strategy. Finally, synthesis of a panel of oligosaccharides capped with α- and β-linked tyvelose and analysis of their ability to be recognized by the anti-tyvelose monoclonal antibodies revealed, surprisingly, that the tyvelose has β stereochemistry in the ES antigens (Ellis *et al.*, 1997).

Recent studies have shown that anti-tyvelose monoclonal antibodies are able to prevent larval invasion of MDCK tissue cultured cells, thus establishing that tyvelose-specific antibodies are protective *in vitro* and providing further support for the hypothesis that tyvelose plays a role in *in vivo* tissue invasion (McVay *et al.*, 1998). Muscle-stage larvae (L1) are also rich in tyvelose but it is not known whether glycans play a role in de-differentiation. Nevertheless it is interesting that a tyvelosylated 43 kDa glycoprotein secreted by the muscle-stage larva locates exclusively to the nurse cell cytoplasm from day 12 to day 15 of nurse cell development (Vassilatis *et al.*, 1996). In addition, it has been demonstrated that muscle-stage larvae contain PC-substituted *N*-glycans by the loss of binding of an anti-PC monoclonal antibody (MoAb2) to a family of glycoprotein antigens (TSL-4 antigens) after *N*-glycanase treatment (Ortega-Pierres *et al.*, 1996).

Concluding Remarks

As the amount of structural data from different species of nematodes increases, a number of principles are emerging. Firstly, oligomannose structures are abundant in many different nematodes, as are truncated glycans with as few as one or two mannoses attached to the chitobiose core. With respect to the production of truncated structures, nematode *N*-glycans closely resemble insect structures (Altmann *et al.*, 1995; März *et al.*, 1995). Complex and hybrid structures are also major constituents in many nematodes. These may have antennae truncated to a single GlcNAc, a situation that is relatively rare in mammals. Non-truncated antennae commonly have backbones that are composed of lacdiNAc (GalNAcβ1-4GlcNAc), as observed in the *N*-glycans of *D. immitis* (Kang *et al.*, 1993) as well as *Trichinella* and *Haemonchus* (see above). LacdiNAc is a relatively rare building block in higher animals, which more commonly use lacNAc (Galβ1-4GlcNAc). This variation could be due to the lack of a functional β-galactosyl transferase in some nematodes, though the suggestion that this enzyme is not functional in all nematodes (Nyame *et al.*, 1998) has recently been contradicted in *Dictyocaulus viviparus* glycoproteins (Haslam *et al.*, 2000). Fucose is a dominant modifying sugar and, as exemplified by *Haemonchus*, it can occur in unusual locations. The contribution of methyl groups to immunodominancy is a common theme – for example, as integral components of fucose and tyvelose and as substituents on the *Toxocara* *O*-glycans. Indeed, the presence of hydrophobic non-reducing structures is in contrast to those normally observed on mammalian glycans, which commonly contain hydrophilic sialic acid residues. No convincing evidence for the presence of sialic acid on any nematode glycoconjugate has been presented. Finally, phosphorylcholine is an important hapten in many glycan epitopes.

Knowing the structures of glycans implicated in nematode infections is an important first step in elucidating their roles. Much progress has already been made in this area and we anticipate a rapid growth in the number of species examined and new structures reported. Synthetic work will clearly be an important component of parasite glycobiology, because the tiny amounts of material that can normally be isolated as pure compounds are rarely sufficient for functional studies. In this regard we expect the *C. elegans* genome project (Bürglin *et al.*, 1998), which has already facilitated the identification of an α1,3 fucosyltransferase (DeBose-Boyd *et al.*, 1998), *N*-acetylglucosaminetransferase I (Chen *et al.*, 1999) and a family of polypeptide *N*-acetylgalactosaminyl transferases (Hagen and Nehrke, 1998), to play an increasingly important role in the identification of parasitic nematode glycosyl transferases of potential utility for glycan synthesis.

Acknowledgements

We are grateful to the Wellcome Trust and the Biotechnology and Biological Sciences Research Council for financial support for our research on nematode glycobiology.

References

Adam, R., Kaltmann, B., Rudin, W., Friedrich, T., Marti, T. and Lucius, R. (1996) Identification of chitinase as the immunodominant filarial antigen recognized by the sera of vaccinated rodents. *Journal of Biological Chemistry* 271, 1441–1447.

Altmann, F., Schwihla, H., Staudacher, E., Glössl, J. and März, L. (1995) Insect cells contain an unusual, membrane-bound β-*N*-acetylglucosaminidase probably involved in the processing of protein *N*-glycans. *Journal of Biological Chemistry* 270, 17344–17349.

Ashman, K., Mather, J., Wiltshire, C., Jacobs, H.J. and Meeusen, E. (1995) Isolation of a larval surface glycoprotein from *Haemonchus contortus* and its possible role in evading host immunity. *Molecular and Biochemical Parasitology* 70, 175–179.

Bakkers, J., Semino, C.E., Stroband, H., Kijne, J.W., Robbins, P.W. and Spaink, H.P. (1997) An important developmental role for oligosaccharides during early embryogenesis of cyprinid fish. *Proceedings of the National Academy of Sciences USA* 94, 7982–7986.

Blaxter, M.L., Page, A.P., Rudin, W. and Maizels, R.M. (1992) Nematode surface coats: actively evading immunity. *Parasitology Today* 8, 243–247.

Britton, C., Canto, G.J., Urquhart, G.M. and Kennedy, M. (1993) Characterization of excretory-secretory products of adult *Dictyocaulus viviparus* and the antibody-response to them in infection and vaccination. *Parasite Immunology* 15, 163–174.

Bürglin, T.R., Lobos, E. and Blaxter, M.L. (1998) *Caenorhabditis elegans* as a model for parasitic nematodes. *International Journal for Parasitology* 28, 395–411.

Chen, S.H., Zhou, S.H., Sarkar, M., Spence, A.M. and Schachter, H. (1999) Expression of three *Caenorhabditis elegans* N-acetylglucosaminyltransferase I genes during development. *Journal of Biological Chemistry* 274, 288–297.

Cummings, R.D. and Nyame, A.K. (1996) Glycobiology of schistosomiasis. *FASEB Journal* 10, 838–848.

DeBose-Boyd, R.A., Nyame, A.K. and Cummings, R.D. (1998) Molecular cloning and characterization of an α1,3 fucosyltransferase, CEFT-1 from *Caenorhabditis elegans*. *Glycobiology* 8, 905–917.

Dell, A., Haslam, S.M., Morris, H.R. and Khoo, K.-H. (1999a) Immunogenic glyco-conjugates implicated in parasitic nematode diseases. *Biochimica et Biophysica Acta* 1455, 353–362.

Dell, A., Morris, H.R., Easton, R., Haslam, S.M., Panico, M., Sutton-Smith, M., Reason, A.J. and Khoo, K.-H. (1999b) Structural analysis of oligosaccharides: FAB-MS, ES-MS and MALDI-MS. In: Ernst, B., Sinaÿ, P. and Hart, G. (eds) *Oligosaccharides in Chemistry and Biology – a Comprehensive Handbook.* Wiley-VCH, Weinheim (in press).

Ellis, L.A., McVay, C.S., Probert, M.A., Zhang, J., Bundle, D.R. and Appleton, J.A. (1997) Terminal β-linked tyvelose creates unique epitopes in *Trichinella spiralis* glycan antigens. *Glycobiology* 7, 383–390.

Fuhrman, J.A. (1995) Filarial chitinases. *Parasitology Today* 11, 259–261.

Gems, D. and Maizels, R.M. (1996) An abundantly expressed mucin-like protein from *Toxocara canis* infective larvae: the precursor of the larval surface coat glycoproteins. *Proceedings of the National Academy of Sciences USA* 93, 1665–1670.

Gerdt, S., Lochnit, G., Dennis, R.D. and Geyer, R. (1997) Isolation and structural analysis of three neutral glycosphingolipids from a mixed population of *Caenorhabditis elegans* (Nematoda: Rhabditdida). *Glycobiology* 7, 265–275.

Geremia, R.A., Mergaert, P., Geelen, D., Van Montague, M. and Holsters, M. (1994) The NODC protein of *Azorhizobium-caulinodans* is an N-acetylglucosaminyl-transferase. *Proceedings of the National Academy of Sciences USA* 91, 2669–2673.

Gilleard, J.S., Duncan, J.L. and Tait, A. (1995) An immunodominant antigen on the *Dictyocaulus viviparus* L3 sheath surface-coat and a related molecule in other strongylid nematodes. *Parasitology* 11 193–200.

Gillespie, S.H. (1988) The epidemiology and pathogenesis of zoonotic toxocariasis. *Parasitology Today* 4, 180–182.

Hagen, F.K. and Nehrke, K. (1998) cDNA cloning and expression of a family of UDP-*N*-acetyl-D galactosamine: polypeptide *N*-acetylgalactosaminyltransferase. Sequence homologs from *Caenorhabditis elegans*. *Journal of Biological Chemistry* 273, 8268–8277.

Harnett, W. and Parkhouse, R.M.E. (1995) Nature and function of parasite nematode surface and excretory–secretory antigens. In: Sood, M.L. and Kapur, J. (eds) *Perspectives in Nematode Physiology and Biochemistry.* Narendra Publishing House, Delhi, pp. 207–242.

Harnett, W., Houston, K.M., Amess, R. and Worms, M.J. (1993) *Acanthocheilonema viteae*: phosphorylcholine is attached to the major excretory-secretory product via a N-linked glycan. *Experimental Parasitology* 77, 498–502.

Haslam, S.M., Coles, G.C., Munn, E.A., Smith, T.S., Smith, H.F., Morris, H.R. and Dell, A. (1996) *Haemonchus contortus* glycoproteins contain *N*-linked oligo-saccharides with novel highly fucosylated core structures. *Journal of Biological Chemistry* 271, 30561–30570.

Haslam, S.M., Khoo, K.-H., Houston, K.M., Harnett, W., Morris, H.R. and Dell, A. (1997) Characterisation of the phosphorylcholine-containing *N*-linked oligo-saccharides in the excretory-secretory 62 kDa glycoprotein of *Acanthocheilonema viteae*. *Molecular and Biochemical Parasitology* 85, 53–66.

Haslam, S.M., Coles, G.C., Reason, A.J., Morris, H.R. and Dell, A. (1998) The novel core fucosylation of *Haemonchus contortus* *N*-glycans is stage specific. *Molecular and Biochemical Parasitology* 93, 143–147.

Haslam, S.M., Houston, K.M., Harnett, W., Reason, A.J., Morris, H.R. and Dell, A. (1999) Structural studies of phosphorylcholine-substituted *N*-glycans of filarial parasites: conservation amongst species and discovery of novel chito-oligomers. *Journal of Biological Chemistry* 274, 20953–20960.

Haslam, S.M., Coles, G.C., Morris, H.R. and Dell, A. (2000) Structural characterisa-tion of the *N*-glycans of *Dictyocaulus viviparus*: discovery of the Lewis[x] structure in a nematode. *Glycobiology* 10, 223–229.

Houston, K.M., Cushley, W.C. and Harnett, W. (1997) Studies on the site and mech-anism of attachment of phosphorylcholine to a filarial excretory-secretory product. *Journal of Biological Chemistry* 272, 1527–1533.

Jasmer, D.P., Perryman, L.E., Conder, G.A., Crow, S. and McGuire, T. (1993) Pro-tective immunity to *Haemonchus contortus* induced by immunoaffinity isolated antigens that share a phylogenetically conserved carbohydrate gut surface epitope. *Journal of Immunology* 151, 5450–5460.

Kang, S., Cummings, R.D. and McCall, J.W. (1993) Characterization of the *N*-linked oligosaccharides in glycoproteins synthesised by microfilariae of *Dirofilaria immitis*. *Journal of Parasitology* 79, 815–828.

Kennedy, M.W., Maizels, R.M., Meghji, M., Young, L., Qureshi, F. and Smith, H.V. (1987) Species-specific and common epitopes on the secreted and surface antigens of *Toxocara cati* and *Toxocara canis* infective larvae. *Parasite Immunology* 9, 407–420.

Kennedy, M.W., Qureshi, F., Fraser, E.M., Haswell-Elkins, M.R., Elkins, D.B. and Smith, H.V. (1989) Antigenic relationships between the surface-exposed, secreted and somatic materials of the nematode parasites *Ascaris lumbricoides, Ascaris suum* and *Toxocara canis. Clinical and Experimental Immunology* 75, 493–500.

Khoo, K.-H., Maizels, R.M., Page, A.P., Taylor, G.W., Rendell, N.B. and Dell, A. (1991) Characterization of nematode glycoproteins: the major O-glycans of *Toxocara* excretory-secretory antigens are O-methylated trisaccharides. *Glycobiology* 1, 163–171.

Khoo, K.-H., Morris, H.R. and Dell, A. (1993) Structural characterisation of the major glycans of *Toxocara canis* ES antigens. In: Lewis, J.W. and Maizels, R.M. (eds) *Toxocara and Toxocariasis: Clinical, Epidemiological and Molecular Perspectives*. Institute of Biology, London, pp. 133–140.

Khoo, K.-H., Chatterjee, D., Caulfield, J.P., Morris, H.R. and Dell, A. (1997a) Structural characterization of the glycosphingolipids from the eggs of *Schistosoma mansoni* and *Schistosoma japonicum*. *Glycobiology* 7, 653–661.

Khoo, K.-H., Chatterjee, D., Caulfield, J.P., Morris, H.R. and Dell, A. (1997b) Structural mapping of the glycans from the egg glycoproteins of *Schistosoma mansoni* and *Schistosoma japonicum*. Identification of novel core structures and terminal sequences. *Glycobiology* 7, 663–677.

Kornfeld, R. and Kornfeld, S. (1985) Assembly of asparagine linked oligosaccharides. *Annual Review of Biochemistry* 54, 631–664.

Kubelka, V., Altmann, F., Staudacher, E., Tretter, V., März, L., Hård, K., Kamerling, J.P. and Vliegenthart, F.G. (1993) Primary structure of the N-linked carbohydrate chains from honeybee venom phospholipase A_2. *European Journal of Biochemistry* 213, 1193–1204.

Levery, S.B., Weiss, J.B., Salyan, M.E., Roberts, C.E., Hakomori, S.-I., Magnani, J.L. and Strand, M. (1992) Characterisation of a series of novel fucose-containing glycosphingolipid immunogens from eggs of *Schistosoma mansoni*. *Journal of Biological Chemistry* 267, 5542–5551.

Lochnit, G., Dennis, R.D., Zähringer, U. and Geyer, R. (1997) Structural analysis of neutral glycosphingolipids from *Ascaris suum* adults (Nematoda: Ascaridida). *Glycoconjugate Journal* 14, 389–399.

Lochnit, G., Dennis, R.D., Ulmer, A.J. and Geyer, R. (1998a) Structural elucidation and monkine-inducing activity of two biologically active zwitterionic glycosphingolipids derived from the porcine parasitic nematode *Ascaris suum*. *Journal of Biological Chemistry* 273, 466–474.

Lochnit, G., Nispel, S., Dennis, R.D. and Geyer, R. (1998b) Structural analysis and immunohistochemical location of two glycosphingolipids from the porcine parasitic nematode *Ascaris suum*. *Glycobiology* 8, 891–899.

Maizels, R.M. and Page, A.P. (1990) Surface associated glycoproteins from *Toxocara canis* larval parasites. *Acta Tropica* 47, 355–364.

Maizels, R.M. and Robertson, B.D. (1990) *Toxocara canis*: secreted glycoconjugate antigens in immunobiology and immunodiagnosis. In: Kennedy, M.W. (ed.) *Parasitic Nematodes: Antigens, Membranes and Genes*. Taylor and Francis, London, pp. 95–115.

Maizels, R.M. and Selkirk, M.E. (1988) Immunobiology of nematode antigens. In: Englund, P.T. and Sher, F.A. (eds) *The Biology of Parasitism*. Alan R. Liss, New York, pp. 285–308.

Maizels, R.M., Kennedy, M.W., Meghji, M., Robertson, B.D. and Smith, H.V. (1987a) Shared carbohydrate epitopes on distinct surface and secreted epitopes of the parasitic nematode *Toxocara canis*. *Journal of Immunology* 139, 207–214.

Maizels, R.M., Bianco, A.E., Flint, J.E., Gregory, W.F., Kennedy, M.W., Lim, G.E., Robertson, B.D. and Selkirk, M.E. (1987b) Glycoconjugate antigens from parasitic nematodes. In: MacInnis, A.J. (ed.) *Molecular Paradigms for Eradicating Helminthic Parasites.* Alan R. Liss, New York, pp. 267–279.

Maizels, R.M., Burke, L. and Denham, D.A. (1987c) Phosphorylcholine-bearing antigens in filarial nematode parasites: analysis of somatic extracts, *in-vitro* secretions and infection sera from *Brugia malayi* and *B. pahangi. Parasite Immunology* 9, 49–66.

Makaaru, C., Damian, R.T., Smith, D.F. and Cummings, R.D. (1992) The human blood fluke *Schistosoma mansoni* synthesizes a novel type of glycosphingolipid. *Journal of Biological Chemistry* 267, 2251–2257.

März, L., Altmann, F., Staudacher, E. and Kubelka, V. (1995) Protein glycosylation in insects. In: Montreuil, J., Vliegenthart, J.F.G. and Schachter, H. (eds) *Glycoproteins,* Vol. 29a in Neuberger, A. and Van Deenen, L.L.M. (eds) *New Comprehensive Biochemistry.* Elsevier, Amsterdam, pp. 543–563.

McVay, C.S., Tsung, A. and Appleton, J.A. (1998) Participation of parasite surface glycoproteins in antibody-mediated protection of epithelial cells against *Trichinella spiralis. Infection and Immunity* 66, 1941–1945.

Munn, E.A., Smith, T.S., Graham, M., Tavernor, A.S. and Greenwood, C.A. (1993) The potential value of integral membrane-proteins in the vaccination of lambs against *Haemonchus contortus. International Journal for Parasitology* 23, 261–269.

Nyame, A.K., Pilcher, J.B., Tsang, V.C.W. and Cummings, R.D. (1996) *Schistosoma mansoni* infection in humans and primates induces cytolytic antibodies to surface Le(x) determinants on myeloid cells. *Experimental Parasitology* 82, 191–200.

Nyame, A.K., Pilcher, J.B., Tsang, V.C.W. and Cummings, R.D. (1997) Rodents infected with *Schistosoma mansoni* produce cytolytic IgG and IgM antibodies to the Lewis x antigen. *Glycobiology* 7, 207–215.

Nyame, A.K., Debose-Boyd, R., Long, T.D., Tsang, V.C.W. and Cummings, R.D. (1998) Expression of Le^x antigen in *Schistosoma japonicum* and *S. haematobium* and immune responses to Le^x in infected animals: lack of Le^x expression in other trematodes and nematodes. *Glycobiology* 8, 615–624.

Ortega-Pierres, M.G., Yepez-Mulia, L., Homan, W., Gamble, H.R., Lim, P.L., Takahashi, Y., Wassom, D.L. and Appleton, J.A. (1996) Workshop on the detailed characterisation of *Trichinella spiralis* antigens: a platform for the future studies on antigens and antibodies to this parasite. *Parasite Immunology* 18, 273–284.

Otubu, O.E., Carlisle-Nowak, M.S., McGregor, D.D., Jacobson, R.H. and Appleton, J.A. (1993) *Trichinella spiralis:* the effect of specific antibody on muscle larvae in the small intestines of weaned rats. *Experimental Parasitology* 76, 394–400.

Prenner, C., Mach, L., Glössl, J. and März, L. (1992) The antigenicity of the carbohydrate moiety of an insect glycoprotein, honey-bee (*Apis mellifera*) venom phospholipase A2. The role of α1,3-fucosylation of the asparagine-bound *N*-acetylglucosamine. *Biochemical Journal* 284, 377–380.

Raleigh, J.M. and Meeusen, E.N.T. (1996) Developmentally regulated expression of a *Haemonchus contortus* surface antigen. *International Journal for Parasitology* 26, 673–675.

Raleigh, J.M., Brandon, M.R. and Meeusen, E. (1996) Stage-specific expression of surface molecules by the larval stages of *Haemonchus contortus. Parasite Immunology* 18, 125–132.

Ramirez-Soto, D. and Poretz, R.D. (1991) The (1–3)-linked α-L-fucosyl group of the *N*-glycans of the *Wistaria floribunda* lectins is recognised by a rabbit antiserum. *Carbohydate Research* 213, 27–36.

Reason, A.J., Ellis, L.A., Appleton, J.A., Wisnewski, N., Grieve, R.B., McNeil, M., Wassom, D.L., Morris, H.R. and Dell, A. (1994) Novel tyvelose-containing tri- and tetra-antennary *N*-glycans in the immunodominant antigens of the intracellular parasite *Trichinella spiralis*. *Glycobiology* 4, 593–603.

Rhoads, M.L. and Fetterer, R.H. (1990) Biochemical and immunological character-ization of I-125 labeled cuticle components of *Haemonchus contortus*. *Molecular and Biochemical Parasitology* 42, 155–164.

Rhoads, M.L. and Fetterer, R.H. (1994) Purification and characterization of surface-associated proteins from adult *Haemonchus contortus*. *Journal of Parasitology* 80, 756–763.

Savage, A.V., Donohue, J.J., Koeleman, C.A.M. and van den Eijnden, D.H. (1990) Structural characterization of sialylated tetrasaccharides and pentasaccharides with blood-group H and Le^x activity isolated from bovine submaxillary mucin. *European Journal of Biochemistry* 193, 837–843.

Schachter, H. (1995) Biosynthesis. In: Montreuil, J., Vliegenthart, J.F.G. and Schachter, H. (eds) *Glycoproteins* Vol. 29a in Neuberger, A. and Van Deenen, L.L.M. (eds) *New Comprehensive Biochemistry*. Elsevier, Amsterdam, pp. 123–454.

Semino, C.E. and Robbins, P.W. (1995) Synthesis of NOD-like chitin oligosaccha-rides by the *Xenopus* developmental protein DG42. *Proceedings of the National Academy of Sciences USA* 92, 3498–3501.

Siciliano, R.A., Morris, H.R., McDowell, R.A., Azadi, P., Rogers, M.E., Bennett, H.P.J. and Dell, A. (1993) The Lewis^x epitope is a major non-reducing structure in the sulphated *N*-glycans attached to Asn-65 of bovine pro-opiomelanocortin. *Glycobiology* 3, 225–239.

Smith, H.V., Kusel, J.R. and Grindwood, R.W.A. (1983) The production of human A and B blood group-like substances by *in vitro* maintained second stage *Toxocara canis* larvae: their presence on the outer larval surface and in the excretions/secretions. *Clinical and Experimental Immunology* 54, 625–633.

Smith, T.S., Munn, E.A., Graham, M., Tavernor, A.S. and Greenwood, C.A. (1993) Purification and evaluation of the integral membrane protein H11 as a protec-tive antigen against *Haemonchus contortus*. *International Journal for Parasitology* 23, 271–280.

Smith, T.S., Graham, M., Munn, E.A., Newton, S.E., Knox, D.P., Coadwell, W.J., McMichael-Phillips, D., Smith, H., Smith, W.D. and Oliver, J.J. (1997) Cloning and characterization of a microsomal aminopeptidase from the intestine of the nematode *Haemonchus contortus*. *Biochimica et Biophysica Acta* 1338, 295–306.

Taylor, M.R.H., Keane, C.T., O'Connor, P., Mulvihill, E. and Holland, C. (1988) The expanded spectrum of toxocaral disease. *Lancet* 1, 692–695.

Tretter, V., Altmann, F. and März, L. (1991) Peptide-*N*4-(*N*-acetyl-β-glucosaminyl) asparagine amidase F cannot release glycans with fucose attached α1–3 to the asparagine-linked *N*-acetylglucosamine residue. *European Journal of Biochemistry* 199, 647–652.

Vassilatis, D.K., Polvere, R.I., Despommier, D.D., Gold, A.M. and Van der Ploeg, L.H.T. (1996) Developmental expression of a 43-kDa secreted glycoprotein from *Trichinella spiralis*. *Molecular and Biochemical Parasitology* 78, 13–23.

Velupillai, P. and Harn, D.A. (1994) Oligosaccharide-specific induction of interleukin-10 production by B220+ cells from schistosome-infected mice – a mechanism for regulation of CD4[+] T-cell subsets. *Proceedings of the National Academy of Sciences USA* 91, 18–22.

Weber, A., Schröder, H., Thalberg, K. and März, L. (1987) Specific interactions of IgE antibodies with a carbohydrate epitope of honey-bee venom phospholipase-A2. *Allergy* 42, 464–470.

Wilson, I.B.H., Harthill, J.E., Mullin, N.P., Ashford, D.A. and Altmann, F. (1998) Core α1,3-fucose is a key part of the epitope recognized by antibodies reacting against plant *N*-linked oligosaccharides and is present in a wide variety of plant extracts. *Glycobiology* 8, 651–661.

Wisnewski, N., McNeil, M., Grieve, R.B. and Wassom, D.L. (1993) Characterisation of novel fucosyl- and tyvelosyl-containing glycoconjugates from *Trichinella spiralis* muscle stage larvae. *Molecular and Biochemical Parasitology* 61, 25–36.

Structurally Novel Lipid-binding Proteins

Malcolm W. Kennedy

Division of Infection and Immunity, Institute of Biomedical and Life Sciences, Joseph Black Building, University of Glasgow, Glasgow G12 8QQ, UK

Introduction

One of the lessons gained from the genomics of *Caenorhabditis elegans* is the number of protein types that vertebrates and nematodes have in common, and the large number that they do not (*C. elegans* Consortium, 1998). Enzymes such as glutathione peroxidase, superoxide dismutase and the various classes of proteinase, and structural elements such as collagens and actin, are all readily identifiable from their sequence similarities with vertebrate proteins. It is usually such likenesses that lead to the discovery of the biochemical activities of nematode products, which have attracted attention for their potential relevance to the biochemistry and immunology of host–parasite interactions. Expressed sequence tag (EST) projects have found even more apparently novel protein types from organisms such as *Toxocara canis* and *Brugia malayi* (Blaxter *et al.*, 1996; Blaxter, 1998; Tetteh *et al.*, 1999) and many of these appear to be of types not even represented in the *C. elegans* genome.

About 58% of the proteins encoded in the *C. elegans* genome alone appear to encode entirely novel protein types (*C. elegans* Consortium, 1998), though this proportion may fall with the development of better structural and domain prediction methods. A surprising finding is that the proportion of new types even applies when comparing two different species of nematode, such as the three mentioned above. Some of these novel proteins may carry out biological tasks that are special to nematodes, and some may represent examples of where two structurally different kinds of protein carry out the same or similar functions in different animal groups.

Some forms of protein may have been possessed by the common ancestor of nematodes and vertebrates but subsequently lost in one line; some may be entirely new inventions of either group, modified from an unidentifiable common precursor.

Of all the novel protein types found in nematodes, only two have had biochemical activites ascribed, and these both happen to be lipid-binding proteins (LBPs). This chapter will focus on these, plus those that are structurally similar to those of vertebrates but appear to have nematode-specific modifications to their structures and functions.

The Importance of Lipid-binding Proteins

LBPs are likely to have conventional roles in the energy metabolism and transport of lipids in nematodes for membrane construction, etc. Many parasitic helminths have deficiencies in the synthesis of some lipids and so their lipid acquisition, transport and storage mechanisms clearly need to be specialized and therefore pertinent to the host–parasite relationship (Barrett, 1981). From a practical point of view, lipid transporter proteins may also be important in the delivery of anthelmintic drugs to their target; most anthelmintics are hydrophobic and if they do not distribute to their site of action within the parasites by simple diffusion across and along membranes, then the parasite's own carrier proteins may be involved.

Most interest is likely to be attracted to LBPs for their role in parasitism and disease. For example, small lipids such as retinoids (retinol/vitamin A, retinoic acid) and the eicosanoids (arachidonic acid, leucotrienes, prostaglandins) have important roles in signalling events in inflammation, immune responses, and tissue repair and differentiation in mammals (Kliewer et al., 1999; Kersten et al., 2000). The involvement of retinoids in gene activation and control is particularly well understood, as are the structure and function of the nuclear (hormone) receptors involved in retinoid-induced gene activation (Kliewer et al., 1999; Kersten et al., 2000). Thus, proteins binding any of these signalling lipids could be involved in the control of the tissue environment by parasitic nematodes through sequestration of signalling lipids. Moreover, some nematodes are known to release eicosanoids, which may function to alter local and systemic inflammatory and immune reactions against the parasites (Liu et al., 1992). However, signalling lipids are highly susceptible to degradation and so the parasites may require to release the lipids within specialized carrier proteins. Such proteins may additionally need to interact with specific receptors on the surface of host cells in order to deliver their cargo.

The Different Types of Lipid-binding Protein

Lipid-binding and transport proteins come broadly in two types: those that carry lipids through aqueous compartments such as blood and haemolymph, or in the cytosol; and those that transport lipids across cell membranes. The soluble lipid-binding/transporter proteins come in many forms, some of which may be universal among metazoans and some which are confined to particular groups. The best understood LBP is mammalian serum albumin, which is the most abundant species of protein in blood plasma. It is a large protein (approximately 67 kDa), with multiple binding sites for lipids (chiefly for fatty acids) and several other ligand types, and the binding of drugs by albumin is important to their delivery and pharmacokinetics (He and Carter, 1992; Curry *et al.*, 1998). Serum albumin is considered to be the main plasma protein for the bulk transport of fatty acids; its large size also prevents its loss by the kidneys and limits its diffusion in tissues.

The other major class of extracellular LBPs of mammals is the lipocalins (Flower, 1996). These are approximately 20 kDa, β-sheet-rich proteins, performing functions such as the transport of retinol in plasma or milk, the capture of odorants in olfaction, invertebrate coloration, dispersal of pheromones, and solubilizing the lipids in tears (Flower, 1996). The retinol-binding protein (RBP) of human plasma is found in association with a larger protein, transthyretin, the complex being larger than the kidney threshold and thus not excreted, although the RBP itself may dissociate from the complex to interact with cell surface receptors in the delivery of retinol (Papiz *et al.*, 1986; Sundaram *et al.*, 1998).

The small (approximately 14 kDa) cytoplasmic LBPs (cLBPs) of the FABP/P2/CRBP/CRABP family are involved in the intracellular transport of fatty acids, retinol and retinoic acid and may also be involved in the transport of eicosanoid signalling lipids (Coe and Bernlohr, 1998). They fold as somewhat flattened and crumpled β-barrels (sometimes curiously termed 'β-clam'), which consist of a cage formed of antiparallel β-strands enclosing a binding site for hydrophobic ligands (Sacchettini and Gordon, 1993; Coe and Bernlohr, 1998). They have a small section of helix (the 'lid') adjacent to the point of entry and exit of lipids (the 'portal'). In most of these proteins, as with many of the lipocalins, the ligands are taken into the binding site in their entirety, including the charged carboxylate of fatty acids (Sacchettini and Gordon, 1993; LaLonde *et al.*, 1994). The advantage of this is presumably that the lipid is entirely removed from contact with solvent water; it is thereby protected from degradation and does not alter the interaction of the protein with solvent water or other cellular components. The cLBPs are found widely amongst the animal phyla and have been reported from vertebrates, insects, arachnids, cestodes, trematodes and nematodes (Stewart, 2000). In vertebrates, different tissues have different paralogues/isoforms of cLBP and discrete types have been described from the liver, brain, intestine, heart muscle, epidermis and adipocytes, but many cell types produce more than one isoform (Coe and Bernlohr, 1998; Simpson *et al.*, 1999).

Phospholipid transporter proteins are larger (approximately 21–35 kDa) and are thought to be involved in the transport of phospholipids to support a number of functions, such as the equilibration of phospholipids between membrane bilayers, and processes involved in membrane reorganization. Proteins specific for different classes of phospholipid (e.g. phosphatidylethanolamine, phosphatidylinositol, phosphatidylcholine) have been described. The structure of one of these from yeast is known: the Sec14 phosphatidylinositol-binding protein, which is essential for vesicle budding from the Golgi complex (Sha *et al.*, 1998); this appears to have an unusual external hydrophobic helix, which, it is speculated, acts in a bulldozer-like fashion in the collection and deposition of its ligand (Sha *et al.*, 1998). A phosphatidylethanolamine-binding protein (TES-26, now Tc-PEB-1) has been described as a secreted product of *T. canis* larvae and the encoding transcript for the protein is among the most abundant produced by the larval, but not the adult parasites (see Chapter 12); apparent homologues have been described from *Onchocerca volvulus* (Ov16) and *B. malayi* (Lobos *et al.*, 1991; Gems *et al.*, 1995; Blaxter *et al.*, 1996).

The remaining major classes of water-soluble lipid transporter proteins (other than the polyproteins of nematodes; see below) come from plants and helminths. Plants possess very small (approximately 9 kDa) helix-rich, fatty-acid-binding proteins, the structures of some of which are known (Lerche and Poulsen, 1998). A recently described class comes from cestodes: these are also very small (approximately 8 kDa), presumably intracellular, and helix-rich, and bind anthelmintic drugs in addition to fatty acids (Janssen and Barrett, 1995; Barrett *et al.*, 1997). The only helix-rich small (approximately 14 kDa) lipid transporter from vertebrates is the acetyl-CoA-binding protein (Kragelund *et al.*, 1993).

The best-described lipid transporter proteins are involved in solubilizing and protecting lipids in transit across an aqueous phase in blood or cytosol, but how do lipids cross membranes, which they clearly must do to enter and leave cells and organelles? The simplest mechanism would seem to be exchange between water-soluble transporter proteins and membranes by simple diffusion, where they flip-flop to the other membrane leaflet and are then picked up by transporter proteins on the other side (Kleinfeld *et al.*, 1997). However, two of these steps involve thermodynamically unfavourable entry into an aqueous phase. Another idea is that the transporter proteins interact with membranes by collision or direct contact and the ligand is then exchanged without contact with water (Hsu and Storch, 1996). This idea has considerable support from experimental systems involving lipid transfer to and from artifical vesicles, but even this might eventually be superseded with the discovery of integral membrane proteins that participate in transmembrane lipid transport, and with which water-soluble transporter proteins may interact directly (Dutta-Roy, 2000). Retinoids represent a case in point, where there is strengthening evidence for a continuous chain of transporter proteins to deliver and protect these sensitive lipids from food source to nuclear receptor (Sundaram *et al.*, 1998). Moreover, because retinoids are closely involved in gene regulation, the pathway for their transport must be controlled and directional in order to avoid inappropriate signalling. The situation is less clear with fatty acids, but there is increasingly strong evidence that there are membrane receptor proteins with which extra- and intracellular lipid carrier proteins interact (Dutta-Roy, 2000). Such transmembrane transporters are found in the adipocytes of humans (Abumrad *et al.*, 1993) and a particularly graphic example of their importance is in the human placenta, where trophoblast cells possess a plasma membrane protein (p-FABPpm) that is highly selective of the types of unsaturated fatty acid that it takes from the maternal blood circulation to enrich in the fetal circulation (Campbell *et al.*, 1995).

Nothing is yet known of the transmembrane lipid transporters of nematodes, so this chapter will concentrate on those proteins that appear to be unique to nematodes or those that are of a type familiar in other animal groups (principally vertebrates) but exhibit modifications of structure and function that are unusual to nematodes.

The Polyproteins of Nematodes: NPAs

The term 'polyproteins' is used for two different types of entity. The first refers to precursor polypeptides which are cleaved post-translationally into biologically active proteins or peptides of quite different functions. Examples of these include polyproteins of viruses and some prohormones of vertebrates (reviewed in Kennedy, 2000b). The other type is large proproteins which comprise tandem repetitions of identical or similar polypeptides that are post-translationally cleaved into multiple copies of biochemically similar functional entities. The nematode polyprotein allergens/antigens (NPAs) fall into this class (Fig. 16.1).

The NPAs originally attracted interest because one of their members (the ABA-1 protein of *Ascaris*) is an allergen in infected experimental animals, and there is strong genetic control of the immune response to this and similar proteins in filarial nematodes (Tomlinson *et al.*, 1989; Kennedy *et al.*, 1991; Allen *et al.*, 1995). ABA-1-like proteins are at particularly high concentrations in the pseudocoelomic fluid of large ascaridid nematodes (Kennedy and Qureshi, 1986; Christie *et al.*, 1990). Quite independently, NPAs were also identified as the curious 'ladder proteins' that appear to be present immediately below the cuticle of filarial nematodes, though later work showed that they are widely dispersed within the connective tissue of the nematode body (Maizels *et al.*, 1989; Selkirk *et al.*, 1993; Tweedie *et al.*, 1993). NPAs appear to be extracellular, and secreted by a number of species. They are the subject of recent reviews (Kennedy, 2000a,b) and their essential features are covered here.

NPA Biosynthesis and the Gene

NPAs are produced as large proproteins with a short hydrophobic leader peptide and about 11 tandemly repeated polypeptide units (Fig. 16.1). Each unit is about 133 amino acids long and is post-translationally cleaved from the precursor polyprotein into functional entities of about 15 kDa in mass. The regularly spaced cleavage sites conform to the $^{Lys}/_{Arg}$.Xaa.$^{Lys}/_{Arg}$.Arg consensus site for post-translational processing enzymes (Barr, 1991). It is the progressive processing of the large repetitive polypeptide which results in the ladder-like appearance in SDS-PAGE gels, representing different stages in the cleavage process, which, with time, culminates in large quantities of approximately 15 kDa functional units (Selkirk *et al.*, 1993; Tweedie *et al.*, 1993). The amino acid sequence of the individual repeat units can be very similar to one another, or very different,

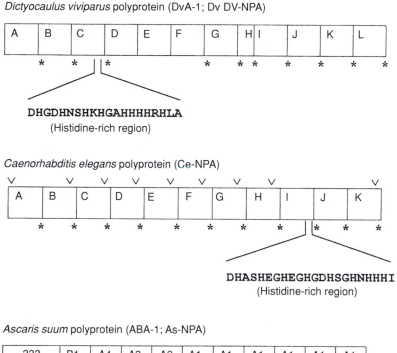

Dictyocaulus viviparus polyprotein (DvA-1; Dv DV-NPA)

DHGDHNSHKHGAHHHHRHLA
(Histidine-rich region)

Caenorhabditis elegans polyprotein (Ce-NPA)

DHASHEGHEGHGDHSGHNHHHI
(Histidine-rich region)

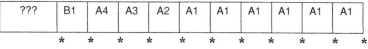

Ascaris suum polyprotein (ABA-1; As-NPA)

depending on the species (reviewed in Kennedy, 2000b), and there is some evidence of allelic differences within a species (Britton *et al.*, 1995).

Biosynthesis as a tandemly repetitive polypeptide which is cleaved into multiple functionally similar proteins is highly unusual, and the only similar examples are the filaggrins of the skin of mammals (Rothnagel and Steinert, 1990). The reason why NPAs and filaggrins are produced in this way remains obscure. If a protein needs to be produced in large quantities rapidly, then a polyprotein would seem advantageous, but it seems inefficient for a genome to encode multiple copies of a protein when one will do, and other abundantly produced proteins are produced in the conventional

---

Fig. 16.1. (Opposite) Organization of the polyprotein of three species of nematode. Only the central repetitive sections of the polyproteins are illustrated, ignoring the short hydrophobic N-terminal (signal?) and C-terminal extension peptides. The full structure of the nematode polyprotein antigen/allergen (NPA) of *D. viviparus* is known from cDNA analysis (Britton *et al.*, 1995), and the *C. elegans* sequence derives from the genome project (cosmids VC5 and F27B10). In the case of the *A. suum* NPA, which is the next best understood, the 5′ region of cDNA is unresolved (Moore *et al.*, 1999). For the NPAs of other species, only fragments of the central region are known. The positions of the consensus $^{Lys}/_{Arg}$-Xaa-$^{Lys}/_{Arg}$-Arg cleavage sites are marked (*). In some species (e.g. *D. viviparus* and *Dirofilaria immitis*) no consensus cleavage site is evident between some of the units (Britton *et al.*, 1995; Poole *et al.*, 1996), suggesting either that another type of enzyme is involved in processing the polyprotein at these points or that double or triple units are produced without further processing; intact multimers have been detected in extracts from whole parasites (Britton *et al.*, 1995; Poole *et al.*, 1996). Following cleavage, the proteins appear to be trimmed to remove the cleavage site amino acids – mass spectrometry on ABA-1 purified from the parasite yielded a value of 14,643.2 ± 1.4, which, within error, corresponds to the calculated mass (14,642.6) of the ABA-1A1 unit with the cleavage site amino acids (Arg-Arg-Arg-Arg) removed (Christie *et al.*, 1993). Unit H in *D. viviparus* is truncated in a manner that has not yet been found in any other species. The histidine-rich stretches that are found in *D. viviparus* and *C. elegans* are positioned as indicated. The positions of the small (40–396 bp) introns in the genomic DNA encoding the *C. elegans* NPA are shown (marked with an arrow, V). For complete sequences, the naming convention for the units is alphabetical from the N terminus/5′ end. The *Ascaris* NPA units were named according to order of discovery and similarity, and should be renamed when the complete DNA sequence becomes available. In this NPA, the A1 to A4 units are grouped because of their closely similar amino acid sequences, which are distinctly different from the single B1 unit (Moore *et al.*, 1999). The units within the NPA arrays of *D. viviparus* and *C. elegans* are quite different from one another, but those of ascaridid species and certain filarial nematodes (e.g. *Brugia malayi*) are more similar (Paxton *et al.*, 1993; Tweedie *et al.*, 1993; Moore *et al.*, 1999). See Kennedy (2000a) for the original and suggested new nomenclatures of NPAs, the individual units within the arrays, and the encoding genes. Figure adapted from Kennedy (2000a), with permission.

manner (e.g. serum albumin). There may also be risks of disruption of the gene by homologous recombination in a tandemly repetitive array, though this could presumably be circumvented by the evolution of sequence diversity amongst the members of the array after the amplification event(s), subject to functional constraints on the encoded protein(s). Because the filaggrins are produced by terminally differentiating cells (keratinocytes), which are undergoing programmed cell death, it has been argued that tandemly repetitive polyproteins, such as the filaggrins, are only produced by apoptotic cells (Rothnagel and Steinert, 1990). The argument is that the biosynthetic and nuclear processes of such cells are compromised or deteriorating, and synthesis of proteins by a means in which mRNA and polypeptide processing can be minimized would be advantageous. The fact that the filaggrins appear to have no introns (although complete genomic DNA sequences are still to appear) in the repetitive region bolsters this idea (Rothnagel and Steinert, 1990). However, there is no evidence that the NPAs are produced by apoptotic cells – in fact, the reverse is the case, since nematodes only undergo apoptosis at early stages of development, and not in adults (Soulston, 1988; Wood, 1988). It is true that this has been shown for only a few species, but some of those for which this holds good produce NPAs (*C. elegans* in particular). So, either the selective advantages that led to the evolution of polyprotein filaggrins and NPAs are different, or the cause/effect relationship with apoptosis is invalid. If the latter, then there might be some unguessed-at commonality about NPAs and filaggrins as polyproteins. For instance, if a protein is required at a very high concentration in the synthesizing cell or site of action, then there is a risk that aggregation and crystallization might occur (Arakawa and Timasheff, 1985). If (even slightly) different forms of the protein are present, then the risk of such aggregation is much reduced. Given the high concentration of filaggrins in differentiating keratinocytes, and the NPAs in nematode pseudocoelomic fluid, such an explanation is worthy of attention.

Of all the nematodes from which cDNA-encoding NPAs have been isolated, only one complete sequence is available: that from the lungworm of cattle, *Dictyocaulus viviparus* (Britton *et al.*, 1995) (Fig. 16.1). Complete genomic DNA sequence is now available for *C. elegans* (across cosmids VC5 and F27B10). The two polyproteins are very similar in the number of units present and in the diversity of amino acid sequences of the repeated units within an array. The characteristics of NPAs are therefore common to free-living and parasitic species. Whilst there is still too little information to generalize, it appears at the moment that only in parasites are there examples of stretches of units within an array being very similar, such as in *Ascaris* and *Brugia* (Paxton *et al.*, 1993; Moore *et al.*, 1999).

Structure of NPAs

Secondary structural predictions about NPAs, and direct biophysical measurements, have demonstrated that the NPAs are rich in α-helix, with no β-structure either predicted from secondary structure prediction algorithms, or detected by circular dichroism (Kennedy *et al.*, 1995b). In this they are the antithesis of the similarly sized cLBPs and lipocalins. The predictions are that each individual NPA unit protein will fold into four main regions of helix, and it has been speculated that the tertiary structure is as a four-bundle helix protein, similar to other invertebrate carrier proteins (Sheriff *et al.*, 1987).

Functions of NPAs

The NPAs bind fatty acids, lysophospholipids and retinol, and constitute the only example of a lipid-binding protein produced as a polyprotein (Kennedy *et al.*, 1995a,b,c). They have yet to be screened for the binding of diglyceride phospholipids, triglycerides, cerebrosides, etc., but they do not appear to bind cholesterol. They are therefore likely to be transport proteins for small lipids, just as are human serum albumin, plasma retinol-binding protein, β-lactoglobulin of bovine milk, and the several different types of cytosolic fatty acids and retinoid transport proteins found widely in metazoans. The NPAs, however, are structurally quite distinct from these protein types, and nothing similar has yet been found in animal groups other than nematodes.

The NPAs are probably involved in the mass transport of small lipids within nematodes, via the pseudocoelomic fluid, and are therefore functionally similar to human serum albumin. NPAs appear to be produced only by cells of the gut (Fig. 16.2) (Xia *et al.*, 1999) and they may be loaded with lipid there and then released for distribution to consuming tissues. At their destination, they may exchange their cargo directly to cell membranes or membrane receptors, or be internalized for stripping before recycling back into the pseudocoelomic fluid. Their role in the secretions of parasites is more speculative. They could be involved in the simple acquisition of host lipids as nutrients, although this would seemingly require a mechanism for their reabsorption, possibly requiring a specific receptor. Alternatively, they could be involved in sequestering or delivering signalling lipids, as mentioned above, and thereby assist the parasites' survival against host defence mechanisms, or to optimize the tissue environment for the survival and reproduction of the parasite.

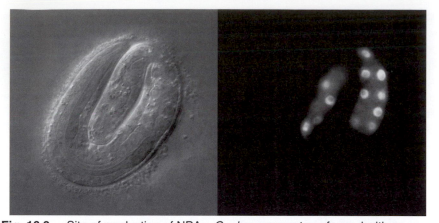

Fig. 16.2.   Site of production of NPAs. *C. elegans* was transformed with a
plasmid containing the putative promoter sequences of *Ce-npa-1* fused to the
green fluorescent protein (GFP) gene. The version of GFP used was engineered
to contain a nuclear localization signal, causing the GFP protein to locate to the
nucleus of the cell in which the promoter was activated. A bright field image of an
egg of a transformant is on the left, and a fluorescence image of the same egg is
on the right. Fluorescence is confined to the nuclei of the gut cells. Although not
shown, a similar expression pattern can be seen in all later developmental
stages. This suggests that the normal site of expression of *Ce-npa-1* is the gut.
A similar conclusion was reached using Northern blotting with cDNA encoding
the NPA of *A. suum* and mRNA from a range of tissues of adult worms (Xia *et al.*,
1999). The eggs of *C. elegans* are approximately 50 μm in length. Construction
of the plasmid for transformation of *C. elegans*, construction of transgenic strains
and microscopy by Moira Watson and Iain Johnstone, Wellcome Centre for
Molecular Parasitology, University of Glasgow.

Polyfunctional Polyproteins?

A recent finding adds to the mystery of the biological functions of these
proteins in that a histidine-rich stretch of amino acids has been found at the
C-terminus of one unit of the array in the two species for which complete
sequences are available (Fig. 16.1). Histidine-rich peptides and proteins are
known to bind certain divalent cations and haem (Morgan, 1981; Burch
and Morgan, 1985; Koide *et al.*, 1986; Wulfing *et al.*, 1994; Sullivan *et al.*,
1996) and synthetic peptides of the NPA sequences have indeed been
found to bind nickel, zinc and copper ions, but not calcium (M.W.
Kennedy, unpublished results). These peptides also bind haemin, in which
they produce a dramatic change in the absorbance spectrum (M.W.
Kennedy, unpublished results). This binding propensity might therefore
indicate that the particular NPA units concerned are involved in either the
transport or salvage of certain metal ions and haems. The packaging of
haem is an important function in blood-feeding arthropods (Oliveira *et al.*,

1999) and so it might be in the pseudocoelomic fluid of blood-feeding nematodes. However, no information on the NPAs of blood feeders such as *Haemonchus*, *Necator* or *Ancylostoma* has yet appeared. A bound metal ion might also act to cross-link two individual His-containing NPA units, thereby producing a larger complex. It is already suspected that NPAs form non-covalent dimers (Britton *et al.*, 1995; Kennedy *et al.*, 1995b,c; Poole *et al.*, 1996) and some adjacent units do not have discernible consensus proteinase cleavage sites. Thus, there may be at least two mechanisms for the formation of larger entities (Fig. 16.1).

The LBP-20 Proteins

The first of these proteins to be investigated was the Ov20 protein of *O. volvulus*, which is an approximately 20 kDa protein originally used to examine the immune responses to onchocerciasis in humans. It is secreted by the parasites (Tree *et al.*, 1995; Nirmalan *et al.*, 1999) and was later found to be, like the NPAs, a fatty acid and retinol-binding protein (Kennedy *et al.*, 1997). It is its retinol-binding activity that particularly strikes a chord, because retinol (vitamin A) and other retinoids have particular roles in many cellular processes, including gene activation, cell differentiation, tissue repair and light detection, and so their sequestration by a protein released by a parasite that causes damage to skin and the eye demands particular attention. Measurements of the dissociation constants for ligand binding of Ov20 indicated that its affinity for retinol was higher than for fatty acids, though it can by no means be considered a retinol-specific binding protein.

Following the description of Ov20, a closely similar protein sequence from *Brugia malayi* has been described and it was found to possess ligand-binding properties that were almost indistinguishable from those of Ov20 (M.W. Kennedy and J.E. Allen, unpublished results). A comparative immunological study on the two has since been published (Nirmalan *et al.*, 1999). There then followed the identification of genes encoding these proteins (now generically termed LBP-20) from a wide range of nematodes parasitic in humans, animals and plants (Blaxter, 1998). A curious finding is that, to date, only one such gene has been found in any parasite, but the genome of *C. elegans* encodes six similar proteins, recombinant forms of which have been demonstrated to have similar binding properties to Ov20 and Bm20 (M.-C. Rollinson, A. Garofalo, M.W. Kennedy and J.E. Bradley, unpublished results). Why *C. elegans* requires so many paralogues of these proteins, whilst the parasitic species do not, is far from clear, but it is quite possible that more LBP-20-like proteins will eventually be found in parasites.

The LBP-20 proteins were predicted to be rich in α-helix, and with very little, if any, β/extended structure, and this has been confirmed by circular dichroism analysis (Kennedy *et al.*, 1997). In this they are like the NPAs, and both therefore represent departures from the richness in β-structure of small LBPs of vertebrates, such as the lipocalins and the cLBPs. The similarity between the ligand-binding activities of the LBP-20s and the NPAs prompts one to ask why nematodes require both types of protein, why the functional units of each type are different in size, and why one type is synthesized as a polyprotein and the other is not. At the very least this indicates that there is much yet to learn about the biology of these proteins, and it may yet transpire that their true binding and transport function *in vivo* is quite different from that deduced to date.

The β-Barrel FABPs

These are members of the FABP/P2/CRBP/CRABP family of cLBPs, which are found widely amongst the animal phyla. Examples from humans include the cytosolic fatty acid transporter proteins of adipocytes, brain, heart muscle, intestinal tissue and liver, and the cytoplasmic retinol- and retinoic acid-binding proteins (Coe and Bernlohr, 1998). Examples from invertebrates include a highly abundant protein from the flight muscles of adult insects, an allergen from tropical dust mites, and important antigens from trematodes and cestodes (Stewart, 2000). The reason they are included here is that nematodes appear to have evolved uses for these proteins that are quite different from other groups. Firstly, whereas these LBPs are universally intracellular in other animals (Stewart, 2000), the nematodes produce forms that are released not only from the synthesizing cell, but also from the organisms themselves (Mei *et al.*, 1997; Plenefisch *et al.*, 2000). Secondly, whilst the sequences of cLBPs are diverse, even between tissue types in mammals, their structures are virtually superimposable, except in nematodes where modifications are apparent (Mei *et al.*, 1997).

As-p18, the First Secreted Cytoplasmic Lipid-binding Protein

The first of the nematode cLBP-like proteins to be examined was the As-p18 protein of *Ascaris suum*. This protein was identified as a major component of the perivitelline fluid surrounding the developing embryo within the egg (Mei *et al.*, 1997) and it has recently been found also to be present in the perienteric fluid of the adult worms (Plenefisch *et al.*, 2000). Biochemical analysis of recombinant As-p18 showed that it binds fatty acids but not retinol or cholesterol, so may be involved in the maintenance of the lipid layer of ascarid eggs or sequestering damaged lipids. If it is indeed involved in maintaining the lipid layer (which is thought to contribute to the environmental robustness of some nematode eggs), then it may bind the unusual ascaroside lipids present in this structure, although this has yet to be tested. Unlike cLBPs from any other source, As-p18 is synthesized with a hydrophobic leader which bears the characteristics of hydrophobic signal peptides of many types of secreted protein, and is cleaved from the protein by the time it appears in the perivitelline fluid (Mei *et al.*, 1997). Similar proteins are also now known from filarial nematodes (Plenefisch *et al.*, 2000) and so, although interest in As-p18 arose from its relevance to a non-parasitic stage of a nematode, it might be able to teach us something about the function of similar proteins in nematodes in direct contact with human tissues.

The protein that is most similar to As-p18 is represented by ESTs from *B. malayi*; this protein also bears a putative signal sequence (Plenefisch *et al.*, 2000). More complete information is available from the *C. elegans* genome, which encodes eight cLBP-like proteins, comprising two distinct clusters according to their amino acid sequence similarities (Fig. 16.3); As-p18 belongs to a cluster most of whose members have signal sequences, and the members of the other cluster are more closely related to the cLBPs found in mammals and trematodes, of which none have signal/leader peptides. The possession of a signal sequence represents a departure so far exclusive to nematodes, and if these proteins are specialized for the maintenance of

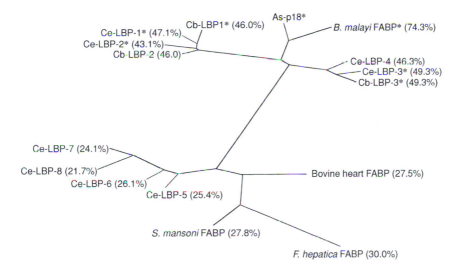

Fig. 16.3. Phylogenetic relationships between the predicted amino acid sequence of As-p18 and cLBPs from other sources. Asterisked (*) proteins possess secretory leader/signal sequences. The percent identity of each protein with As-p18 is shown in parenthesis. Protein sequences from *C. elegans* and *C. briggsae* are prefixed 'Ce' and 'Cb', respectively. It is clear that cLBP-like proteins from nematodes fall into two groups – one group whose members have secretory leaders, and those with no such leader – and cLBPs from vertebrates fall into the latter group. Reproduced from Plenefisch *et al.*, 2000, with permission.

developing embryos, then this clearly has implications both for transmission of the infections and for the pathology of those infections in which it is the larval stages that cause the most damage (such as in onchocerciasis).

Modelling As-p18 on X-ray crystallographically derived structures of mammalian cLBPs indicated that it has essentially the same fold as other cLBPs, but with unusual modifications (Fig. 16.4) (Mei *et al.*, 1997). The ligand-binding propensities of the protein have not yet been found to be fundamentally different from those of other cLBPs but, given the location

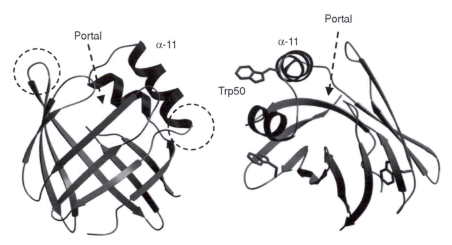

Fig. 16.4. The structure of As-p18. The protein as modelled comprises a β-barrel in which the ligand is held, and two small helices, one of which (α-II), is adjacent to the presumptive portal of ligand entry and exit. The protruding hydrophobic side-chains postulated to be involved in interaction between As-p18 and other proteins or membranes is tryptophan 50 (Trp50) (Kennedy and Beauchamp, 2000; Kennedy *et al.*, 2000). The portal of entry and exit for the ligand is thought to lie between the helix with Trp50 and an opposing loop, as indicated. There are three regions where there are insertions in As-p18 that are not matched in other similar proteins, and two of these are encircled. The image on the left is a general view of the protein to illustrate its overall β-barrel structure, and on the right is a side view of the molecule looking down the axis of helix α-II to illustrate the unusually exposed hydrophobic side-chain of Trp50. Conserved charged residues on the surface of the protein seem to be involved in protein : membrane interaction and have already been shown to influence cLBP : membrane vesicle interaction in mammalian cLBPs (Herr *et al.*, 1995). The other tryptophans in the protein are also shown, illustrating their more concealed position in the protein's tertiary structure. For more discussion on the postulated 'sticky finger' motif, see Kennedy and Beauchamp (2000). The structure is a theoretical model based on the experimentally derived three-dimensional structures of mammalian cytoplasmic fatty acid-binding proteins, as described in Mei *et al.* (1997). The coordinates of the model are deposited in the Protein Data Bank with the accession number 1as1. Ilustration prepared by Dr J. Beauchamp, Protein Crystallography, Department of Chemistry, University of Glasgow.

of As-p19 in the perivitelline fluid and its unusual structural features (Fig. 16.4), its function *in vivo* may be to bind and transport/exchange lipids that are different from those which it can be demonstrated to bind *in vitro*. If As-p18 does indeed bind ascarosides, then it and similar proteins in other species of nematode may be important in the maintenance of the protective lipid layers of the eggshell. Given that eggs of species such as *Ascaris* can survive for several years in soil, then it may also be that the enclosed larvae continuously maintain the lipid layer using As-p18 as a transporter for damaged and replacement lipids.

Sticky Finger Interaction Sites?

Another unusual feature of As-p18 is a prominent, bulky, hydrophobic amino acid side-chain (a tryptophan) which is predicted to project directly away from the protein and into solvent. This projects from an α-helix positioned immediately adjacent to the presumed site of ligand entry and exit (the portal) (Sacchettini and Gordon, 1993; Coe and Bernlohr, 1998; Corsico *et al.*, 1998). This is a highly unusual position for such a side-chain, but a similar feature is evident in mammalian cLBPs that have been shown experimentally to exchange their fatty acid ligand by collision or contact with a donor or acceptor membrane; it is absent in a fatty acid binding protein that does not (reviewed in Kennedy and Beauchamp, 2000). In the former type of cLBP from mammals, the protruding hydrophobic side-chain is usually a phenylalanine or a leucine, both of which are highly hydrophobic and, like tryptophans, are usually directed towards the interior of proteins. Tryptophan and phenylalanines are known to be able to penetrate membranes, and are usually found on the outside of proteins in regions of protein : membrane interaction (reviewed in Kennedy and Beauchamp, 2000).

In other types of protein, protruding hydrophobic side-chains occur frequently in regions of protein : protein interactions; for instance, 23% of the intermolecular contacts between the gp120 protein of the human immunodeficiency virus and its target CD4 protein on lymphocytes is attributed to a single phenylalanine side-chain (Kwong *et al.*, 1998). It is therefore not beyond the bounds of possibility that this 'sticky finger' motif (Kennedy and Beauchamp, 2000) of As-p18 is instead (or additionally?) involved in interactions with other proteins, since it has recently been shown that at least one mammalian cLBP indeed interacts with another cytosolic protein (Shen *et al.*, 1999). Whether As-p18 interacts with lipidic structures and/or other proteins remains to be established, but it shows all the signs of being part of a lipid transporter chain.

No cLBP-like proteins other than those from nematodes and schistosomes have tryptophan side-chains in this sticky finger position (Kennedy and Beauchamp, 2000; Kennedy *et al.*, 2000). This is potentially useful to the structural analysis of these proteins because the indole ring of tryptophan is fluorescent; the fluorescence emission characteristic is environment-sensitive in that it varies in intensity and wavelength depending on the degree of exposure to solvent water. This can be used, for example, to check structural predictions on a particular protein (Eftink and Ghiron, 1976, 1984; Kennedy *et al.*, 2000) by examining the exposure or isolation of a tryptophan to solvent water. Theoretically, fluorescence could also be useful in detecting when and how a protein from which a tryptophan protrudes interacts with membranes or other proteins (Kachel *et al.*, 1995; Yau *et al.*, 1998; Gelb *et al.*, 1999). The nematode proteins, such as As-p18, could therefore be useful in understanding interactions of cLBP-like proteins from all sources, including humans, with other cellular or tissue components.

Concluding Remarks

The various types of novel lipid-binding protein from nematodes can provide a number of lessons.

- There are many more ways to build a lipid transport protein for small lipids than previously known. The NPAs, for instance, have similar ligand-binding propensities to similarly sized LBPs of vertebrates, but are helix-rich rather than being β-rich. The LBP-20-type proteins represent yet another novel type of LBP.
- The binding site environments of certain LBPs (the NPAs in particular) appear to be extremely, and so far inexplicably, apolar.
- Polyproteins (as exemplified by the NPAs) are not the exclusive products of apoptotic cells, so the polyprotein nature of the keratinocyte filaggrins of humans requires a new explanation. The NPAs might provide a model for repetitive-type polyproteins which could impinge on the understanding of polyproteins as a whole.
- Nematodes provide exceptions to the rule that members of the FABP/ P2/CRBP/CRABP family of proteins are universally intracellular. Some of those secreted by nematodes may be involved in specialized functions, such as the protection of developing embryos within shelled eggs. Nematodes also provide examples of where the highly conserved tertiary structure of these proteins has been modified.
- Finally, there are the questions that these novel proteins pose. Do any of these unusual lipid-binding proteins, or modifications of otherwise usual proteins, have any role in parasitism? Are they secreted to the advantage of the parasite and, if so, in what way?

If the answer to the last question is yes, then it may be time for signalling lipids, and the parasite proteins that carry or capture them, to become more of a focus in research on infections with multicellular parasites.

References

Abumrad, N.A., El-Maghrabi, M.R., Amri, E.Z., Lopez, E. and Grimaldi, P.A. (1993) Cloning of a rat adipocyte membrane-protein implicated in binding or transport of long chain fatty acids that is induced during preadipocyte differentiation: homology with human CD36. *Journal of Biological Chemistry* 268, 17665–17668.

Allen, J.E., Lawrence, R.A. and Maizels, R.M. (1995) Fine specificity of the genetically controlled immune-response to native and recombinant gp15/400 (polyprotein allergen) of *Brugia malayi*. *Infection and Immunity* 63, 2892–2898.

Arakawa, T. and Timasheff, S.N. (1985) Theory of protein solubility. *Methods in Enzymology* 114, 49–77.

Barr, P.J. (1991) Mammalian subtilisins: the long-sought dibasic processing endoproteases. *Cell* 66, 1–3.

Barrett, J. (1981) *Biochemistry of Parasitic Helminths*. MacMillan, London.

Barrett, J., Saghir, N., Timanova, A., Clarke, K. and Brophy, P.M. (1997) Characterisation and properties of an intracellular lipid-binding protein from the tapeworm *Moniezia expansa*. *European Journal of Biochemistry* 250, 269–275.

Blaxter, M. (1998) *Caenorhabditis elegans* is a nematode. *Science* 282, 2041–2046.

Blaxter, M.L., Raghavan, N., Ghosh, I., Guiliano, D., Lu, W., Williams, S.A., Slatko, B. and Scott, A.L. (1996) Genes expressed in *Brugia malayi* infective third stage larvae. *Molecular and Biochemical Parasitology* 77, 77–79.

Britton, C., Moore, C., Gilleard, J.S. and Kennedy, M.W. (1995) Extensive diversity in repeat unit sequences of the cDNA encoding the polyprotein antigen/ allergen from the bovine lungworm *Dictyocaulus viviparus*. *Molecular and Biochemical Parasitology* 72, 77–88.

Burch, M.K. and Morgan, W.T. (1985) Preferred heme binding sites of histidine-rich glycoprotein. *Biochemistry* 24, 5919–5924.

C. elegans Consortium (1998) Genome sequence of *Caenorhabditis elegans*: a platform for investigating biology. *Science* 282, 2012–2018.

Campbell, F.M., Gordan, M.J., Taffasse, S. and Dutta-Roy, A.K. (1995) Plasma membrane fatty acid-binding protein from human placenta: identification and characterization. *Biochemical and Biophysical Research Communications* 209, 1011–1017.

Christie, J.F., Dunbar, B., Davidson, I. and Kennedy, M.W. (1990) N-terminal amino acid sequence identity between a major allergen of *Ascaris lumbricoides* and *Ascaris suum*, and MHC-restricted IgE responses to it. *Immunology* 69, 596–602.

Christie, J.F., Dunbar, B. and Kennedy, M.W. (1993) The ABA-1 allergen of the nematode *Ascaris suum*: epitope stability, mass spectrometry, and N-terminal sequence comparison with its homologue in *Toxocara canis*. *Clinical and Experimental Immunology* 92, 125–132.

Coe, N.R. and Bernlohr, D.A. (1998) Physiological properties and functions of intracellular fatty acid-binding proteins. *Biochimica et Biophysica Acta* 1391, 287–306.

Corsico, B., Cistola, D.P., Frieden, C. and Storch, J. (1998) The helical domain of intestinal fatty acid binding protein is critical for collisional transfer of fatty acids to phospholipid membranes. *Proceedings of the National Academy of Sciences USA* 95, 12174–12178.

Curry, S., Mandelkow, H., Brick, P. and Franks, N. (1998) Crystal structure of human serum albumin complexed with fatty acid reveals an asymmetric distribution of binding sites. *Nature Structural Biology* 5, 827–835.

Dutta-Roy, A.K. (2000) Cellular uptake of long chain fatty acids: role of membrane associated fatty acid binding/transport proteins. *Cellular and Molecular Life Sciences* (in press).

Eftink, M.R. and Ghiron, C.A. (1976) Exposure of tryptophanyl residues in proteins: quantitative determination by fluorescence quenching studies. *Biochemistry* 15, 672–679.

Eftink, M.R. and Ghiron, C.A. (1984) Indole fluorescence quenching studies on proteins and model systems: use of the efficient quencher succinimide. *Biochemistry* 23, 3891–3899.

Flower, D.R. (1996) The lipocalin protein family: structure and function. *Biochemical Journal* 318, 1–14.

Gelb, M.H., Cho, W. and Wilton, D.C. (1999) Interfacial binding of secreted phospholipase A_2: more than electrostatics and a major role for tryptophan. *Current Opinion in Structural Biology* 9, 428–432.

Gems, D., Ferguson, C.J., Robertson, B.D., Nieves, R., Page, A.P., Blaxter, M.L. and Maizels, R.M. (1995) An abundant, *trans*-spliced mRNA from *Toxocara canis* infective larvae encodes a 26-kDa protein with homology to phosphatidylethanolamine-binding proteins. *Journal of Biological Chemistry* 270, 18157–18522.

He, X.M. and Carter, D.C. (1992) Atomic structure and chemistry of human serum albumin. *Nature* 358, 209–215.

Herr, F.H., Matarese, V., Bernlohr, D.A. and Storch, J. (1995) Surface lysine residues modulate the collisional transfer of fatty acid from adipocyte fatty acid binding protein to membranes. *Biochemistry* 34, 11840–11845.

Hsu, K.-T. and Storch, J. (1996) Fatty acid transfer from liver and intestinal fatty acid-binding proteins to membranes occurs by different mechanisms. *Journal of Biological Chemistry* 271, 13317–13323.

Janssen, D. and Barrett, J. (1995) A novel lipid-binding protein from the cestode *Moniezia expansa. Biochemical Journal* 311, 49–57.

Kachel, K., Asuncionpunzalan, E. and London, E. (1995) Anchoring of tryptophan and tyrosine analogs at the hydrocarbon polar boundary in model membrane-vesicles – paralax analysis of fluorescence quenching induced by nitroxide-labelled phospholipids. *Biochemistry* 34, 15475–15479.

Kennedy, M.W. (2000a) The nematode polyprotein allergens/antigens. *Parasitology Today* 16, 373–380.

Kennedy, M.W. (2000b) The polyprotein lipid binding proteins of nematodes. *Biochimica et Biophysica Acta* 1472, 149–164.

Kennedy, M.W. and Beauchamp, J. (2000) Sticky finger interaction sites on cytosolic lipid binding proteins? *Cellular and Molecular Life Sciences* 57, 1379–1387.

Kennedy, M.W. and Qureshi, F. (1986) Stage-specific secreted antigens of the parasitic larval stages of the nematode *Ascaris. Immunology* 58, 515–522.

Kennedy, M.W., Fraser, E.M. and Christie, J.F. (1991) MHC class II (I-A) region control of the IgE antibody repertoire to the ABA-1 allergen of the nematode *Ascaris. Immunology* 72, 577–579.

Kennedy, M.W., Allen, J.E., Wright, A.S., McCruden, A.B. and Cooper, A. (1995a) The gp15/400 polyprotein antigen of *Brugia malayi* binds fatty acid and retinoids. *Molecular and Biochemical Parasitology* 71, 41–50.

Kennedy, M.W., Brass, A., McCruden, A.B., Price, N.C., Kelly, S.M. and Cooper, A. (1995b) The ABA-1 allergen of the parasitic nematode *Ascaris suum*: fatty acid and retinoid binding function and structural characterization. *Biochemistry* 34, 6700–6710.

Kennedy, M.W., Britton, C., Price, N.C., Kelly, S.M. and Cooper, A. (1995c) The DvA-1 polyprotein of the parasitic nematode *Dictyocaulus viviparus*: a small helix-rich lipid-binding protein. *Journal of Biological Chemistry* 270, 19277–19281.

Kennedy, M.W., Garside, L.H., Goodrick, L.E., McDermott, L., Brass, A., Price, N.C., Kelly, S.M., Cooper, A. and Bradley, J.E. (1997) The Ov20 protein of the parasitic nematode *Onchocerca volvulus*. A structurally novel class of small helix-rich retinol-binding protein. *Journal of Biological Chemistry* 272, 29442–29448.

Kennedy, M.W., Scott, J.C., Lo, S.J., Beauchamp, J. and McManus, D.P. (2000) The Sj-FABPc fatty acid binding protein of the human blood fluke *Schistosoma japonicum*: structural and functional characterisation and unusual solvent exposure of a portal-proximal tryptophan. *Biochemical Journal* 349, 377–384.

Kersten, S., Desvergne, B. and Wahli, W. (2000) Roles of PPARs in health and disease. *Nature* 405, 421–424.

Kleinfeld, A.M., Chu, P. and Storch, J. (1997) Flip-flop is slow and rate-limiting for the movement of long chain anthroyloxy fatty acids across lipid vesicles. *Biochemistry* 36, 5702–5711.

Kliewer, S.A., Lehmann, J.M. and Willson, T.M. (1999) Orphan nuclear receptors: shifting endocrinology into reverse. *Science* 284, 757–760.

Koide, T., Foster, D.C., Yoshitake, S. and Davie, E.W. (1986) Amino acid sequence of human histidine-rich glycoprotein derived from the nucleotide sequence of its cDNA. *Biochemistry* 25, 2220–2225.

Kragelund, B.B., Andersen, K.V., Madsen, J.C., Knudsen, J. and Poulsen, F.M. (1993) Three-dimensional structure of the complex between acyl-coenzyme A binding protein and palmitoyl-coenzyme A. *Journal of Molecular Biology* 230, 1260–1277.

Kwong, P.D., Wyatt, R., Robinson, J., Sweet, R.W., Sodroski, J. and Hendrickson, W.A. (1998) Structure of an HIV complex with the CD4 receptor and a neutralizing human antibody. *Nature* 393, 648–659.

LaLonde, J.M., Levenson, M.A., Roe, J.J., Bernlohr, D.A. and Banaszak, L.J. (1994) Adipocyte lipid-binding protein complexed with arachidonic acid: titration calorimetry and X-ray crystallographic studies. *Journal of Biological Chemistry* 269, 25339–25347.

Lerche, M.H. and Poulsen, F.M. (1998) Solution structure of barley lipid transfer protein complexed with palmitate. Two different binding modes of palmitate in the homologous maize and barley nonspecific lipid transfer proteins. *Protein Science* 7, 2490–2498.

Liu, L.X., Buhlmann, J.E. and Weller, P.F. (1992) Release of prostaglandin-E2 by microfilariae of *Wuchereria bancrofti* and *Brugia malayi*. *American Journal of Tropical Medicine and Hygiene* 46, 520–523.

Lobos, E., Weiss, N., Karam, M., Taylor, H.R., Ottesen, E.A. and Nutman, T.B. (1991) An immunogenic *Onchocerca volvulus* antigen: a specific and early marker of infection. *Science* 251, 1603–1605.

Maizels, R.M., Gregory, W.F., Kwan-Lim, G.-E. and Selkirk, M.E. (1989) Filarial surface antigens: the major 29 kilodalton glycoprotein and a novel 17–200 kilodalton complex from adult *Brugia malayi* parasites. *Molecular and Biochemical Parasitology* 32, 213–228.

Mei, B., Kennedy, M.W., Beauchamp, J., Komuniecki, P.R. and Komuniecki, R. (1997) Secretion of a novel, developmentally regulated fatty acid binding protein into the perivitelline fluid of the parasitic nematode, *Ascaris suum*. *Journal of Biological Chemistry* 272, 9933–9941.

Moore, J., McDermott, L., Price, N.C., Kelly, S.M., Cooper, A. and Kennedy, M.W. (1999) Sequence-divergent units of the ABA-1 polyprotein array of the nematode *Ascaris* have similar fatty acid and retinol binding properties but different binding site environments. *Biochemical Journal* 340, 337–343.

Morgan, W.T. (1981) Interactions of the histidine-rich glycoprotein of serum with metals. *Biochemistry* 20, 1054–1061.

Nirmalan, N., Cordeiro, N.J.V., Kläger, S.L., Bradley, J.E. and Allen, J.E. (1999) Comparative analysis of glycosylated and non-glycosylated filarial homologues of the 20-kilodalton retinol binding protein from *Onchocerca volvulus* (Ov20). *Infection and Immunity* 67, 6239–6334.

Oliveira, M.F., Silva, J.R., Dansa-Petreski, M., do Souza, W., Lins, U., Braga, C.M.S., Masuda, H. and Oliveira, P.L. (1999) Haem detoxification by an insect. *Nature* 400, 517–518.

Papiz, M.Z., Sawyer, L., Eliopoulos, E.E., North, A.C.T., Findlay, J.B.C., Sivaprasadarao, R., Jones, T.A., Newcomer, M.E. and Kraulis, P.J. (1986) The structure of β-lactoglobulin and its similarity to plasma retinol-binding protein. *Nature* 324, 383–385.

Paxton, W.A., Yazdanbakhsh, M., Kurniawan, A., Partono, F., Maizels, R.M. and Selkirk, M.E. (1993) Primary structure and IgE response to the repeat subunit of gp15/400 from human lymphatic filarial parasites. *Infection and Immunity* 61, 2827–2833.

Plenefisch, J., Xiao, H., Mei, B., Geng, J., Komuniecki, P.R. and Komuniecki, R. (2000) Secretion of a novel class of iFABPs in nematodes: coordinate use of the *Ascaris/Caenorhabditis* model systems. *Molecular and Biochemical Parasitology* 105, 223–236.

Poole, C.B., Hornstra, L.J., Benner, J.S., Fink, J.R. and Mcreynolds, L.A. (1996) Carboxy-terminal sequence divergence and processing of the polyprotein antigen from *Dirofilaria immitis*. *Molecular and Biochemical Parasitology* 82, 51–65.

Rothnagel, J.A. and Steinert, P.M. (1990) The repeating structure of the units for mouse filaggrin and a comparison of the repeating units. *Journal of Biological Chemistry* 265, 1862–1865.

Sacchettini, J.C. and Gordon, J.I. (1993) Rat intestinal fatty acid binding protein: a model system for analyzing the forces that can bind fatty acids to proteins. *Journal of Biological Chemistry* 268, 18399–18402.

Selkirk, M.E., Gregory, W.F., Jenkins, R.E. and Maizels, R.M. (1993) Localization, turnover and conservation of gp15/400 in different stages of *Brugia malayi*. *Parasitology* 107, 449–457.

Sha, B., Phillips, S.E., Bankaitis, V.A. and Luo, M. (1998) Crystal structure of the *Saccharomyces cervisiae* phosphatidylinositol-transfer protein. *Nature* 391, 506–510.

Shen, W.-E., Sridhar, K., Bernlohr, D.A. and Kraemer, F.B. (1999) Interaction of rat hormone-sensitive lipase with adipocyte lipid-binding protein. *Proceedings of the National Academy of Sciences USA* 96, 5528–5532.

Sheriff, S., Hendrickson, W.A. and Smith, J.L. (1987) Structure of myohemery-thrin in the azidomet state at 1.7/1.3Å. *Journal of Molecular Biology* 197, 273–296.

Simpson, M.A., LiCata, V.J., Coe, N.R. and Bernlohr, D.A. (1999) Biochemical and biophysical analysis of the intracellular lipid binding proteins of adipocytes. *Molecular and Cellular Biochemistry* 192, 33–40.

Soulston, J. (1988) Cell lineage. In: Wood, W.B. (ed.) *The Nematode Caenorhabditis elegans.* Cold Spring Harbor Laboratory Press, Cold Spring Harbor, New York, pp. 123–155.

Stewart, J.M. (2000) The cytoplasmic fatty acid binding proteins: thirty years and counting. *Cellular and Molecular Life Sciences* (in press).

Sullivan, D.J., Gluzman, I.Y. and Goldberg, D.E. (1996) Plasmodium hemozoin formation mediated by histidine-rich proteins. *Science* 271, 219–222.

Sundaram, M., Sivaprasadarao, A., DeSousa, M.M. and Findlay, J.B.C. (1998) The transfer of retinol from serum retinol-binding protein to cellular retinol-binding protein is mediated by a membrane receptor. *Journal of Biological Chemistry* 273, 3336–3342.

Tetteh, K.A., Loukas, A., Tripp, C. and Maizels, R.M. (1999) Identification of abundantly expressed novel and conserved genes from the infective larval stage of *Toxocara canis* by an expressed sequence tag strategy. *Infection and Immunity* 67, 4771–4779.

Tomlinson, L.A., Christie, J.F., Fraser, E.M., Mclaughlin, D., Mcintosh, A.E. and Kennedy, M.W. (1989) MHC restriction of the antibody repertoire to secretory antigens, and a major allergen, of the nematode parasite *Ascaris. Journal of Immunology* 143, 2349–2356.

Tree, T.I.M., Gillespie, A.J., Shepley, K.J., Blaxter, M.L., Tuan, R.S. and Bradley, J.E. (1995) Characterisation of an immunodominant glycoprotein antigen of *Onchocerca volvulus* with homologues in other filarial nematodes and *Caenorhabditis elegans. Molecular and Biochemical Parasitology* 69, 185–195.

Tweedie, S., Paxton, W.A., Ingram, L., Maizels, R.M., McReynolds, L.A. and Selkirk, M.E. (1993) *Brugia pahangi*: a surface-associated glycoprotein (gp15/400) is composed of multiple tandemly repeated units and processed from a 400-kDa precursor. *Experimental Parasitology* 76, 156–164.

Wood, W.B. (1988) Introduction to *C. elegans* biology. In: Wood, W.B. (ed.) *The Nematode Caenorhabditis elegans.* Cold Spring Harbor Laboratory, Cold Spring Harbor, New York, pp. 1–16.

Wulfing, C., Lombardero, J. and Pluckthun, A. (1994) An *Escherichia coli* protein consisting of a domain homologous to FK506-binding proteins (FKBP) and a new metal binding motif. *Journal of Biological Chemistry* 269, 2895–2901.

Xia, Y., Spence, H.J., Moore, J., Heaney, N., McDermott, L., Cooper, A., Watson, D.G., Mei, B., Komuniecki, R. and Kennedy, M.W. (1999) The ABA-1 allergen of *Ascaris lumbricoides*: sequence polymorphism, stage and tissue-specific expression, lipid binding function, and protein biophysical properties. *Parasitology* 120, 211–224.

Yau, W.M., Wimley, W.C., Gawrisch, K. and White, S.H. (1998) The preference of tryptophan for membrane interfaces. *Biochemistry* 37, 14713–14718.

T Helper Cell Cytokine Responses During Intestinal Nematode Infection: Induction, Regulation and Effector Function

David Artis and Richard K. Grencis

Immunology Group, School of Biological Sciences, Stopford Building, University of Manchester, Manchester M13 9PT, UK

Introduction

The global prevalence of gastrointestinal helminthiases highlights the major public health significance and economic impact that this group of pathogens represents in both human and animal populations. Helminth infections are amongst the most prevalent of all chronic human diseases, with at least one-quarter of the world's population harbouring intestinal nematode infections. Recent estimates suggest that more than 1000 million people worldwide are infected with each of the four major pathogen species of humans: *Trichuris trichiura*, *Ascaris lumbricoides* and the hookworms *Necator americanus* and *Ancylostoma duodenalis* (World Health Organisation, 1996; Chan, 1997; Albonico *et al.*, 1999). These infections are characteristically chronic and exhibit an overdispersed distribution within infected populations (reviewed in Bundy and Cooper, 1989; Behnke *et al.*, 1992). The most severe clinical symptoms of infection, including anaemia, protein-losing enteropathy, chronic dysentery and rectal prolapse, occur in the minority of heavily infected individuals (reviewed in Symons, 1969; Bundy and Cooper, 1989; Grencis and Cooper, 1996). However, significant detrimental clinical outcomes also occur in moderately infected individuals, including impaired nutritional status, growth retardation and lower educational achievement (Cooper *et al.*, 1992; Nokes and Bundy, 1994). Within the livestock industry, *Haemonchus* spp., *Trichostrongylus* spp. and *Ostertagia ostertagi* are the main disease-causing intestinal nematode infections, contributing significantly to reduced productivity and representing a substantial economic concern, with anthelmintic treatment costing approximately £1000 million annually (Newton and Munn, 1999).

Strong temporal stability of nematode populations exists in endemic areas due to the high frequency and long-lived nature of infective larval stages. As a result, reinfection rates are high and chemotherapeutic intervention, although successful, provides only a short-term benefit to infected people and livestock. In addition, the development of anthelmintic resistance and the growing costs of developing new anthelmintic drugs (and concern over drug residues entering the food chain through livestock: Emery and Wagland, 1991) require the development of more long-lived immunological intervention strategies. The apparent success of antinematode vaccines in the livestock industry (Emery, 1996) and growing evidence from human immunoepidemiology studies suggesting that immunity to helminth infection can operate (Maizels *et al.*, 1993) suggest that a vaccine approach is a necessary and obtainable goal. However, more detailed knowledge of the immunoregulatory processes operating during intestinal nematode infection is required to allow the development of effective and commercially viable vaccine strategies that elicit long-term reductions in the intensity and prevalence of infection.

The development of immunologically well-defined laboratory models of intestinal nematode infection has allowed significant advances to be made in understanding the immunological basis of susceptibility and resistance to infection under controlled laboratory conditions. Therefore, considerations, such as infection and co-infection history, nutritional status and behavioural and environmental differences within communities, that have made human studies difficult to interpret can be overcome. The utilization of rodent models also has the advantage of a wealth of immunological reagents with which to manipulate and analyse responses during infection. Four infection models in particular have allowed key developments in the understanding of immunity to nematode infection: *Trichuris muris*, *Heligomosoides polygyrus*, *Nippostrongylus brasiliensis* and *Trichinella spiralis*. Important differences in life-cycle strategies exist between these model systems. *T. muris* inhabits the caecum and proximal colon, while the latter three reside in the small intestine. *T. muris* and *T. spiralis* share an unusual microenvironmental niche for metazoan pathogens, being partially or entirely embedded within host enterocytes, depending on the life-cycle stage. Infective stages of *T. muris*, *T. spiralis* and *H. polygyrus* are administered orally, while infective *N. brasiliensis* larvae penetrate the skin, migrate via the lymphatics to the lungs, and enter the alimentary tract via the tracheal–oesophageal route. Primary *T. spiralis* and *N. brasiliensis* infections are expelled from most hosts within weeks, while expulsion of *T. muris* is host–strain dependent. In contrast, most experiments using the *H. polygyrus* model involve drug clearance of normally persistent primary infection and analysis of the host response to subsequent challenge. These life-cycle characteristics particular to different nematode species will influence the generation of anti-nematode immune responses and require different effector mechanisms to mediate worm expulsion.

Experiments using these four nematode infections have allowed analysis of the cellular and molecular interactions involved in the generation and regulation of immune responses during infection, and have identified common immunological events that are triggered during all these nematode infections. In addition, these studies have provided insights into the nature of protective effector responses operating in the gut microenvironment, which confer worm expulsion and host resistance. This review will provide a synopsis of our current understanding of the induction and regulation of immunity to intestinal nematode infection drawn predominantly from studies using these four model systems and, where relevant, highlight areas of the immune response to nematodes that require further research.

T Cell and Cytokine Regulation of Host Protective Immunity

The critical importance of T cells in the generation of host protective immunity to intestinal nematode parasites was initially shown in nude mice and rats (which lack mature thymus-derived T cells) (Prowse *et al.*, 1978; Vos *et al.*, 1983; Ito, 1991; McKay *et al.*, 1995). Subsequently, CD4+ T cells were shown to be the critical subset in resistance by adoptive transfer of fractionated T cells during *N. brasiliensis* and *T. spiralis* infections (Grencis *et al.*, 1985; Katona *et al.*, 1988; Ramaswany *et al.*, 1994) and anti-CD4 monoclonal antibody treatment during *T. muris* and *H. polygyrus* infections (Urban *et al.*, 1991a; Koyama *et al.*, 1995). The seminal work of Mosmann *et al.* (1986), in which two distinct CD4+ T cell subsets were defined by differential secretion of cytokines, has revolutionized understanding of the regulatory mechanisms underlying resistance and susceptibility to nematode infection. T helper type 1 (Th1) cells produce the type 1 cytokines IFN-γ, lymphotoxin and interleukin (IL)-2, stimulating immunoglobulin (Ig)G2a production and cell-mediated effector responses. The type 2 cytokines IL-4, IL-5, IL-6, IL-9 and IL-13 are secreted by T helper type 2 (Th2) cells in the relative absence of type 1 cytokines, and promote mastocytosis, eosinophilia and the production of IgE and IgG1 (reviewed by Abbas *et al.*, 1996; Mosmann and Sad, 1996; O'Garra, 1998). These T cel lsubsets were first discovered following *in vitro* differentiation of murine T cell clones (Mosmann *et al.*, 1986). Subsequent studies identified the existence of human and murine CD4+ and CD8+ T cells with the capacity to produce type 1 and type 2 cytokines *in vivo* (Romagnani, 1991; Seder *et al.*, 1992). Numerous *in vitro* and *in vivo* studies have demonstrated that, following antigenic stimulation, naive T cells pass through an intermediate stage in their development (Th0 cells) in which a mixed profile of cytokines is produced before becoming polarized into committed and cross-regulatory cell populations promoting type 1 or type 2 responses (reviewed in Fitch *et al.*, 1992; Abbas *et al.*, 1996; O'Garra, 1998). The classification of immune responses as type 1 or type 2 based on the dominant cytokines being produced is likely to be a simplified model, but it has provided an excellent framework with which to characterize the cellular and molecular regulation of immunity to invading pathogens.

Resistance to infection requires a type 2 response

While resistance to a range of intracellular pathogens, including *Leishmania major* (Scott, 1991), *Mycobacterium leprae* (Yamamura *et al.*, 1991), *Cryptosporidium parvum* (Ungar *et al.*, 1991) and *Toxoplasma gondii* (Gazinelli *et al.*, 1991), requires the induction of a type 1 response to elicit host protection, protective immune responses and expulsion of nematodes in murine hosts are universally coincident with the generation of a type 2 response (Grencis *et al.*, 1991; Else *et al.*, 1992; Wahid *et al.*, 1994; Lawrence *et al.*, 1996; Finkelman *et al.*, 1997). Although there are currently no published studies on T cell and cytokine responses during intestinal nematode infection of humans, a number of studies support observations made in murine models, showing the presence of type 2 cytokine-controlled responses during infection, including elevated IgE and peripheral eosinophilia (Pritchard *et al.*, 1990, 1995; Needham *et al.*, 1994; Quinnell *et al.*, 1995). Similar responses indicative of dominant type 2 cytokine production are observed during intestinal nematode infection of livestock (reviewed in Miller, 1984; Buddle *et al.*, 1992). Indeed, prototypic type 2 responses composed of peripheral and tissue eosinophilia, elevated IgE levels and intestinal mastocytosis are common responses in all intestinal helminth infections (Lobos, 1997).

The *in vivo* manipulation of specific type 2 cytokines using anti-cytokine monoclonal antibodies, or mouse strains with targeted deletions in cytokine and/or cytokine receptor genes, has proved a fruitful approach in identifying the importance of individual cytokines and the responses that they control in contributing to host resistance. These studies have identified important roles for IL-4, IL-9 and IL-13 in host protection against nematode infection, though the relative importance of each cytokine appears to be nematode-species dependent.

A cardinal role for IL-4 in host protection against intestinal nematode infection was first shown in the *H. polygyrus* challenge model. Worm expulsion was delayed following treatment with anti-IL-4 or anti-IL-4 receptor monoclonal antibodies, while control treated animals successfully cleared infection (Urban *et al.*, 1991b). Blockade of the IL-4 receptor effectively prevents the *in vivo* function of IL-4 and IL-13, as these two cytokines share the IL-4 receptor α-chain for signalling functions (Lin *et al.*, 1995). In addition, *in vivo* administration of IL-4 complex (IL-4C), which enhances the half-life of the cytokine *in vivo* (Finkelman *et al.*, 1993), facilitated expulsion of primary infection in normally permissive hosts. Similar experiments in the *T. muris* model demonstrated that blockade of the IL-4 receptor in resistant mouse strains resulted in the production of a predominantly type 1 response, with a chronic infection developing (Else *et al.*, 1994). Conversely, the administration of IL-4C to susceptible mouse strains resulted in the expansion of a type 2 response and clearance of infection (Else *et al.*, 1994). It is interesting that, in the latter case, delivery of IL-4C late in infection also facilitated clearance of an established worm infection.

These studies were confirmed in mice with a targeted deletion in their IL-4 gene (IL-4 knockout, KO). These animals did not generate protective type 2 responses during both *T. muris* and *H. polygyrus* infection and hence failed to clear infection (Finkelman *et al.*, 1997; Bancroft *et al.*, 1998). In the case of *T. muris*, the expulsion phenotype of IL-4 KO mice has been found to be dependent on the background strain of mouse used, with C57BL/6 IL-4 KO mice developing chronic unresolving infections, while BALB/c IL-4 KO mice cleared infection (Bancroft *et al.*, 1998; Artis *et al.*, 1999a; A.J. Bancroft, unpublished). Expulsion of primary *T. spiralis* infection has also been reported to be delayed in IL-4 KO mice (Lawrence *et al.*, 1998). A role for IL-4 (and perhaps IL-13) in expulsion of *T. spiralis* was confirmed in studies in which immunocompetent animals were treated with anti-IL-4 receptor monoclonal antibody, resulting in prolonged adult infections and higher muscle larvae burdens (Finkelman *et al.*, 1997).

In the case of *N. brasiliensis* infection, a role for IL-4 and IL-13 has been identified. While treatment with IL-4C cured chronic infection in anti-CD4 monoclonal antibody treated or SCID mice (which lack B and T cells) (Urban *et al.*, 1995), anti-IL-4 treated mice and IL-4 KO mice cleared infection with similar kinetics to wild-type controls, which suggests that IL-4 was not important in host resistance (Madden *et al.*, 1991; Lawrence *et al.*, 1996). Critically, mice deficient in IL-4 receptor signalling failed to clear infection, as did anti-IL-4 receptor treated IL-4 KO mice (Barner *et al.*, 1998; Urban *et al.*, 1998). In addition, mice deficient in signal transducer and activation of transcription (Stat)-6 molecules also failed to expel *N. brasiliensis* (Urban *et al.*, 1998). As stated earlier, IL-4 and IL-13 share the IL-4 receptor α-chain. Signalling through this receptor occurs via Stat6 activation and is critical in the development of type 2 cytokine responses (Kaplan *et al.*, 1996; Takeda *et al.*, 1996). IL-4 and IL-13 are the only two cytokines currently known to activate this signalling pathway. Therefore, expulsion of *N. brasiliensis* was IL-4 receptor and Stat6 dependent (but independent of IL-4), suggesting an important role for IL-13 in resistance. Definitive evidence of such a function came from McKenzie *et al.* (1998a,b), demonstrating delayed expulsion of *N. brasiliensis* in IL-13 KO mice despite these mice mounting robust type 2 responses. Supporting this, blockade of IL-13 function during infection in immunocompetent and IL-4 KO mice by *in vivo* administration of a soluble IL-13 receptor α2-human IgG-Fc fusion protein (A25) completely blocked worm expulsion and confirmed the importance of IL-13 in host resistance (Urban *et al.*, 1998). The IL-4 KO mice treated with A25 had higher worm burdens than immunocompetent animals given the same treatment, however. This observation, coupled with the fact that mice doubly deficient in IL-4 and IL-13 displayed more severely impaired host resistance than mice deficient in IL-13 alone (McKenzie *et al.*, 1999), suggested that IL-4 can play at least a partial role in host protection.

A critical and IL-4-independent role for IL-13 has also been identified in host resistance to *T. muris*. Despite generating strong and equivalent type 2 responses to wild-type mice, mice deficient in IL-13 production failed to clear infection (Bancroft *et al.*, 1998). Expulsion in BALB/c IL-4 KO mice was completely blocked following treatment with A25, confirming an important role for IL-13 in resistance to *T. muris* (A.J. Bancroft *et al.*, unpublished observations) (Table 17.1). In addition, IL-4 receptor KO mice (lacking IL-4 and IL-13 functions) are completely susceptible to *T. muris* (A.J. Bancroft, unpublished observations).

We have recently identified a novel role for TNF-α in the regulation of IL-13-mediated expulsion of *T. muris*. *In vivo* treatment of normally resistant immunocompetent strains with anti-TNF-α monoclonal antibody prevented worm expulsion without significantly altering the magnitude of the type 2 cytokine response compared with control treated mice, which cleared infection (Artis *et al.*, 1999a). In addition, administration of anti-TNF-α to *T. muris* infected BALB/c IL-4 KO mice (in which expulsion is IL-13 mediated) prevented worm expulsion (Fig. 17.1). These data show that TNF-α, a cytokine normally associated with type 1 mediated phenomena including inflammation, autoimmunity and control of intracellular pathogens (reviewed in Vassilli, 1992), plays a critical role in IL-13 mediated anti-*T. muris* effector responses (see below).

Table 17.1.   Mean worm burdens in IL-13 knockout (KO), IL-4 KO and wild-type mice at day 35 post-*Trichuris muris* infection.

	Mean worm burden (± SEM)
IL-13 KO	186.29 ± 24.30
Wild-type	0
BALB/c IL-4 KO + A25	126.5 ± 19.71
BALB/c IL-4 KO + control Ig	8.0 ± 0.58

IL-13 KO mice failed to clear infection despite mounting equivalent type 2 cytokine responses to wild-type mice. Further evidence of a role for IL-13 in expulsion is shown in studies utilizing BALB/c IL-4 KO mice. While control mice cleared infection, treatment with A25 (a soluble IL-13 receptor alpha 2-human IgG-Fc fusion protein) prevented worm expulsion. (Data adapted from Bancroft *et al.*, 1998; A.J. Bancroft, unpublished.)

The role of IL-9, another type 2 cytokine, has received rather less attention than IL-4 and IL-13, despite being produced during infection with the nematode species discussed in this review. IL-9 is produced predominantly by T cells – although a non-T cell source does exist (Svetic *et al.*, 1993) – and acts as a mast cell and T cell growth factor in addition to potentiating the production of IL-4, IgE and IgG (Uyttenhove *et al.*, 1988; Eklund *et al.*, 1993; Petit-Frere *et al.*, 1993; Louahed *et al.*, 1995). Studies in which IL-9 levels have been enhanced *in vivo* have identified an important role for IL-9 in resistance to *T. spiralis*. Therefore, IL-9 transgenic mice (which constitutively over-express this cytokine) displayed enhanced parasite-specific IgG1 and profound intestinal mastocytosis, resulting in accelerated worm expulsion (Faulkner *et al.*, 1997). *In vivo* elevation of IL-9 levels during *T. muris* infection also resulted in enhanced type 2 responses and accelerated worm expulsion (Faulkner *et al.*, 1998). Indeed, IL-9 mRNA was detected prior to IL-4 production in the draining mesenteric lymph node of resistant strains, suggesting that the protective effect of IL-9 may be operating through the promotion and expansion of Th2 cells and enhancement of their protective effector function. Recent studies have also found that *in vivo* treatment with anti-IL-9 monoclonal antibody significantly delayed expulsion of *T. muris* in normally resistant immuno-competent strains and in BALB/c IL-4 KO mice. This suggests that IL-9, either directly or indirectly, is also critical in host resistance (Humphreys and R.K. Grencis, unpublished observations). There have been no studies

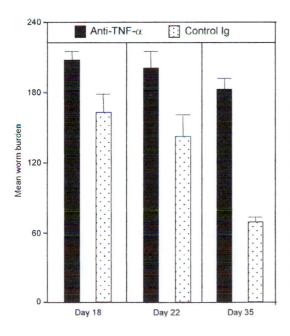

Fig. 17.1. Treatment of normally resistant BALB/c IL-4 KO mice with anti-TNF-α monoclonal antibody prevents IL-13-mediated expulsion of *T. muris*. Mice were infected on day 0 with 200 *T. muris* eggs and worm burdens (± SEM) from four mice per group determined on days 18, 22 and 35 post-infection. (Data adapted from Artis *et al.*, 1999a.)

as yet on the role of IL-9 in protection against *H. polygyrus* and *N. brasiliensis*, despite elevated levels of this cytokine being detected following infection (Svetic *et al.*, 1993; Katona *et al.*, 1995).

IL-3 and IL-5 are also produced during the four intestinal nematode infections of rodents discussed in this chapter (Grencis *et al.*, 1991; Madden *et al.*, 1991; Else *et al.*, 1992; Wahid *et al.*, 1994), though neither cytokine appears to be essential in host protection in any model system. Treatment with anti-IL-3 monoclonal antibody or infection in IL-3 KO mice resulted in no impairment of host protection to *T. muris*, *H. polygyrus*, *N. brasiliensis* or *T. spiralis* (Madden *et al.*, 1991; Betts and Else, 1999; R.K. Grencis and Tybulewicz, unpublished observations). However, IL-3 is important in the development of intestinal mastocytosis and host protection against the related pathogen *Strongyloides stercoralis* (Lantz *et al.*, 1998). Conflicting evidence exists on the role of IL-5 in protection against intestinal nematode infection. Treatment with anti-IL-5 monoclonal antibody during *T. muris* (Betts and Else, 1999), *H. polygyrus* (Urban *et al.*, 1991), *N. brasiliensis* (Coffman *et al.*, 1989) or *T. spiralis* infections (Herndon and Kayes, 1992) had no effect on the kinetics of worm expulsion, suggesting that this cytokine did not play a major role in protection. Studies in transgenic mice over-expressing IL-5 suggest that this cytokine might play a protective role against invading *N. brasiliensis* larvae (Dent *et al.*, 1999) (see below).

Type 1 cytokine responses promote susceptibility

Mouse strains that are naturally susceptible to *T. muris* respond to infection with the production of antigen-specific IFN-γ, IL-12 and IgG2a (characteristic of a strong type 1 response) and develop unresolving infections (Else *et al.*, 1992; D. Artis and R.K. Grencis, unpublished observations). The functional involvement of type 1 cytokines in promoting susceptibility to infection was shown in studies where the *in vivo* depletion of IFN-γ (Else *et al.*, 1997) or IL-12 (A.J. Bancroft, unpublished) in normally susceptible strains allowed the expansion of type 2 responses and effective clearance of infection. These results are supported by the kinetics of worm expulsion in IFN-γ KO and IFN-γ receptor KO mice, which display very early production of type 2 cytokines following infection and more rapid worm expulsion than wild-type control animals (D. Artis, Boeuf, and R.K. Grencis, unpublished observations). Furthermore, *in vivo* administration of recombinant IL-12 to normally resistant mice resulted in the generation of a predominantly type 1 response (through its effects on IFN-γ production) and the development of a chronic infection (Bancroft *et al.*, 1997).

Similarly, the administration of IFN-γ or IL-12 at the time of inoculation with *N. brasiliensis* significantly prolonged the course of infection, with the effects of IL-12 being IFN-γ dependent (Urban *et al.*, 1993; Finkelman *et al.*, 1994). It is interesting that the effects of administration of IL-12 during *T. muris* and *N. brasiliensis* infection were different, with IL-12 treatment early during *T. muris* infection of resistant strains inducing the development of a polarized type 1 response and resulting in chronic infection (Bancroft *et al.*, 1997). However, administration of IL-12 during *N. brasiliensis* infection was only effective in blocking expulsion for the duration of treatment, with protective type 2 responses developing and worm expulsion occurring after cessation of cytokine treatment (Finkelman *et al.*, 1994). Therefore, it would appear that IL-12 treatment induces a loss in IL-4 responsiveness during *T. muris* but not *N. brasiliensis* infection, although differential expression of cytokine receptor expression on lymphocyte populations was not examined in these experiments.

In 1995 a new cytokine, interferon-γ-inducing factor (IGIF) or IL-18, was identified with an important role in promoting type 1 responses (Okamura *et al.*, 1995). IL-18 has been found to share a number of biological functions with IL-12, such as inducing IFN-γ production by lymphocytes and Th1 clones and enhancing natural killer cell activity, though most of the effects of IL-18 are dependent on synergy with IL-12 (Okamura *et al.*, 1995; Kohno *et al.*, 1997; Yoshimoto *et al.*, 1997). Recently, IL-12-independent induction of IFN-γ responses have been identified (Magram *et al.*, 1996; Schijns *et al.*, 1998; Takeda *et al.*, 1998), but the role of IL-18 in inducing IFN-γ production during nematode infection, and so in regulating host susceptibility to infection, has yet to be investigated.

Nematode Cytokines?

As discussed above, it appears that administration of the Th1-inducing cytokine IL-12 to resistant mouse strains results in the apparent loss of IL-4 responsiveness and the development of chronic infection, while IL-12 is only effective at preventing expulsion of *N. brasiliensis* for the duration of cytokine administration, suggesting IL-4 responsiveness is not lost in this case. One possible explanation for the differences in effect of IL-12 treatment on infection outcome in these two models relates to the identification of a potentially important immunomodulatory molecule secreted by *T. muris*. This 43 kDa *T. muris*-derived protein has been found to share cross-reactive epitopes with host IFN-γ, binding IFN-γ receptors on host lymphocytes and mediating cellular changes similar to those induced by IFN-γ itself (Grencis and Entwistle, 1997). Immuno-gold staining using anti-IFN-γ antibodies has localized this epitope-sharing molecule to the cuticle, stichosome and the bacillary band (Fig. 17.2). The latter is interesting as this structure is a major interface between host and parasite and appears from ultrastructural studies to have a secretory function (Bughdadi and R.K. Grencis, unpublished). This parasite-derived IFN-γ homologue may be critical in potentiating type 1 responses and so promoting the development of chronic infection. Following prolongation of infection by IL-12

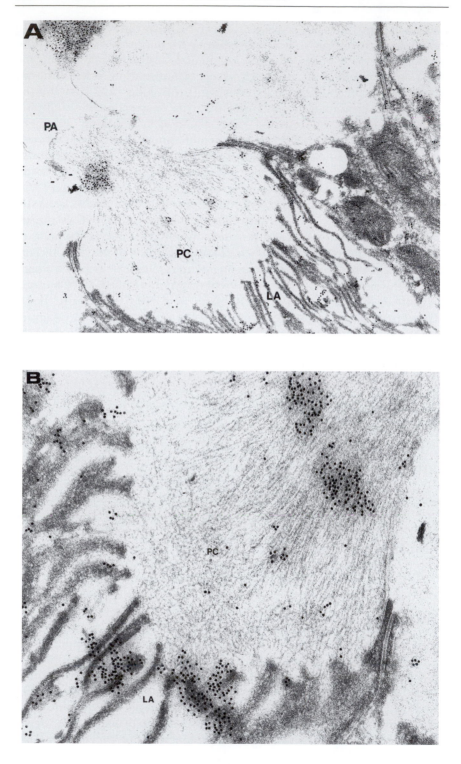

treatment, a protective type 2 response may be prevented from developing due to the secretion of this immunomodulatory protein by the parasite.

Regulation of Th Responses During Nematode Infection

It is clear that the type 2 cytokines IL-4, IL-9 and IL-13 play an obligatory role in host resistance to nematode infection whereas type 1 responses promote host susceptibility. Therefore, given that susceptibility to nematode infection is not due to a lack of responsiveness *per se*, but rather the development of an inappropriate response, it is important to understand the factors that influence the induction and expansion of Th subset responses and so control infection outcome. Studies in nematode models and other systems have addressed these questions and identified the importance of host genetic factors, the nature of the antigen and the antigen presenting cell, co-stimulatory molecules on these cells, and the cytokine and chemokine environment immediately following induction of the response.

Genetic control of Th responses

Studies in the *T. muris* and *T. spiralis* models have investigated genetically determined differences in immune responsiveness that exist during infection. Effects of genes within the mouse major histocompatibility complex (MHC) (H-2) on host resistance were examined using B10 mice. Within different B10 congenic strains a gradation of resistance to *T. muris* infection was observed, from the relatively resistant B10 (H-2^b), B10.G (H-2^q) and B10.D2/n (H-2^d) strains through to the B10.BR (H-2^k) strain, which was completely susceptible to infection (Else and Wakelin, 1988). In both BALB and B10 congenic strains, a correlation between *b* and *q* alleles within the I-A locus was associated with host resistance while *d* and *k* alleles were associated with susceptibility (Else and Wakelin, 1988). Similarly, the H-2^k haplotype had a greater association with relative susceptibility than the H-2^q haplotype during *T. spiralis* infection (Wakelin and Donochie, 1983). It is clear, therefore, that the H-2 complex plays a significant role in controlling resistance and susceptibility to intestinal nematode infection. However, studies using H-2 recombinant strains have also shown the importance of background genes in determining host resistance. Expression of the *q* allele has been shown to have different effects on resistance to infection

Fig. 17.2. (Opposite) Immuno-gold localization of a *T. muris*-derived IFN-γ homologue to the bacillary band and cuticular pore. (A) Transmission electron micrograph of bacillary band showing the pore chamber (PC), pore aperture (PA) and lamellar apparatus (LA) (×22,000). (B) High-power localization of antibody staining (black dots) to the lamellar apparatus (LA) and pore chamber (PC) (×36,000). (Courtesy of F. Bughdadi.)

when expressed on a variety of genetic backgrounds, and mouse strains sharing the same H-2 alleles differ significantly in their resistance to *T. muris* infection (Else *et al.*, 1990, 1992).

Using a serial backcross approach, the loci linked with resistance to *L. major* infection have been mapped to six distinct chromosomal regions (Beebe *et al.*, 1997; Guler *et al.*, 1997). The loci important in determining host susceptibility and resistance to intestinal nematode infections have not yet been identified. It is clear, however, that both H-2 and non-H-2 genes influence the nature of the immune response during nematode infection, with the former likely to be having a regulatory effect on antigen presentation to T cells and the latter exerting effects at other levels, including the antigen-presenting cell, cells of the myeloid lineage and the induction of inflammatory mediators.

Professional antigen-presenting cells

Polarization of CD4+ T cell subsets is also influenced by the cell type presenting the antigen. In *in vitro* studies using ovalbumin-specific T cell clones, adherent spleen cells (presumed macrophages) were shown to promote predominantly type 1 responses whereas presentation by B cells enhanced type 2 responses (Gajewski *et al.*, 1991) (but whether different antigen-presenting cell populations induce different responses under physiological conditions remains a contentious issue). Although there have been no studies investigating the effects of antigen presentation by these cell populations during nematode infection, an important role for dendritic cells has been identified in the *T. muris* model. Current understanding of antigen presentation in the gut suggests that, in Peyer's patches of the small intestine and lymphoid follicles of the large intestine, antigen is taken up by resident dendritic cells and other antigen-presenting cells following antigen sampling by specialized M cells that overlie lymphoid tissue in the gut (reviewed in Neutra *et al.*, 1996). Antigen presentation to T cells occurs in these induction sites and in the draining mesenteric lymph nodes following migration of antigen-presenting cells. Thereafter, antigen-specific lymphocytes recirculate to the site of original antigen exposure (Springer, 1994; Mowat and Viney, 1997). Using bone marrow-derived dendritic cells from susceptible and resistant strains, studies have shown that following exposure to antigen *in vitro*, both these populations can prime for enhanced host resistance to *T. muris* after transfer into normally resistant animals (Meredith and R.K. Grencis, unpublished). This suggests that the cytokine environment in which antigen presentation takes place (see below) is more important than the inherent properties of dendritic cell populations from different mouse strains. The importance of dendritic cells in generating immunity to *T. muris* was also shown in studies in which dendritic cell populations were expanded *in vivo* using Flt3 ligand.

This haemopoietic growth factor is known to expand both myeloid- and lymphoid-derived dendritic cells (Pulendran *et al.*, 1997). An expansion of the dendritic cell population was observed in the intestinal mucosa and draining mesenteric lymph following Flt3 ligand treatment, and resulted in enhanced type 2 cytokine responses in the draining lymph node and accelerated worm expulsion following primary *T. muris* infection (Meredith and R.K. Grencis, unpublished observations).

Recent studies have identified two distinct dendritic cell subsets derived from human peripheral blood following *in vitro* culture under different conditions. 'Myeloid' dendritic cells (termed DC1) develop from peripheral blood monocytes in the presence of IL-4 and GM-CSF, whereas 'lymphoid' dendritic cells (DC2) develop from $CD4^+CD3^-CD11c^-$ plasmacytoid cells after culture in the presence of IL-3. These two populations were found to differ in a number of ways, in particular the cytokines they produce during antigen presentation to T cells. The DC1 population produced large amounts of IL-12, inducing the differentiation of Th1 cells, while DC2 promoted the development of Th2 cells producing little IL-12 (Rissoan *et al.*, 1999). Interestingly, DC2 induction of Th2 cells was independent of IL-4. Precursors of the DC2 lineage have also been shown to act as specialized leucocytes ('natural IFN-producing cells') (Siegal *et al.*, 1999). Therefore, cells of this lineage perform different immune functions depending on their maturation state, with immature precursors producing large amounts of IFN-α (important in innate resistance to bacteria and viruses), while mature DC2 populations stimulate antigen-specific type 2 responses by the adaptive immune system. DC1 and DC2 populations have not yet been identified in mice, but functional heterogeneity does exist in murine dendritic cell populations (Suss and Shortman, 1996). It will be of interest to determine whether such DC1 and DC2 populations exist in mouse strains, whether Flt3 treatment selectively enhances the DC2 population, and whether these populations are responsible for the differential induction of type 1 and type 2 responses observed during some nematode infections.

Non-professional antigen-presenting cells

The role of non-professional antigen-presenting cell populations (i.e. cells that are not macrophages, B cells or dendritic cells) in initiation of responses during intestinal nematode infection has also not been studied. This is surprising, given the growing evidence from *in vitro* studies that enterocytes can process and present antigen to T cells (Bland and Warren, 1986; Mayer and Schlien, 1987; Bland and Whiting, 1989; Kaiserlian *et al.*, 1989). Differentiated rodent and human enterocytes of the small intestine are known to 'constitutively' express MHC class II, which is upregulated in relatively undifferentiated enterocytes in the small intestine and enterocytes in the colon during inflammation (Wilson *et al.*, 1990; Bland and Whiting, 1992). Certainly, our unpublished observations have demonstrated a dramatic upregulation of class II expression on enterocytes of the caecum during *T. muris* infection. Due to their apparent lack of co-stimulatory molecules, however, class II expression on intestinal epithelium is thought to perform a predominantly peptide receptor and shuttling function (discussed in Bland, 1999) rather than antigen presentation. The functional significance of elevated class II expression on enterocytes during intestinal nematode infection remains to be elucidated.

In common with all nucleated cells, enterocytes also constitutively express class I molecules and, in addition, express the class I-like molecule CD1 (Bleicher *et al.*, 1990; Blumberg *et al.*, 1991). CD1 proteins are a heterologous group of non-polymorphic molecules encoded outside the MHC and capable of binding peptide and non-peptide antigens (reviewed in Bland, 1999). The functional importance of CD1 has been shown in studies where enterocyte-induced T cell proliferation and cytolysis can be blocked following treatment with anti-CD1 monoclonal antibodies (Panja *et al.*, 1993; Sydora *et al.*, 1993). T cells bearing the γδ T cell receptor are found in abundance in the gut (Klein, 1996) and are known to be activated following antigen presentation in association with CD1 (Kaufmann, 1996). Therefore, CD1-restricted antigen presentation by enterocytes during nematode infection may be important in initiation of responses against nematode infection.

Co-stimulatory molecules

Clonal expansion of antigen-specific T cells following T cell receptor ligation to the peptide : MHC complex on antigen-presenting cells requires a second activation signal through the interaction of co-stimulatory molecules on the antigen-presenting cell with their receptors on T cells. Interactions of the co-stimulatory molecules B7-1 and B7-2 with CD28 and CTLA-4 on naive and activated T cells are well characterized in terms of influencing the polarization of Th subset development. Studies carried out in a number of murine systems demonstrated that stimulation of T cells through interaction with B7-1 preferentially induced type 1 responses while stimulation through B7-2 promoted type 2 responses (Kuchroo *et al.*, 1995; Brown *et al.*, 1997; Subramanian *et al.*, 1997; Tsuyuki *et al.*, 1997). Studies in the *T. muris* model have shown that blockade of B7-2 prevented the induction of protective type 2 responses during nematode infection in normally resistant mice, resulting in the generation of a predominantly type 1 response and impaired clearance of infection (Fig. 17.3). Therefore B7-2 interaction with CD28 is critical in host protection in this model. Interestingly, CD28 expression on T cells also appears to be down-regulated in normally susceptible mouse strains during *T. muris* infection (Soltys *et al.*, 1999).

The importance of co-stimulation through B7 in the generation of type 2 responses during *H. polygyrus* infection has also been shown in studies where administration of CTLA-4Ig (a chimeric fusion protein that blocks interaction between CD28 and B7 molecules) prevented the generation of type 2 responses (Lu *et al.*, 1994). Blockade of B7-2 alone was not sufficient to block the generation of this response during *H. polygyrus* infection, demonstrating that signalling through both B7-1 and B7-2 could stimulate type 2 responses in this model (Greenwald *et al.*, 1997). However, B7-2 KO mice had lower type 2 cytokine-mediated effector responses at 2 weeks post-primary *H. polygyrus* infection compared with wild-type animals. As a result, mice deficient in B7-2 expression had significantly higher adult worm egg production (although no difference in worm burden) at day 24 post-infection (Greenwald *et al.*, 1999).

The importance of B7 interactions in the generation of type 2 responses during *N. brasiliensis* infection has been shown in studies where CTLA-4Ig treatment resulted in reduced generation of type 2 cytokines, although it did not alter worm expulsion (Harris *et al.*, 1999). In addition, blockade of CTLA-4 interactions using monoclonal antibodies resulted in enhanced type 2 responses and accelerated expulsion of *N. brasiliensis*, suggesting that CTLA-4 was competing with B7 in interacting with CD28 (McCoy *et al.*, 1997). Therefore the balance of interaction between B7 molecules and their receptors, CD28 and CTLA-4, are critical in determining the development of protective type 2 responses.

The development of effector T cells following activation by antigen-presenting cells also requires the interaction of another receptor-ligand pair. CD40 molecules on antigen-presenting cells are required to bind

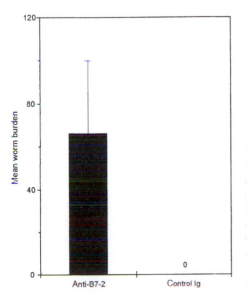

Fig. 17.3. *In vivo* blockade of B7-2 using monoclonal antibodies impairs expulsion of *T. muris* from normally resistant BALB/k mice. Mice were infected on day 0 with 200 *T. muris* eggs and worm burdens (± SEM) from four to five mice per group determined on day 35 post-infection. (K.J. Else, A.J. Bancroft and R.K. Grencis, unpublished.)

CD40L on the surface of T cells for effective T cell activation (Grewel *et al.*, 1995). Following CD40 ligation on the surface of B cells and dendritic cells, a dramatic upregulation of IL-12 production by the antigen-presenting cell was observed (Cella *et al.*, 1996). Indeed, blockade of this interaction resulted in a marked reduction in type 1 responses (Cella *et al.*, 1996). Preliminary experiments in which CD40–CD40L interactions were blocked during *T. muris* infection of normally susceptible mice resulted in lower IgG isotype responses and a reduction in the number of adult worms compared with control treated mice (Fig. 17.4), suggesting that the magnitude of the type 1 response (promoting susceptibility) was reduced (H.F. Faulkner and R.K. Grencis, unpublished). Subsequently, a number of *in vitro* and *in vivo* studies have demonstrated that signalling through CD40L is important in the induction and maximization of IL-4 production by T cells, and so in influencing the expansion of type 2 cytokine responses (van Essen *et al.*, 1995; Blotta *et al.*, 1996; Poudrier *et al.*, 1998). However, no studies have investigated the role of this co-stimulatory pathway in the development of resistance to intestinal nematode infection.

Nature of antigens

In addition to the type of cell and co-stimulatory molecules expressed on that cell, the nature of the antigen being presented is important in influencing the development of Th responses. Although relatively little is known about the nature of many dominant nematode antigens, studies have shown that the inherent allergenic nature of these proteins can influence host responses. Indeed, it has been suggested that allergenic

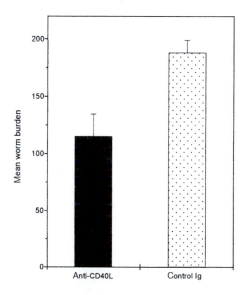

Fig. 17.4. Reduced establishment of chronic *T. muris* infection in normally susceptible AKR mice following treatment with anti-CD40L monoclonal antibody. Mice were infected on day 0 with 200 *T. muris* eggs and worm burdens (± SEM) from five mice per group determined on day 35 post-infection. (H.F. Faulkner and R.K. Grencis, unpublished.)

nematode antigens with enzymatic properties can cleave low affinity IgE receptors on B cells and so enhance type 2 responses (discussed in Pritchard *et al.*, 1997). In order to enhance their survival, however, secretion of antigens that preferentially induce type 1 responses would be more evolutionarily advantageous. As discussed above, preliminary data suggests the existence of a *T. muris*-secreted antigen with such properties. *In vivo* evidence of a role for *T. muris*-derived antigens in influencing the Th response comes from studies in which drug clearance of primary infection in normally susceptible strains was carried out on various days post-infection. The degree of immunity to subsequent challenge infection increased with the duration of primary infection until day 21 post-infection, after which stage-specific expression of an antigen released around this time resulted in enhanced host susceptibility to challenge infection (Else *et al.*, 1989).

Antigen dose and affinity

In vitro studies using T cell receptor transgenic CD4$^+$ T cells have also demonstrated the importance of the antigen dose in influencing the polarization of Th responses, with low and moderate doses of antigen preferentially inducing type 1 responses, whereas very high or very low doses of antigen induced type 2 responses (Constant *et al.*, 1995; Hosken *et al.*, 1995). The underlying mechanisms of these effects remain to be determined but the effects of the type of antigen-presenting cell and the frequency of T cell receptor engagement with different doses of antigen are thought to be important. Studies with the *L. major* model confirmed that low doses of infection given to susceptible strains that normally mount a type 2 response resulted in a dominant type 1 response and resistance to infection (Bretscher *et al.*, 1992). Results from the *T. muris* system confirmed that this phenomenon was also important in determining the development of Th responses during nematode infection, with low numbers of parasites inducing a type 1 response and a chronic infection in mouse strains that readily clear infection under normal conditions (Bancroft *et al.*, 1994). The MHC binding affinity of antigen can also alter the development of Th responses, in that peptides with low binding affinity preferentially induce type 2 responses (Kumar *et al.*, 1995).

Cytokine environment

The most clearly defined factors in determining the differentiation of Th responses are the cytokines present during T cell receptor engagement of the peptide : MHC complex. IL-12 and IFN-γ are important in the development of type 1 responses (Hsieh *et al.*, 1993; Seder *et al.*, 1993), while IL-4 is important in the induction of type 2 responses (Abishira-Amar *et al.*, 1992; Seder *et al.*, 1992). The importance of these cytokines in influencing the development of Th responses and so determining the outcome of nematode infection has been discussed above. The cellular source of IL-4 and IL-12 at the initiation of the response has been an area of intense interest. Macrophages and dendritic cells have been shown to produce large amounts of IL-12 upon activation, though the early source of IL-4 has been more difficult to identify. Some studies have suggested that NK1.1[+] CD4[+] $\alpha\beta$ T cells (which can recognize antigen in the absence of MHC) or non-B non-T FcγR1[+] cells may provide the early IL-4 that would drive the development of type 2 responses (Conrod *et al.*, 1990; Yoshimoto *et al.*, 1995). It has also been suggested that T cells themselves might be an early source of IL-4 (Schmitz *et al.*, 1994). However, there is a lack of definitive evidence demonstrating that any of these cell types are the early source of IL-4 that is important in inducing type 2 responses during nematode infection.

Generic pathogen recognition

The presence of evolutionarily conserved pattern recognition receptors (PRRs) on cells of the innate immune system (including dendritic cells, macrophages, mast cells and basophils) is a possible pathway whereby these populations could become activated and act as a source of early cytokines that are important in determining the nature of the Th response (Medzhitov and Janeway, 1997). PRRs such as scavenger receptors, integrins and lectin-like receptors recognize 'structural signatures' or pathogen-associated molecular patterns (PAMPs). Examples of PAMPs include LPS on the surface of Gram-negative bacteria, and double-stranded RNA characteristic of RNA viruses (discussed in Medzhitov and Janeway, 1999; Ulevitch and Tobias, 1999). It is possible that nematodes share common PAMPs that activate cells of the innate immune system, which then provide signals about the nature and origin of the antigen and so instruct the host's adaptive immune system on the type of effector response required. It will be important to examine nematode PAMPs that are involved in activation of the innate immune system during infection, and to identify the cell populations (and PRRs on those cells) that are involved in these processes.

Enterocytes may also play an important role in the balance of type 1 and type 2 responses following infection, as this population is also known to express PRRs (Medzhitov and Janeway, 1999). Studies *in vivo* and using human and rodent intestinal epithelial cell lines have also shown that enterocytes are a potent source of cytokines, including IL-1, IL-6, IL-7, IL-8, IL-15, IL-18, TNF-α and TGF-β (Barnard *et al.*, 1993; Eckmann *et al.*, 1993; Scheuerermaly *et al.*, 1994; Jung *et al.*, 1995; Stadnyck *et al.*, 1995; Watanabe *et al.*, 1995; Reinecker *et al.*, 1996; Takeuchi *et al.*, 1997). Therefore, these cells may be activated through PRRs and provide important early signals following infection that can influence the development of Th subsets. The expression of PRRs on enterocytes may also provide a mechanism of cell invasion by nematodes, such as *T. muris* and *T. spiralis*, that inhabit intracellular niches.

As mentioned above, γδ T cells are numerous in the gut and can produce cytokines in the absence of conventionally presented antigen (Kaufmann, 1996). These cells have been shown to produce IL-4 or IFN-γ (Ferrick *et al.*, 1995) and are potentially an important source of these cytokines in the intestinal microenvironment following nematode infection. Supporting this suggestion, IL-4 mRNA has been found in γδ T cells from the caecum of resistant mice immediately following *T. muris* infection (Lukaszewski and R.K. Grencis, unpublished observations).

Chemokines

Chemokines are a superfamily of proteins involved in leukocyte recruitment and activation (Baggiolini *et al.*, 1997). Recently, murine and human polarized Th1 and Th2 cells have been shown to respond to different chemokines, with Th1 cells exclusively expressing the chemokine receptors CXCR3 and CCR5, and Th2 cells expressing CCR3, CCR4 and CCR8 (Sallusto *et al.*, 1997; Siveke and Hamann, 1998; Zingoni *et al.*, 1998). There has been some suggestion that chemokines may be directly involved in the polarization of Th cells (and so in determining the balance of type 1 and type 2 responses) through their signalling and activation functions on T cells (Chensue *et al.*, 1996; Karpus *et al.*, 1997; Lukacs *et al.*, 1997). Notwithstanding this possibility, it is clear that the local production of chemokines can directly influence the nature of tissue responses through selective recruitment of Th1 or Th2 cells. In addition, chemokine receptor sharing between polarized Th cell subsets and different effector cells promotes their encounter and orchestrates the appropriate effector response. For example, common chemokine receptor expression is found on Th2 cells, eosinophils and basophils, all of which contribute to type 2 cytokine-mediated effector responses (Sallusto *et al.*, 1998).

Given that epithelial cells have been shown to be a source of chemokines at mucosal sites (Li *et al.*, 1999; Santy *et al.*, 1999; Song *et al.*, 1999), it will be important to define the expression and function of chemokines and their receptors in the intestinal microenvironment during nematode infection, and to determine what effects, if any, these have on the polarization or regulation of anti-nematode responses.

Immune Effector Mechanisms

As mentioned above, a prototypic type 2 response is mounted against nematode infection in most cases (consisting of elevated IgE, eosinophilia and intestinal mastocytosis), with a number of coexisting variables likely to be influencing the response under physiological conditions. Following the initiation and expansion of a protective type 2 response, the classical model of nematode expulsion from the gut has been the 'leak-lesion hypothesis'. This commonly quoted model of host protection suggested that the initial phase of the anti-parasite response consists of antigen-specific T cells inducing IgE production by B cells, which then sensitize mast cells by binding the surface of the cell through FcεR1 receptors. Following subsequent exposure to antigen, IgE-mediated degranulation of mast cells occurs, inducing type 1 hypersensitivity in the gut through the release of pharmacological mediators such as histamine, proteases, leukotrienes and prostaglandins. This non-specific inflammatory phase was thought to make the intestinal microenvironment unsuitable for nematode feeding and survival or directly to damage the worm, hence inducing expulsion (discussed in Wakelin, 1978, 1993).

Although in some cases antibody responses do appear to be important in expulsion, the availability of specific monoclonal antibodies and gene-targeted mice has suggested that antibody and mast cells are not involved in expulsion of a number of intestinal nematode parasites, certainly during a primary infection, although these (and other host responses) often cause immunopathology (reviewed in Grencis, 1997). Interestingly, a situation in which antigen-specific T cells and their products induce expulsion through non-specific inflammatory responses and alterations in gut physiology is still a likely scenario based on a number of recent findings.

Immune effectors: antibody

Studies in the *T. muris* model have shown that passive transfer of immune serum containing either predominantly IgA or IgG1 could enhance resistance to infection (Else *et al.*, 1990b; Roach *et al.*, 1991). Clearance of infection in resistant strains has also been shown to be coincident with a peak in the number of IgG1 and IgA producing B cells in the draining lymph node (Koyama *et al.*, 1999). However, a number of studies in which humoral immune responses have been either depressed or eliminated have shown that antibody is not essential for expulsion of *T. muris*. The transfer of a highly purified population of CD4+ T cells into normally susceptible SCID mice demonstrated that expulsion of *T. muris* (operating against larval stages of the parasite) could occur in the complete absence of antibody (Else and Grencis, 1996). Supporting this, μMT mice (which lack B cells) expel *T. muris* with efficiency and kinetics comparable to wild-type controls (Blackwell and K.J. Else, unpublished). Furthermore, Fc receptor γ-chain-deficient mice (lacking Fcγ receptors and Fcε receptors) (Takai *et al.*, 1994) are also completely resistant to *T. muris* infection, demonstrating that antibody-dependent cellular cytotoxicity mechanisms are not likely to play a major role in host protection in this model (Betts and Else, 1999).

In the case of *H. polygyrus*, passive transfer of purified IgG1 has implicated this isotype in host protection (Pritchard *et al.*, 1983). However, IgE does not appear to be important, because host protection is not blocked following treatment with anti-IgE monoclonal antibodies or in mice that lack IgE receptors (Finkelman *et al.*, 1997). In view of the evidence that the relative resistance of different mouse strains to *H. polygyrus* infection does not correlate well with the ability of sera from these animals to protect naive animals from infection (Williams and Behnke, 1983), it has been argued that antibody-dependent mechanisms are not a key component in protection against *H. polygyrus*. This also appears to be the case during *N. brasiliensis* infection, with depletion of antibody responses having no effect on worm expulsion (Jacobson *et al.*, 1977).

Immunity to *T. spiralis* can be passively transferred following either primary infection or vaccination with parasite antigens (Jarvis and Pritchard, 1992; Robinson *et al.*, 1995). Studies in the rat have also identified an important role for IgE and IgG1 in the rapid expulsion phenomenon following challenge infection (Love *et al.*, 1976; Dessein *et al.*, 1981; Appleton and McGregor, 1987; Ahmad *et al.*, 1991). However, it is clear that, in mice, primary infection can be efficiently expelled in Fc receptor γ-chain-deficient mice, demonstrating that antibody-independent expulsion can occur (R.K. Grencis and J.V. Ravetch, unpublished observations).

Overall, it appears that humoral responses during nematode infection may contribute to host resistance under normal physiological circumstances, although in most cases it is not essential for expulsion. Antibody responses may be important in challenge infections, however, as appears to be the case during *H. polygyrus* infection (Finkelman *et al.*, 1997). The evidence suggests that antibody is not a principal mechanism of host protection but the potential role of 'blocking antibodies' in promoting susceptibility to infection has not been investigated. In this context, mouse strains that are susceptible to *T. muris* have strikingly higher antibody levels and altered patterns of antigen recognition compared with resistant strains (Else and Wakelin, 1989).

Immune effectors: mast cells

A variety of type 2 cytokines and the non-lymphoid cell-derived growth factor, stem cell factor (SCF) (Hültner *et al.*, 1989; Copeland *et al.*, 1990; Huang *et al.*, 1990; Thompson-Snipes *et al.*, 1991; reviewed in Befus, 1995), are known to be important in the development of intestinal mast cell responses associated with nematode infection but the contribution of this population to host protection depends on the nematode species in question.

Mast cells do not appear to play a direct role in host protection against primary *T. muris* infection, as no correlation was found between intestinal mast cell responses and expulsion kinetics in different inbred strains of mice (Lee and Wakelin, 1982). *In vivo* ablation of mast cell responses using anti-SCF monoclonal antibody (causing a 99% reduction in numbers of intestinal mast cells) did not alter the kinetics of worm expulsion in normally resistant mice (Betts and Else, 1999), demonstrating that mast cells were not important in the expression of host resistance. Likewise, expulsion of *N. brasiliensis* occurs independently of mast cell responses. In this case, treatment with a combination of anti-IL-3 and anti-IL-4 monoclonal antibodies (resulting in an 85% decrease in mastocytosis) or infection in W/W^v mice (which have a natural mutation in SCF receptor, *c-kit*, and so have drastically impaired intestinal mast cell responses) (Galli *et al.*, 1994) did not alter worm expulsion (Crowle and Reed, 1981; Madden *et al.*, 1991).

Mast cells may be important in resistance to primary *H. polygyrus* infection. Although primary infection in most mouse strains is chronic, some strains do eventually clear infection, with the production of IL-3 and IL-9, and the development of intestinal mastocytosis coincident with worm expulsion (Behnke *et al.*, 1993; Wahid *et al.*, 1994). Indeed, coinfection studies with *H. polygyrus* and *T. spiralis* have identified the ability of *H. polygyrus* to selectively down-regulate mast cell responses (Dehlawi *et al.*, 1987).

The involvement of mast cells in host protection against nematode infection is well characterized in *T. spiralis* infection. W/W^v mice exhibited a significant delay in worm expulsion, and treatment with either anti-SCF or anti-SCF receptor monoclonal antibody dramatically inhibited mast cell responses and expulsion of *T. spiralis* for the duration of treatment (Donaldson *et al.*, 1996). W/W^v mice also lack interstitial cells of cajal and intraepithelial γδ T cells (Maeda *et al.*, 1992; Puddington *et al.*, 1994), which may contribute to the impaired response in these animals (see below). However, supporting evidence of a role for mast cells in protection against *T. spiralis* comes from studies in which overexpression of IL-9 in mice (which is known to influence the mast cell responses; see above) resulted in an extremely rapid mast cell-dependent expulsion of *T. spiralis* (Faulkner *et al.*, 1997).

The mechanisms whereby mast cells enhance host protection to *H. polygyrus* and *T. spiralis* (and whether these are related to the leak-lesion hypothesis) have not yet been fully defined. Certainly, mast cells contribute to intestinal inflammation during infection through the secretion of a range of cytokines (Gordon *et al.*, 1990) and vasoactive substances (see above). In addition, the release of mast cell proteases are known to increase enterocyte permeability to macromolecules in the rat intestine (Scudamore *et al.*, 1995) and regulate epithelial cell functions at other mucosal sites (Cairns and Walls, 1996).

Mast cell-dependent alterations in smooth muscle contractility and intestinal permeability may also be important in expulsion of *H. polygyrus* and *T. spiralis*. Increases in smooth muscle contractility (with associated changes in peristalsis) are observed during the protective response to *H. polygyrus* challenge infections. These mast cell-dependent effects were blocked following treatment with anti-IL-4 receptor monoclonal antibody, or induced in naive animals by administration of IL-4C (discussed in Finkelman *et al.*, 1997). Increased intestinal permeability and decreased fluid absorption in response to glucose during *H. polygyrus* challenge infection were also found to be dependent on IL-4 and intestinal mast cells (Finkelman *et al.*, 1997). Therefore a combination of enhanced fluid secretion and altered peristalsis may function to 'flush' worms out of the intestine.

Similar changes in fluid absorption and gut motility may also be important in expulsion of *T. spiralis*. Increases in smooth muscle contractility are observed coincident with expulsion of primary (Weisbrodt *et al.*, 1994; Vallance *et al.*, 1997) and challenge *T. spiralis* infections (Palmer and Castro, 1986). Indeed, studies in which smooth muscle contraction, granulocyte-dependent inflammation (measured by myeloperoxidase activity) and expulsion of primary *T. spiralis* infection were investigated found that genetically determined differences in the ability of inbred mouse strains to expel *T. spiralis* reflected the magnitude of the increase in smooth muscle contraction rather than the degree of mucosal inflammation (Vallance *et al.*, 1997). The pathophysiological changes in smooth muscle contractility associated with expulsion of *T. spiralis* were subsequently found to be dependent on haemopoetic MHC class II expression and CD4+ T cells (Vallance *et al.*, 1998). W/W^v mice were also found to have reduced expulsion and impaired increases in smooth muscle contraction following *T. spiralis* infection, compared with control animals, although whether this is due to a defect in the mast cell or interstitial cells of cajal populations (which regulate propulsive activity) is unknown (Vallance, 1999). Disruption of fluid secretion is also associated with rapid expulsion of challenge *T. spiralis* infection in rats (Castro *et al.*, 1979). Furthermore, artificial increases in fluid secretion through administration of serotonin reduced the establishment of *T. spiralis* infection in naive rats (Zhang and Castro, 1990). Whether these responses are essential for expulsion of *H. polygyrus* and *T. spiralis*, or only contribute to the overall unfavourable environment for the worm in the intestine, is unclear at present.

Immune effectors: eosinophils

Eosinophilia is a hallmark of intestinal nematode infection and is known to be under the control of IL-5 (Finkelman *et al.*, 1992). As discussed above, treatment with anti-IL-5 monoclonal antibody (and so ablation of eosinophilia) had no effect on expulsion of *T. muris, H. polygyrus, N. brasiliensis* or *T. spiralis* infections, suggesting that either redundant mechanisms operate under these circumstances or that eosinophils are not a critical component of effector responses operating against most murine intestinal nematode infections. However, mouse strains that constitutively overexpress IL-5 (and have high numbers of peripheral and tissue eosinophils) exhibited enhanced resistance to primary *N. brasiliensis* infection (Dent *et al.*, 1999). These studies showed that IL-5 transgenic mice had fewer numbers of adult worms in their intestine following infection, with those worms that did migrate there being smaller, apparently malnourished, and less fecund than worms in wild-type control animals (Dent *et al.*, 1997, 1999). The peak intestinal worm burden in these animals (although significantly lower than in wild-types) was also later (day 5 in transgenic animals compared with day 3 in controls), suggesting that IL-5 and/or eosinophils were operating against larval stages of the parasite to impede their migration to the intestine. Therefore, under certain conditions eosinophils may be important in host resistance, but an IL-5-dependent, eosinophil-independent mechanism of enhanced resistance cannot be ruled out in these transgenic animals.

Using the same transgenic animals, experiments using the *T. spiralis* model suggested that over-expression of IL-5 (and high numbers of eosinophils) may be an advantage to the parasite, with the number of tissue larvae being higher in transgenic animals at day 10 post-infection (Dent *et al.*, 1997). This may be a result of prolonged survival of adult worms in the intestine, generating more larvae, or reduced killing of larvae in the muscle tissue. The latter suggestion is not supported by *in vitro* studies, which have shown that eosinophils can kill larval stages of *T. spiralis* (Kazura and Aikawa, 1987), presumably through the release of cationic or superoxide radicals.

Although definitive evidence of a role for eosinophils in protection against intestinal nematode parasites is not currently available, it is clear that this population does have at least a partially protective function against migratory larvae of tissue-dwelling nematodes, including *Angiostrongylus cantonensis* (Yoshida *et al.*, 1996) and *S. stercoralis* (Korenaga *et al.*, 1994). It is also important to note that a fundamental difference exists between murine and human eosinophils (the former lacking high affinity IgE receptors) and so this may be the basis of the apparent lack of function of eosinophils in murine models of nematode infection, although perhaps not in human and livestock infection. At present it is clear that expulsion of a number of species of intestinal nematodes can occur in the absence of eosinophils, although further analysis of the protective role of this population against different life-cycle stages of other intestinal nematode species is required.

Immune effectors: cytokines and enterocytes

It is clear from studies in the *T. muris*, *H. polygyrus* and *N. brasiliensis* systems that administration of IL-4C to infected SCID mice can mediate expulsion of adult stages of these nematodes in the absence of an adaptive immune system (Urban *et al.*, 1995; K.J. Else, unpublished). Therefore, the hypothesis that type 2 cytokines can mediate their protective effects in the intestine through regulating the function of host enterocytes is an attractive one.

Dramatic changes in cell proliferation and turnover within the intestinal epithelium have been observed during a number of nematode infections, resulting in crypt hyperplasia and/or villus atrophy (Symons, 1965, 1978; Manson-Smith *et al.*, 1979; Hoste, 1989; Lawrence *et al.*, 1998; Artis *et al.*, 1999b), with these changes suggested to be a result of either parasite-derived products or the host response to infection. In the case of *T. muris*, we have recently shown that enterocyte hyperproliferation observed during chronic infection is regulated by IFN-γ (Artis *et al.*, 1999b). Interestingly, we have observed clear qualitative differences in the patterns of enterocyte proliferation between susceptible and resistant mouse strains during primary *T. muris* infection – in particular, significant changes in the patterns of proliferation within the epithelium coincident with worm expulsion in resistant strains. Whether cytokine regulation of enterocyte proliferation is a precipitating event in worm expulsion or a response to it remains to be determined. As mentioned above, we have also identified a novel role for TNF-α in regulating expulsion of *T. muris*. As TNF-α has been shown to regulate enterocyte proliferation and apoptosis in other systems (Kaiser and Polk, 1997; Piguet *et al.*, 1998), it is possible that this cytokine is mediating its protective effects through regulation of enterocyte function and physiology (see also Chapter 18).

In addition to disruption in enterocyte proliferation, changes in enterocyte differentiation patterns are observed during nematode infection. Although the signals that regulate differentiation of goblet cells from enterocyte stem cells are unknown (Leblond and Messier, 1958; Paulus *et al.*, 1993), a marked goblet cell hyperplasia is observed during *T. muris* (D. Artis and R.K. Grencis, unpublished), *N. brasiliensis* (Ishikawa *et al.*, 1993) and *T. spiralis* infection (Garside *et al.*, 1992). Evidence exists to support a role for goblet cells and the mucins they secrete in expulsion of *N. brasiliensis*, and in the CD4+ T cell control of this response. Treatment with anti-CD4 monoclonal antibody significantly reduced the numbers of goblet cells and amount of mucins secreted during infection, resulting in a delay in worm expulsion (Khan *et al.*, 1995). IL-13 plays a critical role in the goblet cell response to *N. brasiliensis*, with IL-13 KO mice having lower numbers of goblet cells during infection and delayed worm expulsion (McKenzie *et al.*, 1998).

It is the quality rather than the quantity of the goblet cell response that is important in resistance to nematode infection, as even strains that are susceptible to *T. muris* (harbouring chronic infections) have a dramatic increase in the number of goblet cells and amount of mucus secreted during infection. Indeed, detailed analysis of the biochemical nature of mucins secreted by goblet cells during *T. muris* infection has found that differences do exist between susceptible and resistant strains (Khan and Grencis, unpublished observations). Although the mechanisms of protection afforded by goblet cells are unknown, it is likely that mucins impede worm feeding and/or the ability of the worm to locate within its preferred niche, and so enhance expulsion (see also Chapter 18).

T cell and cytokine regulation of enterocyte apoptosis may also be important in the expulsion of nematodes, in particular *T. spiralis* and *T. muris*, which inhabit an intracellular niche. Certainly an increase in the number of apoptotic cells within the epithelium is observed around the period of expulsion of *T. muris* in resistant mouse strains (D. Artis, C.S. Potten and R.K. Grencis, unpublished). Apoptosis of host enterocytes may dislodge the nematode or perhaps expose vital feeding organs to immune attack, and so enhance expulsion. Whether enterocyte apoptosis results from the burrowing action of the worms or a tissue repair mechanism, or is involved in expulsion, remains to be investigated.

Concluding Remarks

Significant advances in our understanding of immune regulation of intestinal nematode infection have been made in recent years but a number of questions remain to be addressed. In particular, what are the early molecular events following infection that determine infection outcome? The design of anti-nematode vaccine candidates will be aided by the identification of the critical cytokines that influence the development of protective Th responses during infection, with a view to developing these molecules as adjuvants of potential vaccines. In addition, examination of the PRRs on enterocytes and other cell populations (and the molecular structures common to nematodes that they recognize) will be important in the development of potential vaccines, providing essential information on the nature of nematode antigens that initiate (and potentially modulate) immune responses. The precise effector mechanisms induced by type 2 cytokines in the intestinal microenvironment also remain unknown. In this regard it is necessary to define the target cells of protective cytokines and what aspects of altered gut physiology are important in worm expulsion. Although caution must be exercised in extrapolating the results of murine studies to nematode infections of humans and livestock, laboratory models of infection have undoubtedly allowed rigorous characterization and functional analysis of the cellular and molecular basis of immunity to intestinal nematode infection. In addition, these studies now provide a solid framework within which the host and nematode molecules important in determining infection outcome can be manipulated in the design of rational immunotherapies.

Acknowledgements

We would like to thank all our colleagues in Manchester for allowing us to discuss their unpublished work. Research in this laboratory has received funding from the BBSRC, the British Medical Research Council, and The Wellcome Trust.

References

Abbas, A.K., Murphy, K.M. and Sher, A. (1996) Functional diversity of helper T lymphocytes. *Nature* 383, 787–793.

Abishira-Amar, O., Gilbert, M., Jolly, M., Theze, J. and Jankovic, D.I. (1992) IL-4 plays a dominant role in the differential development of Th0 into Th1 and Th2 cells. *Journal of Immunology* 148, 3820–3829.

Ahmad, A., Wang, C.H. and Bell, R.G. (1991) A role for IgE in intestinal immunity. Expression of rapid expulsion of *Trichinella spiralis* in rats transfused with IgE and thoracic lymphocytes. *Journal of Immunology* 146, 3563–3570.

Albonico, M., Crompton, D.W.T. and Savioli, L. (1999) Control strategies for human intestinal nematode infections. *Advances in Parasitology* 42, 277–341.

Appleton, J.A. and McGregor, D.D. (1987) Characterisation of the immune mediator of rapid expulsion of *Trichinella spiralis* in suckling rats. *Immunology* 62, 477–484.

Artis, D., Humphreys, N.H., Bancroft, A.J., Rothwell, N.J., Potten, C.S. and Grencis, R.K. (1999a) TNF-α is a critical component of IL-13-mediated protective T helper cell type 2 responses during helminth infection. *Journal of Experimental Medicine* 190, 953–962.

Artis, D., Potten, C.S., Else, K.J., Finkelman, F.D. and Grencis, R.K. (1999b) *Trichuris muris*: host intestinal epithelial cell hyperproliferation during chronic infection is regulated by interferon-γ. *Experimental Parasitology* 92, 144–153.

Baggiolini, M., Dewald, B. and Moser, B. (1997) Human chemokines: an update. *Annual Review of Immunology* 15, 675–705.

Bancroft, A.J., Else, K.J. and Grencis, R.K. (1994) Low level infection of *Trichuris muris* significantly affects the polarisation of the CD4 response. *European Journal of Immunology* 24, 3113–3118.

Bancroft, A.J., Else, K.J., Sypek, J. and Grencis, R.K. (1997) IL-12 promotes a chronic intestinal nematode infection. *European Journal of Immunology* 27, 866–870.

Bancroft, A.J., McKenzie, A.N.J. and Grencis, R.K. (1998) A critical role for IL-13 in resistance to intestinal nematode infection. *Journal of Immunology* 160, 3453–3461.

Barnard, A.J., Warwick, G.J. and Gold, C.I. (1993) Localisation of TGFα isoforms in the normal murine small and large intestine. *Gastroenterology* 105, 67–73.

Barner, M., Mohrs, M., Brombacher, F. and Kopf, M. (1998) Differences between IL-4Rα-deficient and IL-4-deficient mice reveal a role for IL-13 in the regulation of Th2 responses. *Current Biology* 8, 669–672.

Beebe, A.M., Mauze, S., Schork, N.J. and Coffman, R.L. (1997) Serial backcross mapping of multiple loci associated with resistance to *Leishmania major* in mice. *Immunity* 6, 551–557.

Befus, A.D. (1995) The immunophysiology of mast cells in intestinal immunity and symbiosis. In: Blaser, M.J., Smith, P.D., Ravdin, J.I., Greenberg, H.B. and Guerrant, R.L. (eds) *Infections of the Gastrointestinal Tract.* Raven, New York, pp. 227–236.

Behnke, J.M., Barnard, C.J. and Wakelin, D. (1992) Understanding chronic nematode infections: evolutionary considerations, current hypotheses, and the way forward. *International Journal for Parasitology* 22, 861–907.

Behnke, J.M., Wahid, F.N., Grencis, R.K., Else, K.J., Ben-Smith, A. and Goyal, P.K. (1993) Immunological relationships during primary infection with *Heligmosomoides polygyrus* (*Nematospiroides dubius*): down-regulation of specifc cytokine secretion (IL-9 and IL-10) correlates with poor mastocytosis and chronic survival of adult worms. *Parasite Immunology* 15, 415–421.

Betts, C.J. and Else, K.J. (1999) Mast cells, eosinophils, and antibody-mediated-cellular-cytotoxicity are not critical in resistance to *Trichuris muris*. *Parasite Immunology* 21, 45–52.

Bland, P.W. (1999) Mucosal T cell–epithelial cell interactions. *Chemical Immunology* 71, 40–63.

Bland, P.W. and Warren, L.G. (1986) Antigen presentation by epithelial cells of the rat small intestine. I. Kinetics, antigen specificity and blocking by anti-Ia antisera. *Immunology* 58, 1–7.

Bland, P.W. and Whiting, C.V. (1989) Antigen processing by rat intestinal villus enterocytes. *Immunology* 68, 497–502.

Bland, P.W. and Whiting, C.V. (1992) Induction of MHC class II gene products in rat intestinal epithelium during graft-versus-host disease and effects on the immune function of the epithelium. *Immunology* 75, 366–371.

Bleicher, P.A., Balk, S.P., Hagen, S.J., Blumberg, R.S., Flotte, T.J. and Terhorst, C. (1990) Expression of murine CD1 on gastrointesinal epithelium. *Science* 250, 679–682.

Blotta, M.H., Marshall, J.D., De Kruyff, R.H. and Umetsu, D.T. (1996) Cross-linking of the CD40 ligand on human CD4+ T lymphocytes generates a co-stimulatory signal that upregulates IL-4 synthesis. *Journal of Immunology* 156, 3133–3140.

Blumberg, R.S., Terhorst, C., Bleicher, P., Macdermott, F.V., Allan, C.H., Landan, S.B., Trier, J.S. and Balk, S.P. (1991) Expression of a non-polymorphic MHC class I-like molecule, CD1d, by human intestinal epithelial cells. *Journal of Immunology* 147, 2518–2524.

Bretscher, P.A., Wei, G., Menon, J.N. and Bielefeldt, O.H. (1992) Establishment of stable, cell-mediated immunity that makes susceptible mice resistant to *Leishmania major. Science* 257, 539–542.

Brown, J.A., Titus, R.G., Nabavi, N. and Glimcher, L.H. (1997) Blockade of CD86 ameiliorates *Leishmania major* infection by downregulating the Th2 response. *Journal of Infectious Diseases* 174, 1303–1308.

Buddle, B.M., Jowett, G., Green, R.S., Dough, P.G.C. and Risdon, P.L. (1992) Association of blood eosinophilia with the expression of resistance in Romney lambs to nematodes. *International Journal for Parasitology* 22, 955–960.

Bundy, D.A.P. and Cooper, E.S. (1989) *Trichuris* and trichuriasis in humans. *Advances in Parasitology* 28, 107–173.

Cairns, J.A. and Walls, A.F. (1996) Mast cell tryptase is a mitogen for epithelial cells. *Journal of Immunology* 156, 275–283.

Castro, G.A., Hessel, J.J. and Whalen, G. (1979) Altered intestinal fluid movement in response to *Trichinella spiralis* in immunized rats. *Parasite Immunology* 1, 259–266.

Cella, M., Scheidegger, D., Plamer-Lehmann, K., Lane, P., Lanzavecchia, A. and Alber, G. (1996) Ligation of CD40 on dendritic cells triggers production of high levels of interleukin-12 and enhances T cell stimulatory capacity: T-T help via APC activation. *Journal of Experimental Medicine* 184, 747–752.

Chan, M.S. (1997) The global burden of intestinal nematode infections – fifty years on. *Parasitology Today* 13, 438–443.

Chensue, S.W., Warmington, K.S., Ruth, J.H., Sanghi, P.S., Lincoln, P. and Kunkel, S.L. (1996) Role of monocyte chemoattractant protein-1 (MCP-1) in Th1 (mycobacterial) and Th2 (schistosomal) antigen-induced granuloma formation. *Journal of Immunology* 157, 4602–4608.

Coffman, R.L., Seymour, B.W.P., Hudak, S., Jackson, J. and Rennick, D. (1989) Antibody to IL-5 inhibits helminth-induced eosinophilia in mice. *Science* 245, 308–310.

Conrod, D.H., Ben-Sasson, S.Z., Le Gros, G., Finkelman, F.D. and Paul, W.E. (1990) Injection with anti-IgD antibodies markedly enhances Fc receptor mediated interleukin 4 production by non B, non T cells. *Journal of Experimental Medicine* 171, 1497–1508.

Constant, S., Pfeiffer, C., Woodard, A., Pasqualini, T. and Bottomly, K. (1995) Extent of T cell receptor ligation can determine the functional differentiation of naive CD4+ T cells. *Journal of Experimental Medicine* 182, 1591–1596.

Cooper, E.S., Whyte-Alleng, C.A.M., Finzi-Smith, J.S. and McDonald, T.T. (1992) Intestinal nematode infections in children: the pathophysiological price paid. *Parasitology* 104, S91–S103.

Copeland, N.G., Gilbert, D.J., Cho, B.C., Donovan, P.J., Jenkins, N.A., Cosman, D., Anderson, D., Lyman, S.D. and Williams, D.E. (1990) Mast cell growth factor maps near the steel locus on mouse chromosome 10 and is deleted in a number of steel alleles. *Cell* 63, 175–183.

Crowle, P.K. and Reed, N.D. (1981) Rejection of the parasite *Nippostrongylus brasiliensis* by mast cell deficient W/W^v anaemic mice. *Infection and Immunity* 33, 54–58.

Dehlawi, M.S., Wakelin, D. and Behnke, J.M. (1987) Suppression of mucosal mastocytosis by infection with the intestinal nematode *Nematospiroides dubius*. *Parasite Immunology* 12, 561–566.

Dent, L.A., Daly, C., Geddes, A., Cormie, J., Finlay, D.A., Bignold, L., Hagan, P., Parkhouse, R.M.E., Garate, T., Parsons, J. and Mayrhofer, G. (1997) Immune responses of IL-5 transgenic mice to parasites and aeroallergins. *Memoirs of the Institute Oswaldo Cruz* 92, 45–54.

Dent, L.A., Daly, C.M., Mayrhofer, G., Zimmerman, T., Hallett, A., Bignold, L.P., Creaney, J. and Parsons, J.C. (1999) Interleukin-5 transgenic mice show enhanced resistance to primary infections with *Nippostrongylus brasiliensis* but not primary infections with *Toxocara canis*. *Infection and Immunity* 67, 989–993.

Dessein, A.F., Parker, W.L., James, S.L. and David, J.R. (1981) IgE antibody and resistance to infection. I. Selective suppression of the IgE antibody response to *Trichinella spiralis* infection. *Journal of Experimental Medicine* 153, 423–436.

Donaldson, L.E., Schmitt, E., Huntey, J.F., Newlands, G.F.J. and Grencis, R.K. (1996) A critical role for stem cell factor and c-kit in host protective immunity to an intestinal helminth. *International Immunology* 8, 559–567.

Eckmann, C., Jung, H.C., Schurer-Maly, C., Panja, A., Wroblewska, E.M. and Kagnoff, M. (1993) Differential cytokine expression by human intestinal epithelial cell lines: regulated expression of IL-8. *Gastroenterology* 105, 1689–1697.

Eklund, K.K., Ghildyal, N., Austen, K.F. and Stevens, R.L. (1993) Induction by IL-9 and suppression by IL-3 and IL-4 of the levels of chromosome 14-derived transcripts that encode late expressed mouse mast cell proteases *Journal of Immunology* 151, 4266–4273.

Else, K.J. and Grencis, R.K. (1996) Antibody-independent effector mechanisms in resistance to the intestinal nematode parasite *Trichuris muris*. *Infection and Immunity* 64, 2950–2954.

Else, K.J. and Wakelin, D. (1988) The effects of H-2 and non H-2 genes on the expulsion of the nematode *Trichuris muris* from inbred and congenic mice. *Parasitology* 96, 543–550.

Else, K.J. and Wakelin, D. (1989) Genetic variation in the humoral immune response of mice to the nematode *Trichuris muris*. *Parasite Immunology* 11, 77–90.

Else, K.J., Wakelin, D. and Roach, T.I.A. (1989) Host predisposition to trichuriasis: the mouse–*T. muris* model. *Parasitology* 98, 275–282.

Else, K.J., Wakelin, D., Wassom, D.L. and Hauda, K.M. (1990a) The influence of genes mapping within the major histocompatibility complex on resistance to *Trichuris muris* infections in mice. *Parasitology* 101, 61–67.

Else, K.J., Wakelin, D., Wassom, D.L. and Hauda, K.M. (1990b) MHC-restricted antibody responses to *Trichuris muris* excretory/secretory (E/S) antigen. *Parasite Immunology* 12, 509–527.

Else, K.J., Hültner, L.H. and Grencis, R.K. (1992) Modulation of cytokine production and response phenotypes in murine trichuriasis. *Parasite Immunology* 14, 441–449.

Else, K.J., Finkelman, F.D., Maliszewski, C.R. and Grencis, R.K. (1994) Cytokine-mediated regulation of chronic intestinal helminth infection. *Journal of Experimental Medicine* 179, 347–351.

Emery, D.L. (1996) Vaccination against worm parasites of animals. *Veterinary Parasitology* 64, 31–45.

Emery, D.L. and Wagland, B.M. (1991) Vaccines against gastrointestinal nematode parasites of ruminants. *Parasitology Today* 7, 347–349.

Faulkner, H.F., Humphreys, N.H., Renauld, J.C., Van Snick, J. and Grencis, R.K. (1997) Interleukin-9 is involved in host protective immunity to intestinal nematode infection. *European Journal of Immunology* 27, 2536–2540.

Faulkner H.F., Renauld, J.C., Van Snick, J. and Grencis, R.K. (1998) Interleukin-9 enhances resistance to the intestinal nematode *Trichuris muris*. *Infection and Immunity* 66, 3832–3840.

Ferrick, D.A., Schrenzel, M.D., Mulvania, T., Hsieh, B., Ferlin, W.G. and Lepper, H. (1995) Differential production of interferon-γ and interleukin 4 in response to Th1 and Th2 stimulating pathogens by $\gamma\delta$ T cells *in vivo*. *Nature* 373, 265–257.

Finkelman, F.D., Pearce, E.J., Urban, J.F. and Sher, A. (1992) Regulation and biological function of helminth-induced cytokine responses. *Immunoparasitology Today*, A62–A66.

Finkelman, F.D., Madden, K.B., Morris, S.C., Holmes, J.M., Boiani, N., Katona, I.M. and Maliszewski, C.R. (1993) Anti-cytokine antibodies as carrier proteins: prolongation of *in vivo* effects of exogenous cytokines by injection of cytokine–anti-cytokine antibody complexes. *Journal of Immunology* 151, 1235–1244.

Finkelman, F.D., Madden, K.B., Cheever, A.W., Katona, I.M., Morris, S.C., Gately, M.K., Hubbard, B.R., Gause, W.C. and Urban, J.F. (1994) Effects of IL-12 on immune responses and host protection in mice infected with intestinal nematode parasites. *Journal of Experimental Medicine* 179, 1563–1572.

Finkelman, F.D., Shea-Donohue, T., Goldhill, J., Sullivan, C.A., Morris, S.C., Madden, K.B., Gause, W.C. and Urban, J.F. (1997) Cytokine regulation of host defense against parasitic gastrointestinal helminths: lessons from studies with rodent models. *Annual Review of Immunology* 15, 505–533.

Fitch, F.W., McKisick, M.D., Lanck, D.W. and Gajweski, T.F. (1992) Differential regulation of murine T lymphocyte subsets. *Annual Review of Immunology* 11, 29–48.

Gajewski, T.F., Pinnas, M., Wong, T. and Fitch, F.W. (1991) Murine Th1 and Th2 clones proliferate optimally in response to distinct antigen presenting cell populations. *Journal of Immunology* 146, 1750–1758.

Galli, S.J., Geissler, E.N. and Zsebo, K.M. (1994) The kit ligand, stem cell factor. *Advances in Immunology* 55, 1–96.

Garside, P., Grencis, R.K. and McI Mowat, A.M. (1992) T lymphocyte dependent enteropathy in murine *Trichinella spiralis* infection. *Parasite Immunology* 14, 217–225.

Gazinelli, R.T., Hakim, F.T., Hieny, S., Shearer, G.M. and Sher, A. (1991) Synergistic role of CD4+ and CD8+ T lymphocytes in IFN-γ production and protective immunity induced by an attenuated *Toxoplasma gondii* vaccine. *Journal of Immunology* 146, 286–292.

Gordon, J.R., Burd, P.R. and Galli, S.J. (1990) Mast cells as a source of multifunctional cytokines. *Immunology Today* 11, 458–464.

Greenwald, R., Lu, P., Zhou, X.D., Nguyen, H., Chen, S.J., Perrin, P.J., Madden, K.B., Morris, S.C., Finkelman, F.D., Peach, R., Linsley, P.S., Urban, J.F. and Gause, W.C. (1997) Effects of blocking B7-1 and B7-2 interactions during a type 2 *in vivo* immune response. *Journal of Immunology* 158, 4088–4096.

Greenwald, R.J., Urban, J.F., Ekkens, M.J., Chen, S.J., Nguyen, D., Fang, H., Finkelman, F.D., Sharpe, A.H. and Gause, W.C. (1999) B7-2 is required for the progression but not the initiation of the type 2 immune response to a gastrointestinal nematode parasite. *Journal of Immunology* 162, 4133–4139.

Grencis, R.K. (1997) Enteric helminth infection: immunopathology and resistance during intestinal nematode infection. *Chemical Immunology* 66, 41–61.

Grencis, R.K. and Cooper, E.S. (1996) *Enterobius, Trichuris, Capillaria,* and hookworm including *Ancylostoma caninum. Gastroenterology Clinics of North America* 25, 579–597.

Grencis, R.K. and Entwistle, G.M. (1997) Production of an interferon-gamma homologue by an intestinal nematode: functionally significant or interesting artefact? *Parasitology* S101–S105.

Grencis, R.K., Reidlinger, J. and Wakelin, D. (1985) L3T4 positive lymphoblasts are responsible for transfer of immunity to *Trichinella spiralis* in mice. *Immunology* 56, 213–218.

Grencis, R.K., Hültner, L. and Else, K.J. (1991) Host protective immunity to *Trichinella spiralis* in mice: activation of Th cell subsets and lymphokine secretion in mice expressing different response phenotypes. *Immunology* 74, 329–332.

Grewel, I.S., Xu, J. and Flavell, R.A. (1995) Impairment of antigen-specific T cell priming in mice lacking CD40 ligand. *Nature* 378, 617–620.

Guler, M.L., Gorham, J.D., Hsieh, C.S., Mackey, A.J., Steen, R.G., Dietrich, W.F. and Murphy, K.M. (1996) Genetic susceptibility to *Leishmania*: IL-12 responsiveness in Th1 development. *Science* 271, 984–987.

Harris, N.L., Peach, R.J. and Ronchese, F. (1999) CTLA-4-Ig inhibits optimal T helper cell development but not protective immunity or memory response to *Nippostrongylus brasiliensis*. *European Journal of Immunology* 29, 311–316.

Herdon, F.J. and Kayes, S.G. (1992) Depletion of eosinophils by anti-IL-5 monoclonal antibody treatment of mice infected with *Trichinella spiralis* does not alter parasite burden or immunologic resistance to re-infection. *Journal of Immunology* 149, 3642–3647.

Hosken, N.A., Shibuya, K., Heath, A.W., Murphy, K.M. and O'Garra, A. (1995) The effect of antigen dose on CD4+ T cell phenotype development in an $\alpha\beta$ TCR-transgenic mouse model. *Journal of Experimental Medicine* 182, 1579–1584.

Hoste, H. (1989) *Trichostrongylus colubriformis*: epithelial cell kinetics in the small intestine of infected rabbits. *Experimental Parasitology* 68, 99–104.

Hsieh, C.S., Macatonia, S.E., Tripp, C.S., Wolf, S.F., O'Garra, A. and Murphy, K.M. (1993) Development of Th1 CD4+ T cells through IL-12 produced by *Listeria*-induced macrophages. *Science* 260, 547–549.

Huang, E., Nocka, K., Beier, D.R., Chu, T.Y., Buck, J., Lahm, H.W., Wellner, D., Leder, P. and Besmer, P. (1990) The haemopoetic growth factor KL is encoded by the *Sl* locus and is the ligand of the *c-kit* receptor, the gene product of the *w* locus. *Cell* 63, 225–233.

Hültner, L., Moeller, J., Schmitt, E., Jäger, G., Reisbach, G., Ring, J. and Dörmer, P. (1989) Thiol-sensitive mast cell lines derived from mouse bone marrow respond to a mast cell growth-enhancing activity different from both IL-3 and IL-4. *Journal of Immunology* 142, 3440–3446.

Ishikawa, N., Horii, Y. and Nawa, Y. (1993) Immune mediated alteration of the terminal sugars of goblet cell mucins in the small intestine of *Nippostrongylus brasiliensis* infected rats. *Immunology* 78, 303–307.

Ito, Y. (1991) The absence of resistance in congenitally athymic nude mice toward infection with the intestinal nematode, *Trichuris muris*: resistance restored by lymphoid cell transfer. *International Journal of Parasitology* 21, 65–69.

Jacobson, R.H., Reed, N.D. and Manning, D.D. (1977) Expulsion of *Nippostrongylus brasiliensis* from mice lacking antibody production potential. *Immunology* 32, 867–874.

Jarvis, L.M. and Pritchard, D.I. (1992) An evaluation of the role of carbohydrate epitopes in immunity to *Trichinella spiralis*. *Parasite Immunology* 14, 489–501.

Jung, H.C., Eckmann, L., Yang, S.K., Panja, A., Fierer, J., Morzyckawroblewska, E. and Kagnoff, M.F. (1995) A distinct array of proinflammatory cytokines is expressed in human colon epithelial cells in response to bacterial invasion. *Journal of Clinical Investigation* 95, 55–65.

Kaiser, G.C. and Polk, D.B. (1997) Tumour necrosis factor-α regulates proliferation in a mouse intestinal cell line. *Gastroenterology* 112, 1231–1240.

Kaiserlian, D., Vidal, K. and Revillar, J.P. (1989) Murine enterocytes can present soluble antigen to specific class II restricted CD4+ T cells. *European Journal of Immunology* 19, 1513–1516.

Kaplan, M.H., Schindler, U., Smiley, S.T. and Grusby, M.J. (1996) Stat6 is required for mediating responses to IL-4 and for the development of Th2 cells. *Immunity* 4, 313–319.

Karpus, W.J., Lukas, N.W., Kennedy, K.J., Smith, W.S., Hurst, S.D. and Barrett, T.A. (1997) Differential C-C chemokine-induced enhancement of T helper cell cytokine production. *Journal of Immunology* 158, 4129–4136.

Katona, I.M., Urban, J.F. and Finkelman, F.D. (1988) The role of L3T4+ and Ly2+ T cells in the IgE response and immunity to *Nippostrongylus brasiliensis. Journal of Immunology* 140, 3206–3211.

Katona, I.M., Urban, J.F., Finkelman, F.D., Gause, W.C. and Madden, K.B. (1995) Cytokine regulation of intestinal mastocytosis in *Nippostrongylus brasiliensis* infection. In: Mestecky, J. *et al.* (eds) *Advances in Mucosal Immunology.* Plenum Press, New York, pp. 971–973.

Kaufmann, S.H.E. (1996) γδ and other unconventional T lymphocytes: what do they see and what do they do? *Proceedings of the National Academy of Sciences USA* 93, 2272–2279.

Kazura, J.W. and Aikawa, M. (1980) Host defence mechanisms against *Trichinella spiralis* infection in the mouse: eosinophil mediated destruction of newborn larvae *in vitro. Journal of Immunology* 124, 355–361.

Khan, W.I., Abe, T., Ishikawa, N., Nawa, Y. and Yoshimura, K. (1995) Reduced amount of intestinal mucus after treatment with anti-CD4 antibody interferes with the spontaneous cure of *Nippostrongylus brasiliensis* infection in mice. *Parasite Immunology* 17, 485–491.

Klein, J.R. (1996) Whence the intestinal intraepithelial lymphocyte? *Journal of Experimental Medicine* 184,1203–1206.

Kohno, K., Kataoka, J., Ohtsuki, T., Suemoto, Y., Okamoto, I., Usui, M., Ikeda, M. and Kurimoto, M. (1997) IFN-γ-inducing factor (IGIF) is a costimulator factor on the activation of Th1 but not Th2 cells and exerts its effect independently of IL-12. *Journal of Immunology* 158, 1541–1550.

Korenaga, M., Hitoshi, Y., Takatsu, K. and Tada, I. (1994) Regulatory effect of anti-interleukin-5 monoclonal antibody on intestinal worm burden in a primary infection with *Strongyloides venezuelensis. Journal of Parasitology* 24, 951–957.

Koyama, K., Tamanchi, H. and Ito, Y. (1995) The role of CD4+ and CD8+ T cells in protective immunity to the murine parasite *Trichuris muris. Parasite Immunology* 17, 161–165.

Koyama, K., Tamauchi, H., Tomita, M., Kitajima, T. and Ito, Y. (1999) B cell activation in the mesenteric lymph nodes of resistant BALB/c mice infected with the murine nematode parasite *Trichuris muris. Parasitology Research* 85, 194–199.

Kuchroo, K., Das, M.P., Brown, J.A., Ranger, A.M., Zamvil, S.S., Sbel, R.A., Weiner, H.L., Nabavi, N. and Glimcher, L.H. (1995) B7-1 and B7-2 costimulatory molecules activate differentially the Th1/Th2 developmental pathways: application to autoimmune disease therapy. *Cell* 80, 707–718.

Kumar, V., Bhardwaj, V., Soares, L., Alexander, J., Sette, A. and Sercaz, E. (1995) Major histocompatibility complex binding affinity of an antigenic determinant is crucial for the differential secretion of interleukin 4/5 or interferon-γ by T cells. *Proceedings of the National Academy of Sciences USA* 92, 9510–9514.

Lantz, C.S., Boesiger, J., Song, C.H., Mach, N., Kobayashi, T., Mulligan, R.C., Nawa, Y., Dranoff, G. and Galli, S.J. (1998) Role for IL-3 in mast cell and basophil development and in immunity to parasites. *Nature* 393, 90–93.

Lawrence, C.E., Paterson, J.C.M., Higgins, L.M., McDonald, T.T., Kennedy, M.W. and Garside, P. (1998) IL-4 regulated enteropathy in an intestinal nematode infection. *European Journal of Immunology* 28, 2672–2684.

Lawrence, R.A., Gray, C.A., Osbourne, J. and Maizels, R. (1996) *Nippostrongylus brasiliensis*: cytokine responses and nematode expulsion in normal and IL-4 deficient mice. *Experimental Parasitology* 84, 65–73.

Leblond, C.P. and Messier, B. (1958) Renewal of chief cells and goblet cells in the small intestine as shown by radioautography after injection of thymidine-H3 into mice. *Anatomical Record* 132, 49–58.

Lee, T.D.G. and Wakelin, D. (1982) The use of host strain variation to assess the significance of mucosal mast cells in the spontaneous cure response of mice to the nematode *Trichuris muris*. *International Archives of Allergy and Applied Immunity* 67, 302–305.

Li, L., Xia, Y., Nguyen, A., Lai, Y.H., Feng, L., Mosmann, T.R. and Lo, D. (1999) Effects of Th2 cytokines on chemokine expression in the lung: IL-13 potently induces eotaxin expression by airway epithelial cells. *Journal of Immunology* 162, 2477–2487.

Lin, J.X., Migone, T.S., Tsang, M., Friedman, M., Weatherbeem, J.A., Zhou, L., Yamauchi, A., Bloom, E.T., Mietz, J., John, S. and Leonard, W.J. (1995) The role of shared receptor motifs and common stat proteins in the generation of cytokine pleiotropy and redundancy by IL-2, IL-4, IL-7, IL-13, and IL-15. *Immunity* 2, 331–339.

Lobos, E. (1997) The basis of IgE responses to specific antigenic determinants in helminthiasis. *Chemical Immunology* 66, 1–25.

Louahed, J., Kermouni, A., van Snick, J. and Renauld, J.C. (1995) IL-9 induces expression of granzymes and high affinity IgE receptor in murine T helper cell clones. *Journal of Immunology* 154, 5061–5070.

Love, R.J., Ogilvie, B.M. and McLaren, D.J. (1976) The immune mechanism which expels the intestinal stage of *Trichinella spiralis* from rats. *Immunology* 30, 7–15.

Lu, P., di Zhou, X., Chen, S.J., Moorman, M., Morris, S.C., Finkelman, F.D., Linsley, P., Urban, J.F. and Gause, W.C. (1994) CTLA-4 ligands are required to induce an *in vivo* interleukin 4 response to a gastrointestinal nematode parasite. *Journal of Experimental Medicine* 180, 693–698.

Lukacs, N.W., Chensue, S.W., Karpus, W.J., Lincoln, P., Keefer, C., Strieter, R.M. and Kunkel, S.L. (1997) C-C chemokines differentially alter IL-4 production from lymphocytes. *American Journal of Pathology* 150, 1861–1868.

Madden, K.B., Urban, J.F., Ziltener, H.J., Schrader, J.W., Finkelman, F.D. and Katona, I.M. (1991) Antibodies to IL-3 and IL-4 suppress helminth induced intestinal mastocytosis. *Journal of Immunology* 147, 1387–1391.

Maeda, H., Yamagata, A., Nishikawa, S., Yoshinaga, K., Kobayashi, S., Nishi, K. and Nishikawa, S. (1992) Requirement of c-kit for development of intestinal pacemaker system. *Development* 116, 369–375.

Magram, J., Connaughton, S.E., Warrier, R.R., Carvajal, D.M., Wu, C.Y., Ferrante, J., Stewart, C., Sarmiento, U., Faherty, D.A. and Gately, M.K. (1996) IL-12-deficient mice are defective in IFN-γ production and type 1 cytokine responses. *Immunity* 4, 471–481.

Maizels, R.M, Bundy, D.A.P., Selkirk, M.E., Smith, D.F. and Anderson, R.M. (1993) Immunological modulation and evasion by helminth parasites in human populations. *Nature* 365, 797–805.

Manson-Smith, D.F., Bruce, R.G. and Parrot, D.M.V. (1979) Villus atrophy and expulsion of intestinal *Trichinella spiralis* are mediated by T cells. *Cellular Immunology* 47, 285–293.

Mayer, L. and Schlien, R. (1987) Evidence for function of Ia molecules on gut epithelial cells in man. *Journal of Experimental Medicine* 166, 1471–1483.

McCoy, K., Camberis, M. and Le Gros, G. (1997) Protective immunity to nematode infection is induced by CTLA-4 blockade. *Journal of Experimental Medicine* 186, 183–187.

McKay, D.M., Benjamin, M., Baca-Estrada, M., D'Inca, R., Croitoru, K. and Perdue, M.H. (1995) Role of T lymphocytes in secretory response to an enteric nematode parasite. Studies in athymic rats. *Digestive Disease Sciences* 40, 331–337.

McKenzie, G.J., Emson, C.L., Bell, S.E., Anderson, S., Fallon, P., Zuraweski, G., Murray, R., Grencis, R.K. and McKenzie, A.N.J. (1998a) Impaired development of Th2 cells in IL-13-deficient mice. *Immunity* 9, 423–432.

McKenzie, G.J., Bancroft, A.J., Grencis, R.K. and McKenzie, A.N.J. (1998b) A distinct role for interleukin-13 in Th2-cell-mediated immune responses. *Current Biology* 8, 339–342.

McKenzie, G.J., Fallon, P.G., Emson, C.L., Grencis, R.K. and McKenzie, A.N.J. (1999) Simultaneous disruption of interleukin (IL)-4 and IL-13 defines individual roles in T helper type-2-mediated responses. *Journal of Experimental Medicine* 189, 1565–1572.

Medzhitov, R. and Janeway, C.A. (1997) Innate immunity: the virtues of a nonclonal system of recognition. *Cell* 91, 295–298.

Medzhitov, R. and Janeway, C.A. (1999) Innate immunity: impact on the adaptive immune response. *Current Opinion in Immunology* 9, 4–9.

Miller, H.R.P. (1984) The protective mucosal response against gastrointestinal nematodes in ruminants and laboratory animals. *Veterinary Immunology and Immunopathology* 6, 167–259.

Miller, H.R.P. (1996) Prospects for the immunological control of ruminant gastrointestinal nematodes: natural immunity, can it be harnessed? *International Journal for Parasitology* 26, 801–811.

Mosmann, T. and Sad, S. (1996) The expanding universe of T cell subsets: Th1, Th2 and more. *Immunology Today* 17, 138–146.

Mosmann, T.R., Cherwinski, H., Bond, M.W., Giedlin, M.A. and Coffman, R.L. (1986) Two types of murine helper T cell clone. I. Definition according to profiles of lymphokine activities and secreted proteins. *Journal of Immunology* 136, 2348–2357.

Mowat, A.M. and Viney, J.L. (1997) The anatomical basis of intestinal immunity. *Immunological Reviews* 156, 145–166.

Needham, C.S., Lillywhite, J.E., Didier, J.M., Bianco, A.E. and Bundy, D.A.P. (1994) Temporal changes in *Trichuris trichiura* infection and serum isotype responses in children. *Parasitology* 109, 197–200.

Neutra, M.R., Pringault, E. and Kraehenbuhl, J.P. (1996) Antigen sampling across epithelial barriers and induction of mucosal immune responses. *Annual Review of Immunology* 14, 275–300.

Newton, S.E. and Munn, E.A. (1999) The development of vaccines against gastro-intestinal nematode parasites, particularly *Haemonchus contortus. Parasitology Today* 15, 116–122.

Nokes, C. and Bundy, D.A.P. (1994) Does helminth infection affect mental process-ing and educational achievement? *Parasitology Today* 10, 1–18.

O'Garra, A. (1998) Cytokines induce the development of functionally hetero-geneous T helper cell subsets. *Immunity* 8, 275–283.

Okamura, H., Tsutsui, H., Komatsu, T., Yutsudo, M., Hakura, A., Tanimoto, T., Torigoe, K., Okura, T., Nukada, Y., Hattori, K., Akita, K., Namba, M., Tanabe, F., Konishi, K., Fukuda, S. and Kurimoto, M. (1995) Cloning of a new cytokine that induces IFN-γ production by T cells. *Nature* 378, 88–91.

Palmer, J.M. and Castro, G.A. (1986) Anamnestic stimulus-specific myoelectric responses associated with intestinal immunity in the rat. *American Journal of Physiology* 250, G266–G273.

Panja, A., Blumberg, R.S., Balk, S.P. and Mayer, L. (1993) CD1d is involved in T cell : epithelial cell interactions. *Journal of Experimental Medicine* 178, 1115–1119.

Paulus, U., Loeffler, M., Zeidler, J., Owen, G. and Potten, C.S. (1993) The differenti-ation and lineage development of goblet cells in the murine small intestinal crypt: experimental and modelling studies. *Journal of Cell Science* 106, 473–484.

Petit-Frere, C., Dugas, B., Braquet, P. and Mencia-Huerta, J.M. (1993) Interleukin 9 potentiates the IL-4 induced IgE and IgG1 release from murine B lymphocytes. *Immunology* 79, 146–151.

Piguet, P.F., Vesin, C., Guo, J., Donati, Y. and Barazzone, C. (1998) TNF-induced enterocyte apoptosis in mice is mediated by the TNF receptor 1 and does not require p53. *European Journal of Immunology* 28, 3499–3505.

Poudrier, J., van Essen, D., Morlaes-Alcelay, S., Leanderson, T., Bergthorsdottir, S. and Gray, D. (1998) CD40 ligand signals optimize T helper cell cytokine production: role in Th2 development and induction of germinal centres. *European Journal of Immunology* 28, 3371–3383.

Pritchard, D.I., Williams, D.J., Behnke, J.M. and Lee, T.D.G. (1983) The role of IgG1 hypergammaglobinaemia in immunity to the gastrointestinal nematode *Nematospiroides dubius.* The immunochemical purification, antigen specificity, *in vivo* anti-parasite effect of IgG1 from immune serum. *Immunology* 49, 353–365.

Pritchard, D.I., Quinnell, R.J., Slater, A.F.G., McKean, P.G., Dale, D.D.S., Raiko, A. and Keymer, A. (1990) Epidemiology and immunology of *Necator americanus* infection in a community in Papua New Guinea: humoral responses to excretory–secretory and cuticular collagen antigens. *Parasitology* 100, 317–326.

Pritchard, D.I., Quinnell, R.J. and Walsh, E.A. (1995) Immunity in humans to *Necator americanus*: IgE, parasite weight, and fecundity. *Parasite Immunology* 17, 71–75.

Pritchard, D.I., Hewitt, C. and Moqbel, R. (1997) The relationship between immunological responsiveness controlled by T helper 2 lymphocytes and infections with parasitic helminths. *Parasitology* S33–S44.

Prowse, S.J., Mitchell, G.F., Ey, P.L. and Jenkin, C.R. (1978) *Nematospiroides dubius*: susceptibility to infection and the development of resistance in hypothymic (nude) BALB/c mice. *Australian Journal of Experimental Biology and Medical Science* 56, 561–570.

Puddington, L., Olson, S. and Lefrancois, L. (1994) Interactions between stem cell factor and c-kit are required for intestinal immune homeostasis. *Immunity* 1, 733–739.

Pulendran, B., Linggappa, J., Kennedy, M.K., Smith, J., Teepe, M., Rudensky, A., Maliszewski, C.R. and Maraskovsky, E. (1997) Developmental pathways of dendritic cells *in vivo*: distinct function, phenotype, and localisation of dendritic cell subsets in FLT3 ligand-treated mice. *Journal of Immunology* 159, 2222–2231.

Quinnell, R.J., Woolhouse, M.E.J., Walsh, E.A. and Pritchard, D.I. (1995) Immunoepidemiology of human necatoriasis: correlations between antibody responses and parasite burdens. *Parasite Immunology* 17, 313–318.

Ramaswamy, K., Goodman, R.E. and Bell, R.G. (1994) Cytokine profile of protective anti-*Trichinella spiralis* CD4 Ox22– and non-protective CD4+ Ox22+ thoracic duct cells in rats: secretion of IL-4 alone does not determine protective capacity. *Parasite Immunology* 16, 435–445.

Reinecker, H.C., MacDermott, R.P., Mirua, S., Dignass, A. and Podolsky, D.K. (1996) Intestinal epithelial cells express and respond to IL-15. *Gastroenterology* 111, 1706–1713.

Rissoan, M.C., Soumelis, V., Kadowaki, N., Grouard, G., Briere, F., Malefyt, R.D.W. and Liu, Y.J. (1999) Reciprocal control of T helper cell and dendritic cell differentiation. *Science* 283, 1183–1186.

Roach, T.I.A., Else, K.J., Wakelin, D., McLaren, D.J. and Grencis, R.K. (1991) *Trichuris muris*: antigen recognition and transfer of immunity in mice by IgA monoclonal antibodies. *Parasite Immunology* 13, 1–12.

Robinson, K., Bellaby, T. and Wakelin, D. (1995) Immunity to *Trichinella spiralis* transferred by serum from vaccinated mice not protected by immunisation. *Parasite Immunology* 17, 85–90.

Romagnani, S. (1991) Human Th1 and Th2 subsets: doubt no more. *Immunology Today* 12, 256–257.

Sallusto, F., Mackay, C. and Lanzavecchia, A. (1997) Selective expression of the eotaxin receptor CCR3 by human T helper 2 cells. *Science* 277, 2005–2007.

Sallusto, F., Lanzavecchia, A. and Mackay, C.R. (1998) Chemokine and chemokine receptors in T-cell priming and Th1/Th2-mediated responses. *Immunology Today* 12, 569–574.

Santy, A., Dziejman, M., Taha, R.A., Iarossi, A.S., Neote, K., Garcia-Zepeda, E.A., Hamid, Q. and Luster, A.D. (1999) The T cell specific CXC chemokines IP-10, Mig, and I-TAC are expressed by activated human bronchial epithelial cells. *Journal of Immunology* 162, 3549–3558.

Scheuerermaly, C.C., Eckmann, L., Kagnoff, M.F., Falco, M.T. and Maly, F.E. (1994) Colonic epithelial cell lines as a source of interleukin-8 stimulation by inflammatory cytokines and bacterial lipopolysaccharide. *Immunology* 81, 85–91.

Schijns, V.E.C.J., Haagmans, B.L., Wierda, C.M.H., Kruithof, B., Heijnen, I.A.F.M., Alber, G. and Horzinek, M.C. (1998) Mice lacking IL-12 develop polarised Th1 cells during viral infection. *Journal of Immunology* 160, 3958–3964.

Schmitz, J., Thiel, A., Kühn, R., Rajewsky, K., Müller, W., Assenmacher, M. and Radbruch, A. (1994) Induction of interleukin (IL)-4 expression in T helper (Th) cells is not dependent on IL-4 from non-Th cells. *Journal of Experimental Medicine* 179, 1349–1353.

Scott, P. (1991) IFN-γ modulates the early development of Th1 and Th2 responses in a murine model of cutaneous leishmaniasis. *Journal of Immunology* 147, 3149–3155.

Scudamore, C.L., Thornton, E.M., McMillan, L., Newlands, G.F.J. and Miller, H.R.P. (1995) Release of the mucosal mast cell granule chymase, rat mast cell protease II during anaphylaxis is associated with the rapid development of paracellular permeability to macromolecules in rat jejunum. *Journal of Experimental Medicine* 182, 1871–1881.

Seder, R.A., Boulay, J.L., Finkelman, F.D., Barbier, S., Ben-Sasson, S.Z., Le Gros, G. and Paul, W.E. (1992a) CD8+ T cells can be primed *in vitro* to produce IL-4. *Journal of Immunology* 148, 1652–1656.

Seder, R.A., Paul, W.E., Davis, M.M. and Farekas de St Groth, B. (1992b) The presence of interleukin 4 during *in vitro* priming determines the lymphokine producing potential of CD4+ T cells from T cell receptor transgenic mice. *Journal of Experimental Medicine* 176, 1091–1098.

Seder R.A., Gazzinelli, R., Sher, A. and Paul, W.E. (1993) IL-12 acts directly on CD4+ T cells to enhance priming for IFN-γ production and diminishes IL-4 inhibition of such priming. *Proceedings of the National Academy of Sciences USA* 90, 10188–10192.

Siegal, F.P., Kadowaki, N., Shodell, M., Fitzgerald-Bocarsly, P.A., Shah, K., Ho, S., Antonenko, S. and Liu, Y.J. (1999) The nature of principal type 1 interferon-producing cells in human blood. *Science* 284, 1835–1837.

Siveke, J.T. and Hamann, A. (1998) T helper 1 and T helper 2 cells respond differentially to chemokines. *Journal of Immunology* 160, 550–554.

Soltys, J., Goyal, P.K. and Wakelin, D. (1999) Cellular immune responses in mice infected with the intestinal nematode *Trichuris muris*. *Experimental Parasitology* 92, 40–47.

Song, F., Ito, K., Denning, T.L., Kuninger, D., Papaconstantinou, J., Gourley, W., Klimpel, G., Balish, E., Hokanson, J. and Ernst, P.B. (1999) Expression of the neutrophil chemokine KC in the colon of mice with enterocolitis and by intestinal epithelial cell lines: effects of flora and proinflammatory cytokines. *Journal of Immunology* 162, 2275–2280.

Springer, T.A. (1994) Traffic signals for lymphocyte recirculation and leukocyte emigration: the multistep paradigm. *Cell* 76, 301–314.

Stadnyk, A.W., Sisson, G.R. and Waterhouse, C.C.M. (1995) IL-1α is constitutively expressed in the rat intestinal epithelial cell line IEC-6. *Experimental Cell Research* 220, 298–303.

Subramanian, G., Kazura, J.W., Pearlman, E., Jia, X., Malhorta, I. and King, C.L. (1997) B7-2 requirement for helminth induced granuloma formation and CD4 type 2 T helper cell cytokine expression. *Journal of Immunology* 158, 5914–5920.

Suss, G. and Shortman, K. (1996) A subclass of dendritic cells kills CD4 T cells via Fas/Fas-ligand-induced apoptosis. *Journal of Experimental Medicine* 183, 1789–1796.

Svetic, A., Madden, K.B., Zhou, X., Lu, P., Katona, I.M., Finkelman, F.D., Urban, J.F. and Gause, W.C. (1993) A primary intestinal helminthic infection rapidly induces a gut associated elevation of Th2-associated cytokines and IL-3. *Journal of Immunology* 150, 3434–3441.

Sydora, B.C., Mixter, P.F., Holcombe, H.R., Eghtesady, P., Williams, K., Amaral, M.C., Nel, A. and Kronenberg, M. (1993) Intestinal intraepithelial lymphocytes are activated and cytolytic but do not proliferate as well as other T cells in response to mitogenic signals. *Journal of Immunology* 150, 2179–2191.

Symons, L.E.A. (1965) Kinetics of epithelial cells and morphology of villi and crypts in the jejunum of the rat infected with *Nippostrongylus brasiliensis*. *Gastroenterology* 49, 158–168.

Symons, L.E.A. (1969) Pathology of gastrointestinal helminthiases. *International Review of Tropical Medicine* 3, 49–100.

Symons, L.E.A. (1978) Epithelial cell mitosis and morphology in worm free regions of the intestine of the rat infected with *N. brasiliensis*. *Journal of Parasitology* 64, 958–959.

Takeda, K., Tanake, T., Shi, W., Matsumoto, M., Minami, M., Kashiwamura, S., Nakanishi, K., Yoshida, N., Kishimoto, T. and Akira, S. (1996) Essential role of Stat6 in IL-4 signalling. *Nature* 380, 627–630.

Takeda, K., Tsutsui, H., Yoshimoto, T., Adachi, O., Yoshida, N., Kishimoto, T., Okamura, H., Nakanishi, K. and Akira, S. (1998) Defective NK cell activity and Th1 response in IL-18-deficient mice. *Immunity* 8, 383–390.

Takeuchi, M., Nishizaki, Y., Sano, O., Ohta, T., Ikeda, M. and Kurimoto, M. (1997) Immunohistochemical and immuno-electron-microscopic detection of interferon-gamma-inducing factor ('interleukin 18') in mouse intestinal epithelial cells. *Cell and Tissue Research* 289, 499–503.

Takia, T., Li, M., Sylvestre, D., Clynes, R. and Ravetch, J.V. (1994) FcRγ chain deletion results in pleiotropic effector cell defects. *Cell* 76, 519–529.

Thompson-Snipes, L., Dhar, V., Bond, M.W., Mosmann, T.R., Moore, K.W. and Rennick, D.M. (1991) Interleukin 10: a novel stimulatory factor for mast cells and their progenitors. *Journal of Experimental Medicine* 173, 507–510.

Tsuyuki, S., Tsuyuki, J., Einsle, K., Kopf, M. and Coyle, A.J. (1997) Co-stimulation through B7-2 (CD86) is required for the induction of a lung mucosal T helper cell 2 (Th2) immune response and altered airway responsiveness. *Journal of Experimental Medicine* 185, 1671–1679.

Ulevitch, R.J. and Tobias, P.S. (1999) Recognition of gram-negative bacteria and endotoxin by the innate immune system. *Current Opinion in Immunology* 11, 19–22.

Ungar, B.L.P., Kao, T.C., Burpis, J.A. and Finkelman, F.D. (1991) Cryptosporidium infection in the adult mouse model. Independent roles for IFN-γ and CD4+ T lymphocytes in protective immunity. *Journal of Immunology* 147, 1014–1022.

Urban, J.F., Katona, I.M. and Finkelman, F.D. (1991a) *Heligmosomoides polygyrus*: CD4+ cells but not CD8+ T cells regulate the IgE response and protective immunity in mice. *Experimental Parasitology* 73, 500–511.

Urban, J.F., Katona, I.M., Paul, W.E. and Finkelman, F.D. (1991b) IL-4 is important in protective immunity to a gastrointestinal nematode infection in mice. *Proceedings of National Academy of Sciences USA* 88, 5513–5517.

Urban, J.F., Madden, K.B., Cheever, A.W., Trotta, P.P., Katona, I.M. and Finkelman, F.D. (1993) Interferon inhibits inflammatory responses and protective immunity in mice infected with the nematode parasite, *Nippostrongylus brasiliensis*. *Journal of Immunology* 151, 7086–7094.

Urban, J.F., Maliszewski, C.R., Madden, K.B., Katona, I.M. and Finkelman, F.D. (1995) IL-4 treatment can cure established gastrointestinal nematode infections in immunocompetent and immunodeficient mice. *Journal of Immunology* 154, 4675–4684.

Urban, J.F., Noben-Trauth, N., Donaldson, D.D., Madden, K.B., Morris, S.C., Collins, M. and Finkelman, F.D. (1998) IL-13, IL-4Rα, and stat6 are required for the expulsion of the gastrointestinal nematode parasite *Nippostrongylus brasiliensis*. *Immunity* 8, 255–264.

Uyttenhove, C., Simpson, R.J. and van Snick, J. (1988) Functional and structural characterisation of P40, a mouse glycoprotein with T cell growth factor activity. *Proceedings of the National Academy of Sciences USA* 85, 6934–6938.

Vallance, B.A. (1999) The immunomodulation of intestinal smooth muscle function. PhD thesis, McMaster University, Canada.

Vallance, B.A., Blennerhasset, P.A. and Collins, S.M. (1997) Increased intestinal muscle contractility and worm expulsion in nematode-infected mice. *American Journal of Physiology* 272, G321–G327.

Vallance, B.A., Croitoru, K. and Collins, S.M. (1998) T lymphocyte dependent and independent smooth muscle dysfunction in the *T. spiralis* infected mouse. *American Journal of Physiology* 275, G1157–G1165.

van Essen, D., Kikutani, H. and Gray, D. (1995) CD40 ligand-transduced co-stimulation of T cells in the development of helper function. *Nature* 378, 620–623.

Vassalli, P. (1992) The pathophysiology of tumour necrosis factor. *Annual Review of Immunology* 10, 411–452.

Vos, J.G., Ruitenberg, E.J., Van Basten, N., Buys, J., Elgersma, A. and Kruizinga, W. (1983) The athymic nude rat. IV. Immunocytochemical study to detect T cells, and immunological and histopathological reactions against *Trichinella spiralis*. *Parasite Immunology* 5, 195–215.

Wahid, F.N., Behnke, J.M., Grencis, R.K., Else, K.J. and Ben-Smith, A.W. (1994) Immunological relationships during primary infection with *Heligmosomoides polygyrus*: Th2 cytokines and primary response phenotype. *Parasitology* 108, 461–471.

Wakelin, D. (1978) Immunity to intestinal parasites. *Nature* 273, 617–620.

Wakelin, D. (1993) Allergic inflammation as a hypothesis for the expulsion of worms from tissues. *Parasitology Today* 9, 115–116.

Wakelin, D. and Donochie, A.M. (1983) Genetic control of immunity to *Trichinella spiralis*: influence of H-2-linked genes on immunity to the intestinal phase of infection. *Immunology* 48, 343–350.

Watanabe, Y., Sudo, T., Minato, N., Ohnishi, A. and Katsura, Y. (1995) IL-7 preferentially supports the growth of γδ T cells from foetal thymocytes *in vitro*. *International Immunology* 3, 1067–1075.

Weisbrodt, N.W., Lai, M., Bowers, R.L., Harari, Y. and Castro, G.A. (1994) Structural and molecular changes in intestinal smooth muscle induced by *Trichinella spiralis* infection. *American Journal of Physiology* 266, G856–G862.

Williams, D.J. and Behnke, J.M. (1983) Host protective antibodies and serum immunoglobulin isotypes in mice chronically infected or repeatedly immunized with the nematode parasite *Nematospiroides dubius*. *Immunology* 48, 37–47.

Wilson, A.D., Bland, P.W. and Stokes, C.R. (1990) Expression and distribution of Ia antigen in the murine small intesine: influence of environment and cholera toxin. *International Archives of Allergy and Applied Immunology* 91, 348–353.

World Health Organisation (1996) *The World Health Report 1996: Fighting Disease, Fostering Development*. WHO, Geneva.

Yamamura, M., Uyemura, K., Deans, R.J., Weinberg, K., Rea, T.H., Bloom, B.R. and Modlin, R.L. (1991) Defining protective responses to pathogens: cytokine profiles in leprosy lesions. *Science* 254, 277–279.

Yoshida, T., Ikuta, K., Sugaya, H., Maki, K., Takagi, M., Kanazawa, H., Sunaga, S., Kinashi, T., Yoshimura, K., Miyazaki, J.I., Takaki, S. and Takatsu, K. (1996) Defective B-1 cell development and impaired immunity against *Angiostrongylus cantonensis* in IL-5Rα deficient mice. *Immunity* 4, 483–494.

Yoshimoto, T., Bendelac, A., Watson, C., Hu-li, J. and Paul, W.E. (1995) Role of NK1.1 T cells in a Th2 response and in immunoglobulin E production. *Science* 270, 1845–1847.

Yoshimoto, T., Okamura, H., Tagawa, Y., Iwakura, Y. and Nakanishi, K. (1997) Interleukin 18 together with interleukin 12 inhibits IgE production by induction of interferon-γ production from activated B cells. *Proceedings of the National Academy of Sciences USA* 94, 3948–3953.

Zhang, S. and Castro, G.A. (1990) Involvement of type 1 hypersensitivity in rapid rejection of *Trichinella spiralis* from adult rats. *International Archives of Allergy and Applied Immunology* 93, 272–279.

Zingoni, A., Soto, H., Hedrick, J.A., Stoppacciaro, A., Storlazzi, C., Sinigaglia, F., D'Ambrosia, D., O'Garra, A., Robinson, D., Rocchi, M., Santoni, A., Zlotnik, A. and Napolitano, M. (1998) The chemokine receptor CCR8 is preferentially expressed in Th2 but not Th1 cells. *Journal of Immunology* 161, 547–551.

Gut Immunopathology in Helminth Infections – Paradigm Lost?

Catherine E. Lawrence,[1] Malcolm W. Kennedy[1] and Paul Garside[2]

[1]Division of Infection and Immunity, Institute of Biomedical and Life Sciences, University of Glasgow, Glasgow G12 8QQ, UK; [2]Department of Immunology, University of Glasgow, Glasgow G11 6NT, UK

Introduction

Gastrointestinal (GI) nematode infections in humans and domestic animals cause significant morbidity and mortality throughout the developing world, yet the existence and nature of protective mechanisms against these parasites remain unclear. Although a long history of experimental studies in laboratory rodents has provided a detailed knowledge of the immunology of protective responses against GI nematodes, the precise mechanisms that bring about loss of worms from the gut have yet to be defined. The consensus that has arisen from many experimental studies is that protective responses are dependent upon the pathological changes that infection induces. This idea was first formulated in 1975 by Larsh and Race in their concept of allergic inflammation (Larsh, 1975). It is crucial to know whether or not this hypothesis is correct, for if vaccines against intestinal nematodes are to be developed, it will be important to ensure that they provide protection without inducing unacceptable pathological responses.

Nematodes occupy a variety of niches within the intestine – luminal (e.g. *Ascaris* spp., *Nippostrongylus brasiliensis*), mucosal surface (e.g. hookworms), intraepithelial (e.g. *Trichinella spiralis*, *Trichuris* spp.) or tissue-penetrating (e.g. *Heligmosomoides polygyrus*, strongyles) – and the pathology associated with them varies according to the particular host–parasite combination. The intestinal pathology (enteropathy) may have several causes, including direct damage from the attachment, migration, burrowing and feeding activities of the worm, or secondary damage resulting from opportunistic bacterial infection, or the host's immune response against the parasite or bacteria.

The relative contributions of parasite and immune response to enteropathy remain unclear in many infections, as does the requirement for intestinal inflammation in protection. The immune response may actually provide no benefit to the host in terms of limiting parasite survival or fecundity, instead resulting in damaging pathological reactions. Thus, there are several scenarios: the parasite may be expelled by an immune response that is temporally associated with a mucosal lesion (*T. spiralis* in responder strains of mouse) (Garside *et al.*, 1992a; Lawrence *et al.*, 1998); the parasite may survive in the face of such a reaction (*Ascaris suum* in pigs; hookworms; *T. spiralis*-infected hamsters) (Stephenson *et al.*, 1980; Behnke *et al.*, 1994a,b; Prociv, 1997); the parasite may be expelled with little or no apparent pathology (*Trichuris muris* in high-responder strains of mouse) (Else and Grencis, 1991; Else *et al.*, 1992); or the parasite may survive without an obvious intestinal lesion (low-level infections with *T. trichiura* in humans and *T. muris* in low-responder mouse strains) (Else and Grencis, 1991; MacDonald *et al.*, 1991; Else *et al.*, 1992). While much effort has concentrated on dissecting the protective immune response to a variety of GI nematodes, there has been relatively little attention paid to the mechanisms controlling the accompanying intestinal pathology, and to whether or not this is an important and necessary component of protection.

In those infections that are associated with enteropathy (exemplified by *T. spiralis*), no experimental manipulation has, until recently, been able to separate enteropathy and immune expulsion – if one is abrogated, so is the other. This chapter illustrates how the two processes can be separated, and discusses implications of this for understanding immune expulsion of gut nematodes and the prospects for anti-nematode vaccines that cause no ill effects at either the initial induction of immunity or the expression of protective responses. The definition of that which consitututes enteropathy may vary between authors, but we take as our primary definition the most destructive and quantifiable changes in intestinal tissue that are associated with expulsion, villus atrophy and crypt hyperplasia.

Pathology of Nematode Infections

The morbidity and mortality that are often associated with human GI helminth infections reflect in part the nutritional consequences of diarrhoea and malabsorption, and the resulting malnutrition that can accentuate the effects of infection by suppressing the protective immune response as well as compromising intestinal repair (Ferguson *et al.*, 1980; Keymer and Tarlton, 1991; Cooper *et al.*, 1992). In experimental rodents the pathology associated with infection is characterized by villus atrophy, crypt hyperplasia, goblet cell hyperplasia and infiltration of the mucosa by a variety of inflammatory cells, of which eosinophils and mast cells are prominent (Manson-Smith *et al.*, 1979; Wang *et al.*, 1990; Garside *et al.*, 1992a).

A requirement for gross intestinal pathology in the expulsion of nematodes has been widely accepted because the two phenomena are usually coincident in immunologically normal hosts. Two explanations for the association between expulsion and pathology have been proposed. Firstly, that pathological changes create an unfavourable environment for the parasites, which are thereby forced from their preferred niches or, secondly, that the local inflammatory response (through increased mucosal permeability) results in increased exposure to components of the systemic immune system (Wakelin, 1978). Studies in nude, thymectomized or cyclosporin A-treated mice have established the T cell dependency of parasite expulsion and much of the accompanying intestinal pathology (Manson-Smith *et al.*, 1979; Garside *et al.*, 1992a). Cell transfer studies further demonstrated that protection against gastrointestinal nematodes is mediated by CD4+ T helper cells (Grencis *et al.*, 1985; Riedlinger *et al.*, 1986). More recently (as described in Chapter 17) protective responses have been associated with production of Th2 cytokines, including IL-3, IL-4, IL-5, IL-9 and IL-13, and anti-parasite IgE and IgG1 antibody responses (Grencis *et al.*, 1991; Svetic *et al.*, 1993; Else *et al.*, 1994; Faulkner *et al.*, 1997). Th1-mediated events appear unimportant or antagonistic in terms of protection (Urban *et al.*, 1993; Else *et al.*, 1994; Ishikawa *et al.*, 1998).

Enteropathies of Other Aetiologies

Intestinal inflammation, regardless of specific initiating events, shares common immunologically mediated pathways of tissue injury and repair. Activation of immune, mesenchymal and epithelial cells, recruitment of circulating effector cells, and tissue damage and repair are the consequence of a complex balance of cytokines that are produced by a variety of cells within the intestine. Cytokines play a key role in the pathogenesis of inflammatory bowel diseases but have not been investigated in detail in the enteropathy accompanying the expulsion of gastrointestinal nematode parasites.

The nature of the pathology associated with immune rejection of many intestinal nematodes shows gross similarities to a number of immunologically mediated enteropathies such as graft-versus-host disease (GvHD) and inflammatory bowel disease (IBD) (MacDonald and Spencer, 1988; Garside, 1999). Cytokine and T cell receptor-deficient mice have proved useful in determining the role of the immune response in such changes (Kuhn *et al.*, 1993; Kramer *et al.*, 1995; Mizoguchi *et al.*, 1996). For example, in GvHD, small intestinal damage appears to progress through a proliferative phase, comprising crypt hyperplasia, enhanced natural killer cell activity, increased expression of major histocompatibility complex (MHC) class II antigens, mucosal mast cell (MMC) hyperplasia, lymphocytic infiltration and goblet cell hyperplasia, to a destructive phase characterized by villus atrophy, immunosuppression, malabsorption and diarrhoea. However, whereas immune rejection of the majority of gastrointestinal nematodes is a Th2-dependent process, intestinal pathology in GvHD and IBD is usually associated with Th1-type cytokines and particularly TNF-α (Piguet et al., 1987; Mowat, 1989; Garside et al., 1993; Murch et al., 1995). Since many pathological features are common to parasite-induced lesions and Th1-mediated enteropathies, there may be mechanisms in common. This has been studied by using mice defective for cytokines (or their receptors) to investigate cytokine regulation of both immunopathology and parasite expulsion.

The Paradox

Whilst the expulsion of intestinal nematodes is absolutely Th2-mediated (Chapter 17 and below), the associated enteropathy is indistinguishable from that previously attributed to Th1-like responses. The solution to this paradox, which is addressed below, may therefore reside in the recognition of common mediators of tissue destruction, remodelling and repair, induced by both Th1 and Th2 cytokine-dependent processes.

Pathology and Nematode Expulsion

We have used a sensitive microdissection technique (MacDonald and Ferguson, 1977) to analyse the effects of *T. spiralis* infection on the development of enteropathy and the role the latter plays in parasite expulsion. In this, the vital statistics of the functional unit of the small intestine are determined as the length of villus and the depth of the crypts, along with the number of mitotic figures per crypt as an indicator of cell proliferation. Initial studies demonstrated that the kinetics of parasite expulsion and development of intestinal pathology were not precisely synchronous, suggesting that intestinal pathology and the expulsion of these parasites were not necessarily interdependent (Lawrence et al., 1998). Figure 18.1 shows that the development of intestinal pathology, as assessed by a decrease in villus height and increases in crypt depth and the proliferative index of the crypts, precedes the loss of the parasite and is beginning to resolve by the time the parasite is lost.

Immune expulsion of *T. spiralis* is clearly Th2- and, specifically, IL-4-dependent. However, contrary to expectations, enteropathy (as assessed by changes in villus/crypt ratios) is regulated by IL-4 and not IFN-γ (Lawrence et al., 1998). Moreover, the usual severe pathology is not induced in p55 TNF receptor (TNF-R1) gene-deficient mice, which nevertheless expelled

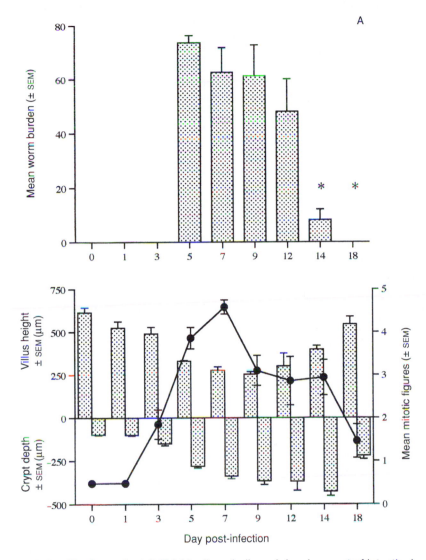

Fig. 18.1. Burdens of adult *Trichinella spiralis* and development of intestinal pathology. BALB/c were infected with 400 *T. spiralis* muscle larvae on day 0. (A) Adult worm burdens are represented as mean ± SEM, five mice per group (*, significantly different from day 5 p.i., $P < 0.01$). (B) Crypt lengths, villus lengths and number of mitotic figures per crypt were measured at intervals throughout the infection. Bars represent villus and crypt lengths; filled circles represent number of mitotic figures per crypt. Results are shown as mean ± SEM for five mice per group. Villus lengths were significantly reduced compared with uninfected controls from day 3 until day 14 p.i. ($P < 0.05$). Crypt lengths and number of mitotic figures were significantly greater than controls from day 3 until day 18 p.i. ($P < 0.01$). Reproduced from Lawrence *et al.* (1998), with permission.

their parasites (Fig. 18.2). The results were even more surprising as tumour necrosis factor (TNF) is usually considered a Th1-associated cytokine and down-regulated by IL-4 (Mosmann and Coffman, 1989; Joyce and Steer, 1996; Bennett *et al.*, 1997).

In contrast with other enteropathies, there is no evidence that IFN-γ plays an important role in either protection or pathology in *T. spiralis* infections. Infections in mice treated with anti-IFN-γ antibody or in IFN-γR-deficient mice (Rose *et al.*, 1994; Lawrence *et al.*, 1998) were comparable in time course and pathology to those in controls. However, crypt hyperplasia was reduced following immunodepletion of IFN-γ in SCID mice infected

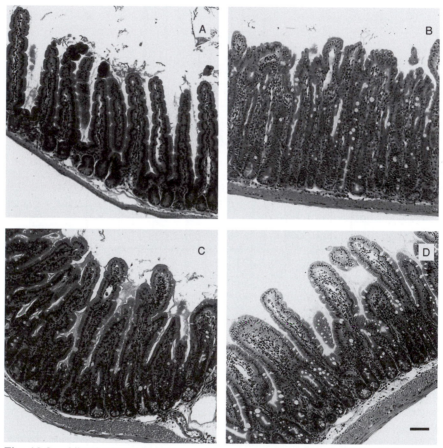

Fig. 18.2. Histological appearance of haematoxylin and eosin-stained jejunum of: (A) uninfected control, with villi and crypts of normal lengths; (B) day 13 p.i. *T. spiralis*-infected control, showing crypt hyperplasia and villus atrophy; (C) day 13 p.i. *T. spiralis*-infected TNF-R1$^{-/-}$ mice, showing normal appearance, similar to (A); (D) day 13 p.i. *T. spiralis*-infected IL-4$^{-/-}$ mice, again showing a normal appearance. Bar represents 100 μm. Reproduced from Lawrence *et al.* (1998), with permission.

with *T. muris*, although no assessment of effects on protection could be made since the parasites are not expelled in these mice (Artis *et al.*, 1999b).

The results of these studies lead to two main conclusions that prompt re-examination of the roles of immunopathology in the expulsion of gastrointestinal nematode parasites and of IL-4 in enteropathies. Firstly, whilst expulsion of *T. spiralis* was temporally associated with severe intestinal pathology in normal mice, immune expulsion occurred with minimal pathology in TNF-R1-deficient mice. Secondly, a novel role for IL-4 was indicated by its regulation of the enteropathy accompanying *T. spiralis* infection. Thus, although protection against the parasite is dependent upon an IL-4- and Th2-mediated response, the accompanying pathology is TNF-mediated but IL-4-dependent.

Tumour Necrosis Factor Induces Pathology via iNOS

It has been shown in other parasite infections that TNF mediates pathological effects via the production of inducible nitric oxide synthase (iNOS) (James, 1995). Furthermore, neutralization of iNOS in other models of intestinal pathology can ameliorate enteropathy (Garside *et al.*, 1992b; Hogaboam *et al.*, 1995) and it has been demonstrated that NO can protect against a number of helminth infections (Oswald *et al.*, 1994; Rajan *et al.*, 1996, Taylor *et al.*, 1996). Following infection of iNOS-deficient mice with *T. spiralis*, we were able to demonstrate that, like TNF, iNOS is involved in the pathogenic but not in the protective response to *T. spiralis* (Lawrence *et al.*, 2000).

An overproduction of TNF-α by epithelial cells has been suggested to be responsible for the pathology induced in a number of models of IBD (Jung *et al.*, 1995; Neurath *et al.*, 1997; Corazza *et al.*, 1999). Neutralization with antibody, inhibitors or the use of TNF- or TNF-R-deficient mice also ameliorates enteropathy in a number of models (Deckert-Schluter *et al.*, 1998; Peschon *et al.*, 1998; Higgins *et al.*, 1999). However, the source of this cytokine and its mechanism of action have yet to be determined. TNF-α has the potential to be involved at a number of levels in the development of intestinal inflammation. Locally, TNF induces upregulation of adhesion molecule and chemokine expression and changes in intestinal permeability (Hirribarren *et al.*, 1993; Haraldsen *et al.*, 1996; Hornung *et al.*, 2000). TNF also acts directly on cells of the epithelium to induce their death by either necrosis or apoptosis (Guy-Grand *et al.*, 1998; Piguet *et al.*, 1998). TNF promotes the production of inflammatory mediators such as NO (Drapier *et al.*, 1988) and tissue-degrading enzymes such as matrix metalloproteases (Pender *et al.*, 1997) in addition to potentiating cytokine production by cells including monocytes, macrophages, lymphocytes, eosinophils, neutrophils and mast cells (Reinecker *et al.*, 1991; Lukacs *et al.*, 1995; Cope *et al.*, 1997; Furuta *et al.*, 1997; Corazza *et al.*, 1999).

Another possible role for TNF in GI helminth infection is the induction of Th2 responses. Although some components of the Th2 response appear to be reduced in TNF-R1-deficient mice infected with *T. spiralis* (e.g. mucosal mastocytosis and Th2-dependent antibodies) the response appears, nonetheless, to be sufficient to induce a protective but not a pathological response. The increased crypt cell turnover in TNF-R1-deficient mice, accompanied by changes in intestinal physiology, may be sufficient to expel *T. spiralis*; indeed, it has been suggested previously that there is considerable redundancy in the mechanisms leading to worm expulsion (Urban *et al.*, 1995, Finkelman *et al.*, 1997). This emphasizes the finding that not all components of intestinal pathology are necessary for protection against *T. spiralis*, but the relative roles of TNF and IL-4, and the mechanisms regulated by each, require further study. It is interesting to note that TNF-R1- or TNF-R2-deficient mice infected with *T. muris*, a parasite expelled by an IL-13-dependent, mast cell-independent mechanism (Bancroft *et al.*, 1998; Betts and Else, 1999), were unable to expel the parasite and the Th2 response was substantially reduced (Artis *et al.*, 1999a). *In vivo* blockade of TNF-α, although not impairing the Th2 response, could also delay worm expulsion (Artis *et al.*, 1999a). It was therefore suggested that TNF-α played a role in regulating Th2 effector activity, possibly by regulating IL-4 and IL-13 receptor expression on cells in the intestinal microenvironment (Lugli *et al.*, 1997).

Th2 Involvement in Pathological Responses in Other Helminth Infections

Examination of pathology and protection in non-intestinal helminth infections has also suggested that Th2 responses may primarily be pathological. For instance, IL-4-deficient mice infected with *Onchocerca volvulus* showed reduced corneal damage (Pearlman *et al.*, 1995). In murine *Schistosoma mansoni* infections, where TNF-α plays a crucial role in the pathology associated with pulmonary granulomata and Th2 responses are protective, IL-4-defective mice had significant reduced pathology (Pearce *et al.*, 1996). Mast cell products have also been shown to induce epithelial injury at villus tips and this may be one mechanism by which villus atrophy is induced (Perdue *et al.*, 1989). (The absence of a mastocytosis in TNF-R1- and IL-4-deficient animals may explain why there is no villus atrophy despite a crypt hyperplasia in TNF-R1-deficient mice.)

The reason for the long-standing association between intestinal pathology and protection against GI nematodes, and the apparent discrepancy between Th2-mediated protection and Th1-mediated pathology, can be rationalized through a central role of IL-4 in both processes. Our data (Lawrence *et al.*, 1998) show that, in *T. spiralis* infections, pathology is regulated by IL-4, although TNF is also necessary. Interestingly, there is gathering evidence of a role for Th2 cytokines in enteropathies of other aetiologies, elevated levels of IL-4, IL-5, IL-6 and IL-10 and decreased levels of IFN-γ and IL-2 having been demonstrated in a number of intestinal pathologies (Boirivant *et al.*, 1998; Iijima *et al.*, 1999; Mizoguchi *et al.*, 1999).

Effector Cells in Induction of Enteropathy

In order to understand fully the relationship between protective and pathological Th2 responses in GI helminth infection, it will be necessary to identify the sources of the mediators that have been implicated in each phenomenon. It has been established that MMCs play a role in a range of inflammatory and immunological events. Furthermore, mast cells are thought to be crucial effector cells in expulsion of nematode parasites and this role has been well documented in infections with *N. brasiliensis*, *Strongyloides venezuelensis* and *T. spiralis* (Befus and Bienenstock, 1979; Abe and Nawa, 1988; Khan *et al.*, 1993; Donaldson *et al.*, 1996; Lantz *et al.*, 1998). However, this point is debatable because there is also evidence suggesting that these cells are not essential for worm expulsion even in some of the above species, and appear unimportant in responses against other species such as *T. muris* (Betts and Else, 1999). Infection of mast cell-deficient W/Wᵛ mice or the use of antibodies against IL-3 or IL-4 (which block mastocytosis) have failed to prevent expulsion of *N. brasiliensis* (Crowle and Reed, 1981; Madden *et al.*, 1991). Furthermore, while Stat6 knockout mice generate a substantial mucosal mastocytosis, they fail to expel *N. brasiliensis* (Stat6 is activated following the binding of IL-4 or IL-13 to the IL-4Rα chain, and hence these mice can be considered Th2 knockout) (Urban *et al.*, 1998). Conversely, mucosal mastocytosis and parasite expulsion are both inhibited in the same Stat6-deficient mice infected with *T. spiralis* (Urban *et al.*, 2000). This suggests that, although Stat6 signalling is required for expulsion of these parasites, multiple mechanisms can effect protection, and possibly also pathology.

Upon activation, mast cells release numerous mediators, including vasoactive amines, proteases, pro-inflammatory cytokines (e.g. IL-1β, IL-6, IL-18 and TNF-α) and also regulatory Th2 cytokines (e.g. IL-4, IL-10 and IL-13) (Burd *et al.*, 1989; Gordon and Galli, 1990; Marietta *et al.*, 1996; Toru *et al.*, 1998; Aoki *et al.*, 1999; Lorentz *et al.*, 2000). Therefore, the mastocytosis in the infected mucosa represents an immunopathological rather than a protective response. Indeed, our studies have shown that expulsion of *T. spiralis* from TNF-R1⁻/⁻ or iNOS⁻/⁻ mice was achieved in the absence of a substantial mastocytosis and subsequent amelioration of enteropathy (Lawrence *et al.*, 1998, 2000).

Infection of W/Wᵛ mast-cell-deficient mice with *T. muris* showed that although mast cells were not important for protection they appeared to be important for the generation of the Th2 responses (Koyama and Ito, 2000).

Additionally, mast cells appear to be involved in the regulation of T and B cell function, and several studies have provided evidence that mast cells can act as antigen-presenting cells *in vitro*, a function that is augmented by IL-4 and abrogated by IFN-γ (Frandji *et al.*, 1993, 1995; Bhattacharyya *et al.*, 1998; Inamura *et al.*, 1998; Aoki *et al.*, 1999). Activated mast cells also migrate to local lymph nodes, indicating their potential to influence naive T cell activation and differentiation *in vivo* (Huels *et al.*, 1995; Frandji *et al.*, 1996; Wang *et al.*, 1998). Thus, there are a number of mechanisms by which mast cells can influence the development of protective and pathological responses or modulate ongoing responses, both in the periphery and the gut, rather than merely functioning as endpoint effector cells. This point is illustrated by the demonstration that, while infection of W/W^v mast cell-deficient mice with *T. spiralis* resulted in delayed expulsion, it was also associated with decreased enteropathy and Th2 responses (Fig. 18.3). Therefore, the part that mast cells and their products play in both protection and pathology in *T. spiralis* infection, as well as their role in generating a Th2 response, warrants further investigation.

Whilst eosinophils appear unimportant in the induction of protective responses to GI helminths, they are present in large numbers in the inflamed gut and it has therefore been suggested that they play a part in the induction of enteropathy. Moreover, eosinophils have been implicated in the induction of intestinal inflammation eosinophilic gastroenteritis, ulcerative colitis and Crohn's disease. However, IL-5-deficient mice, or GM-CSF transgenic mice (which typically have a blood eosinophilia of approximately 25%) infected with *T. spiralis* did not show a significant exacerbation or amelioration of either protective or pathological responses (C.E. Lawrence, unpublished observation).

Pathophysiological Effects of Helminth-induced Cytokines

Although it appears that severe IL-4-regulated enteropathy is not required for immune expulsion of *T. spiralis*, it is still possible that Th2 cytokines can act in a direct fashion to create an environment unfavourable for intestinal parasites. It remains to be shown directly whether these effects are sufficient to expel parasites. Indeed, there is considerable evidence to support a variety of pathophysiological effects of IL-4 and/or TNF on the gut. These effects may be mediated by factors including cytokines and mast-cell products (e.g. leukotrienes and 5-hydroxytryptamine). *T. spiralis* infections result in increased fluid and mucus secretion into the lumen as well as increased intestinal propulsive activity and more rapid intestinal transit (Castro *et al.*, 1979; Russell, 1986; Vermillion and Collins, 1988; Vermillion *et al.*, 1991; Weisbrodt *et al.*, 1994; Barbara *et al.*, 1997). The increased contractility of radial and longitudinal muscle is greater in high-

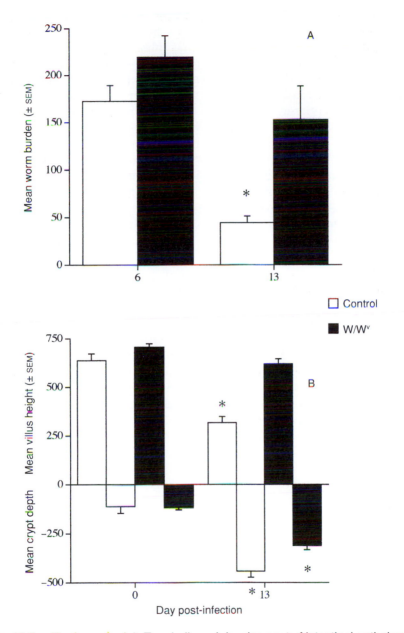

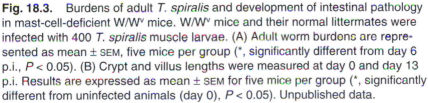

Fig. 18.3. Burdens of adult *T. spiralis* and development of intestinal pathology in mast-cell-deficient W/W^v mice. W/W^v mice and their normal littermates were infected with 400 *T. spiralis* muscle larvae. (A) Adult worm burdens are represented as mean ± SEM, five mice per group (*, significantly different from day 6 p.i., $P < 0.05$). (B) Crypt and villus lengths were measured at day 0 and day 13 p.i. Results are expressed as mean ± SEM for five mice per group (*, significantly different from uninfected animals (day 0), $P < 0.05$). Unpublished data.

than in slow-responder mice and is ameliorated in SCID and CD4+ deficient mice, suggesting a role for CD4+ T lymphocytes. More recently it has been suggested that both IL-4 and IL-5 may play a role in the intestinal neuromuscular dysfunction in *T. spiralis* infection (Vallance *et al.*, 1999a,b). These alterations have also been shown to be mediated by TNF and iNOS in other models of intestinal motility dysfunction (Hogaboam *et al.*, 1995, 1997). However, these changes may be independent of the hyperplasia and hypertrophy of the muscularis, as we have observed no substantial differences in the thickening of the muscularis in infected wild-type, TNF-R1- or IL-4-deficient mice (Lawrence *et al.*, 1998). The effects of infection on smooth muscle contractility in IFN-γR, TNF-R1, IL-4 and iNOS gene-deficient mice therefore require examination. The role of mast cells in the induction of these responses could also be determined by analysing the effects of infection on intestinal neuromuscular function in mast cell-deficient W/W^v mice.

Goblet Cells in Enteropathy

Increased numbers of goblet cells (GCs) and qualitative changes in mucus secretions are coincident with infection with a number of nematode parasites and it has been proposed that mucin proteins mediate this response by enveloping the parasites and/or interrupting attachment (Nawa *et al.*, 1994). However, the role of GCs and mucus in the generation of a protective response versus its role in resolving intestinal inflammation following infection with GI nematode parasites remains unresolved.

Although GC hyperplasia is mediated by a CD4+ Th2 immune response, it appears to be IL-4 independent. The most convincing role for GCs in the expulsion of parasites is seen in *N. brasiliensis* infection. Using IL-13$^{-/-}$ mice, it has been demonstrated that, unlike wild-type and IL-4$^{-/-}$ mice, IL-13$^{-/-}$ animals fail to clear *N. brasiliensis* efficiently, despite developing a Th2 response (McKenzie *et al.*, 1998). Importantly, IL-13$^{-/-}$ animals also failed to generate the GC hyperplasia that is coincident with worm expulsion. It was concluded that IL-13 may induce the production of GC hyperplasia and intestinal mucus, which facilitates expulsion of *N. brasiliensis* (McKenzie *et al.*, 1998). These and other studies further suggest that IL-13 may be the key cytokine involved in the GC response (Grunig *et al.*, 1998). However, GCs do not appear to play a role in the expulsion of other nematode parasites, such as *T. spiralis*, where IL-13 is not important in expulsion of the parasite. In our studies, wild-type, IFN-γR$^{-/-}$, TNF-R1$^{-/-}$ and IL-4$^{-/-}$ mice infected with *T. spiralis* demonstrated significant GC hyperplasia and increased production of mucin glycoprotein (Shekels *et al.*, 2000). *In situ* hybridization showed that this was accompanied by significant increases in GC mucin Muc2 mRNA and Muc3 mRNA at the tips of the intestinal villi. These data indicate that intestinal mucin protein levels,

Muc2 and Muc3, and mucin mRNA are coordinately upregulated in response to *T. spiralis* infection and may form the basis of an innate mucosal response independent of specific IFN-γ, TNF and IL-4 cytokines. Importantly, this study also demonstrated that goblet cell hyperplasia and upregulated mucin secretion are not essential components of the protective immune response to GI helminths.

It has been proposed that changes in mucin levels during intestinal inflammation, due to either a foreign agent or the host response, alter the protective capability of the mucosal layer. Mucin proteins are important in the protection of epithelium from pH extremes, proteases, toxins and pathogenic organisms (Miller, 1987). Polymerization of Muc2 and Muc3 mucins contributes to the viscosity of the mucous gel that lines the epithelium and provides a physical barrier preventing pathogens from interacting directly with the underlying epithelium; in addition, sloughing of the mucus layer allows potentially invasive organisms to be swept out of the intestine (Ho *et al.*, 1993, 1995). However, some pathogens are able to subvert this response; for example, *Entamoeba histolytica* binds to colonic mucins via an adherence lectin and produces either a mucin secretagogue or mucinase to deplete the GC stores of mucin and thus weaken the mucus protective layer (Belley *et al.*, 1996). Similarly, intraepithelial GI parasites such as *T. spiralis* and *T. muris* may employ such strategies to disrupt the integrity of the mucus layer and permit their entry into epithelial cells.

Mucins are also thought to act in cooperation with trefoil proteins in the protection and repair of the epithelium (Kindon *et al.*, 1995). Trefoil factors are expressed along the GI tract and increased levels are noted near sites of inflammation and ulcerative lesions (Babyatsky *et al.*, 1996). Furthermore, it has been demonstrated that mouse intestinal trefoil factor may play a role in the alteration of the physicochemical nature of GC mucins during *N. brasiliensis* infection (Tomita *et al.*, 1995). Perhaps in GI nematode parasite infection mucins are not aiding in the host's protective expulsion of the parasite, but rather are functioning in the repair of the damaged intestinal epithelium.

Matrix Metalloproteases in Enteropathy

T cell activation in the lamina propria is associated with epithelial cell shedding, leading to loss of villi. It has been postulated that this is mediated by increased production of matrix metalloproteases (MMP), which, by degrading the lamina propria matrix, represent a major pathway by which T cells cause injury in the gut (Pender *et al.*, 1997). Production of MMPs also facilitates movement of cells out of the vasculature into sites of inflammation and contributes substantially to the degradation of connective tissue during inflammatory disease (Stetler-Stevenson, 1996). Furthermore, MMPs are required for the release of soluble TNF-α from its membrane precursor form (Gearing *et al.*, 1994) and elevated concentrations of MMP in inflamed intestine may contribute by increasing levels of TNF in the intestinal tissue. More recently, it has been demonstrated that TNF may cause tissue injury in the gut by stimulating mucosal mesenchymal cells to secrete MMPs, since the neutralization of TNF inhibited the production of MMPs (Pender *et al.*, 1998). Therefore, it will be pertinent to analyse MMP levels in the intestine of *T. spiralis*-infected mice and the role that cytokines and mast cells play in the production of these enzymes.

Role of Microbial Flora in the Development of Enteropathy

The intestinal pathology seen in infections with GI parasites may not necessarily be mediated wholly by the parasite but also by commensal bacteria in the gut. Disruption of epithelial integrity, either by the direct actions of the parasite feeding or invading epithelial cells, or indirectly by cytokines induced in response to the parasite infection, may permit entry of bacteria or bacterial products such as LPS to intestinal tissue. This hypothesis is supported by data in which conventionally reared (CR), antibiotic-treated, specific pathogen-free or gnotobiotic pigs have been infected with *Trichuris suis*. These studies demonstrated that absence or removal of bacteria prevented pathology in the colon except in the area immediately surrounding adult worms. Bacteria were also found in large numbers in lymphoglandular complexes, enterocytes, submucosae and muscularis of CR pigs infected with *T. suis*, whereas this was not observed in normal uninfected pigs (Rutter and Beer, 1975; Hall *et al.*, 1976; Mansfield and Urban, 1996). It is therefore conceivable that disruption of epithelial integrity by parasite infection, either by the direct actions of the parasite or indirectly by induction of a Th2 cytokines response, allows bacteria to invade tissue, leading to enhanced pathology, possibly through induction of the TNF and iNOS responses. However, there is also recent evidence that certain bacterial infections may ameliorate the pathological effects of GI parasite infection: mice co-infected with *Lactobacillus casei* and *T. spiralis* showed reduced enteropathy with enhanced resistance to the parasite (Bautista-Garfias *et al.*, 1999). Conversely, parasite infections have been shown to ameliorate the pathology induced by *Helicobacter felis* (Fox *et al.*, 2000). These studies suggest that the mediators of pathology in GI parasite infections may be multifactorial and warrant further investigation.

Parasite Advantage in Induction of Pathology

Gastrointestinal parasites may actually utilize the induction of pathology to increase the generation of the next life-cycle stage. For instance,

IL-4-deficient mice (in which both the protective and pathological responses are ablated) showed, contrary to expectations, decreased numbers of *T. spiralis* muscle worm burdens compared with controls, despite the increased duration of the adult parasite in the gut. Furthermore, induction of intestinal pathology by the administration of TNF had no effect on the expulsion of adult parasites while increasing the numbers of muscle larvae (Table 18.1). However, it could be suggested that the loss of the Th2 responses in these animals resulted in enhanced Th1 responses, which have been suggested to be important in controlling the immune response to the larval stage of the parasite (Pond *et al.*, 1989). It is therefore conceivable that the pathophysiological effects of Th2 cytokines on intestinal epithelium results in a breakdown of epithelial integrity that allows the newborn larvae easier access to the underlying intestinal tissue and hence to continue their life cycle and become encapsulated in muscle. Maintenance of epithelial integrity may prevent newborn larvae accessing the intestinal mucosa and thus decrease the number of larvae able to establish in the muscle. Ablation of the responses that induce enteropathy in parasite infections may also interfere with the life cycle and thus also reduce the pathological effects induced by encapsulated larvae in the muscle.

Concluding Remarks

Studies with a variety of genetically modified mice have shed new light on the complex relationship between the protective and pathological immune responses controlling parasite infections. TNF and NO are important components of the pathological response accompanying the expulsion of a gastrointestinal nematode parasite. In the absence of TNF-R1 or iNOS, mice do not develop the severe villus atrophy and mucosal mastocytosis that usually accompany infection with *T. spiralis*, but their ability to expel the

Table 18.1. Role of intestinal pathology in establishment of *Trichinella spiralis* muscle larvae (unpublished data).

Mouse strain	Expulsion of intestinal stage	Intestinal pathology	Number of muscle larvae establishing
Wild-type	Yes	Yes	$29,133 \pm 4,774$
Wild-type + TNF	Yes	Yes	$29,333 \pm 6,912$
IL-4 KO	No*	No*	$10,433 \pm 2,397$*
IL-4 KO + TNF	No*	Yes	$18,433 \pm 5,395$

IL-4 KO mice and wild-type controls were infected with 400 *T. spiralis*. One group of each was treated with 1×10^5 U rTNF-α i.p. on days 4 and 7 p.i. Adult worm burdens and enteropathy were assessed at days 6 and 13. Muscle worm burdens were assessed at day 30 p.i. and expressed as mean $\pm$ SEM.
*Significantly different from wild-type ($P < 0.01$).

parasite is unaffected. TNF and iNOS may therefore contribute to parasite expulsion by promoting a Th2 response, other components of which are then effective. In biological terms, limitation of excessive TNF and/or NO production and the resulting tissue damage, without loss of protective function, would be of general benefit to the host, but this option is not normally available *in vivo*. Selective inhibitors of TNF or iNOS, currently under consideration for the treatment of inflammatory bowel diseases, may prove beneficial for both Th1- and Th2-mediated enteropathies without compromising Th2 protective responses (Mohler *et al.*, 1994; Muscara and Wallace, 1999) and may therefore have potential in manipulating host responses against GI nematodes. It remains important to elucidate the protective mechanisms involved in the expulsion of GI helminths, which are still unclear. Allergic inflammation does not appear to induce parasite expulsion but enteropathy may instead reflect the extreme activation of the immune system in the gut. Research into this relationship is still relevant in understanding mucosal inflammation in both parasitic and inflammatory bowel disorders, but the new approaches now being taken to examine immune responses to intestinal nematode infections clearly have much to teach us about the relationships between pathology and canonical Th1 and Th2 responses.

References

Abe, T. and Nawa, Y. (1988) Worm expulsion and mucosal mast cell response induced by repetitive IL-3 administration in *Strongyloides ratti*-infected nude mice. *Immunology* 63, 181–185.

Aoki, I., Itoh, S., Yokota, S., Tanaka, S.I., Ishii, N., Okuda, K., Minami, M. and Klinman, D.M. (1999) Contribution of mast cells to the T helper 2 response induced by simultaneous subcutaneous and oral immunization. *Immunology* 98, 519–524.

Artis, D., Humphreys, N.E., Bancroft, A.J., Rothwell, N.J., Potten, C.S. and Grencis, R.K. (1999a) Tumor necrosis factor alpha is a critical component of interleukin 13-mediated protective T helper cell type 2 responses during helminth infection. *Journal of Experimental Medicine* 190, 953–962.

Artis, D., Potten, C.S., Else, K.J., Finkelman, F.D. and Grencis, R.K. (1999b) *Trichuris muris*: host intestinal epithelial cell hyperproliferation during chronic infection is regulated by interferon-gamma. *Experimental Parasitology* 92, 144–153.

Babyatsky, M.W., deBeaumont, M., Thim, L. and Podolsky, D.K. (1996) Oral trefoil peptides protect against ethanol- and indomethacin-induced gastric injury in rats. *Gastroenterology* 110, 489–497.

Bancroft, A.J., McKenzie, A.N. and Grencis, R.K. (1998) A critical role for IL-13 in resistance to intestinal nematode infection. *Journal of Immunology* 160, 3453–3461.

Barbara, G., Vallance, B.A. and Collins, S.M. (1997) Persistent intestinal neuromuscular dysfunction after acute nematode infection in mice. *Gastroenterology* 113, 1224–1232.

Bautista-Garfias, C.R., Ixta, O., Orduna, M., Martinez, F., Aguilar, B. and Cortes, A. (1999) Enhancement of resistance in mice treated with *Lactobacillus casei*: effect on *Trichinella spiralis* infection. *Veterinary Parasitology* 80, 251–260.

Befus, A.D. and Bienenstock, J. (1979) Immunologically mediated intestinal mastocytosis in *Nippostrongylus brasiliensis*-infected rats. *Immunology* 38, 95–101.

Behnke, J.M., Dehlawi, M.S., Rose, R., Spyropoulos, P.N. and Wakelin, D. (1994a) The response of hamsters to primary and secondary infection with *Trichinella spiralis* and to vaccination with parasite antigens. *Journal of Helminthology* 68, 287–294.

Behnke, J.M., Rose, R. and Little, J. (1994b) Resistance of the hookworms *Ancylostoma ceylanicum* and *Necator americanus* to intestinal inflammatory responses induced by heterologous infection. *International Journal of Parasitology* 24, 91–101.

Belley, A., Keller, K., Grove, J. and Chadee, K. (1996) Interaction of LS174T human colon cancer cell mucins with *Entamoeba histolytica*: an *in vitro* model for colonic disease. *Gastroenterology* 111, 1484–1492.

Bennett, B.L., Cruz, R., Lacson, R.G. and Manning, A.M. (1997) Interleukin-4 suppression of tumor necrosis factor alpha-stimulated E-selectin gene transcription is mediated by STAT6 antagonism of NF-kappaB. *Journal of Biological Chemistry* 272, 10212–10219.

Betts, C.J. and Else, K.J. (1999) Mast cells, eosinophils and antibody-mediated cellular cytotoxicity are not critical in resistance to *Trichuris muris*. *Parasite Immunology* 21, 45–52.

Bhattacharyya, S.P., Drucker, I., Reshef, T., Kirshenbaum, A.S., Metcalfe, D.D. and Mekori, Y.A. (1998) Activated T lymphocytes induce degranulation and cytokine production by human mast cells following cell-to-cell contact. *Journal of Leukocyte Biology* 63, 337–341.

Boirivant, M., Fuss, I.J., Chu, A. and Strober, W. (1998) Oxazolone colitis: a murine model of T helper cell type 2 colitis treatable with antibodies to interleukin 4. *Journal of Experimental Medicine* 188, 1929–1939.

Burd, P.R., Rogers, H.W., Gordon, J.R., Martin, C.A., Jayaraman, S., Wilson, S.D., Dvorak, A.M., Galli, S.J. and Dorf, M.E. (1989) Interleukin 3-dependent and -independent mast cells stimulated with IgE and antigen express multiple cytokines. *Journal of Experimental Medicine* 170, 245–257.

Castro, G.A., Hessel, J.J. and Whalen, G. (1979) Altered intestinal fluid movement in response to *Trichinella spiralis* in immunized rats. *Parasite Immunology* 1, 259–266.

Cooper, E.S., Whyte-Alleng, C.A., Finzi-Smith, J.S. and MacDonald, T.T. (1992) Intestinal nematode infections in children: the pathophysiological price paid. *Parasitology* 104, S91–103.

Cope, A.P., Liblau, R.S., Yang, X.D., Congia, M., Laudanna, C., Schreiber, R.D., Probert, L., Kollias, G. and McDevitt, H.O. (1997) Chronic tumor necrosis factor alters T cell responses by attenuating T cell receptor signaling. *Journal of Experimental Medicine* 185, 1573–1584.

Corazza, N., Eichenberger, S., Eugster, H.P. and Mueller, C. (1999) Nonlymphocyte-derived tumor necrosis factor is required for induction of colitis in recombination activating gene (RAG)2($^{-/-}$) mice upon transfer of CD4($^+$)CD45RB(^{hi}) T cells. *Journal of Experimental Medicine* 190, 1479–1492.

Crowle, P.K. and Reed, N.D. (1981) Rejection of the intestinal parasite *Nippostrongylus brasiliensis* by mast cell-deficient W/W^v anemic mice. *Infection and Immunity* 33, 54–58.

Deckert-Schluter, M., Bluethmann, H., Rang, A., Hof, H. and Schluter, D. (1998) Crucial role of TNF receptor type 1 (p55), but not of TNF receptor type 2 (p75), in murine toxoplasmosis. *Journal of Immunology* 160, 3427–3436.

Donaldson, L.E., Schmitt, E., Huntley, J.F., Newlands, G.F. and Grencis, R.K. (1996) A critical role for stem cell factor and c-kit in host protective immunity to an intestinal helminth. *International Immunology* 8, 559–567.

Drapier, J.C., Wietzerbin, J. and Hibbs, J.B. Jr (1988) Interferon-gamma and tumor necrosis factor induce the L-arginine-dependent cytotoxic effector mechanism in murine macrophages. *European Journal of Immunology* 18, 1587–1592.

Else, K.J. and Grencis, R.K. (1991) Cellular immune responses to the murine nematode parasite *Trichuris muris*. I. Differential cytokine production during acute or chronic infection. *Immunology* 72, 508–513.

Else, K.J., Hultner, L. and Grencis, R.K. (1992) Cellular immune responses to the murine nematode parasite *Trichuris muris*. II. Differential induction of TH-cell subsets in resistant versus susceptible mice. *Immunology* 75, 232–237.

Else, K.J., Finkelman, F.D., Maliszewski, C.R. and Grencis, R.K. (1994) Cytokine-mediated regulation of chronic intestinal helminth infection. *Journal of Experimental Medicine* 179, 347–351.

Faulkner, H., Humphreys, N., Renauld, J.C., Van Snick, J. and Grencis, R. (1997) Interleukin-9 is involved in host protective immunity to intestinal nematode infection. *European Journal of Immunology* 27, 2536–2540.

Ferguson, A., Logan, R.F. and MacDonald, T.T. (1980) Increased mucosal damage during parasite infection in mice fed an elemental diet. *Gut* 21, 37–43.

Finkelman, F.D., Shea-Donohue, T., Goldhill, J., Sullivan, C.A., Morris, S.C., Madden, K.B., Gause, W.C. and Urban, J.F. Jr (1997) Cytokine regulation of host defense against parasitic gastrointestinal nematodes: lessons from studies with rodent models. *Annual Review in Immunology* 15, 505–533.

Fox, J.G., Beck, P., Dangler, C.A., Whary, M.T., Wang, T.C., Shi, H.N. and Nagler-Anderson, C. (2000) Concurrent enteric helminth infection modulates inflammation and gastric immune responses and reduces helicobacter-induced gastric atrophy. *Nature Medicine* 6, 536–542.

Frandji, P., Oskeritzian, C., Cacaraci, F., Lapeyre, J., Peronet, R., David, B., Guillet, J.G. and Mecheri, S. (1993) Antigen-dependent stimulation by bone marrow-derived mast cells of MHC class II-restricted T cell hybridoma. *Journal of Immunology* 151, 6318–6328.

Frandji, P., Tkaczyk, C., Oskeritzian, C., Lapeyre, J., Peronet, R., David, B., Guillet, J.G. and Mecheri, S. (1995) Presentation of soluble antigens by mast cells: upregulation by interleukin-4 and granulocyte/macrophage colony-stimulating factor and downregulation by interferon-gamma. *Cellular Immunology* 163, 37–46.

Frandji, P., Tkaczyk, C., Oskeritzian, C., David, B., Desaymard, C. and Mecheri, S. (1996) Exogenous and endogenous antigens are differentially presented by mast cells to CD4+ T lymphocytes. *European Journal of Immunology* 26, 2517–2528.

Furuta, G.T., Schmidt-Choudhury, A., Wang, M.Y., Wang, Z.S., Lu, L., Furlano, R.I. and Wershil, B.K. (1997) Mast cell-dependent tumor necrosis factor alpha production participates in allergic gastric inflammation in mice. *Gastroenterology* 113, 1560–1569.

Garside, P. (1999) Cytokines in experimental colitis. *Clinical and Experimental Immunology* 118, 337–339.

Garside, P., Grencis, R.K. and Mowat, A.M. (1992a) T lymphocyte dependent enteropathy in murine *Trichinella spiralis* infection. *Parasite Immunology* 14, 217–225.

Garside, P., Hutton, A.K., Severn, A., Liew, F.Y. and Mowat, A.M. (1992b) Nitric oxide mediates intestinal pathology in graft-vs.-host disease. *European Journal of Immunology* 22, 2141–2145.

Garside, P., Bunce, C., Tomlinson, R.C., Nichols, B.L. and Mowat, A.M. (1993) Analysis of enteropathy induced by tumour necrosis factor alpha. *Cytokine* 5, 24–30.

Gearing, A.J., Beckett, P., Christodoulou, M., Churchill, M., Clements, J., Davidson, A.H., Drummond, A.H., Galloway, W.A., Gilbert, R., Gordon, J.L. *et al.* (1994) Processing of tumour necrosis factor-alpha precursor by metalloproteinases. *Nature* 370, 555–557.

Gordon, J.R. and Galli, S.J. (1990) Mast cells as a source of both preformed and immunologically inducible TNF-alpha/cachectin. *Nature* 346, 274–276.

Grencis, R.K., Riedlinger, J. and Wakelin, D. (1985) L3T4-positive T lymphoblasts are responsible for transfer of immunity to *Trichinella spiralis* in mice. *Immunology* 56, 213–218.

Grencis, R.K., Hultner, L. and Else, K.J. (1991) Host protective immunity to *Trichinella spiralis* in mice: activation of Th cell subsets and lymphokine secretion in mice expressing different response phenotypes. *Immunology* 74, 329–332.

Grunig, G., Warnock, M., Wakil, A.E., Venkayya, R., Brombacher, F., Rennick, D.M., Sheppard, D., Mohrs, M., Donaldson, D.D., Locksley, R.M. and Corry, D.B. (1998) Requirement for IL-13 independently of IL-4 in experimental asthma. *Science* 282, 2261–2263.

Guy-Grand, D., DiSanto, J.P., Henchoz, P., Malassis-Seris, M. and Vassalli, P. (1998) Small bowel enteropathy: role of intraepithelial lymphocytes and of cytokines (IL-12, IFN-gamma, TNF) in the induction of epithelial cell death and renewal. *European Journal of Immunology* 28, 730–744.

Hall, G.A., Rutter, J.M. and Beer, R.J. (1976) A comparative study of the histopathology of the large intestine of conventionally reared, specific pathogen free and gnotobiotic pigs infected with *Trichuris suis*. *Journal of Comparative Pathology* 86, 285–292.

Haraldsen, G., Kvale, D., Lien, B., Farstad, I.N. and Brandtzaeg, P. (1996) Cytokine-regulated expression of E-selectin, intercellular adhesion molecule-1 (ICAM-1), and vascular cell adhesion molecule-1 (VCAM-1) in human microvascular endothelial cells. *Journal of Immunology* 156, 2558–2565.

Higgins, L.M., McDonald, S.A., Whittle, N., Crockett, N., Shields, J.G. and MacDonald, T.T. (1999) Regulation of T cell activation *in vitro* and *in vivo* by targeting the OX40-OX40 ligand interaction: amelioration of ongoing inflammatory bowel disease with an OX40-IgG fusion protein, but not with an OX40 ligand-IgG fusion protein. *Journal of Immunology* 162, 486–493.

Hiribarren, A., Heyman, M., L'Helgouac'h, A. and Desjeux, J.F. (1993) Effect of cytokines on the epithelial function of the human colon carcinoma cell line HT29 cl19A. *Gut* 34, 616–620.

Ho, S.B., Niehans, G.A., Lyftogt, C., Yan, P.S., Cherwitz, D.L., Gum, E.T., Dahiya, R. and Kim, Y.S. (1993) Heterogeneity of mucin gene expression in normal and neoplastic tissues. *Cancer Research* 53, 641–651.

Ho, S.B., Shekels, L.L., Toribara, N.W., Kim, Y.S., Lyftogt, C., Cherwitz, D.L. and Niehans, G.A. (1995) Mucin gene expression in normal, preneoplastic, and neoplastic human gastric epithelium. *Cancer Research* 55, 2681–2690.

Hogaboam, C.M., Jacobson, K., Collins, S.M. and Blennerhassett, M.G. (1995) The selective beneficial effects of nitric oxide inhibition in experimental colitis. *American Journal of Physiology* 268, G673–684.

Hogaboam, C.M., Snider, D.P. and Collins, S.M. (1997) Cytokine modulation of T-lymphocyte activation by intestinal smooth muscle cells. *Gastroenterology* 112, 1986–1995.

Hornung, F., Scala, G. and Lenardo, M.J. (2000) TNF-alpha-induced secretion of C-C chemokines modulates C-C chemokine receptor 5 expression on peripheral blood lymphocytes. *Journal of Immunology* 164, 6180–6187.

Huels, C., Germann, T., Goedert, S., Hoehn, P., Koelsch, S., Hultner, L., Palm, N., Rude, E. and Schmitt, E. (1995) Co-activation of naive CD4+ T cells and bone marrow-derived mast cells results in the development of Th2 cells. *International Immunology* 7, 525–532.

Iijima, H., Takahashi, I., Kishi, D., Kim, J.K., Kawano, S., Hori, M. and Kiyono, H. (1999) Alteration of interleukin 4 production results in the inhibition of T helper type 2 cell-dominated inflammatory bowel disease in T cell receptor alpha chain-deficient mice. *Journal of Experimental Medicine* 190, 607–615.

Inamura, N., Mekori, Y.A., Bhattacharyya, S.P., Bianchine, P.J. and Metcalfe, D.D. (1998) Induction and enhancement of FcERI-dependent mast cell degranulation following coculture with activated T cells: dependency on ICAM-1- and leukocyte function-associated antigen (LFA)-1-mediated heterotypic aggregation. *Journal of Immunology* 160, 4026–4033.

Ishikawa, N., Goyal, P.K., Mahida, Y.R., Li, K.F. and Wakelin, D. (1998) Early cytokine responses during intestinal parasitic infections. *Immunology* 93, 257–263.

James, S.L. (1995) Role of nitric oxide in parasitic infections. *Microbiology Review* 59, 533–547.

Joyce, D.A. and Steer, J.H. (1996) IL-4, IL-10 and IFN-gamma have distinct, but interacting, effects on differentiation-induced changes in TNF-alpha and TNF receptor release by cultured human monocytes. *Cytokine* 8, 49–57.

Jung, H.C., Eckmann, L., Yang, S.K., Panja, A., Fierer, J., Morzycka-Wroblewska, E. and Kagnoff, M.F. (1995) A distinct array of proinflammatory cytokines is expressed in human colon epithelial cells in response to bacterial invasion. *Journal of Clinical Investigation* 95, 55–65.

Keymer, A.E. and Tarlton, A.B. (1991) The population dynamics of acquired immunity to *Heligmosomoides polygyrus* in the laboratory mouse: strain, diet and exposure. *Parasitology* 103 Pt 1, 121–126.

Khan, A.I., Horii, Y., Tiuria, R., Sato, Y. and Nawa, Y. (1993) Mucosal mast cells and the expulsive mechanisms of mice against *Strongyloides venezuelensis*. *International Journal of Parasitology* 23, 551–555.

Kindon, H., Pothoulakis, C., Thim, L., Lynch-Devaney, K. and Podolsky, D.K. (1995) Trefoil peptide protection of intestinal epithelial barrier function: cooperative interaction with mucin glycoprotein. *Gastroenterology* 109, 516–523.

Koyama, K. and Ito, Y. (2000) Mucosal mast cell responses are not required for protection against infection with the murine nematode parasite *Trichuris muris*. *Parasite Immunology* 22, 21–28.

Kramer, S., Schimpl, A. and Hunig, T. (1995) Immunopathology of interleukin (IL) 2-deficient mice: thymus dependence and suppression by thymus-dependent cells with an intact IL-2 gene. *Journal of Experimental Medicine* 182, 1769–1776.

Kuhn, R., Lohler, J., Rennick, D., Rajewsky, K. and Muller, W. (1993) Interleukin-10-deficient mice develop chronic enterocolitis. *Cell* 75, 263–274.

Lantz, C.S., Boesiger, J., Song, C.H., Mach, N., Kobayashi, T., Mulligan, R.C., Nawa, Y., Dranoff, G. and Galli, S.J. (1998) Role for interleukin-3 in mast-cell and basophil development and in immunity to parasites. *Nature* 392, 90–93.

Larsh, J.E. Jr (1975) Allergic inflammation as a hypothesis for the expulsion of worms from tissues: a review. *Experimental Parasitology* 37, 251–266.

Lawrence, C.E., Paterson, J.C., Higgins, L.M., MacDonald, T.T., Kennedy, M.W. and Garside, P. (1998) IL-4-regulated enteropathy in an intestinal nematode infection. *European Journal of Immunology* 28, 2672–2684.

Lawrence, C.E., Paterson, J.C., Wei, X.Q., Liew, F.Y., Garside, P. and Kennedy, M.W. (2000) Nitric oxide mediates intestinal pathology but not immune expulsion during *Trichinella spiralis* infection in mice. *Journal of Immunology* 164, 4229–4234.

Lorentz, A., Schwengberg, S., Sellge, G., Manns, M.P. and Bischoff, S.C. (2000) Human intestinal mast cells are capable of producing different cytokine profiles: role of IgE receptor cross-linking and IL-4. *Journal of Immunology* 164, 43–48.

Lugli, S.M., Feng, N., Heim, M.H., Adam, M., Schnyder, B., Etter, H., Yamage, M., Eugster, H.P., Lutz, R.A., Zurawski, G. and Moser, R. (1997) Tumor necrosis factor alpha enhances the expression of the interleukin (IL)-4 receptor alpha-chain on endothelial cells increasing IL-4 or IL-13-induced Stat6 activation. *Journal of Biological Chemistry* 272, 5487–5494.

Lukacs, N.W., Strieter, R.M., Chensue, S.W., Widmer, M. and Kunkel, S.L. (1995) TNF-alpha mediates recruitment of neutrophils and eosinophils during airway inflammation. *Journal of Immunology* 154, 5411–5417.

MacDonald, T.T. and Ferguson, A. (1977) Hypersensitivity reactions in the small intestine. III. The effects of allograft rejection and of graft-versus-host disease on epithelial cell kinetics. *Cell and Tissue Kinetics* 10, 301–312.

MacDonald, T.T. and Spencer, J. (1988) Evidence that activated mucosal T cells play a role in the pathogenesis of enteropathy in human small intestine. *Journal of Experimental Medicine* 167, 1341–1349.

MacDonald, T.T., Choy, M.Y., Spencer, J., Richman, P.I., Diss, T., Hanchard, B., Venugopal, S., Bundy, D.A. and Cooper, E.S. (1991) Histopathology and immunohistochemistry of the caecum in children with the *Trichuris* dysentery syndrome. *Journal of Clinical Pathology* 44, 194–199.

Madden, K.B., Urban, J.F. Jr, Ziltener, H.J., Schrader, J.W., Finkelman, F.D. and Katona, I.M. (1991) Antibodies to IL-3 and IL-4 suppress helminth-induced intestinal mastocytosis. *Journal of Immunology* 147, 1387–1391.

Mansfield, L.S. and Urban, J.F. Jr (1996) The pathogenesis of necrotic proliferative colitis in swine is linked to whipworm induced suppression of mucosal immunity to resident bacteria. *Veterinary Immunology and Immunopathology* 50, 1–17.

Manson-Smith, D.F., Bruce, R.G. and Parrott, D.M. (1979) Villous atrophy and expulsion of intestinal *Trichinella spiralis* are mediated by T cells. *Cellular Immunology* 47, 285–292.

Marietta, E.V., Chen, Y. and Weis, J.H. (1996) Modulation of expression of the anti-inflammatory cytokines interleukin-13 and interleukin-10 by interleukin-3. *European Journal of Immunology* 26, 49–56.

McKenzie, G.J., Bancroft, A., Grencis, R.K. and McKenzie, A.N. (1998) A distinct role for interleukin-13 in Th2-cell-mediated immune responses. *Current Biology* 8, 339–342.

Miller, H.R. (1987) Gastrointestinal mucus, a medium for survival and for elimination of parasitic nematodes and protozoa. *Parasitology* 94, S77–100.

Mizoguchi, A., Mizoguchi, E., Chiba, C., Spiekermann, G.M., Tonegawa, S., Nagler-Anderson, C. and Bhan, A.K. (1996) Cytokine imbalance and auto-antibody production in T cell receptor-alpha mutant mice with inflammatory bowel disease. *Journal of Experimental Medicine* 183, 847–856.

Mizoguchi, A., Mizoguchi, E. and Bhan, A.K. (1999) The critical role of interleukin 4 but not interferon gamma in the pathogenesis of colitis in T-cell receptor alpha mutant mice. *Gastroenterology* 116, 320–326.

Mohler, K.M., Sleath, P.R., Fitzner, J.N., Cerretti, D.P., Alderson, M., Kerwar, S.S., Torrance, D.S., Otten-Evans, C., Greenstreet, T., Weerawarna, K. *et al.* (1994) Protection against a lethal dose of endotoxin by an inhibitor of tumour necrosis factor processing. *Nature* 370, 218–220.

Mosmann, T.R. and Coffman, R.L. (1989) Heterogeneity of cytokine secretion patterns and functions of helper T cells. *Advances in Immunology* 46, 111–147.

Mowat, A.M. (1989) Antibodies to IFN-gamma prevent immunologically mediated intestinal damage in murine graft-versus-host reaction. *Immunology* 68, 18–23.

Murch, S.H., Braegger, C.P., Walker-Smith, J.A. and MacDonald, T.T. (1995) Distribution and density of TNF immunoreactivity in chronic inflammatory bowel disease. *Advances in Experimental Medicine and Biology* 30, 1327–1330.

Muscara, M.N. and Wallace, J.L. (1999) Nitric oxide. V. Therapeutic potential of nitric oxide donors and inhibitors. *American Journal of Physiology* 276, G1313–1316.

Nawa, Y., Ishikawa, N., Tsuchiya, K., Horii, Y., Abe, T., Khan, A.I., Bing, S., Itoh, H., Ide, H. and Uchiyama, F. (1994) Selective effector mechanisms for the expulsion of intestinal helminths. *Parasite Immunology* 16, 333–338.

Neurath, M.F., Fuss, I., Pasparakis, M., Alexopoulou, L., Haralambous, S., Meyer zum Buschenfelde, K.H., Strober, W. and Kollias, G. (1997) Predominant pathogenic role of tumor necrosis factor in experimental colitis in mice. *European Journal of Immunology* 27, 1743–1750.

Oswald, I.P., Wynn, T.A., Sher, A. and James, S.L. (1994) NO as an effector molecule of parasite killing: modulation of its synthesis by cytokines. *Comparative Biochemistry, Physiology, Pharmacology, Toxicology and Endocrinology* 108, 11–18.

Pearce, E.J., Cheever, A., Leonard, S., Covalesky, M., Fernandez-Botran, R., Kohler, G. and Kopf, M. (1996) *Schistosoma mansoni* in IL-4-deficient mice. *International Immunology* 8, 435–444.

Pearlman, E., Lass, J.H., Bardenstein, D.S., Kopf, M., Hazlett, F.E. Jr, Diaconu, E. and Kazura, J.W. (1995) Interleukin 4 and T helper type 2 cells are required for development of experimental onchocercal keratitis (river blindness). *Journal of Experimental Medicine* 182, 931–940.

Pender, S.L., Tickle, S.P., Docherty, A.J., Howie, D., Wathen, N.C. and MacDonald, T.T. (1997) A major role for matrix metalloproteinases in T cell injury in the gut. *Journal of Immunology* 158, 1582–1590.

Pender, S.L., Fell, J.M., Chamow, S.M., Ashkenazi, A. and MacDonald, T.T. (1998) A p55 TNF receptor immunoadhesin prevents T cell-mediated intestinal injury by inhibiting matrix metalloproteinase production. *Journal of Immunology* 160, 4098–4103.

Perdue, M.H., Ramage, J.K., Burget, D., Marshall, J. and Masson, S. (1989) Intestinal mucosal injury is associated with mast cell activation and leukotriene generation during *Nippostrongylus*-induced inflammation in the rat. *Digestive Disease Science* 34, 724–731.

Peschon, J.J., Torrance, D.S., Stocking, K.L., Glaccum, M.B., Otten, C., Willis, C.R., Charrier, K., Morrissey, P.J., Ware, C.B. and Mohler, K.M. (1998) TNF receptor-deficient mice reveal divergent roles for p55 and p75 in several models of inflammation. *Journal of Immunology* 160, 943–952.

Piguet, P.F., Grau, G.E., Allet, B. and Vassalli, P. (1987) Tumor necrosis factor/cachectin is an effector of skin and gut lesions of the acute phase of graft-vs.-host disease. *Journal of Experimental Medicine* 166, 1280–1289.

Piguet, P.F., Vesin, C., Guo, J., Donati, Y. and Barazzone, C. (1998) TNF-induced enterocyte apoptosis in mice is mediated by the TNF receptor 1 and does not require p53. *European Journal of Immunology* 28, 3499–3505.

Pond, L., Wassom, D.L. and Hayes, C.E. (1989) Evidence for differential induction of helper T cell subsets during *Trichinella spiralis* infection. *Journal of Immunology* 143, 4232–4237.

Prociv, P. (1997) Pathogenesis of human hookworm infection: insights from a 'new' zoonosis. *Chemical Immunology* 66, 62–98.

Rajan, T.V., Porte, P., Yates, J.A., Keefer, L. and Shultz, L.D. (1996) Role of nitric oxide in host defense against an extracellular, metazoan parasite, *Brugia malayi*. *Infection and Immunity* 64, 3351–3353.

Reinecker, H.C., Steffen, M., Doehn, C., Petersen, J., Pfluger, I., Voss, A. and Raedler, A. (1991) Proinflammatory cytokines in intestinal mucosa. *Immunology Research* 10, 247–248.

Riedlinger, J., Grencis, R.K. and Wakelin, D. (1986) Antigen-specific T-cell lines transfer protective immunity against *Trichinella spiralis in vivo*. *Immunology* 58, 57–61.

Rose, M.E., Wakelin, D. and Hesketh, P. (1994) Interactions between infections with *Eimeria* spp. and *Trichinella spiralis* in inbred mice. *Parasitology* 108, 69–75.

Russell, D.A. (1986) Mast cells in the regulation of intestinal electrolyte transport. *American Journal of Physiology* 251, G253–262.

Rutter, J.M. and Beer, R.J.S. (1975) Synergism between *Trichuris suis* and the microbial flora of the large intestine causing dysentery in pigs. *Infection and Immunity* 11, 395–404.

Shekels, L., Anway, R.E., Lin, J., Kennedy, M.W., Garside, P., Lawrence, C.E. and Ho, S.B. (2000) Coordinated Muc2 and Muc3 mucin gene expression in *Trichinella spiralis* infection in wild-type and cytokine deficient mice. *Infection and Immunity* (submitted).

Stephenson, L.S., Pond, W.G., Nesheim, M.C., Krook, L.P. and Crompton, D.W. (1980) *Ascaris suum*: nutrient absorption, growth, and intestinal pathology in young pigs experimentally infected with 15-day-old larvae. *Experimental Parasitology* 49, 15–25.

Stetler-Stevenson, W.G. (1996) Dynamics of matrix turnover during pathologic remodeling of the extracellular matrix. *American Journal of Pathology* 148, 1345–1350.

Svetic, A., Madden, K.B., Zhou, X.D., Lu, P., Katona, I.M., Finkelman, F.D., Urban, J.F. Jr and Gause, W.C. (1993) A primary intestinal helminthic infection rapidly induces a gut-associated elevation of Th2-associated cytokines and IL-3. *Journal of Immunology* 150, 3434–3441.

Taylor, M.J., Cross, H.F., Mohammed, A.A., Trees, A.J. and Bianco, A.E. (1996) Susceptibility of *Brugia malayi* and *Onchocerca lienalis* microfilariae to nitric oxide and hydrogen peroxide in cell-free culture and from IFN gamma-activated macrophages. *Parasitology* 112, 315–322.

Tomita, M., Itoh, H., Ishikawa, N., Higa, A., Ide, H., Murakumo, Y., Maruyama, H., Koga, Y. and Nawa, Y. (1995) Molecular cloning of mouse intestinal trefoil factor and its expression during goblet cell changes. *Biochemistry Journal* 311, 293–297.

Toru, H., Pawankar, R., Ra, C., Yata, J. and Nakahata, T. (1998) Human mast cells produce IL-13 by high-affinity IgE receptor cross-linking: enhanced IL-13 production by IL-4-primed human mast cells. *Journal of Allergy and Clinical Immunology* 102, 491–502.

Urban, J.F. Jr, Madden, K.B., Cheever, A.W., Trotta, P.P., Katona, I.M. and Finkelman, F.D. (1993) IFN inhibits inflammatory responses and protective immunity in mice infected with the nematode parasite, *Nippostrongylus brasiliensis*. *Journal of Immunology* 151, 7086–7094.

Urban, J.F. Jr, Maliszewski, C.R., Madden, K.B., Katona, I.M. and Finkelman, F.D. (1995) IL-4 treatment can cure established gastrointestinal nematode infections in immunocompetent and immunodeficient mice. *Journal of Immunology* 154, 4675–4684.

Urban, J.F. Jr, Noben-Trauth, N., Donaldson, D.D., Madden, K.B., Morris, S.C., Collins, M. and Finkelman, F.D. (1998) IL-13, IL-4R alpha, and Stat6 are required for the expulsion of the gastrointestinal nematode parasite *Nippostrongylus brasiliensis*. *Immunity* 8, 255–264.

Urban, J.F. Jr, Schopf, L., Morris, S.C., Orekhova, T., Madden, K.B., Betts, C.J., Gamble, H.R., Byrd, C., Donaldson, D., Else, K. and Finkelman, F.D. (2000) Stat6 signaling promotes protective immunity against *Trichinella spiralis* through a mast cell- and T cell-dependent mechanism. *Journal of Immunology* 164, 2046–2052.

Vallance, B.A., Blennerhassett, P.A., Deng, Y., Matthaei, K.I., Young, I.G. and Collins, S.M. (1999a) IL-5 contributes to worm expulsion and muscle hypercontractility in a primary *T. spiralis* infection. *American Journal of Physiology* 277, G400–408.

Vallance, B.A., Galeazzi, F., Collins, S.M. and Snider, D.P. (1999b) CD4 T cells and major histocompatibility complex class II expression influence worm expulsion and increased intestinal muscle contraction during *Trichinella spiralis* infection. *Infection and Immunity* 67, 6090–6097.

Vermillion, D.L. and Collins, S.M. (1988) Increased responsiveness of jejunal longitudinal muscle in *Trichinella*-infected rats. *American Journal of Physiology* 254, G124–129.

Vermillion, D.L., Ernst, P.B. and Collins, S.M. (1991) T-lymphocyte modulation of intestinal muscle function in the *Trichinella*-infected rat. *Gastroenterology* 101, 31–38.

Wakelin, D. (1978) Immunity to intestinal parasites. *Nature* 273, 617–620.

Wang, C.H., Korenaga, M., Greenwood, A. and Bell, R.G. (1990) T-helper subset function in the gut of rats: differential stimulation of eosinophils, mucosal mast cells and antibody-forming cells by OX8⁻OX22⁻ and OX8⁻OX22⁺ cells. *Immunology* 71, 166–175.

Wang, H.W., Tedla, N., Lloyd, A.R., Wakefield, D. and McNeil, P.H. (1998) Mast cell activation and migration to lymph nodes during induction of an immune response in mice. *Journal of Clinical Investigation* 102, 1617–1626.

Weisbrodt, N.W., Lai, M., Bowers, R.L., Harari, Y. and Castro, G.A. (1994) Structural and molecular changes in intestinal smooth muscle induced by *Trichinella spiralis* infection. *American Journal of Physiology* 266, G856–862.

Immunomodulation by Filarial Nematode Phosphorylcholine-containing Glycoproteins

William Harnett[1] and Margaret M. Harnett[2]

[1]Department of Immunology, University of Strathclyde, Glasgow G4 0NR, UK; [2]Department of Immunology, University of Glasgow, Glasgow G12 8QQ, UK

Introduction

Phosphorylcholine (PC) (Fig. 19.1) is an abundant component of eukaryotes (both vertebrates and invertebrates), where it is found as the polar head group of the phospholipids, phosphatidylcholine and sphingomyelin. However, in many lower invertebrates and also in certain fungi and bacteria, PC has been found in a different association: specifically, attached to carbohydrate (reviewed by Harnett and Harnett, 1999). As many of the organisms that contain PC attached in this manner constitute important human pathogens, this association has been subject to significant investigation. It has been particularly well studied in the Gram-positive bacterium *Streptococcus pneumoniae*, the Gram-negative

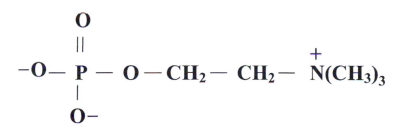

Fig. 19.1. Structure of phosphorylcholine (PC). PC is involved in phosphodiester linkage to carbohydrate in a number of lower organisms, including filarial nematodes (linking sugar appears to be *N*-acetylglucosamine on species examined to date).

©CAB *International* 2001. *Parasitic Nematodes*
(eds M.W. Kennedy and W. Harnett)

bacterium *Haemophilus influenzae* and in filarial nematodes. Thus PC is known to be attached to teichoic acid (Brundish and Baddiley, 1968) and lipoteichoic acid (Briles and Tomasz, 1973) in *S. pneumoniae*, to lipopolysaccharide in *H. influenzae* (Weiser *et al.*, 1997) and to both glycolipids (Wuhrer *et al.*, 2000) and glycoproteins (reviewed by Houston and Harnett, 1999a) of the nematodes. Attachment of PC to the nematode glycoproteins has been shown to be via *N*-type glycans (Harnett *et al.*, 1993), and PC-*N*-glycan structures (see Fig. 15.5, Chapter 15) have been found on a number of glycoproteins, some of which are secreted (Haslam *et al.*, 1999). A major puzzle in filarial nematode research is the role of PC on these latter proteins.

Finding a Role for PC on Secreted Filarial Nematode Glycoproteins

Studies with the rodent filarial nematode *Acanthocheilonema viteae* (Houston and Harnett, 1996) and the feline filarial nematode *Brugia pahangi* (Nor *et al.*, 1997b) indicate that PC-containing glycoproteins are abundant internal components of filarial nematodes. It is almost certain that these are distinct from the molecules that are secreted, because they are dissimilar with respect to molecular mass, rate of synthesis and susceptibility of their PC-glycans to cleavage by *N*-glycosidase F (Houston and Harnett, 1996; Nor *et al.*, 1997a). Thus PC, following attachment to glycoproteins (or glycolipids), clearly must have a role to play *within* filarial nematodes. Studies on bacteria indicate that PC has important physiological roles, being essential for maintenance of cell shape and size and for promoting growth and cell division (reviewed by Harnett and Harnett, 1999). Whether a similar role is played in filarial nematodes awaits investigation but PC has also been found attached to internal glycoproteins of the free-living model nematode *Caenorhabditis elegans* (Gerdt *et al.*, 1999; K.M. Houston and W. Harnett, unpublished results) and hence it can be assumed that its role within worms is likely to be physiological rather than, say, an adaptation to parasitism.

Returning to bacteria, PC also appears to play a role in infection of humans by pathogenic strains/species by allowing colonization and invasiveness due to interaction with appropriate receptors on host endothelial cells (reviewed by Harnett and Harnett, 1999). This may act as a double-edged sword, however, as the PC on the surface of the bacteria can be targeted by both the innate and adaptive immune responses and indeed such responses appear to play a role in the control of *H. influenzae* and *S. pneumoniae*, respectively, in humans (reviewed by Harnett and Harnett, 1999).

PC can also be detected on the surface of L3 infective stages of filarial nematodes and there is some indication from studies on rodent models that, as with bacteria, the exposed PC can represent a target for a protective IgM antibody response (AlQaoud *et al.*, 1998). There may thus be a strong advantage in the worms secreting PC-containing molecules to act as decoys for such a response. However, sequence and functional analysis of ES-62 (Harnett *et al.*, 1999b), the major PC-containing glycoprotein of *A. viteae* (Harnett *et al.*, 1989) and the one PC-containing filarial molecule (due to ready availability) to have been studied in any real depth, strongly suggests that the protein component of PC-containing molecules is not designed simply to act as a carrier for PC. ES-62 may actually function as an aminopeptidase, which raises the question as to why it has PC attached. As outlined below, PC in fact appears to have a number of immunomodulatory properties and hence its attachment to proteins that are actively secreted by parasitic organisms may have evolved due to offering alternative or additional survival advantages. The relationship, if any, between the functional activities of the PC and protein components of secreted PC-containing filarial glycoproteins such as ES-62 awaits unravelling.

Immunological Defects Associated with Human Filarial Nematode Infection: a Role for PC?

Infection of humans (or animals) with filarial nematodes is long-term, with individual worms surviving for in excess of 5 years (Vanamail *et al.*, 1996). The consensus of opinion amongst workers in this area of filariasis research is that such longevity reflects suppression or modulation of the host immune system. 'Defects' in immune responsiveness have been revealed in infected individuals (Maizels and Lawrence, 1991; Kazura *et al.*, 1993) but the exact nature of such defects is uncertain, in that there appears to be a lack of uniformity in findings from the many studies undertaken. Nevertheless, in general, the defects incorporate impairment of lymphocyte proliferation and bias in production of both cytokines, e.g. reduced IFN-γ; increased IL-10 and IgG subclasses – greatly elevated IgG4 (an antibody of little value in eliminating pathogens, due to an inability to activate complement or bind with high affinity to phagocytic cells); and decreases in other IgG subclasses. Overall, therefore, the picture is of an immune response demonstrating a somewhat suppressed, anti-inflammatory or 'Th-2 type' phenotype.

At present, the cause of this immunological bias is probably the most frequently addressed question in research into the immunology of filariasis. We were led to the idea that a contributing factor might be PC on secreted parasite glycoproteins, as there was some reference in the literature to this substance possessing immunomodulatory properties, such as being able to interfere with antibody responses (Mitchell and Lewers, 1976) and driving immune responses in a Th-2 direction (Bordmann *et al.*, 1998). Furthermore, PC-containing secreted glycoproteins are readily detected in the bloodstream of many people harbouring filarial nematodes (reviewed by Harnett *et al.*, 1998b). We therefore investigated the immunological properties of ES-62, in particular focusing on its PC moiety.

Inhibition of Lymphocyte Proliferation by PC on ES-62

It has been known for almost two decades that a percentage of B lymphocytes (murine and human) expresses a non-immunoglobulin receptor for PC (Bach *et al.*, 1983). Interaction of this receptor with PC-containing *S. pneumoniae* results in the polyclonal secretion of antibody from the cells (Beckmann and Levitt, 1984). It may therefore be predicted that interaction of PC-containing molecules such as ES-62 with B lymphocytes would also result in polyclonal antibody production. This in fact may well be the case, since incubation of ES-62 with small resting splenic murine B lymphocytes results in cell activation, as measured by DNA synthesis (Harnett and Harnett, 1993). However, this effect of ES-62 was noted when 'high' concentrations were employed (25–50 μg ml^{-1}). Interestingly, at concentrations 10- to 100-fold less, i.e. within the range at which PC-containing molecules can be found in the human bloodstream (Lal *et al.*, 1987), ES-62 does not cause polyclonal stimulation but, rather, acts to prevent (by up to 60%) proliferation of B lymphocytes associated with ligation of the antigen receptor (Harnett and Harnett, 1993). This effect of ES-62 is almost certainly due to PC, as it can be mimicked by PC conjugated to bovine serum albumin (BSA) or even PC alone. Furthermore, weekly injections of PC-BSA (10 μg) into mice was found to reduce the ability of recovered splenic B lymphocytes to be activated via the antigen receptor (Harnett *et al.*, 1999a). These latter inhibitory effects of ES-62/PC on subsequent activation of B lymphocytes suggests that exposure to low levels of PC may actually anergize B cells. This will be discussed in some detail later.

It was also found that ES-62 was able to inhibit polyclonal activation via the antigen receptor of the human T cell line Jurkat (Harnett *et al.*, 1998a) and again this effect appears to be mainly due to PC (Harnett *et al.*, 1999a). These results are consistent with an earlier finding that PC-containing molecules isolated from a whole worm extract of the human filarial nematode *Brugia malayi* could inhibit activation of human T lymphocytes induced by the mitogen phytohaemagglutinin (Lal *et al.*, 1990). Inhibition of T lymphocyte proliferation during filarial nematode infection has been a much more frequently documented phenomenon than the same effect on its B lymphocyte counterpart. However, we are only currently beginning to investigate whether PC contributes to this *in vivo*.

Mechanism of Action of ES-62/PC

We have spent considerable effort in attempting to elucidate how PC inhibits lymphocyte activation associated with ligation of the antigen receptor. Revealingly, although the concentrations of ES-62 that interfere with activation of B lymphocytes induce no activation of the cells *per se*, as measured by DNA synthesis, they do affect a number of signalling elements associated with the transduction of cellular activation and proliferation. Thus, pretreatment with the parasite molecule has been found to: (i) induce tyrosine phosphorylation and activation of the protein tyrosine kinases (PTKs) Lyn (535% of basal level of tyrosine phosphorylation), Syk (289% basal level) and Blk (172% basal level) but not Fyn (Deehan *et al.*, 1998); (ii) phosphorylate and activate the Erk2 isoform of mitogen-activated protein (MAP) kinase (256% of basal level of tyrosine phosphorylation) (Deehan *et al.*, 1998); (iii) down-regulate the total level and activity (by some 40%) of protein kinase C (PKC) (Harnett and Harnett, 1993); and (iv) modulate the expression of a number of PKC isoforms – for example, whereas α (31% of basal levels), β (37%), ι/λ (41%), δ (49%) and ζ (43%) are down-regulated, expression of γ (492%) and ε (185%) is upregulated (Deehan *et al.*, 1997). That ES-62-induced modulation of these signal transducers requires PC is demonstrated by the fact that the effects on the PTKs, Erk2 and overall PKC levels, at least, are known to be mimicked by PC (Harnett and Harnett, 1993; Deehan *et al.*, 1998).

Importantly, we have also found that pre-exposure of B cells to ES-62 at the low concentrations serves to desensitize the cells almost completely to subsequent activation of the phosphoinositide 3-kinase (PI-3-kinase) and Ras-MAPK pathways via the B cell receptor (BCR) (Deehan *et al.*, 1998). Thus, uncoupling of the BCR from these crucial proliferative pathways could provide a molecular mechanism for the ES-62-mediated inhibition of BCR-driven proliferation of B lymphocytes. More recently (Harnett *et al.*, 1998a) we have found that the ES-62-induced rendering of Jurkat cells anergic to cellular activation via the T cell antigen receptor (TCR) is associated with disruption of TCR coupling to the phospholipase D, PKC, PI-3-kinase and Ras-MAPK signalling cascades but, as with murine B cells (Harnett and Harnett, 1993), not the PLC-mediated generation of inositol phosphates. Again, as for inhibition of BCR signalling, PC appears to be the active component of the parasite molecule as culture with PC or PC-BSA has similar effects to ES-62 on the coupling of the TCR to tyrosine kinase activation (ZAP-70, Lck and Fyn) and the PLC, Ras and MAPkinase signalling cascades (Harnett *et al.*, 1999a).

In B lymphocytes, coupling of the antigen receptors to Erk MAPkinase is protein tyrosine kinase (PTK)-dependent (Pao *et al.*, 1997). Following ligation of the BCR (Fig. 19.2) the PTK, Lyn, tyrosine phosphorylates the immunoreceptor tyrosine-based activation motifs (ITAMs) on the accessory transducing molecules Ig-α and Ig-β, leading to the recruitment and activation of additional PTKs (such as Syk, Lyn, Blk and Fyn) and signalling molecules (PLC-γ, RasGAP) and adaptors (Shc, Grb2) in an SH2- and SH3-domain-dependent manner. Thus, Shc binds to the phosphorylated ITAMs and in turn is phosphorylated by Syk, permitting recruitment of the Grb2Sos complexes required for activation of Ras at the plasma

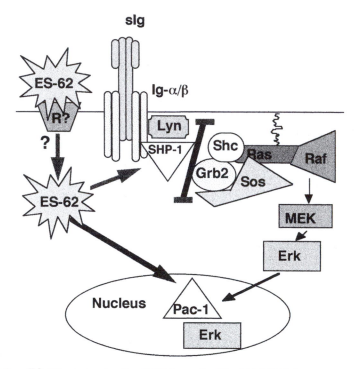

Fig. 19.2. ES-62 uncouples the BCR from the RasErk MAPkinase cascade. Following ligation of the BCR the PTK, Lyn, tyrosine phosphorylates the ITAMs on the accessory transducing molecules Ig-α and Ig-β. The Ras adaptor protein, Shc, binds to the phosphorylated ITAMs and in turn is phosphorylated, leading to the recruitment of the Grb2Sos complexes required for activation of Ras. Active Ras initiates the Erk MAPkinase cascade by binding and activating Raf, leading to stimulation of MEK and consequent activation and nuclear translocation of Erk. ES-62/PC binds to an unknown receptor (R) and, by either subversion of immune receptor signalling or internalization, appears to target two major negative regulatory sites in the control of BCR-coupling to the Ras MAPkinase cascade. Firstly, ES-62 induces the activation of SHP-1 tyrosine phosphatase to prevent initiation of BCR signalling by maintaining the ITAMs in a resting, dephosphorylated state and hence prevents recruitment of the ShcGrb2Sos complexes required to activate Ras. Secondly, ES-62 recruits the nuclear MAPkinase dual phosphatase, Pac-1, to terminate any ongoing Erk signals. This dual-pronged mechanism results in a rapid and profound desensitization of BCR-coupling to the RasErk MAPkinase cascade.

membrane. Following Sos-driven guanine nucleotide exchange and generation of the GTP-bound form of Ras, Ras binds and de-represses Raf-ser/thr kinase, triggering stimulation of MEK (MAPkinasekinase) and consequent activation of MAPkinase.

Although we have found that ES-62 profoundly suppresses BCR-stimulated tyrosine events, it does not appear to uncouple the BCR from MAPkinase activation by disrupting activation of the BCR-associated protein tyrosine kinases such as Lyn, Syk, Blk or Fyn (Deehan *et al.*, 1998). Conversely, we have found that ES-62/PC appears to target two major negative regulatory sites in the control of BCR-coupling to the Ras MAPkinase cascade (Fig. 19.2). Specifically, it induces the activation of SHP-1 tyrosine phosphatase and the MAPkinase phosphatase, Pac-1 (M.R. Deehan *et al.*, 2000, unpublished results). Activation of the former results in dephosphorylation of ITAMs on Ig-β and hence loss of recruitment of other signalling molecules; activation of the latter results in dephosphorylation and hence inactivation of MAPKinase.

How then is PC able to exert these effects on lymphocyte signal transduction pathways? Examination of the literature reveals clear evidence from a number of studies for PC playing a role in cellular proliferation. For example, many human tumours have elevated levels of PC (Daly *et al.*, 1987) and Ras-transformed cell lines produce increased levels of PC which are necessary for cell proliferation (Ratnam and Kent, 1995). Furthermore, it has recently been found that PC can exert mitogenic and co-mitogenic effects on fibroblast cell lines *in vitro* (Crilly *et al.*, 1998) and this is associated with activation of MAPkinase (Jimenez *et al.*, 1995). Thus, since it has been shown for B lymphocytes that ES-62 does not itself induce the generation of cellular PC (Deehan *et al.*, 1998), it is conceivable that the PC component of PC could result in partial activation of B cells, rendering them desensitized to subsequent activation via the BCR. However, evidence has been produced indicating that the mitogenic effects of PC can occur not only intracellularly but also following extracellular interaction with PC (Huang *et al.*, 1999). With respect to ES-62, we cannot at this stage state whether the PC moiety acts at the cell surface or following internalization. The ability of PC to activate certain protein tyrosine kinases that are associated with receptors found at the plasma membrane (Deehan *et al.*, 1998) may be an argument in favour of the former but it is also worth considering that it may be possible for ES-62 to become readily internalized by lymphocytes, perhaps following interaction with a specific receptor. Certainly the PAF receptor is utilized by PC-containing *S. pneumoniae* to enter human endothelial cells during infection (Cundell *et al.*, 1995) and it is known that B lymphocytes express PAF receptors at the plasma membrane (Mazer *et al.*, 1991). Whether the PAF receptor is the uncharacterized receptor referred to earlier (Bach *et al.*, 1983) awaits elucidation.

Regardless of where ES-62 acts, what is interesting about the activation of MAPkinase induced by the parasite product is that it appears to take place in the absence of Ras activation (Deehan *et al.*, 1998). Recent studies in our laboratories indicate that ES-62/PC actually blocks Ras activation by preventing association between Shc and the Ras guanine nucleotide exchange factor Sos (M.R. Deehan *et al.*, unpublished results). The alternative mechanism by which MAPkinase is activated remains to be established but, whatever it is, as mentioned earlier, it clearly fails to result in cell proliferation.

Effects of PC on Antibody and Cytokine Responses

The observation that the effect of PC on B lymphocytes *in vitro* varies with respect to concentration suggests that its effects *in vivo* may be phenotypically diverse. Thus, it is possible to predict that infected individuals exposed to low levels of PC might have suppressed antibody responses to parasite molecules, whereas those exposed to high levels might be subject to increased polyclonal production of immunoglobulin. Review of the literature as it currently stands suggests that we have some way to go before we can fully characterize the effects of filarial nematode infection on host antibody production, but certainly there is evidence for both types of individual existing (Marley *et al.*, 1995; Ottesen, 1995; Wanni *et al.*, 1997). The situation is complicated by the finding that people harbouring filarial nematodes often show a bias in antibody class/subclass production. In particular, it is frequently noted that IgG4 and IgE levels (both specific and non-specific) are greatly elevated, whereas levels of IgG1, IgG2 and IgG3 may be reduced (reviewed by King and Nutman, 1991; Maizels *et al.*, 1995). Intriguingly, we have recently discovered that the combination of IL-4 and ES-62, the latter in concentrations that render B cells anergic to activation via the antigen receptor, actually synergize to produce B lymphocyte activation *in vitro* (Harnett *et al.*, 1999a). This may relate to our earlier finding that IL-4 can overcome the down-regulatory effects of ES-62 on PKC-α and ι/λ expression in B cells (Deehan *et al.*, 1997), as these PKC isoforms are considered to transduce key activation signals (Toker *et al.*, 1994; Pelech, 1996). Since IgG4 and IgE are promoted by IL-4, a similar synergistic activation occurring *in vivo* might offer an explanation for the increased production of these types of antibody. Having said all of this, it is important to note that we currently have no real proof that PC is modulating antibody responses during filarial nematode infection. Nevertheless, we have recently obtained some evidence confirming that the presence of PC on ES-62 modulates the antibody response to other epitopes on the molecule (K.M. Houston, unpublished results). However, as shown below, the mechanism responsible for this effect does not seem to be dependent on IL-4.

In spite of its apparent ability to desensitize lymphocytes, jirds naturally infected with *A. viteae* mount an IgG antibody response to ES-62 (Harnett *et al.*, 1990). BALB/c mice subjected to subcutaneous exposure to ES-62 also mount an antibody response (Harnett *et al.*, 1999a). When this is examined with respect to the Th-1 and Th-2 signature IgG subclasses (IgG2a and IgG1, respectively), it is found that only the latter is produced (Harnett *et al.*, 1999a). Thus, ES-62 induces a Th-2 antibody response. That this is dependent on IL-4 was shown by its absence in the IL-4 knockout (KO) mouse (K.M. Houston, 2000, unpublished results). ES-62 induces the production of a little IL-12 by murine spleen cells (W. Harnett *et al.*, 1999, unpublished results). This cytokine is essential for induction of the Th-1 phenotype but induction does not occur in the presence of IL-4, as the latter cytokine inhibits Th-1 cell development by down-regulating the β-chain of the IL-12 receptor (Szabo *et al.*, 1997). It might therefore be predicted that the IL-4 KO mouse would make a compensatory IgG2a response, as has been observed in the response to adult *B. malayi* (Lawrence *et al.*, 1995). No IgG2a response to ES-62 is detected, however (K.M. Houston, unpublished results), indicating that such a response is not being 'blocked' by IL-4. Thus, either ES-62 simply does not induce a Th-1 response or it is being inhibited in some other way.

ES-62 lacking PC can be produced by culturing *A. viteae* in the presence of 1-deoxymannojirimycin (Houston *et al.*, 1997) or hemicholinium-3 (HC-3) (Houston and Harnett, 1999b). The former is a mannose analogue that inhibits an oligosaccharide processing step on glycoproteins (Elbien, 1987) which is necessary for the generation of the substrate for PC addition to ES-62 (Haslam *et al.*, 1997; Houston *et al.*, 1997). The latter is a choline kinase inhibitor and hence prevents synthesis of PC (Hamza *et al.*, 1983). When PC-free ES-62 prepared in either way was injected into BALB/c mice, it was found by ELISA that there was no significant effect on the previously noted IgG1 antibody response to non-PC epitopes of the parasite product (Fig. 19.3). Unlike the results obtained with normal ES-62, the PC-free material was able to induce a substantial IgG2a response (Fig. 19.3). This implicates a role for PC in blocking the IgG2a response. We thus investigated whether the addition of PC to BSA would inhibit any IgG2a antibody response associated with it. Although it was found that the BALB/c IgG2a response to BSA was relatively weak, the presence of PC did indeed appear to be inhibiting it.

It has previously been reported that the PC moiety of filarial nematodes can induce IL-10 production in B1 cells (Palanivel *et al.*, 1996). IL-10 can down-regulate production of IFN-γ, the cytokine necessary for antibody class switching to IgG2a in mice (Snapper and Paul, 1987; Bogdan *et al.*, 1992). Thus, to determine whether PC was blocking production of IgG2a antibodies by promoting production of IL-10, the antibody response to normal ES-62 in IL-10$^{-/-}$ mice was investigated. When ES-62 was injected into these mice, an IgG2a response to the parasite molecule was generated (Fig. 19.4). This result therefore implicates PC-induced IL-10 as playing a role in determining the nature of the IgG subclass response to ES-62.

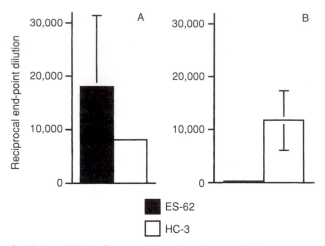

Fig. 19.3. Analysis of (A) IgG1 and (B) IgG2a titres in BALB/c mice exposed to either ES-62, or ES-62 manufactured in the presence of HC-3. Specific IgG2a titres were significantly higher in mice inoculated with ES-62 synthesized in the presence of HC-3 ($P < 0.05$; Mann Whitney U test). Results are expressed as mean reciprocal endpoint dilutions $\pm$ SEM ($n = 4$).

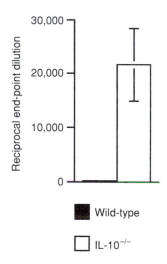

Fig. 19.4. Comparison of the development of IgG2a in IL-10$^{-/-}$ and IL-10$^{+/+}$ mice following 5 weeks exposure to ES-62. Specific IgG2a titres were significantly higher in IL-10$^{-/-}$ ($P < 0.025$; Mann Whitney U test). Values represent the mean reciprocal endpoint dilution $\pm$ SEM ($n = 4$).

As might be expected from the results of the IL-10$^{-/-}$ mouse study, ES-62 is able to induce production of IL-10 in naive BALB/c spleen cells (Harnett *et al.*, 1999a). Although we have not investigated whether this is due to the PC component of the molecule, this would be predicted from the work of Hoerauf and colleagues, who showed that PC causes release of this cytokine from B1 cells (Palanivel *et al.*, 1996; AlQaoud *et al.*, 1998). However, we have also found, as mentioned earlier, that ES-62 promotes the release from naive spleen cells of IL-12 and, in addition, IFN-γ –

cytokines associated with the Th-1 axis of the immune response (Harnett *et al.*, 1999a). Again, we have not investigated whether this is due to PC, but we are describing these results in the context of data obtained by Lochnit *et al.* (1998) which demonstrated that PC on glycolipids of *A. suum* induce human peripheral blood mononuclear cells to release the pro-inflammatory cytokines TNFα, IL-1 and IL-6. Thus, although the immuno-modulatory relevance of all of these observations is uncertain (for example, the *A. suum* glycolipids are internal worm components and hence may not have an opportunity to interact with the host immune system), PC clearly has the potential to be more than simply an inducer of cytokines involved in promoting Th-2 responses. Having said this, we have recently found that although ES-62 can induce a low-level production of IL-12 in macrophages, it also blocks the subsequent substantial production of this cytokine (and also IL-6 and TNFα, but not nitric oxide) induced by LPS/IFN-γ (H. Goodridge *et al.*, 2000, unpublished results).

It has already been noted that the phenotype of an acquired immune response is considered to reflect the early cytokine environment in which naive CD4[+] T cells interact with antigen. Again, it has been suggested, for example, that early exposure to IL-4 can push an immune response in a Th-2 direction (Swain *et al.*, 1990). We therefore investigated (by ELISA) whether ES-62 was able spontaneously to induce IL-4 secretion in naive murine spleen cells (48 h exposure). Ironically, given that the molecule induces a Th-2 antibody response and seems to be able to induce the release of a number of other cytokines, IL-4 was not detected (Harnett *et al.*, 1999a). It was noted, however, that IL-4 was produced by spleen cells from mice that had been pre-exposed to ES-62. This 'established' Th-2 phenotype is consistent with the antibody data.

More recently, we have investigated the ability of ES-62 to prime immune responses via dendritic cells (Whelan *et al.*, 2000). These are specialized antigen-presenting cells required for the priming and activation of CD4[+] T cells and, as such, could potentially direct the subsequent differentiation of T cell function (Banchereau and Steinman, 1998). We employed a well-characterized *in vitro* TH cell assay in which CD4[+] T cells from the DO.11.10 transgenic mouse express a T cell receptor that is specific for an ovalbumin peptide (Hsieh *et al.*, 1993). When these naive CD4[+] T cells are cultured with bone marrow-derived dendritic cells in the presence of the peptide, they secrete both IFN-γ and IL-4. We found that if the dendritic cells were pre-cultured in the presence of ES-62, then there was an increase in the amount of IL-4 produced and a decrease in the amount of IFN-γ (Whelan *et al.*, 2000). Thus ES-62 is found to induce the maturation of dendritic cells with the capacity to induce Th-2 responses. The next step will be to investigate whether this is due to PC. We also need to investigate the molecular mechanism underlying this effect – our initial observations suggest that it is not due to IL-10 or upregulation of known co-stimulatory molecules on the dendritic cell surface.

Concluding Remarks

PC has been suggested as meeting the requirements for a general vaccine targeted against mucosal pathogens (Trolle *et al.*, 1998) and indeed anti-PC antibodies can protect against various bacterial infections (*S. pneumoniae, S. typhimurium*) (Trolle *et al.*, 1998; reviewed by Harnett and Harnett, 1999). Furthermore, filarial nematode infective larvae, a stage possibly unique in expressing PC on their surface, may also be susceptible to host antibody-mediated immunity (AlQaoud *et al.*, 1998). However, it is becoming increasingly apparent that PC has a number of immunomodulatory properties, which dictates that caution should be exercised when considering its use in vaccination. By secreting PC-containing molecules, adult filarial nematodes appear to have utilized these immunomodulatory properties to manipulate the host immune system in certain ways and hence aid their survival in the parasitized host. What could be of interest, therefore, would be to explore the use of PC not for vaccination, but for manipulating immune responses in ways analogous to those being demonstrated by the parasites. One wonders, for example, whether molecules such as PC, by pushing immune responses in an anti-inflammatory, Th-2 direction, might be of value in the treatment of auto-immune diseases such as rheumatoid arthritis. Clearly however, the fact that PC appears to have a plethora of immunomodulatory activities, some of which in fact appear to be pro-inflammatory, dictates that much more investigation has to be carried out on this intriguing molecule, prior to considering its use in therapies.

Acknowledgements

We would like to thank the Edward Jenner Institute for Vaccine Research, the Leverhulme Trust, the Medical Research Council, Tenovus Scotland, the Wellcome Trust and the World Health Organisation for supporting our research on filarial nematodes.

References

AlQaoud, K.M., Fleischer, B. and Hoerauf, A. (1998) The Xid defect imparts susceptibility to experimental murine filariasis – association with a lack of antibody and IL-10 production by B cells in response to phosphorylcholine. *International Immunology* 10, 17–25.

Bach, M.A., Kohler, H. and Levitt, D. (1983) Binding of phosphorylcholine by non-immunoglobulin molecules on mouse B cells. *Journal of Immunology* 131, 365–371.

Banchereau, J. and Steinman, R.M. (1998) Dendritic cells and the control of immunity. *Nature* 392, 245–252.

Beckmann, E. and Levitt, D. (1984) Phosphorylcholine on *Streptococcus pneumoniae* R36a is responsible for *in vitro* polyclonal antibody secretion by human peripheral blood lymphocytes. *Journal of Immunology* 132, 2174–2176.

Bogdan, C., Paik, J., Vodovotz, Y. and Nathan, C. (1992) Contrasting mechanisms for suppression of macrophage cytokine release by transforming growth factor-beta and interleukin-10. *Journal of Biological Chemistry* 267, 23301–23308.

Bordmann, G., Rudin, W. and Favre, N. (1998) Immunization of mice with phosphatidylcholine drastically reduces the parasitaemia of subsequent *Plasmodium chabaudi chabaudi* blood-stage infections. *Immunology* 94, 35–40.

Briles, E.B. and Tomasz, A. (1973) Pneumococcal Forssman antigen. A choline-containing lipoteichoic acid. *Journal of Biological Chemistry* 248, 6394–6397.

Brundish, D.E. and Baddiley, J. (1968) Pneumococcal C-substance, a ribitol teichoic acid containing choline phosphate. *Biochemical Journal* 110, 573–582.

Crilly, K.S., Tomono, M. and Kiss, Z. (1998) The choline kinase inhibitor hemicholinium-3 can inhibit mitogen-induced DNA synthesis independent of its effect on phosphocholine formation. *Archives of Biochemistry and Biophysics* 352, 137–143.

Cundell, D.R., Gerard, N.P., Gerard, C., Idanpaan-Heikkila, I. and Tuomanen, E.I. (1995) *Streptococcus pneumoniae* anchor to activated human cells by the receptor for platelet-activating factor. *Nature* 377, 435–438.

Daly, P.F., Lyon, R.C., Faustino, P.J. and Cohen, J.S. (1987) Phospholipid-metabolism in cancer-cells monitored by P-31 NMR-spectroscopy. *Journal of Biological Chemistry* 262, 14875–14878.

Deehan, M., Harnett, M. and Harnett, W. (1997) A filarial nematode secreted product differentially modulates expression and activation of protein kinase C isoforms in B lymphocytes. *Journal of Immunology* 159, 6105–6111.

Deehan, M.R., Frame, M.J., Parkhouse, R.M., Seatter, S.D., Reid, S.D., Harnett, M.M. and Harnett, W. (1998) A phosphorylcholine-containing filarial nematode-secreted product disrupts B lymphocyte activation by targeting key proliferative signaling pathways. *Journal of Immunology* 160, 2692–2699.

Elbien, A.D. (1987) Inhibitors of the biosynthesis and processing of N-linked oligosaccharide chains. *Annual Review of Biochemistry* 56, 497–534.

Gerdt, S., Dennis, R.D., Borgonie, G., Schnabel, R. and Geyer, R. (1999) Isolation, characterization and immunolocalization of phosphorylcholine-substituted glycolipids in developmental stages of *Caenorhabditis elegans*. *European Journal of Biochemistry* 266, 952–963.

Hamza, M., Lloveras, J., Ribbes, G., Soula, G. and Douste-Blazy, L. (1983) An *in vitro* study of hemicholinium-3 on phospholipid metabolism of Krebs II ascites cells. *Biochemical Pharmacology* 32, 1893–1897.

Harnett, W. and Harnett, M.M. (1993) Inhibition of murine B cell proliferation and down-regulation of protein kinase C levels by a phosphorylcholine-containing filarial excretory–secretory product. *Journal of Immunology* 151, 4829–4837.

Harnett, W. and Harnett, M.M. (1999) Phosphorylcholine: friend or foe of the immune system? *Immunology Today* 20, 125–129.

Harnett, W., Grainger, M., Kapil, A., Worms, M.J. and Parkhouse, R.M.E. (1989) Origin, kinetics of circulation and fate *in vivo* of the major excretory–secretory product of *Acanthocheilonema viteae*. *Parasitology* 99, 229–239.

Harnett, W., Worms, M.J., Grainger, M., Pyke, S.D.M. and Parkhouse, R.M.E. (1990) Association between circulating antigen and parasite load in a model filarial system, *Acanthocheilonema viteae* in jirds. *Parasitology* 101, 435–444.

Harnett, W., Houston, K.M., Amess, R. and Worms, M.J. (1993) *Acanthocheilonema viteae*: phosphorylcholine is attached to the major excretory–secretory product via an N-linked glycan. *Experimental Parasitology* 77, 498–502.

Harnett, M.M., Deehan, M.R., Williams, D.M. and Harnett, W. (1998a) Induction of signalling anergy via the TCR in Jurkat T cells by pre-exposure to a filarial nematode secreted product. *Parasite Immunology* 20, 551–564.

Harnett, W., Bradley, J.E. and Garate, T. (1998b) Molecular and immunodiagnosis of human filarial nematode infections. *Parasitology* 117 (Supplement), S59–71.

Harnett, W., Deehan, M.R., Houston, K.M. and Harnett, M.M. (1999a) Immuno-modulatory properties of a phosphorylcholine-containing secreted filarial glycoprotein. *Parasite Immunology* 21, 601–608.

Harnett, W., Houston, K.M., Tate, R., Garate, T., Apfel, H., Adam, R., Haslam, S.M., Panico, M., Paxton, T., Dell, A., Morris, H. and Brzeski, H. (1999b) Molecular cloning and demonstration of an aminopeptidase activity in a filarial nematode glycoprotein. *Molecular and Biochemical Parasitology* 104, 11–23.

Haslam, S.M., Khoo, K.H., Houston, K.M., Harnett, W., Morris, H.R. and Dell, A. (1997) Characterisation of the phosphorylcholine-containing N-linked oligosaccharides in the excretory–secretory 62 kDa glycoprotein of *Acanthocheilonema viteae*. *Molecular and Biochemical Parasitology* 85, 53–66.

Haslam, S.M., Houston, K.M., Harnett, W., Reason, A.J., Morris, H.R. and Dell, A. (1999) Structural studies of phosphocholine-substituted *N*-glycans of filarial parasites: conservation amongst species and discovery of novel chito-oligomers. *Journal of Biological Chemistry* 274, 20953–20960.

Houston, K.M. and Harnett, W. (1996) Prevention of attachment of phosphorylcholine to a major excretory–secretory product of *Acanthocheilonema viteae* using tunicamycin. *Journal of Parasitology* 82, 320–324.

Houston, K.M. and Harnett, W. (1999a) Attachment of phosphorylcholine to a nematode glycoprotein. *Trends in Glycoscience and Glycotechnology* 11, 43–52.

Houston, K.M. and Harnett, W. (1999b) Mechanisms underlying the transfer of phosphorylcholine to filarial nematode glycoproteins – a possible role for choline kinase. *Parasitology* 118, 311–318.

Houston, K.M., Cushley, W. and Harnett, W. (1997) Studies on the site and mechanism of attachment of phosphorylcholine to a filarial nematode secreted glycoprotein. *Journal of Biological Chemistry* 272, 1527–1533.

Hsieh, C.-S., Macatonia, S.E., Tripp, C.S., Wolf, S.F., O'Garra, A. and Murphy, K.M. (1993) Development of Th1 CD+ T cells through IL-12 produced by *Listeria*-induced macrophages. *Science* 260, 547–549.

Huang, J.S., Mukherjee, J.J., Chung, T., Crilly, K.S. and Kiss, Z. (1999) Extracellular calcium stimulates DNA synthesis in synergism with zinc, insulin and insulin-like growth factor I in fibroblasts. *European Journal of Biochemistry* 266, 943–951.

Jimenez, B., del Peso, L., Montaner, S., Esteve, P. and Lacal, J.C. (1995) Generation of phosphorylcholine as an essential event in the activation of Raf-1 and MAP-kinases in growth factors-induced mitogenic stimulation. *Journal of Cellular Biochemistry* 57, 141–149.

Kazura, J.W., Nutman, T.B. and Greene, B. (eds) (1993) *Filariasis*. Blackwell Scientific Publications, Oxford.

King, C.L. and Nutman, T.B. (1991) Regulation of the immune response in lymphatic filariasis and onchocerciasis. *Immunology Today* 12, A54–58.

Lal, R.B., Paranjape, R.S., Briles, D.E., Nutman, T.B. and Ottesen, E.A. (1987) Circulating parasite antigen(s) in lymphatic filariasis – use of monoclonal-antibodies to phosphocholine for immunodiagnosis. *Journal of Immunology* 138, 3454–3460.

Lal, R.B., Kumaraswami, V., Steel, C. and Nutman, T.B. (1990) Phosphorylcholine-containing antigens of *Brugia malayi* non-specifically suppress lymphocyte function. *American Journal of Tropical Medicine and Hygiene* 42, 56–64.

Lawrence, R.A., Allen, J.E., Gregory, W.F., Kopf, M. and Maizels, R.M. (1995) Infection of IL-4-deficient mice with the parasitic nematode *Brugia malayi* demonstrates that host resistance is not dependent on a T helper 2-dominated immune response. *Journal of Immunology* 154, 5995–6001.

Lochnit, G., Dennis, R.D., Ulmer, A.J. and Geyer, R. (1998) Structural elucidation and monokine-inducing activity of two biologically active zwitterionic glycosphingolipids derived from the porcine parasitic nematode *Ascaris suum*. *Journal of Biological Chemistry* 273, 466–474.

Maizels, R.M. and Lawrence, R.A. (1991) Immunological tolerance: the key feature in human filariasis? *Parasitology Today* 7, 271–276.

Maizels, R.M., Sartono, E., Kurniawan, A., Partono, F., Selkirk, M.E. and Yazdanbakhsh, M. (1995) T-cell activation and the balance of antibody isotypes in human lymphatic filariasis. *Parasitology Today* 11, 50–56.

Marley, S.E., Lammie, P.J., Eberhard, M.L. and Hightower, A.W. (1995) Reduced antifilarial IgG4 responsiveness in a subpopulation of microfilaremic persons. *Journal of Infectious Diseases* 172, 1630–1633.

Mazer, B., Domenico, J., Sawami, H. and Gelfand, E.W. (1991) Platelet-activating factor induces an increase in intracellular calcium and expression of regulatory genes in human B lymphoblastoid cells. *Journal of Immunology* 146, 1914–1920.

Mitchell, G.F. and Lewers, H.M. (1976) Studies on immune responses to parasite antigens in mice IV. Inhibition of an anti-DNP antibody response with antigen, DNP-Ficoll containing phosphorylcholine. *International Archives of Allergy and Applied Immunology* 52, 235–240.

Nor, Z.M., Devaney, E. and Harnett, W. (1997a) The use of inhibitors of N-linked glycosylation and oligosaccharide processing to produce monoclonal anti-bodies against non-phosphorylcholine epitopes of *Brugia pahangi* excretory–secretory products. *Parasitology Research* 83, 813–815.

Nor, Z.M., Houston, K.M., Devaney, E. and Harnett, W. (1997b) Variation in the nature of attachment of phosphorylcholine to excretory–secretory products of adult *Brugia pahangi*. *Parasitology* 114, 257–262.

Ottesen, E.A. (1995) Immune responsiveness and the pathogenesis of human onchocerciasis. *Journal of Infectious Diseases* 171, 659–671.

Palanivel, V., Posey, C., Horauf, A.M., Solbach, W., Piessens, W.F. and Harn, D.A. (1996) B-cell outgrowth and ligand-specific production of IL-10 correlate with TH2 dominance in certain parasitic diseases. *Experimental Parasitology* 84, 168–177.

Pao, L., Carbone, A.M. and Cambier, J.C. (1997) Antigen receptor structure and signalling in B cells. In: Harnett, M.M. and Rigley, K.P. (eds) *Lymphocyte Signalling*. John Wiley & Sons, Chichester, UK, pp. 3–29.

Pelech, S.L. (1996) Kinase connections on the cellular internet. *Current Biology* 6, 551–554.

Ratnam, S. and Kent, C. (1995) Early increase in choline kinase-activity upon induction of the H-Ras oncogene in mouse fibroblast cell-lines. *Archives of Biochemistry and Biophysics* 323, 313–322.

Snapper, C.M. and Paul, W.E. (1987) Interferon-gamma and B cell stimulatory factor-1 reciprocally regulate Ig isotype production. *Science* 236, 944–947.

Swain, S.L., Weinberg, A.D., English, M. and Huston, G. (1990) IL-4 directs the development of Th2-like helper effectors. *Journal of Immunology* 145, 3796–3806.

Szabo, S.J., Dighe, A.S., Gubler, U. and Murphy, K.M. (1997) Regulation of the interleukin (IL)-12R beta 2 subunit expression in developing T helper 1 (Th1) and Th2 cells. *Journal of Experimental Medicine* 185, 817–824.

Toker, A., Meyer, M., Reddy, K.K., Falck, J.R., Aneja, R., Aneja, S., Parra, H., Burns, D.J., Ballas, L.M. and Cantley, L.C. (1994) Activation of protein kinase C family members by the novel polyphosphoinositides PtdIns-3,4-P_2 and PtdIns-3,4,5-P_3. *Journal of Biological Chemistry* 269, 32358–32367.

Trolle, S., Andremont, A. and Fattal, E. (1998) Towards a multipurpose mucosal vaccine using phosphorylcholine as an unique antigen? *STP Pharmasciences* 8, 19–30.

Vanamail, P., Ramaiah, K.D., Pani, S.P., Das, P.K., Grenfell, B.T. and Bundy, D.A.P. (1996) Estimation of the fecund life-span of *Wuchereria bancrofti* in an endemic area. *Transactions of the Royal Society of Tropical Medicine and Hygiene* 90, 119–121.

Wanni, N.O., Strote, G., Rubaale, T. and Brattig, N.W. (1997) Demonstration of immunoglobulin G antibodies against *Onchocerca volvulus* excretory–secretory antigens in different forms and stages of onchocerciasis. *Transactions of the Royal Society of Tropical Medicine and Hygiene* 91, 226–230.

Weiser, J.N., Shchepetov, M. and Chong, S.T. (1997) Decoration of lipo-polysaccharide with phosphorylcholine: a phase-variable characteristic of *Haemophilus influenzae*. *Infection and Immunity* 65, 943–950.

Whelan, M., Harnett, M.M., Houston, K.M., Patel, V., Harnett, W. and Rigley, K.P. (2000) A filarial nematode-secreted product signals dendritic cells to acquire a phenotype that drives development of Th2 cells. *Journal of Immunology* 164, 6453–6460.

Wuhrer, M., Rickhoff, S., Dennis, R.D., Lochnit, G., Soboslay, P.T., Baumeister, S. and Geyer, R. (2000) Phosphocholine-containing, zwitterionic glycosphingo-lipids of adult *Onchocerca volvulus* as highly conserved antigenic structures of parasitic nematodes. *Biochemical Journal* 348, 417–423.

Nematode Neuropeptides **20**

Aaron G. Maule, Nikki J. Marks and David W. Halton

Parasitology Research Group, School of Biology and Biochemistry, The Queen's University of Belfast, Medical Biology Centre, Belfast BT7 1NN, UK

Introduction

Most neuropeptides are intercellular signalling molecules that interact with cell surface receptors to trigger an intracellular transduction pathway. There is huge diversity in peptidic signalling molecules, with well over 200 having been identified in arthropods alone. Not only are they structurally diverse, but also the signalling cascades that they trigger are highly varied and, ultimately, can have a multitude of effects on living cells, from stimulation of ion channel opening to suppression of transcriptional events.

Neuropeptides most probably pre-date the 'classical' neurotransmitters such as acetylcholine (ACh) and 5-hydroxytryptamine (5-HT, serotonin) in that neuropeptide signalling is believed to have evolved prior to that of the classical transmitters (Shaw, 1996; Walker *et al.*, 1996). Classical transmitters have long been recognized as playing an important role in neuronal communication, helped largely by the fact that, relative to the diversity displayed by neuropeptides, they are few in number. Moreover, understanding of the role(s) of neuropeptides is often hindered by the absence of primary sequence information, in that not all of the peptide signalling molecules have been identified and structurally characterized. Also, peptide effects are often subtle, slow in onset and long lasting, making them more difficult to identify and analyse.

The evolution of understanding of the roles of classical neurotransmitters versus those of neuropeptides is perhaps best displayed by research in nematodes. For many years, cutting-edge research on the physiology of nematodes focused on the actions of the classical neurotransmitters, ACh and γ-aminobutyric acid (GABA) (Del Castillo *et al.*, 1963, 1964; Martin, 1980; Stretton *et al.*, 1985, 1992; Martin *et al.*, 1991). Using the large gastrointestinal pig nematode, *Ascaris suum*, models of nematode neuromuscular activity and locomotory behaviour, based on the tonic release of ACh and GABA, were developed (Stretton *et al.*, 1985). Although these models are still valid today, they take no account of the influence of endogenous neuropeptides on nematode locomotory behaviour. This omission was, of course, not deliberate; it was based on the premise that classical transmitters dominated the coordination of nematode behaviour, to the extent that the need for neuropeptide modulation in neuromuscular control was not recognized.

Over the last decade, and triggered by the pioneering work of Stretton and co-workers, numerous nematode neuropeptide sequences have been deduced (Davis and Stretton, 1996). This sequence information has arrived relatively slowly, due to the huge amount of effort required in tissue collection and the ultimate purification of enough peptide to facilitate structural characterization. However, the *Caenorhabditis elegans* genome sequencing project has rapidly provided details on all of the putative neuropeptides in a nematode. The information generated on neuropeptide diversity in *C. elegans* has revolutionized current understanding of nematode neurobiology; simply put, it has revealed a totally unexpected complexity in nematode nervous systems and has highlighted a basic ignorance of the neurobiology of an organism that has only 300 some neurones (Bargmann, 1998).

An ability to recognize and identify potential neuropeptides from the vast quantities of data deposited in the genome sequencing databases is necessarily biased by current perceptions of neuropeptide characteristics. In this respect, a very common feature of neuropeptides is the presence of a C-terminal amide moiety (Bradbury and Smyth, 1991). The reasons that most neuropeptides exhibit this feature are not clear, but it is believed to improve peptide stability in the reduction of carboxypeptidase attack and to be involved in receptor recognition, such that the deamidation of neuropeptides commonly results in loss of function. Searches for novel neuropeptides, therefore, tend to concentrate on search strings comprising two or three C-terminal residues that characterize a known neuropeptide family, a glycyl residue (putative amidation site) and a dibasic cleavage site, usually KR (single basic residues can also occupy the C-terminal position of glycyl amide donor sites). Nevertheless, numerous putative neuropeptides have been identified in *C. elegans* and it is likely that many more await to be discovered.

Although understanding of nematode neuropeptide signalling systems is extremely limited, the last decade has seen the generation of a large body of information on nematode neuropeptide localization, structure and function. The diversity of peptides involved in this signalling system and the variety of action of the identified peptides is so huge that attempts to trawl the existing information can often be disjointed and confusing. In this respect, a number of useful recent reviews have examined the complexity of this system in nematodes (Halton *et al.*, 1994; Geary *et al.*, 1995, 1999; Brownlee *et al.*, 1996a; Davis and Stretton, 1996; Maule *et al.*, 1996a,b,c; Day and Maule, 1999). This chapter aims to present an up-to-date account of nematode neuropeptide systems and attempts to synthesize a rational approach to the development of this research area.

Neuropeptide Localization

The first evidence that neuropeptides occurred in nematodes arose indirectly from histochemical techniques to identify neurosecretory elements in the 1960s (Davey, 1966; Rogers, 1968). The next major breakthrough was the identification of neuropeptide immunostaining in nematodes in the late 1980s (Li and Chalfie, 1986; Leach *et al.*, 1987; Atkinson *et al.*, 1988; Davenport *et al.*, 1988). At this time, the diversity of neuropeptide structure in higher organisms was recognized and had become an ever expanding field of study. These early studies indicated that neuropeptide immunoreactivity was abundant and widespread in nematode nervous systems. Furthermore, evidence for the occurrence of neuropeptides in nematodes was generated for a wide range of species, from free-living to plant- and animal-parasitic forms.

Since these early studies, immunolocalization of neuropeptides has been achieved with a huge array of antisera, all with differing specificities to known vertebrate and invertebrate peptide families (Stretton *et al.*, 1991; Halton *et al.*, 1994; Brownlee *et al.*, 1996a; Davis and Stretton, 1996; Maule *et al.*, 1996c). The information generated indicated that neuropeptides were associated with all of the major neuronal systems in nematodes, including the ganglia, longitudinal nerve cords, motor neurones, interneurones and sensory neurones. Such evidence indicated a diversity in putative function for these neuropeptides. However, the diversity in epitopes identified by the antisera used should be viewed with some caution. Immunocytochemistry is based on a non-competitive antibody–antigen reaction such that there is great potential for non-specific interaction of the antisera employed and consequent false positive results. Therefore, it is unreliable to estimate the number of neuropeptides in nematodes solely by using immunocytochemical criteria. The actual number of neuropeptides present in nematodes can only be deduced from gene or primary sequence information. In this respect, the *C. elegans* genome sequencing project provides a great opportunity to evaluate neuropeptide diversity in nematodes.

Although immunocytochemical screening of neurochemical diversity provides only an indication that a neuropeptide may be present, antisera that interact specifically with known nematode neuropeptides are likely to provide useful information on peptide distribution (Sithigorngul *et al.*, 1989). Even so, when multiple related peptides are present, as is the case in nematodes (see later), the potential for antisera cross-reactivity with numerous peptides is high. Nevertheless, it is possible to generate information on the distribution of specific peptide families, with contrasting structural features, within the nematode nervous system.

Distribution of FMRFamide-related Peptides (FaRPs) in Nematodes

Easily the most widely studied peptide family in nematodes is the FMRFamide-related peptides (FaRPs). Originally, these were classed as having an F/Y-X-R-F.amide motif. However, with increasing reports of primary sequence diversity, most small peptides (= 20 amino acids) that possess a C-terminal R-F.amide, and that are not members of the neuropeptide Y (NPY) family (RXRF/Y.amide), are deemed to be FaRPs. A variety of FaRP-antisera have been used to immunostain nematodes. Moreover, a number of antisera raised against other peptide families (NPY family peptides and SALMF.amide) (Smart *et al.*, 1992a; Brownlee *et al.*, 1996b) were subsequently found to cross-react with nematode FaRPs, such that the immunostaining patterns obtained in nematodes using these can, at least in part, be attributed to endogenous FaRPs.

FaRP distribution in Ascaris suum

In the large enteric pig parasite, *A. suum*, FaRP-immunostaining has been detected in some 60% of its nervous system, including all of its major components (central, peripheral and enteric) and all neuronal classes (motor, sensory and interneurones). Furthermore, the majority of the ganglia associated with the circumpharyngeal nerve ring (the nematode brain) are FaRP-immunopositive (Table 20.1). FaRP-immunostaining is also widespread in nerve fibres, including all the longitudinal nerve cords (ventral, dorsal, lateral and sublateral) and fibres innervating the pharynx, rectum and ovijector (Brownlee *et al.*, 1993a,b, 1994; Cowden *et al.*, 1993; Fellowes *et al.*, 1999).

The interrelationships of the FaRP-immunopositive nerves with pharyngeal and ovijector muscle have been examined using dual wavelength detection confocal microscopy (Fellowes *et al.*, 1999; R.A. Fellowes, D.W. Halton and A.G. Maule, Belfast, 1999, unpublished observations) (Figs 20.1 and 20.2). In these studies, varicose nerve fibres, which were highly FaRP-immunoreactive, were identified in close association with the circular muscle fibres of the ovijector (Fig. 20.1). The varicose nature of these nerve fibres is a well-documented feature of peptidergic nerves

Table 20.1. Occurrence of FMRFamide-related peptide (FaRP)-immunoreactivity in the nervous system *of Ascaris suum* (after Cowden *et al.*, 1993; Fellowes *et al.*, 1999).

Component of nervous system	FaRP-immunoreactive neurones (%)
Circumpharyngeal nerve ring	> 50
Lateral ganglia cells	> 30
Retrovesicular ganglion	> 20
Ventral ganglion	> 50
Dorsal ganglion	100
Subdorsal and subventral ganglia	> 80
Subdorsal sensory cells	80
Tail ganglia cells (females)	> 75
Pharyngeal neurones	90
Ovijector neurones	100

and is believed to represent accumulations of secretory vesicles as they are transported, en masse, along nerve axons. Evidence for the occurrence of neuromuscular synapses requires further studies at the electron microscope level. In the pharynx, FaRP-immunopositive nerve fibres form a number of anastomosing networks which innervate regions of pharyngeal muscle (R.A. Fellowes, D.W. Halton and A.G. Maule, unpublished observations) (Fig. 20.2).

FaRP distribution in Caenorhabditis elegans

The availability of the complete neuronal map of *C. elegans* has greatly facilitated work on its nervous system (White *et al.*, 1986). Immunocytochemical FaRP-screens of the worm showed that some 30 neurones (approximately 10% of its nervous system) were FaRP-immunopositive (Schinkmann and Li, 1992), a significant contrast to the situation in *A. suum*. Such a disparity in the level of immunostaining between these worms is difficult to explain and seems unlikely to be due to small differences in staining technique. Although the distribution of FaRPs in *C. elegans* is more restricted than that reported for *A. suum*, they have similarly been localized to motor, sensory and interneurones. The specific localization of individual FaRPs using immunocytochemical techniques has been restricted by the structural similarity of endogenous FaRPs, making antibody discrimination extremely difficult. It is also unknown if the various FaRP antisera employed in studies on nematodes cross-react with all of the endogenous FaRPs; if not, then reports of FaRP abundance in the nervous systems of nematodes will be underestimates.

Expression of the *C. elegans* FMRFamide-like peptide gene-1 (*flp*-1) has been demonstrated using *flp*-1-*lacZ* reporter constructs in transgenic

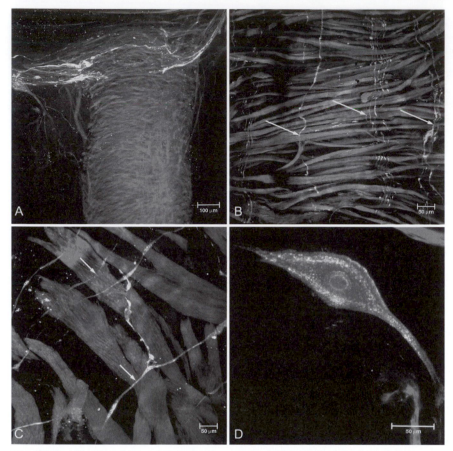

Fig. 20.1. Confocal images of whole mounts of the ovijector region of *A. suum* stained with phalloidin-tetramethylrhodamine isothiocyanate (TRITC) to show muscle and with an anti-RFamide antiserum coupled to fluorescein isothiocyanate (FITC) to show FaRPergic nerves. (A) Main ventral nerve cord encircles opening of ovijector where it meets the body wall and is immunopositive for FaRPs. (B) Flat-fixed preparation of the ovijector showing circular muscles and tracts of parallel FaRPergic nerves (arrows). (C) Detail of the circular muscle of ovijector and associated nerves (arrows). (D) A FaRPergic cell body is localized in the ventral nerve cord at junction with ovijector and provides innervation to ovijector muscle.

animals (Nelson *et al.*, 1998a). Using this technique, the *flp*-1 gene was localized to anterior head neurones, including the AVK, AVA, AVE, RIG, RMG, AIY, AIA and M5 cells. It will be interesting to see how the distribution of FaRPs, as demonstrated using reporter constructs for all the *flp*-genes, compares with that observed using immunocytochemistry.

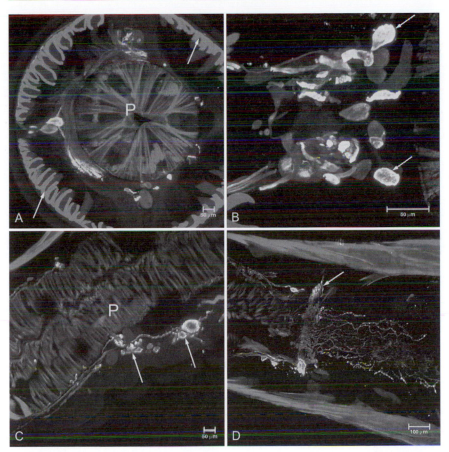

Fig. 20.2. Confocal images of sections of the head region of *A. suum* stained with phalloidin-TRITC for muscle and with an anti-RFamide antiserum-FITC conjugate to show FaRPergic nerves. P, pharynx. (A) T.S. of pharynx at the level of the circumpharyngeal nerve ring. Note FaRPergic nerves and body-wall muscle (arrows). (B) Detail of the ventral ganglion immunostained for FaRPs, showing immunopositive cell bodies (arrows). (C) L.S. of pharynx showing radial organization of muscle and immunostaining in nerves and cell bodies (arrows) in dorsal ganglion. (D) L.S. of pharynx showing FaRP-immunoreactivity in pharyngeal nerves and commissure (arrow).

FaRP distribution in other nematodes

FaRP-immunostaining has been localized to the nervous system of a range of other nematode species, including free-living, animal parasitic and plant parasitic forms (Table 20.2 and references therein).

Table 20.2. Immunocytochemical localization of FMRFamide-related peptides (FaRPs) in nematodes.

Species	Localization of immunostaining	Reference
Ascaris suum	CNS, PNS and ENS	Davenport *et al.*, 1988; Sithigorngul *et al.*, 1990; Cowden *et al.*, 1993; Brownlee *et al.*, 1993a,b, 1994; Fellowes *et al.*, 1999
Brugia pahangi	CNS	Warbrick *et al.*, 1992
Dirofilaria immitis	CNS	Warbrick *et al.*, 1992
Haemonchus contortus	CNS, PNS and ENS	Geary *et al.*, unpublished observations
Onchocerca volvulus	CNS	Geary *et al.*, 1995
Trichostrongylus colubriformis	CNS	Geary *et al.*, 1995
Caenorhabditis elegans	CNS, PNS and ENS	Li and Chalfie, 1986; Atkinson *et al.*, 1988; Schinkmann and Li, unpublished observations
Caenorhabditis vulgaris	CNS	Schinkmann and Li, 1994
Cystidicola faronis	CNS	Wikgren and Fagerholm, 1993
Goodeyus ulmi	CNS	Leach *et al.*, 1987
Heterodera glycines	CNS and PNS	Atkinson *et al.*, 1988
Panagrellus redivivus	CNS, PNS and ENS	Geary *et al.*, unpublished observations
Romanomeris culicivorax	CNS and PNS	Jagdale and Gordon, 1994

CNS, central nervous system; PNS, peripheral nervous system; ENS, enteric nervous system.

Distribution of Other Neuropeptides in Nematodes

A wide range of antisera to vertebrate and invertebrate neuropeptides has given positive immunostaining in nematode nervous systems (Sithigorngul *et al.*, 1990; Brownlee *et al.*, 1993a,b, 1994; Halton *et al.*, 1994). The usefulness of the immunolocalization data to understanding of nematode neuropeptide systems is somewhat limited until the epitopes with which the antisera cross-react have been fully characterized. Nevertheless, of particular interest has been the localization of immunoreactivity to known insect neuropeptides in *A. suum* (Smart *et al.*, 1993, 1995). Such neuropeptides, which differ from those in mammals (presumably more so than those identified by their homology to mammalian neuropeptides), are likely to be more easily targeted by novel chemotherapies against parasitic forms. Such findings suggest that novel drugs that disrupt neuropeptide signalling in nematode and insect pests are a real possibility.

Neuropeptide Structure

The vast majority of neuropeptides that have been structurally characterized from nematodes are FaRPs. One other peptide, TKQELE, has been structurally characterized from *A. suum* (Smart *et al.*, 1992b). TKQELE was identified using an antiserum raised against KGQELE, a fragment of rat chromogranin A, a prohormone that is expressed in the secretory granules of numerous mammalian neuroendocrine cells. This peptide has not been characterized from any other nematode species and its function is unknown. All other nematode neuropeptides that have been structurally characterized are FaRPs. The *C. elegans* genome sequence information has enabled the gene sequences of other putative neuropeptides to be identified. In this respect, there are at least 15 non-FaRP neuropeptide-like encoding genes in *C. elegans* (Bargmann, 1998). However, in the absence of expression and functional data, their assignment as nematode neuropeptides remains equivocal. Indeed, all of the currently available data on nematode neuropeptide expression relates exclusively to FaRPs. All known nematode neuropeptides and candidate neuropeptides have been identified by two methods: isolation and structural characterization of the peptide itself; or by sequencing of the encoding gene and identification of putative products. Most of the available data on nematode neuropeptides have been generated from either *C. elegans* or *A. suum*.

A. suum *neuropeptides*

To date, 20 FaRPs have been structurally characterized from *A. suum* (Table 20.3) (Cowden *et al.*, 1989; Cowden and Stretton, 1993, 1995; Davis and Stretton, 1996). All of these peptides were isolated from acid methanol extracts of *A. suum* heads and tails (which include the circumpharyngeal and perianal ganglia) using reverse phase HPLC. All of these peptides were designated as *Ascaris* FaRPs, AFX, where X is a number given chronologically upon discovery, i.e. AF1 was the first *Ascaris* FaRP to be structurally characterized (Cowden *et al.*, 1989). One of these peptides (FDRDFMHFamide), designated AF17, falls outside the FaRP family specifications in that it possesses a C-terminal HFamide and lacks the signature RFamide motif.

Six of the *A. suum* FaRPs identified by direct purification and sequencing have been found to be co-encoded on a single gene, the *apf*-1 gene (Edison *et al.*, 1997). The occurrence of multiple-copy neuropeptide genes in invertebrates has been recognized for some time and is clearly a feature of *Ascaris* FaRP-encoding genes. This finding shows that single nematode neuropeptide genes can encode a range of related, but structurally unique, neuropeptides. It seems likely that multiple-encoded neuropeptides have arisen as the result of duplication events, and that over time these duplicated peptides have had the opportunity to mutate and assume alternative roles.

Table 20.3. Primary structures (sequence shown using single-letter notation) of *A. suum* FMRFamide-related peptides (Cowden *et al.*, 1989; Cowden and Stretton, 1993, 1995; Davis and Stretton, 1996).

Neuropeptide code	Neuropeptide sequence	Identified molecule(s)
AF1	KNEFIRF . NH$_2$	Peptide
AF2	KHEYLRF . NH$_2$	Peptide
AF3	AVPGVLRF . NH$_2$	Peptide; cDNA
AF4	GDVPGVLRF . NH$_2$	Peptide; cDNA
AF5	SGKPTFIRF . NH$_2$	Peptide
AF6	FIRF . NH$_2$	Peptide
AF7	AGPRFIRF . NH$_2$	Peptide
AF8	KSAYMRF . NH$_2$	Peptide
AF9	GLGPRPLRF . NH$_2$	Peptide
AF10	GFGDEMSMPGVLRF . NH$_2$	Peptide; cDNA
AF11	SDIGISEPNFLRF . NH$_2$	Peptide
AF12	FGDEMSMPGVLRF . NH$_2$	Peptide
AF13	SDMPGVLRF . NH$_2$	Peptide; cDNA
AF14	SMPGVLRF . NH$_2$	Peptide; cDNA
AF15	AQTFVRF . NH$_2$	Peptide
AF16	ILMRF . NH$_2$	Peptide
AF17	FDRDFMHF . NH$_2$	Peptide
AF18	XXXXPNFLRF . NH$_2$	Peptide
AF19	AEGLSSPLIRF . NH$_2$	Peptide
AF20	GMPGVLRF . NH$_2$	Peptide; cDNA

C. elegans *neuropeptides*

Most information on the neuropeptide complement of *C. elegans* has been derived from data generated by the genome sequencing consortium. Prior to this information, a single FaRP-encoding gene had been identified (Rosoff *et al.*, 1992). This gene was designated the FMRFamide-like peptide (*flp*)-1 gene and, like the neuropeptides, was appended a number chronologically upon discovery. With the completion of the *C. elegans* genome sequencing project, many more *flp* genes have been identified. To date, at least 18 *C. elegans* genes have been designated as *flp* genes and the expression of at least 15 of these has been confirmed using PCR methods (Nelson *et al.*, 1998b). Searches of the 'wormpep' facility in the *C. elegans* BLAST database, using known nematode C-terminal FaRP motifs, indicate the occurrence of at least three other putative FaRP-encoding genes. The identification of these genes and confirmation of their expression and their nomenclature assignments have been carried out by Li and co-workers (Boston University). The sequences of 62 different peptides are encoded on these genes, including KPNFMRYG (*flp*-1) and EDGNAPFGTMKFG (*flp*-3), which do not have the characteristic RFamide motif; thus, there are at least 60 known *C. elegans* FaRPs (Table 20.4). The *flp*-1 gene was also characterized from the closely related species, *C. vulgaris* (Schinkmann and Li, 1994). Interestingly, several of these peptides have been identified in other nematode species, including peptides encoded on genes *flp*-1, *flp*-6, *flp*-8, *flp*-11 and *flp*-14. This points to a significant degree of conservation in nematode FaRP structure. Although such conservation of structure may be mirrored by conservation of function, this remains to be tested. Indeed, it seems plausible that those FaRPs that are the most structurally conserved across nematode species include those peptides that have important associated physiology; their key physiological role is likely to have inhibited structural divergence.

Other nematode neuropeptides

FaRPs have been structurally characterized from two other nematode species: the free-living nematode *Panagrellus redivivus* and the sheep gastrointestinal parasite *Haemonchus contortus*. Peptides identified from *P. redivivus* include: SDPNFLRFamide (PF1), SADPNFLRFamide (PF2) (Geary *et al.*, 1992); KSAYMRFamide (PF3/AF8) (Maule *et al.*, 1994a); KPNFIRFamide (PF4) (Maule *et al.*, 1995a); KHEYLRFamide (AF2) (Maule *et al.*, 1994b); and AMRNALVRFamide and NGAPQPFVRFamide (C. Moffett and N.J. Marks, unpublished observations). Of interest here is the fact that all but one (PF4) show complete identity with *C. elegans* peptides; PF4 only differs from the *flp*-1 peptide, KPNFLRFamide, by a single amino acid substitution (I^5 for L^5). Such commonality in FaRP structure may relate to evolutionary relatedness in these two species. However, *H. contortus* peptides include KSAYMRFamide (PF3/AF8) and KHEYLRFamide (AF2), which are present in all four species from which peptides have been structurally characterized (Keating *et al.*, 1995; Marks *et al.*, 1999a). Clearly, some FaRPs are widely distributed across a broad range of nematode species.

Another interesting observation is that AF2 was the most abundant FaRP in each of the species examined (*C. elegans*, *P. redivivus*, *A. suum* and *H. contortus*). Again, this may relate to a key role in nematode biology. It is worth noting that the most abundant neuropeptides in the brains of species as diverse as platyhelminths and humans are neuropeptide Y (NPY)-related peptides (Day and Maule, 1999). Radioimmunometric analyses of nematodes have shown that AF2 cross-reacts with some NPY-family peptide antisera (Maule *et al.*, 1994b). This, coupled with the fact that there are no obvious NPY-family peptides in nematodes (none has been reported in the literature and scans of 'wormpep' in the *C. elegans* BLAST server reveal no hits using C-terminal motifs of known NPY-family peptides as search strings) yet cDNAs with homology to NPY-family receptors have been reported (deBono and Bargmann, 1998), suggests that AF2 could be the nematode equivalent of NPY. Clearly, a detailed understanding of

Table 20.4. *C. elegans* FMRFamide-related peptides.

flp-1	*flp-7*	*flp-14*
KPNFMRYG	TPMQRSSMVRFG	*KHEYLRFG
AGSDPNFLRFG	SPMQRSSMVRFG	*KHEYLRFG
*SQPNFLRFG	SPMQRSSMVRFG	*KHEYLRFG
*ASGDPNFLRFG	SPMQRSSMVRFG	*KHEYLRFG
*SDPNFLRFG	SPMERSAMVRFG	
*AAADPNFLRFG	SPMDRSKMVRFG	*flp-15*
*SADPNFLRFG	TPMQRSSMVRFG	GGPQGPLRFG
KPNFLRFG		GPSGPLRFG
	flp-8	
flp-2	**KNEFIRFG	*flp-16*
LRGEPIRFG	**KNEFIRFG	**AQTFVRFG
SPREPIRFG	**KNEFIRFG	**AQTFVRFG
		GQTFVRFG
flp-3	*flp-9*	
SPLGTMRFG	*KPSFVRFG	*flp-17*
TPLGTMRFG	*KPSFVRFG	KSAFVRFG
SAEPFGTMRFG		KSAFVRFG
NPENDTPFGTMRFG	*flp-10*	KSQYIRFG
ASEDALFGTMRFG	QPKARSGYIRFG	
EDGNAPFGTMKFG		*flp-18*
EAEEPLGTMRFG	*flp-11*	DFDGAMPGVLRFG
SADDSAPFGTMRFG	**AMRNALVRFG	EMPGVLRFG
NPLGTMRFG	ASGGMRNALVRFG	KSVPGVLRFG
	**NGAPQPFVRFG	SVPGVLRFG
flp-4		EIPGVLRFG
PTFIRFG	*flp-12*	SEVPGVLRFG
ASPSFIRFG	*RNKFEFIRFG	DVPGVLRFG
		*SVPGVLRFG
flp-5	*flp-13*	
APKPKFIRFG	SDRPTRAMDSPLIRFG	Peptides from other
AGAKFIRFG	**AADGAPLIRFG	putative *flp* genes
GAKFIRFG	*APEASPFIRFG	
	**AADGAPLIRFG	TKFQDFLRFG
flp-6	APEASPFIRFG	
*KSAYMRFG	ASPSAPLIRFG	AMRNSLVRFG
*KSAYMRFG	SPSAVPLIRFG	
*KSAYMRFG	SAAAPLIRFG	DYDFVRFG
*KSAYMRFG	ASSAPLIRFG	
*KSAYMRFG		DGFVRFG
*KSAYMRFG		
		AFFKNVLRFG

*Peptides that have been isolated and sequenced from *C. elegans*.
**Peptides that have been isolated from nematode species other than *C. elegans*.
Other putative *flp* genes were identified using search strings comprising C-terminal tetrapeptides of known FaRPs plus a C-terminal G (amide donor) residue in searches of the *C. elegans* BLAST server, wormpep 17.

the role of AF2 in nematodes and the endogenous receptors with which it interacts is required before this possibility could be reasonably tested.

Neuropeptide Function

The rapid and relatively recent determination of the structures of so many nematode neuropeptides has resulted in a huge disparity between available structural and functional data; indeed, currently the physiological function of any of the known nematode neuropeptides remains unknown. Most of the accumulated data on nematode neuropeptide activities have been obtained using muscle strips of *A. suum* and, as a result, only provide a glimpse of the putative role for many of the peptides involved.

With respect to the FaRPs, the fundamental problem in resolving physiological function in nematodes stems from the absence of expression data, i.e. where the peptides occur in the worm. Although numerous immunocytochemical studies have localized FaRP expression in a variety of nematodes (see earlier), the distribution of individual FaRPs is unknown. Moreover, the localization of specific receptors for each of the neuropeptides is needed to facilitate rational attempts to evaluate individual peptide roles. At present, no receptor for any of the known nematode neuropeptides has been identified. The identification of the receptors for each of the known neuropeptides and the evaluation of their function are the greatest and most important challenges to scientists working in this area.

The effects of known A. suum *FaRPs*

The first identified *Ascaris* FaRP, AF1 (KNEFIRFamide), and the structurally related peptide, AF2 (KHEYLRFamide), have been found to inhibit locomotory waves when injected into adult worms (Cowden *et al.*, 1989; Cowden and Stretton, 1993). Their effects on body-wall muscle strips are biphasic, comprising a transient relaxation followed by an extended period of increased contractile activity (Maule *et al.*, 1995b; Bowman *et al.*, 1996). When using muscle strips that have had the motor nerve cords removed, only the inhibitory actions of AF1 and AF2 are seen (Maule *et al.*, 1995b). This suggests that the inhibitory phase is due to post-synaptic effects on body-wall muscle in the worm. In contrast, the excitatory effects are nerve-cord dependent and are not observed in muscle strips that have been denervated. Another possibility is that the peptides interact with receptors at the post-synaptic junction – these are also removed in specimens that have had the motor nerve cords removed.

One difference between the effects of AF1 and AF2 is their potency: AF2 is up to 1000-fold more potent than AF1. AF1 has been shown to inhibit both dorsal and ventral inhibitory motor neurones and decreases input resistance in the VI and DI motor neurones. The VI and DI motor neurones are believed to be the site of the excitatory AF1 receptor (Cowden *et al.*, 1989; Davis and Stretton, 1996). The inhibition of inhibitory motor neurones would be expected to result in an excitatory response. The effects of AF2 have also been shown to enhance the effects of the excitatory neurotransmitter, ACh, through the amplification of ACh-induced excitatory junction potentials (Pang *et al.*, 1995). AF2 has also been found to stimulate cAMP levels in *A. suum* body-wall tissue (D.P. Thompson, J.P. Davis and T.G. Geary, Pharmacia and Upjohn Co., 1996, personal communication). These effects do not fully explain the enhanced oscillatory activity of muscle strips exposed to these peptides.

Examination of the structure–activity relationships of AF1 and AF2 provide further evidence of complexity in this signalling system (Bowman *et al.*, 1996). Structural modifications in the form of alanine scan series (replacement of each residue in turn with alanine) and N-terminally truncated analogues of AF1 revealed that both the N- and C-termini were essential for biological activity. One of these analogues, KNEFIAFamide, was found to antagonize the effects of AF1 but not those of AF2. Furthermore, chimeras of AF1 and AF2 (KHEFIRFamide, KNEYIRFamide and KNEFLRFamide) had similar or reduced potencies to that of AF1. These findings suggest, but do not prove, that AF1 and AF2 act at different receptors in the worm.

Six known *Ascaris* FaRPs possess a C-terminal PGVLRFamide (AF3, AF4, AF10, AF13, AF14, AF20). All six of these peptides have been shown to have similar excitatory effects on body-wall muscle of *Ascaris* (Cowden and Stretton, 1995; Davis and Stretton, 1996). Diversity in the N-terminal regions of peptides with a C-terminal PGVLRFamide motif seems to have no obvious effect on observed physiology. One of these peptides, AF3 (AVPGVLRFamide), has also been shown to have excitatory effects on the domestic fowl parasite, *Ascaridia galli* (Trim *et al.*, 1997). Further examination of the mechanism underlying the excitatory effects of AF3 revealed that its actions were independent of the endogenous cholinergic system (Trim *et al.*, 1997). Furthermore, AF3 decreased cAMP levels in both *A. suum* and *A. galli* and inhibited forskolin-stimulation of cAMP levels (Trim *et al.*, 1998). These results show that AF3 has similar effects on muscle contraction and cAMP levels in these two nematodes and points to at least some conservation of neuropeptide physiology across nematode species.

KSAYMRFamide (AF8) has been found to have nerve cord-dependent excitatory effects on ventral and inhibitory effects on dorsal muscle strips of *A. suum* (Maule *et al.*, 1995b). To date, this is the only peptide found to display differential activity on body-wall muscle of *Ascaris*. The segmental oscillator model of locomotion proposed for *A. suum* relies on reciprocal inhibition of opposing effects on dorsal and ventral muscle fields which, with the appropriate time intervals, result in the recognized nematode locomotory wave form (Stretton *et al.*, 1985; Davis and Stretton, 1996). It seems reasonable to hypothesize that AF8 could be involved in the coordination of locomotory behaviour, with simultaneous release on to dorsal and ventral muscles causing a ventral bend. No peptide that has an opposite effect has been identified.

There is a limited amount of information available on the actions of other endogenous *A. suum* FaRPs on body-wall muscle and on four classes of motor neurones. AF5 (SGKPTFIRFamide) was found to have excitatory effects on body-wall muscle and decreases the input resistance of the DE1, DE2, VI and DI motor neurones; AF7 (AGPRFIRFamide) had similar effects on the motor neurones (Davis and Stretton, 1996). AF9 (GLGPRPLRFamide) and AF17 (FDRDFMHFamide) both increased muscle tension in body-wall muscle whereas AF6 (FIRFamide), AF11 (SDIGISEPNFLRFamide) and AF19 (AEGLSSPLIRFamide) decreased muscle tension (Davis and Stretton, 1996). Also, AF11 decreased the input resistance of VI and DI motor neurones and increased the input resistance of DE1 and DE2 motor neurones. AF7 (AGPRFIRFamide) and AF16 (ILMRFamide) were found to have little effect on *Ascaris* body-wall muscle. The evaluation of the physiological role of each of the known *Ascaris* FaRPs and the localization of the specific receptors, as undertaken by workers in the Stretton laboratory (University of Wisconsin-Madison), is central to the understanding of this signalling system. Clearly, the variety of effects of endogenous *Ascaris* FaRPs on body-wall muscle and motor neurones points to a central role for these peptides in locomotory behaviour, with the involvement of a highly complex signalling system.

FaRPs have been implicated not only in the coordination/modulation of locomotory behaviour in *Ascaris*, but also in the activity of two other distinct muscle systems in the worm. The pharynx and the ovijector are highly developed muscular pumping organs involved in the intake and digestion of food and in the release of eggs, respectively. Both of these organs have a rich, well-developed peptidergic neuronal component (Brownlee *et al.*, 1993a, 1994; Fellowes *et al.*, 1999). When active, both the pharynx and the ovijector display waves of coordinated peristaltic activity that have been shown to be modulated by FaRPs. Pharyngeal pumping reponds to AF8 with an initial transient increase in activity followed by a period of inactivity (Brownlee *et al.*, 1995). The ovijector (also known as the vagina vera) displayed a variety of *in vitro* responses when exposed to AF1, AF2, AF3, AF4 or AF8 (Fellowes *et al.*, 1998, 2000). Thus, AF1 caused biphasic responses that comprised an initial contraction followed by a relaxation and a period of inactivity; AF2 and AF8 both reduced contractile activity in the ovijector; AF3 and AF4 had complex multiphasic effects on the ovijector that included a contraction followed by a period of increased activity and then inactivity. The fact that these peptides had distinguishable effects on the ovijector suggests that multiple receptors are present – a situation reminiscent of that seen with body-wall tissue. Clearly, the exact role of these peptides in the pharynx and ovijector are far from clear. Nevertheless, FaRP modulation of pharyngeal pumping and ovijector activity seem highly plausible hypotheses that warrant further evaluation, not least because these sites may well provide important focal points for drug action or drug entry into parasitic nematode species.

The effects of other nematode FaRPs

Not only are known endogenous *Ascaris* neuropeptides active on the worm, but also FaRPs from other species, which may or may not be found to occur in *Ascaris*, have potent and diverse effects on the muscle systems in the worm. Probably the best-characterized signalling system is that employed by PF1 (SDPNFLRFamide) and PF2 (SADPNFLRFamide). These two peptides, originally identified from *P. redivivus* (Geary *et al.*, 1992) and *C. elegans* (Rosoff *et al.*, 1992), have potent nerve cord-independent, Ca^{2+}-dependent inhibitory effects on dorsal and ventral body-wall muscle strips of *A. suum* (Franks *et al.*, 1994; Bowman *et al.*, 1995; Maule *et al.*, 1995b) which cause a long-lasting flaccid paralysis of the tissue. The profound relaxation is accompanied by a small hyperpolarization coupled with no obvious input resistance change in the muscle cells of the body wall (Franks *et al.*, 1994; Bowman *et al.*, 1995). The free radical gas nitric oxide (NO) has been implicated in the actions of PF1 due to the fact that inhibitors of NO synthase (NOS, the enzyme that generates NO) blunt PF1 effects in *Ascaris* muscle strip preparations. Furthermore, the NO donor, sodium nitroprusside, induced muscle strip relaxation similar to that seen with PF1 (Bowman *et al.*, 1995). NOS activity was highest and [^{3}H]PF1 binding was most pronounced in the *Ascaris* hypodermal layer; this layer lies closely associated with the body-wall muscle and surrounds the major nerve cords. This evidence suggests that NOS activity in the hypodermal layer results in the release of NO. NO could easily diffuse to the muscle, where it may act on an as yet unknown intracellular target to induce muscle relaxation. Whether or not this activity involves the stimulation of cytosolic guanylate cyclase and an increase in cytosolic cGMP levels, similar to that seen in the vascular epithelium of mammals, awaits investigation. Nevertheless, the evidence indicates a complex signalling pathway involving several tissue types in the worm and provides a glimpse of some of the nuances that may exist with FaRP signalling in nematodes.

PF1 and PF2 have also been found to modulate the contractile activity of the ovijector of *A. suum* (Fellowes *et al.*, 2000). PF1 and PF2 reduce the inherent contractile activity of the ovijector by an unknown mechanism. It is possible that this inhibitory activity, like that seen in the body wall, relies on NO.

Structure–activity studies with PF1 have indicated that N-terminally truncated and alanine-scan peptides have reduced inhibitory effects (des [Ser^1]-, des[Ser^{1}Asp^2]-, Ala^1-, Ala^2-, Ala^3-, Ala^4- and Ala^6-PF1) or are inactive (Ala^5-, Ala^7- and Ala^8-PF1) (Geary *et al.*, 1995). The pharmacological

profiles of the nematode FaRPs will help to elucidate functional requirements for the individual peptides and may help to delineate receptor
diversity (see later).

PF4 (KPNFIRFamide) has been found to induce nerve cord-
independent and Cl⁻-dependent relaxation of *Ascaris* body-wall muscle
strips and concomitant hyperpolarization of muscle cells (Maule *et al.*,
1995b; Holden-Dye *et al.*, 1997). The relaxation induced by PF4 is much
more rapid and short-lived than that seen with PF1. The classical nematode
inhibitory-neurotransmitter, γ-aminobutyric acid (GABA), also acts directly
on muscle to open Cl⁻ channels and induce relaxation that is qualitatively
similar to that caused by PF4. The possibility of interaction between GABA
and PF4 was examined using the nematode GABA channel inhibitor,
NCS 281-93. This compound reduced the effects of GABA but did not
alter PF4-induced relaxation of *Ascaris* body-wall muscle (Maule *et al.*,
1995b). These results indicate the occurrence of PF4-receptors that are
independent of the GABA receptor; the possibility remains that other
receptor sub-types interact with both signalling molecules. Structure–
activity studies on PF4 have revealed that removal of the proline residue
from position 2 is detrimental to biological activity of the peptide (Kubiak
et al., 1996). PF4 has also been found to have profound inhibitory effects on
the ovijector of *A. suum* (Fellowes *et al.*, 2000).

The *C. elegans* FaRP, APEASPFIRFamide, has been shown to have
inhibitory effects on *Ascaris* body-wall muscle that are nerve cord-
independent (Marks *et al.*, 1997). The ionic dependence of the effects
of APEASPFIRFamide revealed that its actions could be distinguished
from those of the other known inhibitory FaRPs, PF1, PF2 and PF4.
Interestingly, the *C. elegans* FaRP, KPSFVRFamide, was found to have no
observable effects on dorsal or ventral muscle strip preparations of the
worm (Marks *et al.*, 1999b), even though the structurally related peptide
PF4 (KPNFIRFamide) has potent inhibitory actions. However, the chimeric
peptide KPNFVRFamide had similar inhibitory effects to those of PF4
whereas KPSFIRFamide was also inactive. It appears, therefore, that either
the serine residue in position 3 of these peptides inhibits the activity or the
asparagine residue in position 3 of PF4 is essential for the observed effects.
In either case, at least some of the endogenous FaRP receptors have strict
structural prerequisites on the ligands with which they interact. The ability
for nematode FaRP receptors to delineate between different ligands would
seem to be an essential requirement for receptors faced with such a
complex array of signalling molecules.

FaRP effects on C. elegans

Detailed understanding of the development, cytology and now the genome
of *C. elegans* has greatly promoted the value of this soil nematode. The

complexity of the neuropeptidergic system in the worm is clearly evident from the genome data (see earlier). The ability to manipulate gene expression in *C. elegans* provides an extremely valuable tool in attempts to understand neuropeptide function in nematodes. *C. elegans* knockouts (those that have a deletion in a particular gene) for the neuropeptide gene *flp*-1 have been identified and characterized (Nelson *et al.*, 1998a). These studies revealed that disruption of the *flp*-1 gene causes a series of behavioural defects, including loopy and uncoordinated movement, hyperactivity and wandering, nose-touch insensitivity, absence of osmotic avoidance and absence of serotonin-induced inhibition of locomotion (Nelson *et al.*, 1998a). Evaluation of the effects of gene knockouts for each of the endogenous neuropeptide-encoding genes will provide valuable data on neuropeptide function in nematodes. As evidence of FaRP structural similarity between nematode species grows, so the relevance of the results obtained for *C. elegans* to nematodes in general increases.

FaRP effects on H. contortus

Isotonic muscle contraction was used to measure the effects of selected nematode FaRPs on the body-wall muscle of *H. contortus*. AF2 was found to have inhibitory effects on muscle activity and inhibited acetylcholine (ACh)-induced contractions in the worm whereas AF8 had excitatory effects on the muscle and enhanced ACh-induced contractions (Marks *et al.*, 1999a). There were obvious differences in the methodologies used to evaluate the effects of these peptides on *Haemonchus* muscle compared with those used to examine these peptide effects on *Ascaris*. How comparable the results are has yet to be determined.

Further studies using two different *Haemonchus* isolates to examine the effects of AF1 and AF8 provided evidence of some interaction between the FaRP and cholinergic signalling systems in the worm. The responses of a 'normal' worm isolate, which was sensitive to cholinomimetics including levamisole, were compared with the responses of an isolate (Lawes isolate), which had reduced sensitivity to cholinergic drugs (Marks *et al.*, 1999a). In these studies the Lawes isolate, which had reduced cholinomimetic sensitivity, was equally less sensitive to AF8, i.e. its dose–response curve for cholinomimetics and AF8 was shifted to the right relative to that seen in 'normal' worms. Other FaRP effects were identical in both isolates. Although the exact reasons for this are unknown, it is hypothesized that AF8 has its excitatory effects by enhancing worm muscle responses to ACh. One possibility is that AF8 acts presynaptically (as seems to be the case in *A. suum*) to induce the release of ACh such that worms from the Lawes isolate have a reduced response to cholinomimetics and signalling molecules, such as AF8, which act through this system. This hypothesis awaits experimental confirmation.

flp-Gene Expression and Processing

The sequencing of the *C. elegans* genome and in particular the identification of the multiple *flp*-encoding genes coupled with PCR techniques have made it possible to examine the transcribed products and developmental regulation. In a study by Nelson *et al.* (1998b), reverse-transcribed RNA from mixed-stage worms was amplified with specific primers for each gene. The resulting amplification products were sub-cloned and their DNA sequences determined. By using this method the expression of the genes *flp*-1 through to *flp*-13 was confirmed in *C. elegans*. Previous studies had revealed that *flp*-1 may be alternatively spliced such that two different gene products, differing in one of the encoded peptides, are generated (Rosoff *et al.*, 1992, 1993). In subsequent studies, multiple PCR products were amplified from cDNA from *flps*-2 and 11, suggesting that these genes may also be alternatively spliced (Nelson *et al.*, 1998b). To examine the developmental expression of these genes, RNA was isolated from each stage of the *C. elegans* life cycle (eggs, L1, L2, L3, L4 and adult), reverse-transcribed and amplified with specific primers for each gene. The results of these experiments revealed that amplification products from cDNA from all of the developmental stages, as well as adults, were present for *flp* genes 1–7 and 10–12. In the case of *flp* genes 8, 9 and 13, amplification products were obtained from each of the developmental stages but not from the adult nematode. The results indicate that FaRPs play an important role throughout the life cycle of *C. elegans* and that some of these peptides may have functions in specific stages of the nematode life cycle such that their expression is developmentally regulated.

Nematode Neuropeptide Receptors

At the time of writing, no neuropeptide ligand/neuropeptide receptor pair has been unequivocally characterized from a nematode. Many putative neuropeptide receptors have been identified by the genome sequencing project but little progress has been made on ligand identification, partly due to the difficulties in establishing a stable *in vitro* expression system suitable for ligand identification coupled with the isolation of sufficient ligand to facilitate structural characterization. The putative nematode neuropeptide receptors have been identified by their homology to known neuropeptide receptors, classically comprising seven-transmembrane domains and a G-protein coupling cytosolic segment. These receptors have a wide variety of activities mediated by G-protein interaction with, and regulation of, cytosolic proteins and enzymes. Although most of the known peptide receptors are G-protein coupled, the vast majority are endogenous to higher vertebrates and as such may not be representative of the situation in nematodes.

To date, two FaRP receptors have been characterized from molluscs: one from *Helix aspersa* that is a directly ligand-gated Na⁺ channel (Lingueglia *et al.*, 1995) and one from *Lymnaea stagnalis* that is a G-protein coupled receptor (GPCR) (Tensen *et al.*, 1998). This is significant as the occurrence of both types of FaRP receptors would increase the variety of potential responses; should this situation also occur in nematodes, then this is likely to be a factor in the often complex and diverse physiology observed for FaRPs in nematode tissues.

With the number of known nematode FaRPs exceeding 70, it is possible that each neuropeptide has a specific endogenous receptor. However, it seems more likely that there will not be a different receptor for every FaRP and that those peptides with certain stoichiometric characteristics will be able to interact with a particular receptor subtype. For example, the structurally similar FaRPs PF1 and PF2 have similar actions on both body-wall and ovijector muscle of *A. suum* such that it seems reasonable to assume that they interact with a common receptor; the same may be true for other groups of structurally related neuropeptides, including the *A. suum* PGVLRFamides encoded on *afp*-1. All of these have differing N-termini yet similar effects on body-wall muscle in the worm, suggesting that the N-termini are not critical for the observed physiology. This is in contrast to the different effects noted for peptides with the C-terminal PNFLRFamide motif, e.g. PF1 and PF4. In the latter, the N-terminal sequence profoundly alters the observed activity of the peptide, presumably by directing the ligand to interact with a different receptor.

Unfortunately, not enough is known about the activities and structure–activity requirements of each of the neuropeptides to estimate the number of receptors responsible for observed physiology. Nevertheless, some estimate of the potential number of receptors can be made from the genome sequence information in that over 50 peptide-GPCRs have been identified (Bargmann, 1998). Since there are estimated to be in the order of 60 non-FaRP neuropeptides (Bargmann, 1998) and at least 62 FaRPs in *C. elegans* (Table 20.4), it is evident that there is not a GPCR for each of these neuropeptides. Assuming there is no redundancy in the system whereby some of the encoded peptides have no specific function, then more than one neuropeptide may interact with at least some of the identified receptors. It is likely to be very difficult to determine the specific FaRP ligand for these receptors.

Evidence that at least some nematode FaRPs interact with GPCRs was generated by the fact that some G-protein subunit mutants displayed phenotypes similar to those of *flp*-1 knockouts (Nelson *et al.*, 1998a). A series of double mutant crosses for *flp*-1 and selected G-proteins indicated that peptides encoded on the *flp*-1 gene act upstream of G-proteins to control motor and sensory parameters (Nelson *et al.*, 1998a). On a cautionary note, since one of the two FaRP receptors characterized from molluscs is a directly ligand-gated ion channel, it is possible that at least some nematode neuropeptides may interact not with GPCRs but with ion channel proteins. Although it has been suggested that PF4 directly gates a Cl⁻-channel in *A. suum* muscle membranes, membrane-level studies are needed to confirm this hypothesis.

Recently, a *C. elegans* GPCR (designated *npr*-1) with homology to the NPY-family peptide receptors (designated Y receptors) has been identified and shown to influence social feeding behaviour in the worm (deBono and Bargmann, 1998); NPY in mammals has numerous actions including a role in appetite control. That related receptors influence feeding behaviour in *C. elegans* may reveal that this signalling system has been conserved both structurally and functionally in hugely divergent species. One major problem with this theory is that there are no obvious NPY-family peptides in nematodes (see earlier). Clearly, the examination of ligand-receptor relationships within the neuropeptidergic system of nematodes warrants greater attention.

Concluding Remarks

With nematode neuropeptides numbering in excess of 120, it is evident that the endogenous neuropeptide signalling system is highly developed and central to the functional biology of these worms. The potential for a vast array of intricate and complex signalling pathways, modulated or coordinated by neuropeptides, is apparent in nematodes. The capacity for neuropeptide interaction with other classical signalling pathways coupled to highly controlled expression patterns, including developmental regulation and differential splicing of neuropeptide genes, offers immense potential for complexity and adaptability in the nervous system. Although we are a very long way from understanding the processes involved in the neuropeptidic integration and modulation of nematode behaviour, *C. elegans* offers a unique opportunity for a greater understanding of neuropeptide biology. Undoubtedly, the neuropeptide system of these worms offers exciting potential, not only for the neurobiologist, but also for the parasitologist seeking magic bullets to help to control parasite disease. Reports of ever-increasing anthelmintic resistance demand scientific evaluation of more drug targets in worms. Although the neuropeptide system will undoubtedly offer opportunities for the development of novel chemotherapies, the question remains: how long this will take?

Acknowledgements

The authors thank Roz Fellowes for kindly providing the confocal illustrations. Part of the work reported in this chapter was supported by a

Leverhulme Trust grant to N.J.M. and by a grant from the Pharmacia and Upjohn Company to A.G.M. and D.W.H.

References

Atkinson, H.J., Isaac, R.E., Harris, P.D. and Sharpe, C.M. (1988) FMRFamide-immunoreactivity within the nervous system of the nematodes *Panagrellus redivivus, Caenorhabditis elegans* and *Heterodera glycines. Journal of Zoology London* 216, 663–671.

Bargmann, C.L. (1998) Neurobiology of the *Caenorhabditis elegans* genome. *Science* 282, 2028–2033.

Bowman, J.W., Winterrowd, C.A., Friedman, A.R., Thompson, D.P., Klein, R.D., Davis, J.P., Maule, A.G. and Geary, T.G. (1995) Nitric oxide mediates the inhibitory effects of SDPNFLRFamide, a nematode FMRFamide-like neuropeptide, in *Ascaris suum. Journal of Neurophysiology* 74, 1880–1888.

Bowman, J.W., Friedman, A.R., Thompson, D.P., Ichhpurani, A.K., Kellman, M.F. and Geary, T.G. (1996) Structure–activity relationships in a nematode FMRFamide-like peptide, KNEFIRFamide. *Peptides* 17, 381–387.

Bradbury, A.F. and Smyth, D.G. (1991) Peptide amidation. *Trends in Biochemistry* 16, 112–115.

Brownlee, D.J.A., Fairweather, I. and Johnston, C.F. (1993a) Immunocytochemical demonstration of neuropeptides in the peripheral nervous system of the roundworm *Ascaris suum* (Nematoda, Ascaroidea). *Parasitology Research* 79, 302–308.

Brownlee, D.J.A., Fairweather, I., Johnston, C.F., Smart, D., Shaw, C. and Halton, D.W. (1993b) Immunocytochemical demonstration of neuropeptides in the central nervous system of the roundworm, *Ascaris suum* (Nematoda, Ascaroidea). *Parasitology* 106, 305–316.

Brownlee, D.J.A., Fairweather, I., Johnston, C.F. and Shaw, C. (1994) Immunocytochemical demonstration of peptidergic and serotoninergic components in the enteric nervous system of the roundworm, *Ascaris suum* (Nematoda, Ascaroidea). *Parasitology* 108, 89–103.

Brownlee, D.J.A., Holden-Dye, L., Fairweather, I. and Walker, R.J. (1995) The action of serotonin and the nematode neuropeptide KSAYMRFamide on the pharyngeal muscle of the parasitic nematode, *Ascaris suum. Parasitology* 111, 379–384.

Brownlee, D.J.A., Fairweather, I., Holden-Dye, L. and Walker, R.J. (1996a) Nematode neuropeptides: localisation, isolation and functions. *Parasitology Today* 12, 343–351.

Brownlee, D.J.A., Fairweather, I., Thorndyke, M.C. and Johnston, C.F. (1996b) Cellular and subcellular localization of SALMFamide (S1)-like immunoreactivity within the central nervous system of the nematode *Ascaris suum* (Nematoda, Ascaroidea). *Parasitology Research* 82, 149–156.

Cowden, C. and Stretton, A.O.W. (1993) AF2, an *Ascaris* neuropeptide: isolation, sequence and bioactivity. *Peptides* 14, 423–430.

Cowden, C. and Stretton, A.O.W. (1995) Eight novel FMRFamide-like neuropeptides isolated from the nematode *Ascaris suum. Peptides* 16, 491–500.

Cowden, C., Stretton, A.O.W. and Davis, A.E. (1989) AF1, a sequenced bioactive neuropeptide isolated from the nematode *Ascaris suum. Neuron* 2, 1465–1473.

Cowden, C., Sithigorngul, P., Brackley, P., Guastella, J. and Stretton, A.O.W. (1993) Localisation and differential expression of FMRFamide immunoreactivity in the nematode *Ascaris suum*. *Journal of Comparative Neurology* 333, 455–468.

Davenport, T.R.B., Lee, D.L. and Isaac, R.E. (1988) Immunocytochemical demonstration of a neuropeptide in *Ascaris suum* (Nematoda) using an anti-serum to FMRFamide. *Parasitology* 97, 81–88.

Davey, K.G. (1966) Neurosecretion and moulting in some parasitic nematodes. *American Zoologist* 6, 243–249.

Davis, R.E. and Stretton, A.O.W. (1996) The motornervous system of *Ascaris*: electrophysiology and anatomy of the neurons and their control by neuro-modulators. *Parasitology* 113, S99–S118.

Day, T.A. and Maule, A.G. (1999) Parasitic peptides! The structure and function of neuropeptides in parasitic worms. *Peptides* 20, 999–1019.

deBono, M. and Bargmann, C.I. (1998) Natural variation in a neuropeptide Y receptor homolog modifies social behavior and food response in *C. elegans*. *Cell* 96, 679–689.

Del Castillo, J., De Mello, W.C. and Morales, T. (1963) The physiological role of acetylcholine in the neuromuscular system of *Ascaris lumbricoides*. *Archives Internationale Physiologie Biochimie* 71, 741–757.

Del Castillo, J., De Mello, W.C. and Morales, T. (1964) Inhibitory action of γ-aminobutyric acid (GABA) on *Ascaris* muscle. *Experientia* 20, 141–143.

Edison, A.S., Messinger, L.A. and Stretton, A.O.W. (1997) A gene encoding multiple transcripts of a new class of FMRFamide-like neuropeptides in the nematode, *Ascaris suum*. *Peptides* 18, 929–935.

Fellowes, R.A., Maule, A.G., Marks, N.J., Geary, T.G., Thompson, D.P., Shaw, C. and Halton, D.W. (1998) Modulation of the motility of the vagina vera of *Ascaris suum in vitro* by FMRFamide-related peptides. *Parasitology* 116, 277–287.

Fellowes, R.A., Dougan, P.M., Maule, A.G., Marks, N.J. and Halton, D.W. (1999) Neuromusculature of the ovijector of *Ascaris suum* (Ascaroidea, Nematoda): an ultrastructural and immunocytochemical study. *Journal of Comparative Neurology* 415, 518–528.

Fellowes, R.A., Maule, A.G., Marks, N.J., Geary, T.G., Thompson, D.P. and Halton, D.W. (2000) Nematode neuropeptide modulation of the vagina vera of *Ascaris suum: in vitro* effects of PF1, PF2, PF4, AF3 and AF4. *Parasitology* 120, 79–89.

Franks, C.J., Holden-Dye, L., Williams, R.G., Pang, F.-Y. and Walker, R.J. (1994) A nematode FMRFamide-like peptide SDPNFLRFamide (PF1) relaxes the dorsal muscle strip preparation of *Ascaris suum*. *Parasitology* 108, 229–236.

Geary, T.G., Price, D.A., Bowman, J.W., Winterrowd, C.A., Mackenzie, C.D., Garrison, R.D., Williams, J.F. and Friedman, A.R. (1992) Two FMRFamide-like peptides from the free-living nematode *Panagrellus redivivus*. *Peptides* 13, 209–214.

Geary, T.G., Bowman, J.W., Friedman, A.R., Maule, A.G., Davis, J.P., Winterrowd, C.A., Klein, R.D. and Thompson, D.P. (1995) The pharmacology of FMRFamide-related neuropeptides in nematodes: new opportunities for rational anthelmintic discovery? *International Journal of Parasitology* 25, 1273–1280.

Geary, T.G., Marks, N.J., Maule, A.G., Bowman, J.W., Alexander-Bowman, S.J., Day, T.A., Larsen, M.J., Kubiak, T.M., Davis, J.P. and Thompson, D.P. (1999)

Pharmacology of FMRFamide-related peptides (FaRPs) in helminths. *Annals of the New York Academy of Sciences* 897, 212–227.

Halton, D.W., Shaw, C., Maule, A.G. and Smart, D. (1994) Regulatory peptides in helminth parasites. *Advances in Parasitology* 34, 164–227.

Holden-Dye, L., Brownlee, D.J.A. and Walker, R.J. (1997) The effects of the peptide KPNFIRFamide (PF4) on the somatic muscle cells of the parasitic nematode *Ascaris suum. British Journal of Pharmacology* 120, 379–386.

Jagdale, G.B. and Gordon, R. (1994) Distribution of FMRFamide-like peptide in the nervous system of a mermithid nematode, *Romanomermis culicivorax. Parasitology Research* 80, 467–473.

Keating, C., Thorndyke, M.C., Holden-Dye, L., Williams, R.G. and Walker, R.J. (1995) The isolation of a FMRFamide-like peptide from the nematode *Haemonchus contortus. Parasitology* 111, 515–521.

Kubiak, T.M., Maule, A.G., Marks, N.J., Martin, R.A. and Weist, J.R. (1996) Importance of the proline residue to the functional activity and metabolic stability of the nematode FMRFamide-related peptide, KPNFIRFamide (PF4). *Peptides* 17, 1267–1277.

Leach, L., Trudgill, D.L. and Gahan, P.B. (1987) Immunocytochemical localisation of neurosecretory amines and peptides in the free-living nematode, *Goodeyus ulmi. Histochemical Journal* 19, 471–475.

Li, C. and Chalfie, M. (1986) FMRFamide-like immunoreactivity in *C. elegans. Society for Neuroscience Abstracts* 12, 246.

Lingueglia, E., Champigny, G., Lazdunski, M. and Pascal, B. (1995) Cloning of the amiloride sensitive FMRFamide peptide-gated sodium channel. *Nature* 378, 730–733.

Marks, N.J., Maule, A.G., Geary, T.G., Thompson, D.P., Davis, J.P., Halton, D.W., Verhaert, P. and Shaw, C. (1997) APEASPFIRFamide, a novel FMRFamide-related decapeptide from *Caenorhabditis elegans*: structure and myoactivity. *Biochemical and Biophysical Research Communications* 231, 591–595.

Marks, N.J., Sangster, N.C., Maule, A.G., Halton, D.W., Thompson, D.P., Geary, T.G. and Shaw, C. (1999a) Structural characterisation and pharmacology of KHEYLRFamide (AF2) and KSAYMRFamide (PF3/AF8) from *Haemonchus contortus. Molecular and Biochemical Parasitology* 100, 185–194.

Marks, N.J., Maule, A.G., Li, C., Nelson, L.S., Thompson, D.P., Alexander-Bowman, S., Geary, T.G., Halton, D.W., Verhaert, P. and Shaw, C. (1999b) Isolation, pharmacology and gene organisation of KPSFVRFamide: a neuropeptide from *Caenorhabditis elegans. Biochemical and Biophysical Research Communications* 254, 222–230.

Martin, R.J. (1980) The effect of γ-aminobutyric acid on the input conductance and membrane potential of *Ascaris* muscle. *British Journal of Pharmacology* 71, 99–106.

Martin, R.J., Pennington, A.J., Duittoz, A.H., Robertson, S. and Kusel, J.R. (1991) The physiology and pharmacology of neuromuscular-transmission in the nematode parasite, *Ascaris suum. Parasitology* 102, S41–S58.

Maule, A.G., Shaw, C., Bowman, J.W., Halton, D.W., Thompson, D.P., Geary, T.G. and Thim, L. (1994a) KSAYMRFamide: a novel FMRFamide-related heptapeptide from the free-living nematode, *Panagrellus redivivus*, which is myoactive in the parasitic nematode, *Ascaris suum. Biochemical and Biophysical Research Communications* 200, 973–980.

Maule, A.G., Shaw, C., Bowman, J.W., Halton, D.W., Thompson, D.P., Geary, T.G. and Thim, L. (1994b) The FMRFamide-like neuropeptide AF2 (*Ascaris suum*) is present in the free-living nematode, *Panagrellus redivivus* (Nematoda, Rhabditida). *Parasitology* 109, 351–356.

Maule, A.G., Shaw, C., Bowman, J.W., Halton, D.W., Thompson, D.P., Thim, L., Kubiak, T.M., Martin, R.A. and Geary, T.G. (1995a) Isolation and preliminary biological characterization of KPNFIRFamide, a novel FMRFamide-related peptide from the free-living nematode, *Panagrellus redivivus*. *Peptides* 16, 87–93.

Maule, A.G., Geary T.G., Bowman, J.W., Marks, N.J., Blair, K.L., Halton, D.W., Shaw, C. and Thompson, D.P. (1995b) Inhibitory effects of nematode FMRFamide-related peptides (FaRPs) on muscle-strips from *Ascaris suum*. *Invertebrate Neuroscience* 1, 255–265.

Maule, A.G., Bowman, J.W., Thompson, D.P., Marks, N.J., Friedman, A.R. and Geary T.G. (1996a) FMRFamide-related peptides (FaRPs) in nematodes: occurrence and neuromuscular physiology. *Parasitology* 113, S119–S136.

Maule, A.G., Geary, T.G., Bowman, J.W., Shaw, C., Halton, D.W. and Thompson, D.P. (1996b) The pharmacology of nematode FMRFamide-related peptides. *Parasitology Today* 12, 351–357.

Maule, A.G., Geary, T.G., Marks, N.J., Bowman, J.W., Friedman, A.R. and Thompson, D.P. (1996c) Nematode FMRFamide-related peptide (FaRP)-systems: occurrence, distribution and physiology. *International Journal for Parasitology* 26, 927–936.

Nelson, L.S., Rosoff, M.L. and Li, C. (1998a) Disruption of a neuropeptide gene, *flp*-1, causes multiple defects in *Caenorhabditis elegans*. *Science* 281, 1686–1690.

Nelson, L.S., Kim, K., Memmott, J.E. and Li, C. (1998b) FMRFamide-related gene family in the nematode, *Caenorhabditis elegans*. *Molecular Brain Research* 58, 103–111.

Pang, F.-Y., Mason, J., Holden-Dye, L., Franks, C.J., Williams, R.G. and Walker, R.J. (1995) The effects of the nematode peptide, KHEYLRFamide (AF2), on the somatic musculature of the parasitic nematode *Ascaris suum*. *Parasitology* 110, 353–362.

Rogers, W.P. (1968) Neurosecretory granules in the infective stage of *Haemonchus contortus*. *Parasitology* 58, 657–662.

Rosoff, M.L., Burglin, T.R. and Li, C. (1992) Alternatively spliced transcripts of the *flp*-1 gene encode distinct FMRFamide-like peptides in *Caenorhabditis elegans*. *Journal of Neuroscience* 12, 2356–2361.

Rosoff, M.L., Doble, K.E., Price, D.A. and Li, C. (1993) The *flp*-1 propeptide is processed into multiple, highly similar FMRFamide-like peptides in *Caenorhabditis elegans*. *Peptides* 14, 331–338.

Schinkmann, K. and Li, C. (1992) Localization of FMRFamide-like peptides in *Caenorhabditis elegans*. *Journal of Comparative Neurology* 316, 251–260.

Schinkmann, K. and Li, C. (1994) Comparison of two *Caenorhabditis* genes encoding FMRFamide (Phe-Met-Arg-Phe-NH_2)-like peptides. *Molecular Brain Research* 24, 238–246.

Shaw, C. (1996) Neuropeptides and their evolution. *Parasitology* 113, S35–S46.

Sithigorngul, P., Cowden, C., Guastella, J. and Stretton, A.O.W. (1989) Generation of monoclonal antibodies against a nematode peptide extract: another approach for identifying unknown neuropeptides. *Journal of Comparative Neurology* 284, 389–397.

Sithigorngul, P., Stretton, A.O.W. and Cowden, C. (1990) Neuropeptide diversity in *Ascaris*: an immunocytochemical study. *Journal of Comparative Neurology* 294, 362–376.

Smart, D.S., Shaw, C., Johnston, C.F., Halton, D.W., Fairweather, I. and Buchanan, K.D. (1992a) Chromatographic and immunological characterisation of immunoreactivity towards pancreatic polypeptide and neuropeptide Y in the nematode *Ascaris suum*. *Comparative Biochemistry and Physiology* 102C, 477–481.

Smart, D.S., Shaw, C., Curry, W.J., Johnston, C.F., Thim, L., Halton, D.W. and Buchanan, K.D. (1992b) The primary structure of TE-6: a novel neuropeptide from the nematode *Ascaris suum*. *Biochemical and Biophysical Research Communications* 187, 1323–1329.

Smart, D.S., Johnston, C.F., Shaw, C., Halton, D.W. and Buchanan, K.D. (1993) Use of specific antisera for the localization and quantitation of leucokinin immunoreactivity in the nematode, *Ascaris suum*. *Comparative Biochemistry and Physiology* 106C, 517–522.

Smart, D.S., Johnston, C.F., Maule, A.G., Halton, D.W., Hrckova, G., Shaw, C. and Buchanan, K.D. (1995) Localization of *Diploptera punctata* allatostatin-like immunoreactivity in helminths: an immunocytochemical study. *Parasitology* 110, 87–96.

Stretton, A.O.W., Davis, R.E., Angstadt, J.D., Donmoyer, J.E. and Johnson, C.D. (1985) Neural control of behaviour in *Ascaris*. *Trends in Neuroscience* 8, 294–300.

Stretton, A.O.W., Cowden, C., Sithigorngul, P. and Davis, R.E. (1991) Neuropeptides in the nematode *Ascaris suum*. *Parasitology* 102, S107–S116.

Stretton, A.O.W., Donmoyer, J.E., Davis, R.E., Meade, J.A., Cowden, C. and Sithigorngul, P. (1992) Motor behavior and motor nervous system function in the nematode *Ascaris suum*. *Journal of Parasitology* 78, 206–214.

Tensen, C.P., Cox, K.J.A., Burke, J.F., Leurs, R., van der Schors, R.C., Geraerts, W.P.M., Vreugdenhil, E. and van Heerikhuizen, H. (1998) Molecular cloning and characterization of an invertebrate homologue of a neuropeptide Y receptor. *European Journal of Neuroscience* 10, 3409–3416.

Trim, N., Holden-Dye, L., Ruddell, R. and Walker, R.J. (1997) The effects of the peptides AF3 (AVPGVLRFamide) and AF4 (GDVPGVLRFamide) on the somatic muscle of the parasitic nematodes *Ascaris suum* and *Ascaridia galli*. *Parasitology* 115, 213–222.

Trim, N., Brooman, J.E., Holden-Dye, L. and Walker, R.J. (1998) The role of cAMP in the actions of the peptide AF3 in the parasitic nematodes *Ascaris suum* and *Ascaridia galli*. *Molecular and Biochemical Parasitology* 93, 263–271.

Walker, R.J., Brooks, H.L. and Holden-Dye, L. (1996) Evolution and overview of classical transmitter molecules and their receptors. *Parasitology* 113, S3–S34.

Warbrick, E.V., Rees, H.H. and Howells, R.E. (1992) Immunocytochemical localisation of a FMRFamide-like peptide in the filarial nematodes *Dirofilaria immitis* and *Brugia pahangi*. *Parasitology Research* 78, 252–256.

White, J.G., Southgate, E., Thompson, J.N. and Brenner, S. (1986) The structure of the nervous system of *Caenorhabditis elegans*. *Philosophical Transactions of the Royal Society London B* 314, 1–340.

Wikgren, M.C. and Fagerholm, H.P. (1993) Neuropeptides in sensory structures of nematodes. *Acta Biologia Hungarica* 44, 133–136.

Neurobiology of Nematode Muscle: Ligand-gated Ion-channels and Anti-parasitic Drugs

Richard J. Martin,[1] Alan P. Robertson,[1] Momtchil A. Valkanov,[2] Jenny Purcell[2] and C. Stewart Lowden[2]

[1]Department of Biomedical Sciences, College of Veterinary Medicine, Iowa State University, Ames, IA 50011–1250, USA; [2]Department of Preclinical Veterinary Sciences, Royal (Dick) School of Veterinary Science, Summerhall, The University of Edinburgh, Edinburgh EH9 1QH, UK

Introduction

Muscle cell membranes of nematodes possess ion channel receptors that are opened by neurotransmitters and which are gated by selective therapeutic agents. This chapter is an introduction to the physiology and pharmacology of ligand-gated ion channels of nematode muscle.

Nematode parasites present a serious problem for most animals and for humans in developing countries. These parasites produce various symptoms including ill thrift, poor growth, diarrhoea and, in around 1% of cases, loss of life. The economic forces are such that new drugs for the treatment of nematode parasites have been developed first for animal use and only later for human use. A recent example is the development of the antibiotic anthelmintic, ivermectin, which was introduced first for the treatment of cattle nematode parasites and has subsequently been used to control 'river blindness', an eye condition seen in west Africa caused by larvae of *Onchocerca volvulus*.

There are two main mechanisms of action of important anti-nematode drugs.

1. *Binding to β-tubulin.* The anthelmintic benzimidazoles, such as thiabendazole and albendazole, are drugs that bind selectively to nematode β-tubulin molecules and prevent the formation of parasite microtubules. Microtubules perform a number of vital functions: they are required for the transport of intracellular particles, and for absorption of nutrients by the intestinal cells of the nematode. The site of action of the benzimidazoles is the same as that of a drug colchicine, that stops cell division in metaphase, but the benzimidazole anthelmintics are selective for nematode β-tubulin. The therapeutic effect on the parasite takes 2–4 days to develop following administration of the drug.

2. *Opening membrane ion channels.* These are the ion-channel ligands, which bind to and selectively open ion channels of nematode muscle. Three types of ion channel are the sites of action of this group of anthelmintics:

(i) *Nematode nicotinic acetylcholine channels* are excitatory channels found on nematode neurons and muscle. The ion channels on muscle are the site of action of anthelmintics such as levamisole, pyrantel and oxantel.

(ii) *Glutamate-gated Cl channels* are inhibitory channels found on nematode nerve and pharyngeal muscle and are the site of action of the avermectin antibiotics, such as ivermectin and moxidectin.

(iii) *GABA-gated Cl channels* are inhibitory channels found on nematode somatic muscle and are the site of action of piperazine.

Because the second group of anthelmintics act on membrane ion channels and produce electrical changes, it is particularly valuable to study effects of these drugs using electrophysiological techniques. This chapter will provide an introductory review to the ion channel physiology and pharmacology of the nematode muscle and electrophysiological techniques used to examine the effects of the anthelmintics that affect ion channels.

Ligand-gated Ion Channels

All cell membranes contain transmembrane proteins that form ion channels. These ion channels are usually selectively permeable to particular ions. Some channels, such as GABA-gated ion channels, are permeable to Cl ions and are inhibitory in nature because they make the inside of the nerve or muscle cells more negative as the Cl ions enter. Some ion channels are permeable to the cations Na and K, and an example of this type is the nicotinic acetylcholine-gated channel. Nicotinic channels have an excitatory effect when they open because Na ions enter and K ions leave through these channels. The cell becomes more positive inside and depolarizes. If the cell is a muscle cell, calcium accumulates in the cytoplasm and it contracts. We have found that all over the surface of *Ascaris* muscle there are GABA receptors (Martin, 1980) as well as nicotinic acetylcholine channels (Martin, 1982; Robertson and Martin, 1993).

Figure 21.1 is a schematic diagram of a ligand-gated transmembrane ion channel. The ion channel is shown to be larger than the thickness of the membrane and passes right through it. Two drug-selective binding sites have been found on the extracellular surface of the ion channel. In the case of the nematode nicotinic acetylcholine receptor, the drug-binding sites are available for binding by the anthelmintics levamisole and pyrantel. The unbound ion channel receptor remains closed but, as in Fig. 21.1, the binding of two molecules of agonist can lead to a transient opening of the receptor and a short current pulse, about 1 pA in value, lasting for about 1 ms. The total current produced by the opening of many ion channels in the membrane leads to significant current flow across the membrane and a change in potential of the cell.

Sometimes the agonist molecule can also enter the pore of the ion channel to produce a transient and repeated block, known as a 'flickering block'. This happens with levamisole (Robertson and Martin, 1993); thus the anthelmintic is both an agonist and an antagonist of the receptor. There is another state of the ion channel known as the 'desensitized' state, which is usually produced by a high concentration of agonist and which occurs slowly with time after drug application. The desensitized state is a long-lasting closed state associated with a high concentration of drug.

The opening of *masses* of ion channels in nematode muscle membranes may be detected using the two-microelectrode voltage-clamp technique. In contrast, the opening of *single* ion channels may be recorded using the vesicle preparation and patch-clamp technique. These techniques are both described below.

The two-micropipette current-clamp technique for examining the massed responses of ion channels on Ascaris body muscle

Ion channels that open in response to the application of drugs will produce a change in the conductance of the cell membrane: the longer the individual ion channels are open, the greater will be the conductance of the membrane. This change in the conductance of the membrane, produced by anthelmintic drugs that open ion channels, has been detected in *Ascaris* body muscle (Martin, 1980) and in the *Ascaris* pharyngeal muscle (Martin, 1996).

Figure 21.2 illustrates the use of the two-micropipette technique that is used to measure depolarization and conductance changes produced by levamisole application in *Ascaris* somatic muscle. The voltage electrode records the membrane potential and the voltage changes as current is injected through the current electrode. The levamisole is applied through a micro-catheter placed over the muscle cell body (bag) region of *Ascaris* somatic muscle cell with a time- and pressure-controlled system (Pico-Spritzer). The changes in amplitude of the membrane potential and input

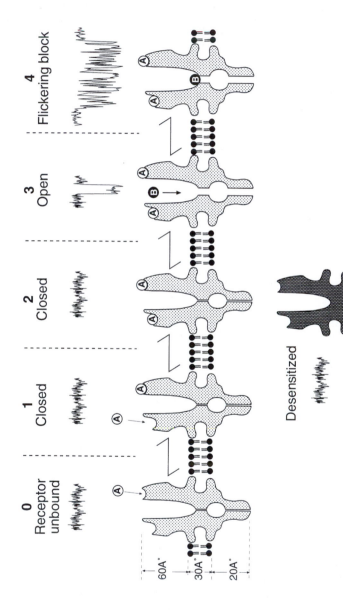

Fig. 21.1. (Opposite) Levamisole gated ion channel (a nematode nicotinic acetylcholine channel) and its opening as a result of the binding of two molecules of agonist. Two agonist molecules (A) (e.g. levamisole) combine sequentially with non-equivalent binding sites on the extracellular surface of the channel and permit the opening of the channel. When the channel opens for a brief period (a few milliseconds) a current pulse of a few picoamps flows in through the channel. A large inorganic cation (B) (including the anthelmintic levamisole) may enter but not pass through the channel pore and produces a 'flickering' channel block as it repeatedly binds and unbinds with a block site deep in the pore. The desensitized state is also shown as a closed non-conducting condition of the channel. Desensitization is produced in a time-dependent manner following the addition of 'high' agonist drug concentrations and is a long closed state.

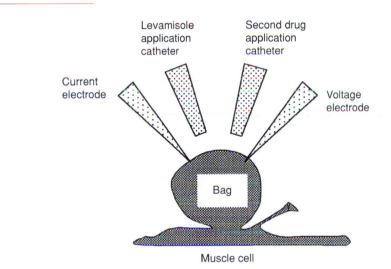

Fig. 21.2. Two-microelectrode current-clamp technique used to observe, in single *Ascaris* body muscle cells in a body-flap preparation, the response to a controlled pulsed application of levamisole. One micropipette, to measure membrane potential, and another micropipette, to inject current, are inserted inside the area of the muscle cell known as the bag region. Levamisole is applied in a time- and pressure-controlled manner from a microcatheter placed over the bag region of the muscle. A second microcatheter is used to apply additional chemical agents (Martin, 1982).

conductance are used to monitor properties of the levamisole receptor ion channels. Figure 21.3 illustrates a representative 15 mV depolarizing response to maintained application of levamisole. As levamisole is applied to the muscle membrane, it opens nematode-specific nicotinic ion channel receptors, the membrane potential changes and the resistance reduces. The advantage of the two-microelectrode current-clamp technique is that the muscle cells of *Ascaris* remain intact and the technique is relatively easy to set up to examine effects of pharmacological agents that open ion

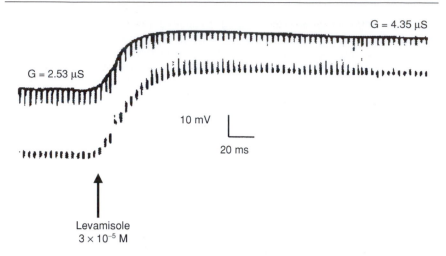

Fig. 21.3. Two-micropipette current-clamp recording and effect of maintained application of 30 µM levamisole, which produces a 15 mV depolarization (upward movement of trace). The downward transients are the result of injected current pulses used to measure membrane conductance. The trace gets narrower as the input conductance increases from 2.35 µS to 4.35 µS as the levamisole ion channels open. The peak amplitude of the membrane potential response and change in input conductance are used as an assay of the number and activity of the levamisole ion channel receptors present in the muscle cell membrane. The response was fully reversible on washing (not shown).

channels. The technique is limited however, because it does not reveal kinetic information about the opening and closing of individual ion channels.

Muscle-vesicle for recording single-channel properties of levamisole receptors

The second technique has been developed to look at the properties of *individual levamisole receptors* at the single-channel level. This is the *Ascaris* muscle-vesicle preparation (Fig. 21.4) (Martin *et al.*, 1990; Robertson and Martin, 1993). We have adapted and use the 'cell-attached' patch and 'inside-out' patch recording technique for recording the properties of single levamisole receptor channels. Figure 21.5 illustrates examples of openings of single levamisole receptors at −75 mV and +75 mV. The record at −75 mV shows brief (mean open-time: 1.1 ms) downward rectangular current pulses of 2.4 pA as the channel steps from the closed-state to the open-state. At +75 mV, the current flow is in the opposite direction. The plot of the amplitude of the current against membrane potential is linear and is also shown in Fig. 21.5: the gradient of the slope is the conductance

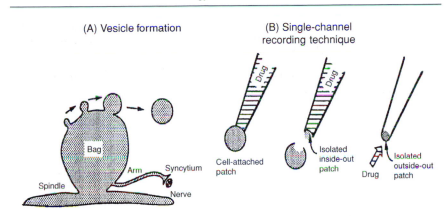

Fig. 21.4. Vesicle formation and patch-clamp techniques used to record levamisole receptor channel currents from *Ascaris* muscle. (A) Muscle membrane vesicles 'bud-off' from the bag membrane following a 10 min collagenase treatment and incubation for 1 h at 37°C in *Ascaris* saline. (B) Levamisole is applied to the outside surface of the membrane to activate receptor channels: *cell-attached* patches are usually used but it is also possible to make *inside-out* and *outside-out* patch recordings.

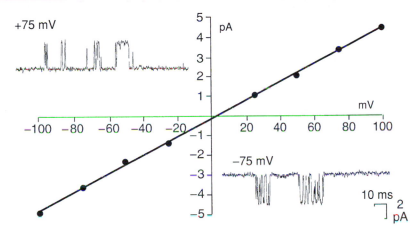

Fig. 21.5. Levamisole-activated single-channel currents activated by 30 μM levamisole in a cell-attached patch and current voltage relationship. The rectangular current pulses were recorded at different patch potentials to determine the relationship between channel current and potential. The slope was linear with a conductance of 34 pS.

of the channel (34 pS ± 1.0). By recording for a long time (2 min) from the levamisole receptor at a fixed potential we can determine the proportion of time the channel is in the open-state (*P-open*). For the patch shown in Fig. 21.5 at −75 mV, *P-open* was 0.004.

The advantage of the patch-clamp technique for recording from single levamisole receptors is that biophysical properties of individual receptors can be measured very precisely. It is possible to measure the single-channel conductance, the mean open-time, *P-open* and other properties of individual receptors, and to observe effects of chemical agents that alter (modulate) the behaviour of the channel. Comparisons can be made of properties of levamisole receptors in nematode parasites that are sensitive and that are resistant to levamisole and related drugs. Isolated inside-out patches are also used for recording from single levamisole channels. This technique exposes the cytoplasmic surface of the receptor to the bathing Ringer solution and chemical agents can be added to the membrane to modulate the behaviour of the channel.

Oesophagostomum dentatum as a Model Nematode for Anthelmintic Resistance

O. dentatum is the nodular intestinal nematode parasite of pigs. Adults are about 1 cm in length with separate males and females (Fig. 21.6A). Infection of pigs may be produced relatively easily by administration of 10,000 infective L3 by stomach tube. About 4 weeks later, mature adults are found in the large intestine. The parasite does not produce clinical symptoms and does not cause a welfare problem. Anthelmintic resistance to levamisole, pyrantel and ivermectin has been produced by sub-therapeutic treatments and collection of eggs from surviving parasites for subsequent infection and passaging into fresh pigs (Verady *et al.*, 1997). After ten generations, it has been possible to produce resistance to normal therapeutic doses of a specific drug, if the concentration of the drug is gradually increased from the sub-therapeutic levels with each passage. As a result we now have isolates of *O. dentatum* that are resistant to levamisole (LEVR), pyrantel (PYRR) and ivermectin (IVMR). We have investigated the properties of the nicotinic ion channel receptors on the muscle cells of this nematode in sensitive (SENS), levamisole-resistant (LEVR), and pyrantel-resistant (PYRR) isolates. (The term 'isolate' is used rather than 'clone' because each of the nematode parasites is produced sexually and not by clonal expansion.)

The female *O. dentatum* parasite is used because it is larger than the male and survives up to 14 days in culture after collection from the pig. The parasite is dissected under the binocular microscope, using a micro-scalpel to cut the worm longitudinally (Fig. 21.6B). It is then gently pinned out and treated with collagenase for 10 min and incubated at 37°C to produce membrane vesicles from muscle cells (Fig. 21.6C). These vesicles are devoid of any extracellular matrix covering the membrane and are suitable for patch-clamp studies. Fire-polished patch-pipettes applied to the vesicle surface, followed by suction, will produce the giga-seal resistance required for recording the activity of single channels with this technique. Single-channel currents may be recorded if levamisole is included in the patch-pipette (Figs 21.5 and 21.6C).

One test that was carried out when trying to compare levamisole receptor properties in SENS and LEVR isolates was to count the numbers of patches of recorded membranes that contained active channels. Figure 21.7 shows the proportion of patches from SENS and LEVR isolates that contained active patches at different concentrations of levamisole. It was

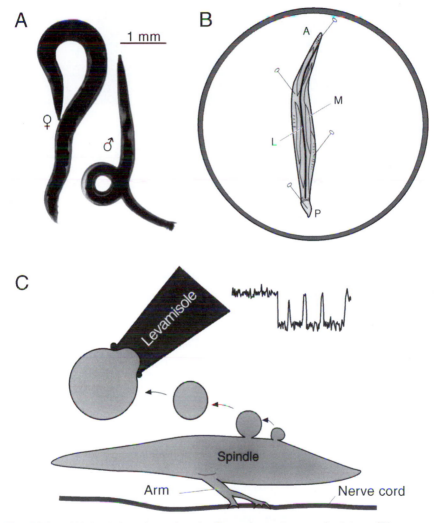

Fig. 21.6. (A) Adult female and male *Oesophagostomum dentatum*. (B) Dissected adult female; the parasite has been cut longitudinally and the gut and reproductive organs removed to form the characteristic 'muscle flap'. (C) Vesicles forming from a muscle cell after treatment with collagenase; the vesicles are then transferred to the experimental chamber, where vesicle-attached patches are formed and single-channel recordings made.

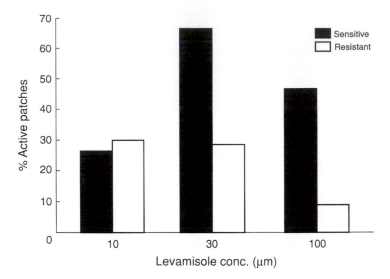

Fig. 21.7. Percentage of active patches at each levamisole concentration for SENS (■) and LEVR (□) parasites. The difference between SENS and LEVR is significant (chi-squared $P < 0.001$).

noticed that the proportion was similar at 10 µM but that there was a greater proportion of active channels in SENS at 30 µM. At 100 µM, although there was a reduction in the proportion of active channels with both isolates, the resistant isolate showed a greater relative decrease in the proportion of active patches. These observations were interpreted as indicating that desensitization was occurring as the concentration of levamisole was increased in the patch-pipette, and this desensitization explained the reduction in the proportion of membrane patches with active receptors at 100 µM levamisole. Thus, in the LEVR isolate, it appears that desensitization is enhanced compared with the SENS isolate. Whatever the explanation for the decrease, it is clear that at 30 µM (near the estimated therapeutic concentration) there are fewer active receptors in the resistant isolate.

The conductance of an individual receptor channel can be measured very precisely. The slope of the I/V plot is linear (Fig. 21.5), mean ± SE 34 pS ± 2 pS. A histogram was constructed of the conductance of the channels that were observed in the SENS and LEVR isolate (Fig. 21.8) and it was found that the channel conductances ranged between 18 pS and 50 pS. This range is much greater than can be explained by experimental error and suggests that there is levamisole receptor channel heterogeneity (receptor subtypes are present). It was found that there was no real difference between the average channel conductance in the SENS isolate (38 pS) and the average of the LEVR isolate (36 pS).

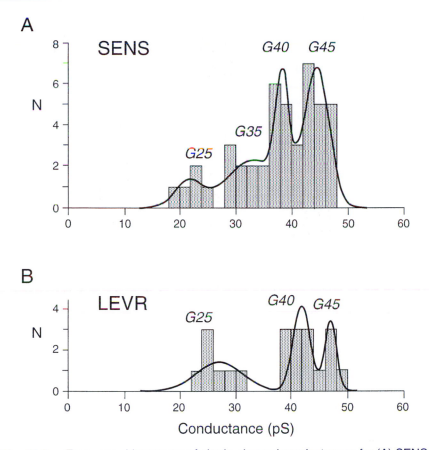

Fig. 21.8. Frequency histograms of single-channel conductances for (A) SENS and (B) LEVR parasites. Gaussian curves were fitted to each distribution using the maximum likelihood procedure. The peaks for the SENS isolate were 21.4 ± 2.3 pS (8% area) labelled G25; 33.0 ± 4.8 pS (31% area) labelled G35; 38.1 ± 1.2 pS (19% area) labelled G40; and 44.3 ± 2.2 pS (42% area) labelled G45. The peaks for the LEVR isolate were 25.2 ± 4.5 pS (21% area) labelled G25; 41.2 ± 1.7 pS (49% area) labelled G40; and 46.7 ± 1.1 pS (30% area) labelled G45.

Receptor Heterogeneity

Variation in conductance

The distribution of the conductance histogram (Fig. 21.8A) could not be described by a normal distribution and showed the presence of several peaks. The SENS conductance distribution was therefore fitted with the sum of four Gaussian distributions that had means of 21.4 ± 2.3 pS (8% area) labelled G25; 33.0 ± 4.8 pS (31% area) labelled G35; 38.1 ± 1.2 pS

(19% area) labelled G40; and 44.3 ± 2.2 pS (42% area) labelled G45. The means for the LEVR isolate were: 25.2 ± 4.5 pS (21% area) labelled G25; 41.2 ± 1.7 pS (49% area) labelled G40; and 46.7 ± 1.1 pS (30% area) labelled G45. Interestingly, the G35 peak was missing from the LEVR isolate. Anthelmintic resistance may arise when there is selection pressure produced by the regular use of the anthelmintic; therefore, the sensitive individuals and their offspring will be eliminated and only the resistant individuals and their offspring will survive. This mechanism is an elimination process: there is no production of mutations giving rise to new variants or alleles.

The different subtypes of levamisole receptor (receptor heterogeneity) were illustrated by the presence, in some patches, of channels with different conductances (Fig. 21.9). The different conductances were seen as channel currents opening to different levels producing different peaks in the scatter

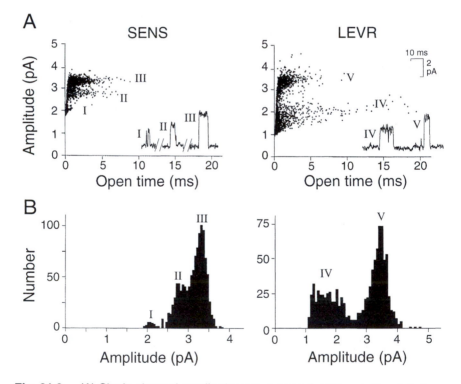

Fig. 21.9. (A) Single-channel amplitude versus open duration scatter plots for SENS and LEVR isolates. Each plot was obtained at +75 mV and demonstrates multiple conductance levels (levels I, II and III for SENS and IV and V for LEVR). Inset: examples of single-channel currents corresponding to each of the conductance levels I, II and III level are shown. (B) Amplitude histograms obtained from the data illustrated in (A). The plots confirm the presence of open-channel currents with multiple conductance levels.

plots (Fig. 21.9A) and current-amplitude histograms (Fig. 21.9B). Again these observations were interpreted as indicating that heterogeneous receptor subtypes are present in individuals of the SENS and LEVR (Robertson *et al.*, 1999). The four peaks seen in the conductance histogram of Fig. 21.9A may be interpreted as indicating the presence of four main channel types and this is represented schematically in Fig. 21.10. The loss of the G35 subtype associated with the LEVR isolates is also illustrated.

Variation in P-open

Figure 21.11A illustrates representative vesicle-attached recordings from a sensitive isolate (SENS), a pyrantel-resistant isolate (PYRR) and a levamisole-resistant isolate (LEVR) produced by 30 µM levamisole. There are differences in mean *P-open* values of receptors between sensitive and resistant isolates that would produce, on average, a reduction in the effect of the anthelmintic. However, one of the puzzling, but very consistent, observations is that there is a big range in the *P-open* values observed in different patches under the same experimental conditions in the same isolate (Fig. 21.11B). For example, the *P-open* values of the sensitive isolate were observed to vary between 0.090 and 0.003 at −50 mV with 30 µM levamisole, a 30-fold difference between the biggest and the smallest *P-open*

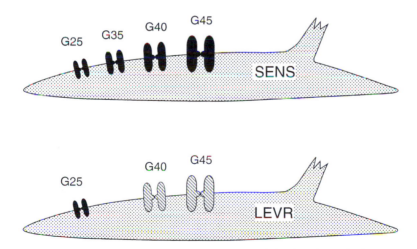

Fig. 21.10. Schematic diagram of the presence of the four levamisole receptor subtypes present in *O. dontatum* in the SENS isolate. The four subtypes are shown to be present in one muscle cell but they may not all occur in the same muscle or animal. In the LEVR isolate three subtypes are shown as present, with the G35 missing. The G40 and G45 are shown as modified because the data on the probability of opening showed that they behaved in a significantly different manner to that of the G40 and G45 subgroups of the SENS isolates.

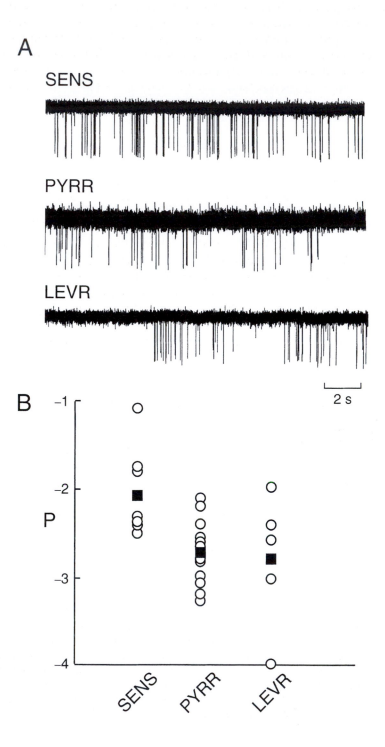

values (Fig. 21.11). This wide variation between *P-open* values of different receptors again indicates receptor heterogeneity. A similar wide range of values occurred in channels observed in patches of levamisole-resistant and pyrantel-resistant isolates. It was also observed that big differences occur between receptors collected from the same muscle-vesicle membrane. This range of *P-open* values is not explained by variations in electrophysiological technique.

Differences between Sensitive and Resistant Isolates

It has already been pointed out that resistant isolates have on average fewer active channels per patch than the sensitive isolates. In addition, the average proportion of time that the levamisole receptors were open (*P-open*) in the LEVR isolates was significantly less than the proportion of time that the levamisole receptors were open in the SENS isolates (Fig. 21.11). Thus, two factors can be identified that lead to a reduction in the current carried across the membrane by levamisole receptors in the resistant isolate when compared with the sensitive isolate. The factors are, again, a reduced density of active channels in LEVR isolates, and a reduced proportion of time that the active channels in the resistant isolate are open. These factors combine so that there is a tenfold reduction in the current carried across the membrane at 30 µM.

Possible Explanations for Receptor Heterogeneity

The distribution of levamisole receptor conductances from the SENS isolate in this study was skewed towards *G45* and fitted by the sum of four Gaussian distributions, suggesting the presence of four main conductance subtypes of levamisole receptor that are referred to as *G25, G35, G40* and *G45*. In addition to the variation in the conductance of the levamisole receptors a large variation was noted between the *P-open* values of individual

---

Fig. 21.11. (Opposite) (A) Representative vesicle-attached patch recordings from SENS, PYRR and LEVR isolates; 30 µM levamisole in the patch-pipette, −50 mV patch potential. Note that the SENS isolate recording contains more openings than the other two patches and so will carry more current across the membrane. (B) Log_{10} plot of the proportion of open-time against the isolate type. Open circles represent individual receptor channels from different preparations at −50 mV with 30 µM levamisole as the agonist. Note that there is a wide spread of greater than tenfold difference between the maximum and minimum values observed for each of the isolates. There is overlap between values of the isolates but the mean (closed square) for SENS is greater than for LEVR and PYRR. The spread between maximum and minimum values suggests that regulation, perhaps by phosphorylation, is present.

receptors with the same conductance. There is both conductance hetero-geneity and *P-open* heterogeneity. This variation may be explained by the presence of multiple alleles.

Multiple alleles

An allele is a sequence variant of a gene specifying a particular trait (Twyman, 1998). Because each *O. dentatum* parasite is produced sexually, the genome will contain an allele from each parent encoding each receptor subunit protein. Therefore, the presence of the different peaks in the distribution of conductances of the levamisole receptors may be explained by the presence of different receptor subunit alleles encoding proteins of different amino acid structure. It is known that all species of parasitic nematode have high nucleotide diversity, with averages of 0.019–0.027 substitutions per nucleotide occurring between individuals from the same population (Blouin *et al.*, 1995). Consider that a 'wild-type' receptor subunit allele is the predominant allele in the population, generally confer-ring the greatest fitness and producing a fully functional gene product. If a mutation occurs, it may alter the gene product and confer a novel function upon the receptor subunit. This novel function may only be apparent in response to adverse environmental conditions, such as levamisole expo-sure. Selection pressure then dictates that the population frequency of this mutant allele increases. The number of receptor subunit alleles in the pop-ulation that are phenotypically distinguishable from the 'wild-type' is not known. Techniques are currently available to detect such uncharacterized gene mutations. Single-strand conformational polymorphism (SSCP) anal-ysis is one technique that has already been used successfully in *O. dentatum* for allele characterization (Gasser *et al.*, 1998; see also Chapter 4). This technique takes advantage of differential electrophoretic mobility in short (200–500 bp) PCR fragments differing by single base mutations (Wallace, 1997). If allelic variation is responsible for receptor heterogeneity, then such analysis may indicate associations of particular alleles with definitive conductance phenotypes. Indeed, further analysis of data will also indicate whether receptor subunit genes are sex-linked, i.e. are present on the sex chromosomes. In this case, there would be an association between the sex of the parasite and the conductance phenotypes. The functional basis of how allelic variation may lead to receptor heterogeneity is not known and the following possible explanations are not necessarily mutually exclusive.

Alleles encoding for stoichiometric changes in the receptor subunits

Caenorhabditis elegans is a soil nematode that has been subjected to detailed genetic analysis. The possible structure of the nematode levamisole receptor, based on the pentameric structure of the nicotinic channel of the *Torpedo* electric organ (Devillers-Thiery *et al.*, 1993), is shown in Fig. 21.12.

A number of genes encoding $nAChR_n$ subunits in *C. elegans* have been described but only three genes encoding $nAChR_n$ subunits are involved in levamisole resistance (Fleming *et al.*, 1997). The three genes are *unc-38*, *unc-29* and *lev-1*; *unc-38* encodes α-like subunits (which possess the vicinal cysteines) while *unc-29* and *lev-1* encode β-like subunits. The molecular structures of levamisole UNC-38 subunit homologues in the parasitic nematodes *Trichostrongylus colubriformis* and *Haemonchus contortus* (Hoekstra *et al.*, 1997; Wiley *et al.*, 1997) are about 90% identical to *C. elegans*, suggesting that information derived from the model nematode may be a useful guide for parasitic nematodes. An explanation for the presence of multiple receptor subtypes within an individual parasite (Martin *et al.*, 1997) based

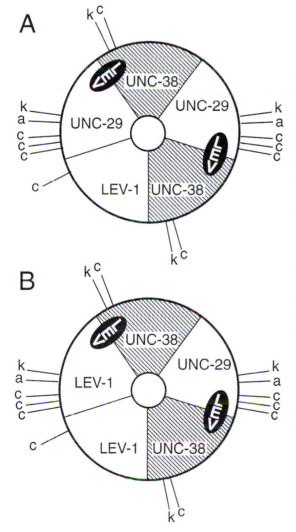

Fig. 21.12. Schematic diagram of the putative structure of a levamisole nicotinic receptor. (A) The receptor is shown composed of five subunits: two *unc-38* subunits, two *unc-29* subunits and one *lev-1* subunit. Phosphorylation sites are also indicated; a, protein kinase A site; c, protein kinase C site; K, tyrosine kinase site. The hole in the centre of the diagram represents the pore of the channel. LEV are the two agonist binding sites shown overlapping *unc-38* and *unc-29*. (B) The levamisole subunit stoichiometry is modified by replacing one *unc-29* subunit with a *lev-1* subunit. The channel will then have different properties and one of the ligand binding sites is modified and unequal to the other of the receptor. Thus variation of the stoichiometry of the receptor will produce receptor heterogeneity.

on possible variations in the stoichiometry of the pentameric subunit composition of the nAChR ion channel in the parasite is shown in Fig. 21.12. The three genes in *O. dentatum* analogous to *C. elegans* genes (*Unc-38*; *lev-1* and *Unc-29*) (Fleming *et al.*, 1997) may contribute one or more of the subunits of the pentameric levamisole nAChR ion channel (McGehee and Role, 1995; Changeaux *et al.*, 1996), each combination producing a discrete channel with distinctive properties. The distribution of channel conductances would then consist of a number of peaks, rather than a single peak, and could be skewed, depending upon the frequency of the subunit combinations. The different subunit combination values may also contribute to the differences in *P-open*.

Alleles encoding for structural changes in individual receptor subunits

An altered amino acid structure of any of the individual receptor subunits may, without altering the stoichiometry of the whole receptor, give rise to receptors with different conductances and *P-open* values.

Alleles encoding for altered receptor phosphorylation sites

Phosphorylation is the most important form of reversible modification of proteins available in animal systems (Tizard, 1996). It involves the use of a high-energy compound (for example, ATP or GTP) to modify a protein, with phosphorylated and non-phosphorylated forms having different functional properties. Protein kinases enzymatically phosphorylate the amino acid residues serine, threonine and tyrosine. Phosphorylation of these three key amino acids plays a critical role in many cellular functions. The most comprehensively studied vertebrate nAChR is that of the *Torpedo* (Hopfield *et al.*, 1988; Huganir and Miles, 1989; Changeaux *et al.*, 1996). The membrane of the *Torpedo* contains protein kinase activity, including cAMP-dependent kinase, a protein kinase C and a tyrosine kinase, all of which can phosphorylate the nicotinic receptor. The experiments have shown that cAMP-dependent phosphorylation markedly alters the channel opening in response to acetylcholine, and affects desensitization (the duration of long closed periods). Less information is available on the structure and function of nematode nicotinic receptors. Consensus regulatory phosphorylation sites for protein kinase C, protein kinase A and tyrosine kinase are also recognized in *C. elegans* (Fleming *et al.*, 1997) (Fig. 21.12). In *Onchocerca volvulus*, cDNA of a $nAChR_n$ subunit (Ajuh *et al.*, 1994) has been used to identify consensus regulatory phosphorylation sites at four places: (i) three PKC sites; and (ii) one PKA site. Analogous phosphorylation sites are also present on $nAChR_n$ subunits of other parasitic nematodes (Hoekstra *et al.*, 1997; Wiley *et al.*, 1997). Phosphorylation of the levamisole receptor protein, then, is one of the most effective means the parasite could have of being able to regulate and modulate the number and activity of the receptors. The presence of mutant alleles encoding alterations in receptor subunit phosphorylation sites provides another functional explanation for receptor heterogeneity in response to levamisole exposure. Phosphorylation is therefore likely to play a role in the development of anthelmintic resistance. Arevalo and Saz (1992) have demonstrated the presence of PKC activity in *Ascaris suum* muscle. The PKC activity was also shown to depend on calcium and 1,2 diacylglyceride and phospholipid. Allelic variation in protein kinases must also be included as a possible cause for regulation of receptor activity and, thus, receptor heterogeneity in response to levamisole exposure. However, protein kinases are ubiquitous in many cellular processes; mutations capable of providing altered receptor function to levamisole exposure may well reduce fitness in other cellular processes and hence should be removed from the population by selection pressure.

Working hypothesis

Our working hypothesis is that subunit stoichiometry or changes in the structure of individual receptor subunits give rise to the different conductance subtypes and that phosphorylation by the different kinase enzymes has major effects on the *P-open* values of the receptor. Should the hypothesis hold, then alterations in the phosphorylation state of the receptor by the different kinase enzymes may (as stated earlier) be important in the development of anthelmintic resistance.

Glutamate-gated Chloride Channels

Ivermectin is a macrocyclic lactone antibiotic anthelmintic produced by *Streptomyces avermitilis*. A series of cloning experiments using *C. elegans* as a source of RNA, and expression in *Xenopus* oocytes, has established that one site of action of avermectins is a group of inhibitory glutamate-gated Cl channels (Cully *et al.*, 1996). This group of inhibitory ion channels is believed to occur only in muscle and nerves of invertebrates such as nematodes, insects and crustacea but not in vertebrates. Expression cloning led to recognition of two *C. elegans* channel subunits, GluClα and GluClβ, that, when expressed together, produced functional ion channels (Cully *et al.*, 1994). Subsequently, PCR-based searches led to recognition of additional GluCl related subunits from *C. elegans*, GluClX and C27H5 (Cully *et al.*, 1996). Another GluCl subunit, *avr-15*, has been recognized from ivermectin resistant *C. elegans* mutants: *avr-15* encodes two alternatively spliced channel subunits present in pharyngeal muscle (Dent *et al.*, 1997).

The GluCl subunits are each approximately 500 amino acids in length. It is assumed that five GluCl subunits come together, as do the subunits of the nAChRs, to produce a pentomeric GluCl ion channel. The stoichiometric arrangement has not yet been determined. It is known that GluClα and GluClβ subunits may form homomeric channels as well as heteromeric ion channels when expressed in *Xenopus* oocytes (Cully *et al.*, 1994), but the

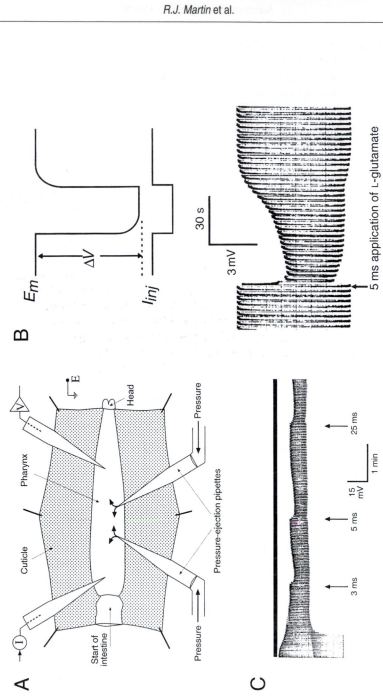

A

B

C

E_m

ΔV

I_{inj}

3 mV

30 s

5 ms application of L-glutamate

Head

Pharynx

Cuticle

Start of intestine

Pressure-ejection pipettes

Pressure

Pressure

E

3 ms

5 ms

25 ms

15 mV

1 min

functions and locations of the native GluCl ion channels in nematodes remain to be determined.

Avermectin-sensitive sites in *A. suum* have been identified on pharyngeal muscle (Martin, 1996) using a two-micropipette current-clamp technique (Fig. 21.13A). Glutamate and avermectins produce hyper-polarization and an increase in Cl conductance when either bath-applied or pressure-ejected on to the pharyngeal preparation (Fig. 21.13B,C). These observations establish that one site of action of the avermectins is the pharyngeal muscle of nematodes and that the avermectins can inhibit feeding in nematodes.

Glutamate-activated Cl single-channel currents can be recorded from nematodes using a vesicle preparation made from the pharyngeal muscle (Addelsberger *et al.*, 1997). The channels (Fig. 21.14) have a conductance in the range 20–50 pS with mean open-times of 5–10 ms. The addition of ivermectin to the preparation dramatically increases the proportion of time that the ion channels are open (Fig. 21.14). Although the single-channel studies are still at an early stage, it can be seen that ivermectin, like glutamate, opens these channels. Questions remain about the location of the receptors in nematodes, their function and the significance of the different genes coding for subunits of the Glu-Cl receptor. The selective action of the avermectins is believed to be due to the fact that vertebrate hosts do not possess the Glu-Cl channel and so are not susceptible to toxicity.

GABA-gated Chloride Channel Receptors

Ascaris somatic muscle cells also possess inhibitory synaptic and extra-synaptic GABA-gated Cl channels (Martin, 1980). Activation of these

Fig. 21.13. (Opposite) (A) Two-micropipette recording from the pharyngeal muscle of *Ascaris suum*. A voltage micropipette (V) is shown and used for recording intracellular potential from the syncytial muscle of the pharynx; the current electrode (I) is shown for injecting the current for measuring input conductance. The two pressure ejection pipettes are used for ejection of glutamate and ivermectin. (B) A diagram of the rectangular 200 ms current pulse (I_{inj}) and the voltage response (ΔV) shown at high time resolution. The injected current produces an exponential change in the membrane potential (*Em*). Below is a series of voltage responses at low resolution time scale and at the voltage calibration shown. During this recording a brief pulse of glutamate is added to the pharyngeal preparation and this leads to a reduction in the width of the trace (increase in conductance as the Glu-Cl channels open) and a hyperpolarization of the membrane potential. (C) Application of ivermectin (during the horizontal bar) results in a hyperpolarization and an increase in the conductance of the membrane. Brief pulses (3 ms, 5 ms and 25 ms) of glutamate marked by the horizontal arrows show that the effect of glutamate is similar to that of ivermectin.

800 µM Glutamate

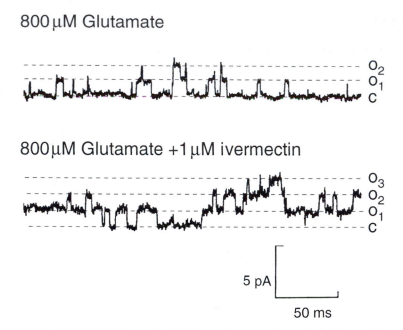

800 µM Glutamate + 1 µM ivermectin

5 pA

50 ms

Fig. 21.14. Isolated inside-out patch from pharyngeal muscle with a patch potential of +50 mV and 800 µM glutamate present in the patch pipette shows the opening of two Glu-Cl channels. The addition of 1 µM ivermectin to the bath and thus to the inside of the patch produces (in the same patch) opening of up to three Glu-Cl channels simultaneously. Ivermectin increases the probability of opening of Glu-Cl channels already opening in the presence of glutamate.

receptors leads to hyperpolarization and relaxation of the body muscle cells. The receptors form part of the inhibitory GABAergic input that controls the excitability of nematode muscle. The anthelmintic piperazine acts as a GABA agonist and will hyperpolarize *Ascaris* muscle (Del Castillo *et al.*, 1964) by increasing the Cl conductance of the muscle membrane. The increase in Cl conductance can be seen following bath application of the piperazine while recording the electrophysiological effect with the two-micropipette recording technique (Martin, 1982). These experiments showed that piperazine is 10–100 times less potent than GABA. Under patch-clamp, the opening of single Cl channels by piperazine was found to produce the same conductance as the GABA channels (Fig. 21.15). However, the openings were much briefer, around 18 ms mean open-time rather than 32 ms mean open-time. Piperazine also produces opening of the ion channels at a lower rate than GABA (Martin, 1982). The explanation for the piperazine being less potent than GABA is that piperazine opens the channels for a shorter proportion of time. However, GABA is an ineffective anthelmintic, because it does not penetrate the cuticle of the nematode. Piperazine,

GABA

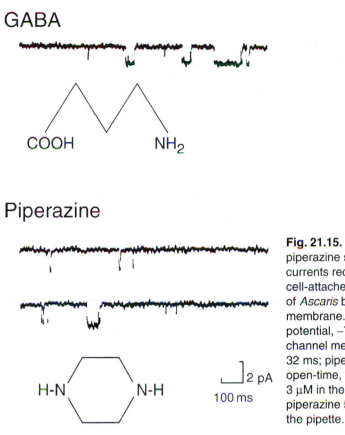

COOH NH₂

Piperazine

H-N N-H

⌐⌐2 pA
100 ms

Fig. 21.15. GABA and piperazine single-channel currents recorded from cell-attached patches of *Ascaris* bag muscle membrane. Trans-patch potential, –75 mV; GABA channel mean open-time, 32 ms; piperazine channel open-time, 18 ms; GABA 3 µM in the pipette; piperazine 500 µM in the pipette.

because it is able to exist in an un-ionized form, is able to penetrate through the cuticle and reach its site of action on the muscle. Piperazine then acts to produce a flaccid paralysis of intestinal nematode parasites.

Concluding Remarks

This chapter has introduced the membrane ion channels and electro-physiological techniques that may be used for exploring the actions and properties of an important group of anthelmintics. It can be seen that these electrophysiological techniques can be used to explore the mechanisms of anthelmintic resistance. It is hoped that in the future new pharmacological approaches may be produced that can reverse the resistance, perhaps by interfering with the phosphorylation of the ion channels.

References

Adelsberger, H., Scheuer, T. and Dudel, J. (1997) A patch clamp study of a glutamatergic chloride channel on pharyngeal muscle of the nematode *Ascaris suum. Neuroscience Letters* 230, 183–186.

Ajuh, P.M., Cowell, P., Davey, M.J. and Shevde, S. (1994) Cloning of a cDNA encoding a putative nicotinic acetylcholine receptor subunit of the human filarial parasite *Onchocerca volvulus. Gene* 144, 127–129.

Arevalo, J.I. and Saz, H.J. (1992) Effects of cholinergic agents on the metabolism of choline in muscle from *Ascaris suum. Journal of Parasitology* 78, 387–392.

Blouin, M.S., Yowell, C.A., Courtney, C.H. and Dame, J.B. (1995) Host movement and the genetic-structure of populations of parasitic nematodes. *Genetics* 141, 1007–1014.

Changeaux, J.-P., Devillers-Thiery, A. and Chemouilli, P. (1996) Acetylcholine receptor: an allosteric protein. *Science* 225, 1335–1345 (abstract).

Cully, D.F., Vassilatis, D.K., Liu, K.K., Paress, P.S., Vanderploeg, L.H.T. and Schaeffer, J.M. (1994) Cloning of an avermectin-sensitive glutamate-gated chloride channel from *Caenorhabditis elegans. Nature* 371, 707–711.

Cully, D.F., Wilkinson, H., Vassilitis, D.K., Etter, A. and Arena, J.P. (1996) Molecular biology and electrophysiology of glutamate-gated chloride channels of invertebrates. *Parasitology* 114, S191–S200.

Dent, J.A., Davis, M.W. and Avery, L. (1997) *Avr-15* encodes a chloride channel subunit that mediates inhibitory glutamatergic neurotransmission and ivermectin sensitivity in *Caenorhabditis elegans. EMBO Journal* 16, 5867–5879.

Devillers-Thiery, A., Galzi, J.L., Eisele, J.L., Bertrand, S., Bertrand, D. and Changeaux, J.P. (1993) Functional architecture of the nictotinic acetylcholine receptor: a prototype of ligand-gated ion channels. *Journal of Membrane Biology* 136, 97–112.

Fleming, J.T., Squire, M.D., Barnes, T.M., Tornoe, C., Matsuda, K., Ahnn, J., Fire, A., Sulston, J.E., Barnard, E.A., Sattelle, D.B. and Lewis, J.A. (1997) *Caenorhabditis elegans* levamisole resistance genes *lev-1, unc-29*, and *unc-38* encode functional nicotinic acetylcholine receptor subunits. *Journal of Neuroscience* 17, 5843–5857.

Gasser, R.B., Woods, W.G. and Bjorn, H. (1998) Distinguishing *Oesophagostomum dentatum* from *Oesophagostomum quadrispinulatum* developmental stages by a single-strand conformation polymorphism method. *International Journal for Parasitology* 28, 1903–1909.

Hoekstra, R., Visser, A., Wiley, L.J., Weiss, A.S., Sangster, N.C. and Roos, M.H. (1997) Characterization of an acetylcholine receptor gene of *Haemonchus contortus* in relation to levamisole resistance. *Molecular and Biochemical Parasitology* 84, 179–187.

Hopfield, J.F., Tank, D.W., Greengard, P. and Huganir, R.L. (1988) Functional modulation of the nicotinic acetylcholine receptor by tyrosine phosphory-lation. *Nature* 336, 677–680.

Huginir, R.L. and Miles, K. (1989) Functional modulation of the nicotinic acetylcholine receptors. *Critical Reviews of Biochemistry and Molecular Biology* 24, 183–215.

Martin, R.J. (1980) The effect of γ-aminobutyric acid in the input conductance and membrane potential of *Ascaris* muscle. *British Journal of Pharmacology* 71, 99–106.

Martin, R.J. (1982) Electrophysiological effects of piperazine and diethylcarbamazine on *Ascaris suum* somatic muscle. *British Journal of Pharmacology* 77, 255–265.

Martin, R.J. (1995) An electrophysiological preparation of *Ascaris suum* pharyngeal muscle reveals a glutamate-gated chloride channel sensitive to the avermectin analogue milbemycin D. *Parasitology* 112, 252.

Martin, R.J., Kusel, J.R. and Pennington, A.J. (1990) Surface-properties of membrane-vesicles prepared from muscle-cells of *Ascaris suum*. *Journal of Parasitology* 76, 340–348.

Martin, R.J., Robertson, A.P., Bjorn, H. and Sangster, N.C. (1997) Heterogeneous levamisole receptors: a single-channels study of nictotinic acetylcholine receptors from *Oesophagostomum dentatum*. *European Journal of Pharmacology* 322, 249–257.

McGehee, D.S. and Role, L.W. (1995) Physiological diversity of nictotinic acetylcholine receptors expressed by vertebrate neurons. *Annual Review of Physiology* 57, 521–546.

Robertson, A.P., Bjorn, H. and Martin, R.J. (1999) Resistance to levamisole resolved at the single-channel level. *FASEB Journal* 13, 749–760.

Robertson, S.J. and Martin, R.J. (1993) Levamisole-activated single-channel currents from muscle of the nematode parasite *Ascaris suum*. *British Journal of Pharmacology* 108, 170–178.

Tizard, I. (1996) In: *Veterinary Immunology*, 5th edn. W.B. Saunders, Philadelphia.

Twyman, R.M. (1998) In: Wisden, W. (ed.) *Advanced Molecular Biology: a Concise Reference*. BIOS Scientific Publishers, Oxford.

Varady, M., Bjorn, H., Craven, J. and Nansen, P. (1997) *In vitro* characterization of lines of *Oesophagostomum dentatum* selected or not selected for resistance to pyrantel, levamisole and ivermectin. *International Journal for Parasitology* 27, 77–81.

Wallace, A.J. (1997) In: Taylor, G.R. (ed.) *Laboratory Methods for the Detection of Mutations and Polymorphisms in DNA*, 1st edn. CRC Press, Boca Raton, Florida.

Wiley, L.J., Weiss, A.S., Sangster, N.C. and Li, Q. (1996) Cloning and sequence analysis of the candidate nicotinic acetylcholine receptor alpha subunit gene *tar*-1 from *Trichostrongylus colubriformis*. *Gene* 182, 97–100.

Index

Note: page numbers in *italics* refer to figures and tables